Praise for
Climate Chaos: Killing People, Places, and the Planet
Edited by John Hans Gilderbloom

From the Foreword:
"John Hans Gilderbloom, a renowned authority in the climate and environmental research field, has crafted a seminal work that revolutionizes our understanding of the critical issues in environmental science and politics. With 29 authors and a message from Pope Francis and Nobel Prize Winner Albert Gore, John Hans Gilderbloom has placed us on a trajectory to reassess what it means to say we are in a crisis not so much of the environment but of human existence. The book's title, *Climate Chaos: Killing People, Places, and the Planet*, says it all. We as citizens must understand that in this monumental work, these scientists are attempting to tell us that we must envision a paradigm shift, a *new worldview*." —**Bobby William Austin**, PhD, President of Neighborhood Associates and author of *Repairing the Breach: Key Ways to Support Family Life, Reclaim Our Streets, and Rebuild Civil Society in America's Communities. Report of the National Task Force on African-American Men and Boys*

ADVANCE PRAISE FOR *CLIMATE CHAOS: KILLING PEOPLE, PLACES, AND THE PLANET*

"While so many predict a coming Dark Age, John Gilderbloom brings light. If we cannot imagine a better future, our future will certainly be darker. But this fine book offers a blueprint for hope—not blind hope, but imaginative hope. It should be a banner carried at the forward edge of a global movement." —**Richard Louv**, author of *The Nature Principle*, *Last Child in the Woods*, *Our Wild Calling*, and *The Web of Life*

"My fact-based story was disclosed in a 60-minute interview with veteran Mike Wallace, then in the movie, *The Insider*, and validated in multiple courts in America. There are several parallel identical events that mimic my experiences in telling the truth to the world about how the tobacco industry harms addicts for profit with what Professor John Gilderbloom is experiencing: a vicious smear campaign, death threats, and even an attempted murder. His truth telling is in a book he has written, *Climate Chaos: Killing People,*

Places, and the Planet, a book that should be required reading for college-level students as it delineates how we can reverse catastrophic climate changes before it is too late." —**Jeffrey Wigand**, MA, PhD, MAT, ScD, highest-ranking former executive to break ranks with the tobacco industry and share the industry's obfuscation of the truth. His testimony led to a $206-billion settlement (MSA) to be paid to the states.

"The imminent publication of *Climate Chaos* was announced by the arrest of an assassin. If people try to kill you for the words you write, it behooves us to read them. Gilderbloom and the contributors to *Climate Chaos* have not only gathered the facts on the cover-up by the corporate hitmen of the planet but lay out a practical program to keep us all from dying. Lancet estimates that nine million people now die prematurely every year because of pollution. This is the first book to focus on the urban battlefront and what thousands of cities are doing that makes a difference. Through shifts in planning, better regulation of effluents, the application of technology, and the practice of a more sustainable way of living in how we build, eat, and get to work, Gilderbloom has gathered a scientific Jeremiad and a practical call to battle that shows us how we might not only slow but reverse, the climate changes pushing us inexorably toward extinction." —**Roger Friedland**, Professor of Religious Studies and Sociology, University of California, Santa Barbara. Author of *Power and Crisis in the City* and *To Rule Jerusalem* (with Richard Hecht).

"Stunning! A powerful action plan for how cities can save our planet from destruction." —**Stephen Roosa**, PhD, MBA, CEM, BEP, CSDP, REP, CBCE, CMVP, LEED AP, Editor-in-Chief of the *International Journal of Strategic Energy and Environmental Planning*

"John Gilderbloom's message couldn't be more timely or important—not only that climate change is the existential challenge of our time but that there is indeed a hopeful path forward. Effective action on climate change cannot help but address and improve our other related challenges, including unequal impacts on health, threats to equitable human development, and declining urban quality of life for too many. In that sense, the growing awareness of climate threats may help to mobilize needed action on these other long-neglected issues." —**Michael W. Mehaffy**, PhD, Executive Director, Lennard Institute/International Conference on Making Cities Livable and author of *Cities Alive: Jane Jacobs, Christopher Alexander, and the Roots of the New Urban Renaissance*

"A must-read for anyone concerned about climate change and what can be done about it." —**Cary Lowe**, PhD, Member of the College of Fellows of the

American Institute of Certified Planners and author of *Becoming American: A Political Memoir*

"This book is an inspiring must-read for anyone who wants to know how to solve the greatest challenge of our time—climate change. World-renowned urban planner John Hans Gilderbloom reveals the untold story of how municipalities, businesses, and private citizens are working together to make a tangible impact. In-depth analysis of how we have the tools and technology to dramatically reduce greenhouse gases and create a sustainable future is the inspiration for our documentary, *Climate of Hope: Cities Saving the World*. The film is filled with inspiring stories from cities all over the world. Together, they are a powerful call to action." —**Chris Sean Nolan**, Director, Climate of Hope, and three-time Emmy Award winner

"This groundbreaking book uncovers the numerous ways climate chaos is killing people, places, and the planet. It makes a powerful contribution by offering a hopeful, practical perspective to reversing destructive climate chaos and reducing greenhouse gases by 80 percent through implementing practical policies that will create a more sustainable, prosperous, and livable Earth." —**Julian Agyeman**, Professor, Tufts University, and Editor, *Local Environment*, and his work has been cited more than 20,000 times

"An excellent overview of the climate crisis facing all of us." —**Scott Cummings**, Professor Emeritus, Saint Louis University and former editor, *Journal of Urban Affairs*

"Dr. Gilderbloom's book, *Climate Chaos: Killing People, Places, and the Planet*, gives an honest and fearless assessment of the problems that manifest in our inner cities and impact the health of residents who are suffering daily. I was with John Hans Gilderbloom when he met Martin Luther King III, who told him that he provided a model for rebuilding livable neighborhoods that his father and mother envisioned. I was also with him when Don Terner, the United States Envoy for South Africa, announced he would meet with President Nelson Mandela and tell him about what Dr. Gilderbloom was doing in West Louisville to rebuild livable neighborhoods." —**La Glenda Reed**, Cofounder of the Center for Sustainable Urban Neighborhoods, West Louisville High School teacher, and long-time community activist

"*Climate Chaos* takes an all-hands-on-deck approach to the climate crisis. It brings the critical issues we confront down to the community level, chronicling the destructive effect of climate change on multiple aspects of life in U.S. cities. Drawing on detailed case studies as well as statistical analysis across cities, *Climate Chaos* documents the differential effects of climate change—both across and within cities—on life expectancy, health, COVID-19, educational outcomes, and even housing values. Drawing on case studies in the United States (such as Portland) and Europe (such as Amsterdam), *Climate Chaos* also makes a strong case for alternatives to the automobile: light rail, walking, and—close to my own heart—bicycling. The solutions offered are systemic, going beyond feel-good calls for planting more trees, arguing instead that cities—where most of the world's population lives—can lead the way. This book is an important read: disturbing in its conclusions, leading to a compelling call for strong, community-based actions." —**Richard Appelbaum**, Distinguished Professor Emeritus and former MacArthur Chair in Sociology and Global & International Studies at the University of California at Santa Barbara (UCSB) and the co-founder of UCSB's pioneering Program in Global Studies

"Climate change will impact all of us in dramatic ways, but at the same time the devastation will be experienced, unequally, across gender, race, income, and geographic positions. Gilderbloom's book draws on empirical evidence to outline the chaotic effects of climate change on health, life expectancy, and overall quality of life. Despite the climate chaos that we face, Gilderbloom's book also identifies practical, imaginative, and proven solutions that can better control pollution and promote a more sustainable life for all, not just societal elites. The book is a comprehensive overview of how climate change will affect us but a book that also offers a thoughtful agenda for change and hope for the future . . . an agenda that can start in our own cities and where we each can play a part." —**Nancy C. Jurik**, Professor Emeritus, Arizona State University and 2019–2020 Fulbright Scholar

"John Hans Gilderbloom is a fearless truth teller. He survived a brutal assault and learned to thrive despite some hearing and eyesight loss and PTSD. This book presents the award-winning research of his 30 colleagues that was originally removed from a university website. But Dr. Gilderbloom rescued and restored and expanded this critical research. People have the right to know the truth." —**Mike Schindler**, Exec Producer, 90000 Feet Productions, US Navy Veteran, three-time author, Co-founder of The VUCA Principle

Climate Chaos

Killing People, Places, and the Planet

Edited by John Hans Gilderbloom

LEXINGTON BOOKS

Lanham • Boulder • New York • London

Lexington Books
Bloomsbury Publishing Inc, 1385 Broadway, New York, NY 10018, USA
Bloomsbury Publishing Plc, 50 Bedford Square, London, WC1B 3DP, UK
Bloomsbury Publishing Ireland, 29 Earlsfort Terrace, Dublin 2, D02 AY28, Ireland
www.rowman.com

British Library Cataloguing in Publication information available

Library of Congress Cataloging-in-Publication Data

Names: Gilderbloom, John Ingram, editor.
Title: Climate chaos : killing people, places, and the planet / edited by
  John Hans Gilderbloom.
Description: Lanham : Lexington Books, [2025] | Includes bibliographical
  references and index. | Summary: "This book combines hard science,
  technology, and progressive planning to reverse climate change, and
  offers a bold yet practical vision for sustainable living" —Provided by
  publisher.
Identifiers: LCCN 2024041448 (print) | LCCN 2024041449 (ebook) |
  ISBN 9781666940497 (cloth) | ISBN 9781666940503 (epub)
Subjects: LCSH: Pollution prevention—Case studies. | Climate change
  mitigation—Case studies. | Sustainable development—Case studies.
  Classification: LCC TD174 .C57 2025 (print) | LCC TD174 (ebook) |
  DDC 363.73/7—dc23/eng/20241115
LC record available at https://lccn.loc.gov/2024041448
LC ebook record available at https://lccn.loc.gov/2024041449

For product safety related questions contact productsafety@bloomsbury.com.

♾™ The paper used in this publication meets the minimum requirements of American
National Standard for Information Sciences—Permanence of Paper for Printed Library
Materials, ANSI/NISO Z39.48-1992.

To Carla, my magnificent lifelong partner who inspired me to push ahead by saying, "Gilderblooms never roll over." So grateful for our journey, support, joy, love, humor, and constant inspiration.

Contents

Foreword

A Human Ecological Worldview

Bobby William Austin

John Hans Gilderbloom, a renowned authority in the climate and environmental research field, has crafted a seminal work that revolutionizes our understanding of the critical issues in environmental science and politics. With 29 authors and a message from Pope Francis and Nobel Prize Winner Albert Gore, John Hans Gilderbloom has placed us on a trajectory to reassess what it means to say we are in a crisis not so much of the environment but of human existence. The book's title, *Climate Chaos: Killing People, Places, and the Planet*, says it all.

These scientists and scholars point us toward a new worldview concerning our narrow understanding of the environment just by perusing the chapter titles alone. The environment they describe is a holistic view of neighborhoods, fields, streams, water supply, housing policies, urban planning, and political goals. Extreme weather events, water scarcity, air pollution, and social unrest assail us daily. In chapter 1, Gilderbloom states,

> It's going to get worse. Look around, and you'll see that no place is safe from the ravages of climate chaos. Every state has experienced the pain of climate chaos: Hawaii, Florida, New York, Arizona, California, and, of course, Kentucky. East to West, North to South, we are witnessing the slow destruction of the Earth as we once knew it. Or look around the world: North Pole, Africa, Middle East, China, Russia, South America. We have a climate crisis that is destroying our way of life with record high temperatures. We are facing a catastrophic crisis that will destroy the Earth as we know it in eight years, according to the United Nations.

This overwhelming complexity appears at first glance to be insurmountable, but Gilderbloom and company are bringing us ways to deal with this situation. We as citizens must understand that in this monumental work, these scientists are attempting to tell us that we must envision a paradigm shift, a *new worldview*. This new worldview, as extensive and comprehensive as the issue itself, includes sustainable urban planning, community-based environmental initiatives, and global cooperation. This book tells us that our public response must be as great as the scientific and scholarly approach. All must understand that a new human ecological viewpoint is our best hope for a comprehensive strategy for citizens and the scientific community to unite to bring hope out of this chaos.

Human ecology first appeared as a course of study at the legendary Department of Sociology at the University of Chicago in the early 1920s. Here, social scientists like Robert Park and Ernest Burgess began to study neighborhoods and how those neighborhoods were affected by their proximity to the city's center and what kind of physical and social impact it had on people and their living conditions (Park et al., 1925). Their sociological work created a theory of concentric zones that emanated from the city core to the suburbs. The most impacted economically, politically, and socially were to be found in the core cities and not the suburbs. Today, this is no longer the case. Suburbs and inner cities experience both environmental problems and social upheaval. As Gilderbloom points out, city cores are still plagued with environmental and industrial pollution.

Gerald G. Marten writes in *Human Ecology: Basic Concepts for Sustainable Development,*

> Ecology is the science of relationships between living organisms and their environment. Human ecology is about relationships between people and their environment. In human ecology, the environment is an ecosystem. An ecosystem is everything in a specified area—the air, soil, water, living organisms and physical structures, including everything built by humans. The living parts of an ecosystem—organisms, plants, animals (including humans)—or biological community. (Marten, 2001, p. 1)

As Gilderbloom and the authors state, the neighborhood is Ground Zero. The American public and politicians must understand that a child's life is in our hands, including the lives of our own children. They are critically affected by the air they breathe, the water they drink, and the food they eat; indeed, they live in their homes and neighborhoods surrounded by this "plague" of pollution, environmental degradation, and social inequality daily.

Martin Luther King Jr.'s description of the World House is the symbol and reality of how we must now envision ourselves. The work of building the World House is the work described in this book. Focus on the fact that we are all heirs to this planet, given to us by God; we are the stewards of this large house and make the critical decisions whether to live or die. He states,

> In a real sense all life is inter-related. All men are caught in an inescapable network of mutuality, tied in a single garment of destiny. Whatever affects one directly, affects all indirectly. I can never be what I ought to be until you are what you ought to be, and you can never be what you ought to be until I am what I ought to be. . . . This is the inter-related structure of reality. (King, 2018)

Gilderbloom and company are expressing the core of a new ecological worldview. The interconnection of mind, body, and soul is no longer a platitude that scientists and scholars have been telling us based on scientific research. King's interconnectedness of all individuals and life on this planet frames the symbiotic relationship that a new human ecological worldview expresses.

E. O. Wilson, world-famous biologist and Pulitzer Prize winner, offered the concept of *Consilience: The Unity of Knowledge.* In the context of Gilderbloom's *Climate Chaos,* consilience means that all disciplines, from science to philosophy to social and urban planning, must come together to address the complex issues of climate change and human ecology. Wilson wrote,

> the search for consilience might seem at first to imprison creativity. The opposite is true. A united system of knowledge is the surest means of identifying the still unexplored domains of reality. It provides a clear map of what is known, and it frames the most productive questions for future inquiry. Historians of science often observe that asking the right question is more important than producing the right answer. The right answer to a trivial question is also trivial, but the right question, even when insoluble in exact form, is a guide to major discovery. And so it will ever be in the future, excursions of science and imaginative flights of the arts. (Wilson, 1999, p. 326)

Gilderbloom and company are on the road to bringing together science, philosophy, and social and urban planning. Therefore, they call for a unified voice of study and public discourse, local and federal governments, religious institutions, and rural and urban America regarding environmental and human ecological issues. The authors emphasize that it is not enough for the scientific and scholarly community to lead the way; it is the responsibility of all citizens, policymakers, and institutions to take action and contribute to the solutions.

Gilderbloom has been fearless in conducting impartial data analysis that uncovers environmental crimes that kill people, places, and the planet. Gilderbloom is persistently fearless in speaking truth to power. Every morning, he is reminded of the effects of a gunshot wound—ringing in his ears, blurred eyesight, and PTSD. He is a survivor who learned to thrive. I salute his bravery as he stands up to the same corporations that promoted dangerous products like DDT and tobacco. We know better now. Climate chaos solutions exist via technology, AI, and the renewable resource of people coming together. Gilderbloom provides a recipe for climate action. The United Nations says we have seven years left before we end in worldwide catastrophe. We still have a small window of opportunity to repair and renew our precious planet by being good stewards of the Earth.

At my youngest daughter's graduation from FISK University in early 2000, the commencement speaker, Astronaut Mae Jamison, made one of the most profound speeches. John Hope Franklin, the distinguished historian, was in the audience, and we both agreed that it was a clear challenge to the new graduates. What impressed us most was the statement she made regarding the environment. I'm paraphrasing. She urged us to be vigilant about the environment, clean it up, and do everything necessary to protect the rainforest and prevent the pollution of our cities and waterways. But she was clear. Rest assured, she said, whether we wish to do it or not, whether we choose to do nothing and thus desert ourselves, nature will take care of herself and transform into her next iteration, with a new species and Earth order with or without us. It is up to us to live and share this beautiful Earth and its joys for generations. If we don't, the Earth will continue with or without us, in one form or another. The loss will be ours. I believe that John Hans Gilderbloom and company are showing us a way out and into a new creation for all of us.

REFERENCES

King, M. L., Jr. (2018). *Letter from Birmingham Jail*. London: Penguin Classics.

Marten, G. G. (2001). *Human Ecology: Basic Concepts for Sustainable Development*. London: Earthscan from Routledge.

Park, R. E., Burgess, E. W., & McKenzie, R. D. (1925). *The City*. University of Chicago Press: Chicago and London.

Wilson, E. O. (1999). *Consilience: The Unity of Knowledge*. Vintage Books A Division of Random House, Inc.: New York.

Our Common Home

A Guide to Caring for Our Living Planet

A Joint Initiative of the Holy See, Pope Francis, and the Stockholm Environment Institute

We must urgently renew our relationship with our living planet. While the scale of this challenge can seem daunting, the good news is that the answers are already at hand—it is only up to us to put them into action. The following pages summarize evidence on issues that are at the heart of our current predicament. The aim is to inform, inspire hope, and stimulate debate and action. This guide is the result of a collaboration between the scientific and spiritual communities, between the Stockholm Environment Institute and the Dicastery for Promoting Integral Human Development. It sets out essential facts and solutions on key topics, along with advice on how communities can respond. It is inspired by Pope Francis' second encyclical, Laudato si': On Care for Our Common Home, which explores our ecological crisis and its roots in over-consumption and current models of economic development. These questions are a priority for Pope Francis, the Catholic Church and all believers, the scientific community, and all citizens, as we answer the call to protect and regenerate our common home.

OUR CLIMATE

"The climate is a common good, belonging to all and meant for all"
Laudato si' (23)

We are living through a climate crisis driven by economic and social systems that run on fossil fuels. Without a far-reaching response, a changing climate will undermine the conditions that have allowed us to thrive on our planet. To tackle this crisis, we need to make fundamental changes to our economies and behavior, to shift our everyday patterns of consumption, and to advance social justice. But the tools we need to take action are in our hands.

The fossil fuel economy is the main cause of global warming. Our planet provided a stable climate that allowed humanity and nature to flourish for thousands of years. But since the start of the Industrial Revolution in the eighteenth century, we have emitted more and more greenhouse gases into our atmosphere that warm the climate by trapping the sun's heat.

These gases are produced by burning coal, oil, and gas to power industry and transport, and to supply our homes and buildings with heat and electricity. We also release greenhouse gases through intensive farming and by cutting down forests.

Climate breakdown is a severe threat to our security and the natural world. Droughts, storms, and floods are already more frequent and severe. Shortages of food and water are causing hunger and conflict. Sea levels are rising: 150 million people currently live on land that will be below the high tide line by mid-century.

Some of our most biodiverse environments, such as coral reefs, have been damaged by warming to a point from which they can't recover. And the world's poor, who are least responsible for climate change, bear an unfair burden of the impacts—though all of us will be affected.

What needs to change? With swift action to rapidly eliminate greenhouse gas emissions, we can limit temperature rise and prevent its most dangerous outcomes. In 2015, nearly 200 countries signed an agreement in Paris to limit temperature rise to 1.5 degrees Celsius, but we have not acted fast enough. The coming decade will be critical.

Nothing less than a complete change in our economies and societies is required. We must:

- halt deforestation
- change how we consume food and how we farm our land
- produce power without fossil fuels through massive shifts to clean energy.

OUR LIVING PLANET

"Because all creatures are connected, each must be cherished with love and respect"
Laudato si' (42)

Industrialization, mass consumption, and transformations in agriculture have brought the living world to a crisis point. We have caused an ongoing mass extinction of species, from mammals down to the smallest plants and bacteria. It's urgent to change course and safeguard those wild ecosystems that remain and revive those that have been degraded. They are the basis of our survival and well-being.

Biodiversity is the foundation for human survival and progress. Healthy ecosystems regulate the climate, supply clean water, provide medicines, clean up pollution, pollinate our crops, and provide vital natural resources. Countless livelihoods directly depend on the services ecosystems provide. Nature is a priceless source of inspiration for art and learning, as well as for cultural and spiritual experiences.

Yet as industry, agriculture, and fisheries reach farther into once-wild landscapes, such as forests, wetlands, and seas, they are destroying the richness of the living world.

Since 1970, global wildlife populations have declined by two-thirds. In the same period, plant and animal life in freshwater has fallen by 84 percent. Bees and other pollinators are in serious decline, directly threatening supplies of important staple foods, fruits, and vegetables.

Without action on climate and deforestation, the Amazon rainforest could collapse into a dry and scrubby landscape within 50 years. Our soil is also under serious threat: without biodiversity in soils, the ecosystems they support cannot survive. Healthy ecosystems also protect against the impacts of climate change, such as extreme heat and storm surges.

OUR WATER

"Access to safe drinkable water is a basic and universal human right"
Laudato si' (30)

We have a moral obligation to ensure everyone has access to enough clean water to meet their fundamental human needs. Yet, increasing numbers of people lack reliable supplies of water, especially the poor. We can ensure access to safe drinking water, sanitation for all, and sustainable use of water in agriculture and industry by treating it as a precious resource and conserving and managing it in a fair and sustainable way.

Access to clean water is a growing challenge. More and more people lack clean water for drinking, washing, and irrigating crops. Sometimes this is because there simply isn't enough water, or because it is overused, wasted, or polluted. Water use has grown at more than twice the rate of population increase in the last century, and more regions are reaching the point where there won't be enough water for people's future needs.

Climate change is also causing shortages and droughts in some areas and flooding in others.

By 2025, two-thirds of the world's people could face water shortages. Around 4 billion people suffer severe water scarcity for at least one month each year—especially in poorer countries and in rural areas. Fast-growing cities around the world also face severe water shortages and often divert freshwater away from rural lands. Water pollution from pesticides and chemical fertilizers is a major problem, while wasteful use of water and global warming also threaten supplies.

Currently, 3.6 billion people live without safely managed sanitation, and 1.7 billion lack even basic sanitation. This exposes them to cholera and other deadly waterborne diseases.

OUR AIR

"Our very bodies are made up of her elements. We breathe her air and we receive life and refreshment from her waters"
Laudato si' (2)

Clean air is everyone's birthright. Yet globally, 9 out of 10 people breathe air that contains high levels of pollutants. At the same time, air pollution contributes to global warming and harms the natural world. Its worst effects are suffered by the poor, especially in cities. Many of the actions we can take to cut air pollution would not only improve health but also bring benefits for our climate, biodiversity, and quality of life.

Outdoor air pollution kills more than 4 million people each year. The main types of outdoor air pollution are smog and soot. These are released from burning fossil fuels in motor vehicles and in industry, and to generate power. In fast-growing cities, open burning of waste is often a major problem.

Breathing the particles released by these activities is a health risk for all, causing heart disease, stroke, and lung cancer. But people in low- and middle-income regions are especially exposed. Some air pollutants are also major contributors to global warming.

Air pollution in the home is also a leading cause of disease—especially among the poor. At least 3 million people die every year from indoor air pollution—mainly from smoke from charcoal, wood, and dung used for cooking.

When people burn these fuels for cooking or to heat their homes, dangerous pollutants are released, such as small particles and carbon monoxide, which damage the health of all members of a household, especially women and children who are often at home the most. Respiratory illnesses, cancer, and eye problems are among the most common illnesses caused by indoor air pollution.

OUR FOOD

"There is a great variety of small-scale food production systems which feed the greater part of the world's peoples"
Laudato si' (129)

The way we produce and consume food is intimately connected with the living world around us, and with climate change, biodiversity, water use, and pollution. Agriculture also increasingly provides raw material for industry and infrastructure. As the world population grows, we need to ensure food security for all as well as safeguard the ecosystems that are the very foundation of agriculture.

Intensive agriculture has reshaped the planet. The growing demand for food in the second half of the last century led to massive changes in agricultural practices around the world. This new model of industrial agriculture was largely driven by mechanization, new breeds of crops, and synthetic pesticides and fertilizers. Initially, these methods brought large increases in yields for some, but in many ways, the system is now unsustainable.

Intensive agriculture has had widespread and devastating environmental and social impacts. Many unique foods and crop varieties have been lost, and with them, bonds of culture in which food plays a vital role.

Agriculture in crisis, people and planet at risk. Acute hunger is rising in more than 50 countries, while a third of all food goes to waste. Intensive agriculture largely relies on fossil fuels; it has turned rich forests into farmland and is responsible for around one-third of greenhouse gas emissions.

As intensive farming methods erode and deplete fertile soil and surrounding biodiversity, they undermine the ability of future generations to grow adequate food. As local and indigenous farmers are displaced, we are quickly losing the deep reserves of skill and knowledge they hold.

OUR CONSUMPTION

"A world of exacerbated consumption is at the same time a world which mistreats life in all its forms"
Laudato si' (230)

By 2050, we would need three planets to support current lifestyles. Since 1970, we have been consuming more than the planet can sustain. What's more, 80 percent of the world's resources are used by only 20 percent of the population. Patterns of production and consumption in the industrialized world strip the Earth of its natural riches, such as forests, fish, minerals, and water. The wasteful ways we produce and consume goods are also highly pol-luting and harm the health of people and the living world, while contributing to global warming.

Each year, around 12 million tons of plastic enter our oceans. Plastic waste is an especially serious symptom of over-consumption. There may now be more than 5 trillion pieces of plastic—both larger pieces and so-called microplastics—floating in our seas, which have devastating effects on marine species and biodiversity. Microplastic particles find their way into our food, water, and even bloodstreams. Half of all plastics ever produced have been made in the last 15 years, and production is growing. At each stage of its life cycle, plastic can harm our health through exposure to plastic particles them-selves and the chemicals used in their manufacture.

OUR SHARED FUTURE

"A true ecological approach always becomes a social approach: it must integrate questions of justice"
Laudato si' (49)

Social justice and human rights are woven through all of today's most press-ing environmental issues. Those who contribute least to environmental harm often suffer its worst effects. Equity and fairness are also often at the heart of solutions. We cannot solve global warming, for example, without facing questions of who is most responsible for it. And women's rights must take a central place: when women are educated and empowered, we see better environmental outcomes.

The environment and social justice cannot be separated. The overuse of natural resources by industrialized nations means that poorer countries have paid a heavy price for the development of the richer world. Climate change is a clear example: richer countries are most responsible for it, but it is the poor and marginalized who are suffering the worst impacts.

Biodiversity and agriculture are also tied to justice and equality. As indus-try and intensive agriculture have advanced, the land rights of people who have long been stewards of our most biodiverse environments—smallscale farmers and indigenous peoples—have often been ignored.

We are all affected—especially the marginalized. Climate impacts, such as sea-level rise, extreme weather, and lower crop yields, will affect the poor the most. By 2050, more than 140 million people in Latin America, sub-Saharan Africa, and Southeast Asia could be forced to migrate because of the climate crisis.

Everyone has the right to clean air and water, yet the poorest in rapidly growing cities are very often deprived of both and are most exposed to pollution. Young people will have to live with the consequences of choices made today, so they need a greater role in decision-making now.

"Many things have to change course"
Pope Francis, Laudato si' (202)

We do not stand separate from the planet we share with other life; we are intimately related to it and so are also responsible for its care. This truth must be recognized in order to meet the challenges of this critical point in history. What is at stake is nothing less than our children's rights to a safe climate, clean water and air, sufficient food, physical security, and the wonders of a planet rich with life.

We cannot overcome the twin crises of climate change and biodiversity loss, nor overcome pollution, resource degradation, poverty, and injustice, without transforming outmoded patterns of behavior, culture, and economics. We must move beyond an exploitative relationship with our planet toward one based on stewardship and care. While such a transformation has already begun, the gravity of the situation demands greater action.

People and communities must join together to ensure that those who make decisions and hold the most responsibility and power understand that action on these issues is a priority for the people they serve and that they are held accountable.

As much as this is a moment of crisis, it is also one of opportunity: we can renew our relationship with the planet so that we not only sustain ourselves but also thrive and flourish.

A Joint Initiative of the Holy See and the Stockholm Environment Institute

The Stockholm Environment Institute is an international non-profit research and policy organization that tackles environmental and development challenges.

The Dicastery, a department of the Vatican, assists the leadership and members of the Church to promote integral human development and care for our common home.

In Laudato si', Pope Francis calls us to develop a "loving awareness" of this home we share and to act on the values we hold dear (Laudato si', 220).

The Laudato si' Action Platform is inspired by the integral ecological vision of Laudato si'. The platform empowers communities, institutions, and groups to become the change that they want to see in the world in the current era of planetary emergency. Learn more about what you can do to bring about positive change.

Visit the Laudato si' Action Platform: laudatosiactionplatform.org

This publication was originally designed as a 20-page printed booklet.

In Laudato si', Pope Francis calls us to develop a "loving awareness" of this home we share and act on the values we hold dear (Laudato si', 220).

The Laudato si' Action Platform is inspired by the integral ecological vision of Laudato si'. The platform empowers communities, institutions, and groups to become the change they want to see in the world in the current era of planetary emergency. Learn more about what you can do to bring about positive change.

Acknowledgments

I extend a big thank you to all who participated in this book journey! This book is a testament to the collaborative efforts of many individuals. We are a team with many different positions, and I am grateful to them all. Over 30 people played a crucial role in shaping this book, a task akin to herding cats—a skill honed during my coaching days (paid and volunteer) in youth baseball, basketball, and football. I still play competitive sports that teach me how to win or how to get back up when I am knocked down. These pursuits have instilled in me a knack for fostering teamwork, joy, fun, work, inquisitiveness, and a passion for intellectual pursuits in students and colleagues. This includes discovering, collecting, analyzing, publishing, and preserving knowledge.

I owe an immense debt of gratitude to my wife, Carla Snyder, whose unwavering support sustained me through the darkest moments; this book is dedicated to her. During the pandemic, Carla and I found solace and inspiration in our long walks through the Arizona desert, covering an average of 7,000 to 12,000 steps five to six days a week. I rest on Sundays with family and friends.

Ellen Slaten deserves special recognition for her meticulous editing suggestions. She took my Introduction to Sociology course at the University of Houston as an undergraduate and went on to earn her Ph.D. in Sociology from the University of Texas at Austin. Her dedication and insights significantly enhanced this book; any errors remain my responsibility alone.

Our research draws on both classic methodologies and statistical techniques pioneered by sociologists such as Emile Durkheim and refined during my tenure at UCSB under the guidance of Harvey Molotch, Richard Appelbaum, Roger Friedland, Bill Bielby, Robert Smith, and visiting scholars like Christopher Jencks and David Harvey from Harvard and the City University of New York, respectively. I should also add that I spent some time in the University of California, Berkeley planning program where I got to work with Dennis Keating, Michael Teitz, Mark Dowie, and Ken Barr. Thanks to Greg Davis, my first college instructor. Their mentorship laid the groundwork for my approach to finding and analyzing data—a critical aspect of impactful research. I also appreciate the help of my brother-in-law, Ronald G. Larsen, who wrote a variety of fiction and non-fiction books, including *The Little Book on Punctuation*. Also, I appreciate the inspiration given to me by Hanneke and Hans Gelderblom of the Hague, Netherlands. My thanks to Tom Fitzgerald, Scott Cummings, Mark Dowie, Donovan Rypkema, and Russ Barnett for the tutoring on sustainable development. My thanks to Nancy Jurik, Ralph Nader, and Andrew Cuomo for supporting environmental stewardship: Thank you to Richard Maraj, Zoe Prowell, James Shields, and Jimmie Scott for spiritual guidance.

This book features contributions from 30 authors, including esteemed professors such as Gregory Squires, Wesley Meares, Chad Frederick, Joshua Ambrosius, Russ Barnett, Robert Friedland, Matt Hanka, and William Riggs, spanning disciplines from medical schools to public health, planning, public administration, sociology, and political science. These chapters were all written by the authors without using AI technology.

I am deeply grateful for the collaboration and contributions of my major co-authors: Wesley Meares, William Riggs, Stephen Roosa, Chad Frederick, Matthew Hanka, and Gregory Squires. Ron Larsen, Cary Lowe, Carla Snyder, and Jon Lorence also played pivotal roles in editing and statistical analysis. Their expertise and dedication enriched these chapters, making them robust, fact-based, and impactful.

I am also grateful for the tremendous amount of funding I raised from the U.S. Department of Education; U.S. Department of State; U.S. Department of Housing and Urban Development; U.S. Environmental Protection Agency; Commonwealth of Kentucky; several local foundations including the Gheens Foundation; the Miller Foundation; my real estate company, Dutch Trading Company; HoyleCohen; Charlene Leung; Carla Snyder; Fidelity Investments; Kentucky Institute for Environment and Sustainable Development; Woody Harrelson; Joe Hickey; David and Elaine Day; Marilyn Melkonian; Bobby Austin; Center for Sustainable Urban Neighborhoods; and Neighborhood Associates. Good writing is always a team effort, and everyone has a position. I am lucky to be the quarterback in this effort, but I also handed it off to others to help me out. Mahalo to all.

I am thankful for the visionary leadership of former Presidents of the University of Louisville (UofL): John Shumaker, James Ramsey, Neeli Bendapudi, and the late Donald Swain, along with Mayor for life Jerry Abramson, who counseled me on getting things done. Early on, the Urban Policy graduate program was ranked number 16 by the U.S. News and World Report. I was grateful to Scott Cumming, the original Chair and Editor of the Journal of Urban Affairs, who brought me to UofL when I was about to accept the offer of an elite Ivy League planning school. I wanted to come to the belly of the beast—the epicenter of horrible environmental policies that kill people, places, and the planet. At that time, we had a powerful faculty made up of Hank Savitch, Bert Useem, Steve Koven, Jerry Vito, and Carrie Donald—most of whom have left or are no longer part of the program. Fighting the good fight to keep our program relevant was Distinguished Professor Russ Weaver, a leading expert on academic freedom, who wrote several published studies of case law interference defending our free speech rights on urban affairs.

My thanks to UofL and the Ph.D. Program in Urban and Public Affairs. In 1988, they supplied me with research assistance, grant assistance, and smart doctoral students who went on to become professors around the world or took powerful positions in government or the private sector. This all came crashing to a halt when the university closed its environmental centers in 2018. I still teach at UofL, and I also hold the position of Senior Fellow at Neighborhood Associates in Washington, D.C.

Special thanks to Bobby Austin, President of Neighborhood Associates, and the late Marilyn Melkonian, CEO of Telesis in Washington D.C., for their unwavering support during my time away from UofL. Their belief in my work was instrumental. Chris Nolan's encouragement has been transformative, urging me to adopt a positive and proactive stance in advocating for cities as pivotal in mitigating climate chaos (https://climateofhopefilm.org/).

I am indebted to the graduate students whose exceptional talent in data research was pivotal. Many of these students co-authored chapters and unearthed critical data, including information leaked from the EPA during challenging times. Chris Bird's bravery in exposing hidden pollution measures during his tenure at the EPA was particularly notable, shedding light on connections between pollution, public health, and societal impacts.

Isaiah Kingsbury and Charles Zhang's contributions to building maps, figures, and tables significantly bolstered our presentation of factual evidence. Moreover, anonymous contributions from a graduate student working for Metro Government and the school district provided crucial insights into the Toxics Release Inventory (TRI) data, pivotal in our award-winning research on the impact of toxic pollution on the health and learning outcomes of schoolchildren and teachers in Louisville.

Carrie Beth Lasley and Matt Hanka's comparative analyses, Zach Kentizer's adeptness in finding biking-related data, and the collective efforts of sociology, public administration, medicine, and political science professors were invaluable in connecting research dots. Porter Stevens' collaboration with us during our study of Portland's urban successes was particularly enlightening.

I extend my gratitude to colleagues from UofL and beyond, including Mark Wright, Rob Mullins, Joshua Ambrosius, Wes Meares, Matt Hanka, Chad Frederick, Billy Riggs, Michael Brazley, Zachary Kenitzer, Kareem Usher, John Markham, Ken Chilton, Karl Bessel, Muthusami Kumaran, Lin Li, Geetha Suresh, Viviana Andreescu, Stella Capek, Annette Allen, Ellen Slaten, David Romero, and others who continue to advance rigorous scientific inquiry in their respective fields. Thanks to my stepson, Gavin Greenstone, who helped me with fixing computer challenges.

Special thanks to Betsy Hatfield of Kentucky Preservation, Steve Wiser of the Louisville Historic League, Edie Bingham, and Joanna Hinton of Preservation Kentucky, Inc., for their unwavering support and contributions. Funding from Joanna Hinton facilitated our research, while Erin House and Tony Lindauer, Donna Hunt, and Jay Mickle of the Jefferson County PVA provided invaluable assistance.

Leah Callahan helped design the book cover and graphics and helped with converting files. We appreciate the cooperation of the Vatican (our visit there will be part of the Climate of Hope documentary) and the Stockholm Environmental Institute for granting permission to print Pope Francis' prayer and his call to connect faith with science to address climate change. I also appreciate my friendship with Al Gore that goes back 30 years and the work I did on the sustainability effort when he was Vice President of the United States. I provided him with our research over the years.

I am indebted to Julian Agyeman, Editor of Local Environment, for giving me his initial support and an outlet to write for his journal, and for letting me reprint revised versions of the articles. Ditto to Stephen Roosa, Editor of the International Journal of Strategic Energy and Environmental Planning, who quickly got five research articles reviewed and published in his outstanding start-up peer-reviewed journal during the pandemic. One of the papers was awarded the highest recognition for scholarship among the 30,000 scientists who are part of the Association of Energy Engineers. This paper shows the strong correlation between high levels of pollution and neighborhoods, and the record low number of students who cannot read or do math. Stephen Roosa also provided a compelling endorsement for the book, reflecting its impactful potential.

The UofL Urban and Public Affairs Program's dedication to scholarship and community service has been recognized globally, thanks to the efforts of numerous individuals honored by *The New York Times*, Martin Luther King III, President Nelson Mandela, Harry Edwards, President Bill Clinton, Andres Duany, Dave Simpson, Jeff Speck, Henry Cisneros, Marc Weiss, Andrew Cuomo, Al Gore, Jerry Brown, and Jerry Abramson.

Despite challenges to our academic freedom and integrity, including smear campaigns and threats of violence, we remain committed to uncovering truths and advocating for environmental justice. These challenges came from those tied to the city's top polluters: liquor, chemical companies, coal, and tobacco. The University of Louisville's decision to defund and dismantle the Kentucky Institute for the Environment and Sustainable Development and the Center for Sustainable Urban Neighborhoods marked a turning point, causing many of the articles that serve as chapters in this book to be destroyed and, in some cases shredded. This necessitated resilience and perseverance in preserving and disseminating our research. I appreciate my Senior Acquisitions Editor, Courtney Morales, and her associate, Breasia Boyd, who saw my writings on climate change in *The New York Times, Harvard Medical Primary Care Review*, and a pre-publication in *Lancet* and wrote a passionate letter inviting me to write this book. Their letter convinced me to bring my draft book over to Lexington (now Bloomsbury) to build what Dr. Bobby Austin calls a "seminal work that revolutionizes our understanding of environmental science and politics." I also appreciate the remarkable group of people working with Bloomsbury/Lexington: Ashleigh Cooke, Arun Rajakumar, and others associated with Deanta. Many thanks for their patience, dedication and 100% effort for making this a change maker book.

Despite numerous efforts to silence me and my team, a brief that defended my right to speak out against climate injustice was developed in collaboration with FIRE (Foundation for Individual Rights in Education), private attorney Buddy Cutler, and Russ Weaver, Distinguished Law Professor at the University of Louisville. Several peer-reviewed journals are documenting this sordid case that was presented at the 2024 International Conference on Free Speech in Budapest (of all places) and the 2024 International Livable Cities Conference. But sadly, the same powerful international corporations are still at work trying to suppress, harass, and discredit scientific climate research. These are the same groups who attacked Rachel Carson when she wrote *Silent Spring* and won the Presidential Medal of Freedom, and Jeffrey Wigand, who as a scientist found ample evidence that smoking is dangerous to health and whose story was documented in the movie *The Insider*. It's an honor to do similar work as these individuals who have been attacked viciously by local and international corporations who kill people, places, and the planet.

As extremist climate denialist groups made efforts to stop the dissemination of scientific, impartial, data-driven research, the interest of three-time Emmy Award-winning documentary filmmaker Chris Nolan was sparked to develop a documentary informing millions of people around the world about this research. He wanted to bring attention to the effort to move away from a gloom-and-doom perspective and promote hope that climate change can be reversed. As this book nears completion, my collaboration with Chris Nolan on the documentary, *Climate of Hope: Cities Saving the World*, marks a milestone in our mission to spark global awareness of irreparable environmental crises and potential solutions.

Together, we aim to mobilize communities worldwide to embrace proactive urban strategies, firmly believing that cities hold the key to a sustainable future. Science won the fight against polio, measles, COVID-19, and AIDS, and we can win the fight against climate chaos.

Chapter 1

Climate Chaos

The Battle Between Science and the Corporate Merchants of Doubt

John Hans Gilderbloom

The following media reports from 2022 to 2024 paint a frightening picture of climate change and the future of the planet. Have we finally reached a tipping point in our civilization at which people will say, "Enough!"?

- Jakarta is the world's most polluted city—the sea is rising so high the entire capital is being moved to higher ground (*CNN*, August 16, 2023)
- As predicted by scientists the intensity of Hurricanes would get even worst as shown in the devastation of Helene and Milton.
- Man found dead in car amid 126-degree heat in Death Valley National Park (*Fox Weather*, July 6, 2023)
- Extreme weather event, resulting in flash flooding leading to over 205 deaths in Valencia, Spain (*New York Times*, November 1, 2024)
- Extreme heat wave bound for Phoenix and Southwest could be worst ever (*Washington Post,* July 7, 2023)
- How hot is South Florida? Beach sand was 137 degrees and playground floor reached 177 (*Miami Herald*, August 10, 2023)
- Earth hit an unofficial record high temperature this week (*AP News*, July 5, 2023)
- East Palestine train derailment killed more than 43,000 fish and animals, officials say (*USA Today*, February 24, 2023)
- Nearly 2 billion at risk from "unprecedented" climate conditions (Axios, May 23, 2023)
- South Florida Ocean Temperature Tops 101 Degrees Fahrenheit, Potentially a Record (CNBC, July 25, 2023)

- Venice faces an Unwelcome Honor: Joining the Endangered Places List (*The New York Times*, August 1, 2023)
- Iran Begins Two-Day Shutdown as Heat Wave Worsens to 140 degrees (*Wall Street Journal*, August 2, 2023)
- Maui wildfire death toll, already highest in modern U.S. history with 117 found dead and 350 still missing many presumed to have drowned, could surge (*NBC News*, August 18, 2023)
- The wildfire caused destruction or damage to 2,200 structures, with 86 percent of them being residential in West Maui. The losses approach $6 billion (*Spectrum News*, August 13, 2023)
- Earth is at its hottest in thousands of years (Washington Post, July 8, 2023)
- For Millions in Mexico City, Water is Disappearing: Drought and Sprawl Push Resources to the Brink (*The New York Times*, May 19, 2024)
- The world's top medical journal, *The Lancet*, found in a review of over 400 studies that places with high pollution are responsible for 9 million premature deaths annually (Beelen et al., 2014).
- "One third of Pakistan underwater, 1,700 people killed" (Al Gore's Speech at the United Nations Climate Change Conference [COP27] in Sharm el-Sheikh, Egypt on November 7, 2022)
- "A heatwave in China that lasted 70 days with the heat above 40 degrees, 104 Fahrenheit over a vast area of China" (Al Gore's Speech at the United Nations Climate Change Conference [COP27] in Sharm el-Sheikh, Egypt on November 7, 2022)
- "Another million displaced in Nigeria two weeks ago. Another million in Chad a few days after that" (Al Gore's Speech at the United Nations Climate Change Conference [COP27] in Sharm el-Sheikh, Egypt on November 7, 2022)
- "Droughts are drying up the mighty Mississippi River . . . the Tigris and the Euphrates in the cradle of civilization, the Poe River in Italy, Loire in France, the Rhine in Germany" (Al Gore's Speech at the United Nations Climate Change Conference [COP27] in Sharm el-Sheikh, Egypt on November 7, 2022)
- Indonesia is sinking, polluted capital is moving to a new city (*AP Wire*, January 26, 2022)

This is not climate warming but climate chaos—unpredictable, savage, deadly, and scary. Who would have known that on August 8, 2023, a deadly firestorm/hurricane in Maui would kill more than a hundred people and destroy more than 2,200 buildings? If it happened in paradise, it could happen anywhere.

Climate chaos is resulting in flooding, fires, starvation, and new wars. How could our leaders be so tone-deaf about climate chaos and the increasingly unlivable places? Have people become brainwashed to believe that nothing should be done for the environment save planting a tree, recycling beer cans and newspapers, or riding a bike on Sundays? How can we prevent further damage to the environment? The answer is simple: strictly regulate pollution, and it will save thousands of lives, places, and the planet.

As Stephen Roosa notes in chapter 6, new environmental problems have surfaced that emphasize the ultimate fragility of our planet's ecosystems. Due to our unresponsive choices, climate change is occurring at a rapidly accelerated pace. Our common future is becoming grim. No matter what actions we take today, the consequences of increasing carbon levels will continue to plague the Earth for generations to come. Never have human-induced environmental changes had such a global impact. Local environmental changes from climate change can be devastating. At this juncture, nothing less than the sustainability of life on our planet is at stake.

The most important issue facing the world is climate chaos and its impact on health, housing, and the livability of the planet. This book is focused on environmental degradation and public health, aiming to show how pollution and other environmental abuses are reducing the quality and length of life and harming social life as well as the economy. Toxins to be examined range from air pollution by industrial smokestacks, chemical polluters, liquor distilleries, coal production, and automobile emissions to ground and water pollution from agricultural runoff, hazardous waste spills, and toxic waste dumps. Much attention will be paid to environmental degradation in cities, where pollution and the resulting threat to dense populations are highest. Also addressed will be the issue of "environmental racism," as we see the most polluted places are home to a far larger percentage of "minorities" (African Americans and Latinos, most noticeably) than white Americans. The roles of both the polluters and the politicians who fail to regulate this activity are discussed. This book will also detail solutions to the problems of pollution.

A growing number of studies in the United States and globally—many of them published in leading medical journals—establish connections between mortality and environmental pollutants. Some of these, particularly relating to fine particle air pollution, focus on the concept of premature death or Years of Life Lost (YLL) in urban areas around the world. Other studies examine everything from fetal damage to cancer, endocrine disruptors, learning disabilities, and a wide variety of chronic health conditions that undermine quality of life, both for children and adults.

Urban air pollution has not received particular attention in the United States. While public attention often focuses on dramatic incidents like Bhopal or Fukushima, a serious source of harmful pollution that is more ordinary and less visible is day-to-day exposure. For example, Samet et al. (2000) studied the effects of five major outdoor-air pollutants on daily mortality rates in 20 of the largest cities and metropolitan areas in the United States from 1987 to 1994, finding "consistent evidence that the levels of fine particulate matter in the air are associated with the risk of death from all causes and from cardio-vascular and respiratory illnesses." Focusing on long-term exposure effects, Pope et al. (2002) concluded that "Fine particulate and sulfur oxide-related pollution were associated with all-cause, lung cancer, and cardiopulmonary mortality." Chay and Greenstone (2003) found that a U.S. economic recession in 1981–1982 reduced the estimated total suspended particulates (TSPs) and caused a decline in infant mortality at the county level, estimating that 2,500 fewer infants died than would have been the case without the TSP reductions. The study also emphasized the important role of fetal exposure and nonlinear effects of TSP pollution. In a study set in Europe, a recent analysis found long-term exposure to fine particulate air pollution to be associated with natural-cause mortality in 22 European cohorts (Beelen et al., 2014). The British medical journal *The Lancet* warns that "it is not just the emerging economies that need to take heed" of the mounting evidence of dangers of small particulate pollution, noting that the United States and Europe have regulatory limits above the World Health Organization's recommendations, and that "Even small reductions in these could confer substantial population health benefits" (Beelen et al., 2014).

Similarly, a recent Organisation for Economic Cooperation and Development (OECD) report, *Environmental Outlook to 2050: The Consequences of Inaction*, projects that unless new policies are instituted, by 2050 air pollution will be the largest contributor to premature mortality, more than doubling current numbers and causing 3.6 million deaths per year (OECD, 2012). The report particularly emphasizes risks to urban residents in Asia, especially India and China; it concludes that the cost of inaction could be "colossal" as far as human health and economic risks, in addition to a projected 50 percent increase in greenhouse gases.

Systematic research literature on environmental justice issues has long established that these hazardous sites—including waste and sources of air pollution—are disproportionately located in minority neighborhoods (Been, 1995; Pastor et al., 2001; Bullard, 1990; Bullard, 2000). Although there has been some debate as to whether hazardous sites have been in or relocated to minority neighborhoods, or whether minority population "move-in" occurred after the location of those sites (Bullard, 2000; Pastor et al., 2001), recent studies indicate that the sites are in minority neighborhoods, not that minorities tend to move to areas where such sites are located (Hipp and Lakon, 2010; Smith, 2009; Swanston, 2012). This can be due to the neighborhood's lack of political power or a weak real estate market (Hamilton, 1995; Banzhaf, 2012). Consistent evidence supporting the disproportionate impact of environmental contaminants on racial and ethnic minority neighborhoods should raise the level of concern about the possible connection to shortened lifespans.

The University of Louisville (UofL) Urban and Public Affairs (UPA) Program, which I helped build, conducted extensive research on how deadly pollution in Louisville, Kentucky shortens lives, destroys neighborhoods, and makes many parts of the Earth uninhabitable (Gilderbloom, 2016; Gilderbloom et al., 2021). Some of the UPA program's findings include the following:

- Louisville has some of the worst and deadliest pollution in the United States (Gilderbloom et al., 2020);
- As predicted by climate scientists, the intensity of hurricanes will get even worse as shown by the devastation of Helene and Milton;
- The poor die five years sooner in Louisville than in California because of pollution (Gilderbloom et al., 2022);
- The lifespan of West Louisville residents is 10 to 12 years shorter than the lifespan of East Louisville residents (Gilderbloom et al., 2020);
- COVID-19 is directly correlated with high levels of four U.S. Environmental Protection Agency (EPA) measures of pollution, with Louisville ranked as having the highest death rates among mid-size cities (Meares et al., 2021);
- Less than 50 percent of schoolchildren living near chemical factories can read or do math at proficiency levels (Gilderbloom et al., 2021);
- The toxic release of chemicals is 12 times higher in West Louisville compared to East Louisville (Gilderbloom and Squires, 2022);
- Asthma, heart disease, and lung disease are up to three times higher in these polluted areas (Gilderbloom et al., 2020); and
- Local foundations have not funded research to counter the propaganda of corporate polluters.

How deadly are these toxins? One accident, one angry employee, or one terrorist attack could wipe out thousands of lives in West Louisville with a chemical leak or explosion. In 1984, perhaps the world's worst chemical accident occurred in Bhopal, India. As many as 3,000 people died instantly after the release of dangerous gases at the Union Carbide industrial plant, and an additional 1,532 people died shortly after the release (Ellis-Petersen, 2019). There were 574,366 injured victims, and upward of 20,000 people suffered from severe respiratory system problems (Ellis-Petersen, 2019). The highly toxic gas that was released, methyl isocyanate, is used as a pesticide, and it is not clear whether the cause of the leak was sabotage or company negligence. Given the relative density of West Louisville, if an accident or terrorist attack occurred at one of the 44 chemical factories or on the train carrying toxins there, it is estimated that roughly 60,000 residents could die instantly and, like Hiroshima, a total of 250,000 could die within a year, according to Russell Barnett, a former official from the Commonwealth of Kentucky Energy and Environmental Cabinet, the recently closed Kentucky

Institute for the Environment and Sustainable Development, and Center for Sustainable Urban Neighborhoods.

Environmental problems in U.S. cities are the most important issue facing America today. The abuse of urban environments is killing the economy, people, places, and the planet. Many books have been written about this crime, and in each one, the blame is directed at different sources: regulations, old technology, the needs of the economy, or the victims themselves. The most recent book by Jeff Goodell, *The Heat Will Kill You First*, ignores the problem of industrial pollution, especially chemical companies located in the rust belt and South (Goodell, 2023). He puts the blame squarely on fossil fuels and most notably gasoline cars. Yet one Louisville chemical company can produce the equivalent of 650,000 cars creating greenhouse gases (McKenna and Bruggers, 2021). Part of the problem is that the impact of pollution is poorly measured. The data exists, but social scientists seem too lazy to analyze it. We have a national emergency, and we need to start looking at this crisis using impartial, rigorous analysis.

The problem of urban toxins is neither an exclusively capitalist nor a socialist problem but a crisis for both political systems that grows every day. Researchers in the United States are just beginning to recognize the connection between location and premature death. For example, the Centers for Disease Control and Prevention estimates that people who live in Alabama have a much shorter lifespan than those who live in Oregon. Thousands have shortened lives in Alabama because of unregulated pollutants compared to Oregon, where the U.S. Environmental Protection Agency (EPA) is not a dirty word. Worse yet, environmental racism is an even bigger problem, with half of African Americans in Alabama never living to their full life expectancy (Gilderbloom et al., 2020b). Within cities, we have seen powerful evidence of how a neighborhood can shape how long someone can live. In Louisville, controlling for race and income, it is a 12-year difference; in Kansas City, public health officials claim it is a 14-year difference. Place matters.

While city officials in Louisville might make disingenuous claims that reduced life expectancy is due to lifestyle choices (diet and lack of exercise) or blame poor education and income, our research counters this by showing that half of the reduced life expectancy is explained by how close a person lives to environmental toxins in the ground, water, and air. None of the books on this subject go directly to the scene of the crime—U.S. cities—or have the courage to point the finger squarely at the agents who are perpetrating this crime—toxic polluters. Like no other book that has come before, we show how much of the disintegration of our social life can be explained by how we treat our urban environment.

This book looks at who is suffering, who is creating this emergency, and, in the most unique aspect of this work, what we are losing as a society. While apologists for toxic industries point the finger at everyone but the polluters and threaten the American people with the loss of jobs, this book illustrates how green living creates jobs, supports long and healthy living, and builds strong communities. It is time to recognize that the belief that allowing unregulated pollution creates a higher standard of living is false and that the exact opposite is true; the intuitive truth is that our standard of living is dependent upon a healthy environment. This book presents the debate in plain terms and makes the reader an offer they cannot refuse: to accept the fact that the economy should work for the people, and that the environment is the only truly precious commodity.

The methods and statistics are grounded in the award-winning and pathbreaking work of Harvey Molotch and Richard Appelbaum. Ironically, the original work on inter-city analysis of planning decisions was published by Praeger over 49 years ago. This methodological approach was later adopted by such scholars as Richard Florida, Roger Friedland, and me (along with my students Wesley Meares, William Riggs, Joshua Ambrosius, Chad Frederick, and Matthew Hanka). We look at the entire universe of all U.S. cities with a population of more than 50,000, not located within 20 miles from another similar size city. We put together over 600 columns of data, from dozens of sources like the EPA, USGS, BLS, CDC, Gallup/Healthways, Places Rated Almanac, National Endowment for the Arts, and the U.S. Census. We utilize sources of data that most urbanists don't use, which go well beyond the standard and tired explanation of urban dynamics and environmental phenomena.

The world's top medical journal, *The Lancet*, found in a review of over 400 studies that places with high pollution are responsible for 9 million premature deaths annually (Beelen et al., 2014). Other pollution research estimates that 1.2 million premature deaths occur annually in Chinese cities due to polluted air (Wong, 2013). Instead of confronting the problem of environmental pollution and its adverse effects on human lifespans, humankind is spending billions of dollars in search of treatments and cures. As the Physicians for Social Responsibility point out, we should try to prevent "what we cannot cure" (Physicians for Social Responsibility, 2013).

Unfortunately, the EPA under President Trump moved in the opposite direction, as it has weakened or eliminated many rules and regulations designed to protect the environment. President Biden is trying to put new teeth into environmental regulation, but all 95 regulations that President Trump dismantled need to be restored. As of April 2023, the EPA acted after seeing our studies, calling for a cut back in emissions and demanding a cut of 6,500 pounds of toxins.

This research can help shape best practices and policies that should be adopted by local, state, and the federal authorities. Even with the availability of an effective COVID-19 vaccination, wearing masks, social distancing, and washing hands continue to be important steps for mitigating the impact of this pandemic. Putting a mask on pollution would save thousands of lives. Reducing the level of air pollution would be a major step. Unfortunately, environmental protection rules were weakened or eliminated by the Trump Administration. Reversing this pattern is one additional step that can be taken to reduce COVID-19 cases and deaths.

Pollution levels, which are the focus here, may not appear at first glance to be something that can be quickly changed. But with appropriate leadership (and perhaps local organizing to pressure leaders), older industrial sites can be remediated in many cases. The EPA could change course and reintroduce what are, in fact, life-saving environmental protection rules. Local and state leaders could act on their own, and many around the world are doing so. But in the meantime, we might follow the guidance of the Physicians for Social Responsibility, who asserted in reference to the costs of pollution: "We must prevent what we cannot cure."

The livability of the Earth worsens every day with droughts, fires, starvation, pandemics, wars, and millions of people dying. What is the future of our great cities? Venice will be gone in a few years, according to the United Nations Educational, Scientific and Cultural Organization (UNESCO). Phoenix, California, Hawaii, and Nevada are experiencing droughts and might run out of water. It's a war between science and the disgusting lies put out by the pollution industry. As Peter Dreier states on Facebook in December 2023,

> The biggest threat to respect for objective, impartial science is corporate America including the fossil fuel, chemical, tobacco. . . . For decades they have spent big bucks trying to undermine scientific research that exposes the dangers of their practices for consumers, workers, public health, and the environment, and raise doubts among the public about the matters like the reality of climate change, the dangers of smoking, the harm caused by unsafe workplaces, etc.

We call it corporate terrorism waged on people and the scientists. Dishonest polluters and government are the "merchants of doubt" who are going after science and want to justify the status quo.

The merchants of doubt have been attacking science since the age of Galileo. While Galileo is celebrated for his magnificent scientific discoveries that changed the world for the better, few know that he spent the last 12 years of his life in home incarceration because he challenged the Catholic Church's teaching with his discovery that the Earth circles the sun. The merchants of doubt attacked Rachel Carson 60 years ago because she warned people about the dangers of chemicals being sprayed on lawns and farmland in her book, *Silent Spring*. Several chemical companies in Louisville funded the attacks on Carson. Thirty years ago, they attacked Jeffrey Wigand, known as "the man who knew too much," for making the truth known to the public via *60 Minutes* about the industry's disregard for health and safety (https://www .jeffreywigand.com). He was vilified by the tobacco industry.

We are at a critical threshold, where action on the environment is required now. This book provides an entirely different perspective and will encourage politicians of all parties to recognize a few simple truths and to legislate accordingly. Instead of pretending that the environment, society, and the economy are in opposition in a zero-sum game, our approach is that they are intimately interconnected and that protecting the environment is a win-win. The operation and function of cities defy intuition. There are many moving parts from multiple scales of government to multiple markets for property, labor, and resources. They function differently at different sizes, densities, and geographies. They require training to be properly understood. There are few, if any, books on urban environments written by scholars of cities. This is the first book to focus on the impact of environmental degradation on the quality of urban social life; it is urgently needed since most of the people in advanced industrial countries live in cities.

A brief outline of the chapters follows. Some of my students did international comparative studies between Amsterdam and U.S. cities.

PART I. NATIONAL STUDIES ON THE IMPACT OF POLLUTION

The Missing Link of Air Pollution: A Closer Look at the Association between Place and Life Expectancy in 146 Mid-Sized Cities

Clean air has long been a priority for environmental planning policy in the United States. From 2016 to 2020, President Trump removed 94 environmental regulations that protected the water, air, and soil. Research has been mixed on the relationships between air pollution and life expectancies; some recent work suggests that these relationships cannot be empirically supported. Our research explores how people living in places with high levels of pollution have lower life expectancies. We test this relationship using four different EPA measures of air quality to analyze the impacts of air pollution. Data for these types of airborne pollution were used to assess the relationships of human mortality rates in 146 semi-isolated, medium-sized U.S. cities.

Accounting for control factors, these data indicate that in areas with high levels of air pollution, men and women in lower income quartiles live for four or five fewer years than in cities with relatively clean air. A more rigorous regression analysis reveals that all four of EPA's measures are statistically significant and have strong coefficients nearly equal to income and gender.

The implications of these findings have very specific implications for public health and planning—air quality matters for more than just climate value; it is associated with shorter human lifespans. Our research shows the importance of how environmental planning measures that reduce air emissions impact life expectancies. It underscores the importance of local policies that reduce emissions from stationary sources and offer incentives to increase the proliferation of clean fuel vehicles.

What Cities Are Most Dangerous to Your Life Expectancy? Toward a Methodology of Livability

Poor air quality is a major worldwide problem, prematurely killing more people than any other disease. This chapter proposes a method for identifying toxic air in cities and how it impacts lifespan, fetal health, education scores, housing values, and labor force participation. After collecting EPA air quality data, we found four standard measures of air quality for medium-sized cities. We use the Molotch/Appelbaum comparative planning method of looking at 142 mid-size cities with populations over 50,000 that are not located within 20 miles (32.2 km) of other similarly sized cities. The method, results, and ranking of mid-size cities from cleanest to dirtiest in terms of air quality will inform citizens on how to address social, health, and economic problems in cities with poor air quality and empower them to support clean air policies. EPA measures are the best available and clash with fraudulent claims of industry insiders that the air quality in specific localities is not toxic. This method aids the medical community in explaining why COVID-19 cases are more prevalent in highly polluted cities.

Pollution and the Pandemic: Explaining Differences in COVID-19 Rates across 146 U.S. Communities

COVID-19 is a nationwide epidemic, but not all communities are hit the same. It is paramount to understand what factors are associated with the varying levels of COVID-19 cases across the counties of the United States. In our sample, the investigators ask why there were no COVID-19 deaths in certain mid-size communities and more than 1,000 deaths in other places. This chapter examines variation in COVID-19 cases and deaths across 146 mid-size communities in the United States to determine the effect of air quality and sociodemographic and political characteristics. It is a cross-sectional design. The study uses a series of predictor variables reflecting exposure and vulnerability to the virus (e.g., share of essential workers, extent of pre-conditions), socio-economic characteristics of these communities (e.g., racial composition, median income, pollution levels), and political factors (e.g., presence of a mask mandate, political party of mayors) were examined. Data was collected from the American Community Survey, John Hopkins, Atmospheric Composition Analysis Group, County Business Patterns, Social Explorer, Various Local Governments, and *The New York Times* Coronavirus data. Bivariate correlations between each of the predictor and two dependent variables were calculated. To assess the independent effects of each predictor, a series of multivariate regression models were constructed.

In evaluating a variety of generally recognized predictors of the number of COVID-19 cases and deaths resulting from the virus, the level of air pollution proved to be a particularly strong and significant predictor. The political affiliation of mayors also was significant, with jurisdictions led by Democratic mayors having fewer cases. When this chapter was written, the vaccine had not yet been developed. The authors mention several steps that can be taken to ameliorate the number of cases and deaths. As is often recommended, wearing masks, social distancing, and washing hands are important steps for mitigating the impact of this pandemic. At the same time, reducing the level of air pollution can be a major step. Unfortunately, environmental protection rules have been weakened rather than strengthened in recent years. Reversing this pattern is one additional step that can be taken in order to reduce COVID-19 cases and deaths. In June of 2023, the EPA, based on our research, estimated that the level of chemical toxins in Louisville and Cancer Alley would be cut down by 6,000 pounds. In Louisville, this is a small but significant step with another 69,000 pounds to go.

Automobile Addiction Kills the Earth: The Need for Multimodality

Urban sprawl has a negative relationship with regional health outcomes, such as high obesity and chronic diseases related to physical inactivity. At the same time, literature has shown that walkable built environments are connected to lower obesity rates and increased physical inactivity. Less understood is the association of transportation multimodality with those health outcomes. Researchers studied variations in a range of public health outcomes to determine their associations with commute mode diversity. Our test measure is the percentage of commuters who use some means other than a single-occupant vehicle. The percentage of such commuters ranges from a low of 11 percent in Houston County, Alabama, to a high of 36 percent in Honolulu County, Hawaii. Using bivariate and multivariate analysis, we show better health outcomes in counties and MSAs with more modal diversity. These findings underscore the positive impact of sustainable transportation policies on community health and open a new direction for public health research and the built environment.

Reducing Greenhouse Gas Emissions and the Need for Renewable Energies

In this chapter, Roosa notes that the environmental changes associated with greenhouse gases have been festering for a long time. Our patterns of energy use since the early 1900s, which have focused on the use of fossil fuels, have contributed to our increasingly dire situation. The results of our past insensitivity to the environment are becoming more noticeable, the consequences less dismissible. Cities in Brazil, Mexico, China, and India are reeling from population growth, poor development choices, and environmental damage. Roosa states that access to energy is key to sustainability and hope for solutions can be found in policies, programs, and technologies.

PART II. LOCAL STUDIES ON THE NEGATIVE IMPACT OF POLLUTION

How to Do a Pollution Audit in Your City

Why do people die 12 years earlier in West Louisville sacrifice zones? In these areas, approximately 75,000 pounds of chemicals are released where 62,000 people live. These people are mostly minorities and low-income persons needing low-cost houses. This chapter describes a typical neighborhood located near chemical companies.

Environmental justice has been defined as the pursuit of equal justice and equal protection under the law for all environmental statutes and regulations without discrimination based on race, ethnicity, and/or socio-economic status. This concept applies to governmental actions at all levels—local, state, and federal—as well as private industry activities. Lower-income communities and minority populations have historically been the targets of many sources of pollution. Air pollution from industrial sites, toxic contamination from incinerators and brownfields, contamination of ground and source water, and lead exposure from aged housing structures are just a few of the environmental hazards that disproportionately affect low-income communities.

"Mama, I Can't Breathe." Louisville's Dirty Air Has Steep Medical and Economic Costs

The call for greater racial equity means cleaning up the air, water, and soil. Poor people needlessly suffer more in Louisville than the same low-income people in West Coast Cities. If Louisville adopted the same tough environmental regulations as its West Coast counterparts, West Louisville would surely bloom instead of slowly die. The unfairness between black and white neighborhoods is stark and vivid. Science and public health officials can show Louisville how to solve some of its most pressing problems, and other cities can learn from its example. If Louisville would find the will to address the pollution of West Louisville, it could prove to be a case study in best practices on how cities can confront environmental and health injustice. Deadly pollution is Louisville's most urgent problem, making many Westside neighborhoods unlivable, unsustainable, unhealthy, and unprosperous. It is the number one cause of environmental racial injustice. The powerful want others to believe that pollution is not a problem. In other words, if they cannot see it, it must not exist. Science, data, facts, and truth tell a much different story. Indeed, pollution is largely invisible because the deadliest pollution, PM2.5, is microscopic, about 1/16th of a human hair, and gets clogged up in a person's lungs, heart, brain, and liver.

Pollution causes shortened lives. An article in a peer-reviewed academic journal, Local Environment, states that shortened lives of 10 to 13 years are due mainly to toxic pollutants in the air, soil, and water impacting 60,000 people in West Louisville (Gilderbloom et al., 2022; Gilderbloom and Squires, 2022; Gilderbloom et al., 2021; Gilderbloom et al., 2016; Gilderbloom et al., 2020; Gilderbloom et al., 2020; Gilderbloom et al., 2020; Meares et al., 2021). A study published in the *Journal of the American Medical Association* and *The New York Times* by Harvard Sociologist and Economist Raj Chetty et al. (2016) found that the overall lifespan of Louisville residents is five years shorter than green West Coast Cities like Santa Barbara and Santa Rosa (Gilderbloom et al., 2022; Chetty et al., 2016; Wong, 2013). Louisville's exploding COVID-19 cases and deaths are strongly correlated with West Louisville polluters. All mid-sized cities with high pollution levels are seeing COVID-19 cases and deaths explode like wildfire (Meares et al., 2021).

Pollution, Place, and Premature Death: Evidence from a Mid-Sized City

Neighborhood life expectancy varies by as much as 10 years in Louisville—perhaps the most extreme difference of any city in America. In 2013, the Greater Louisville Project, funded by government, business, and foundations, argued these differences had little to do with environmental factors. To test these claims, we constructed our own model of neighborhood variation in YPLL by adding two variables testing environmental degradation. We operationalized two different measures of environmental contamination: proximity to brownfield sites and proximity to chemical factories in an industrial park and neighborhood known as "Rubbertown." We conducted several regression analyses, which showed that environmental contaminants in Louisville are a major reason for shortened lives in the neighborhoods within close proximity to a toxic soup of chemicals. Our beta weights challenge the claims made by the Chamber of Commerce, which minimize the impact of nearness to environmental contaminants that cause reduced life expectancy in Louisville neighborhoods.

How Brownfield Sites in Neighborhoods Kill Places and People: An Examination of Neighborhood Housing Values, Foreclosures, Crime, and Lifespan

This study examines the effects of EPA brownfield sites on housing values, foreclosures, and premature deaths in Louisville, Kentucky, between 2000 and 2008. While previous research has focused on the impacts of brownfield sites on neighborhood housing values, little research has been done on the impact of these hazardous sites on foreclosures and premature deaths. We utilize ordinary least squares regression to analyze the net impact of brownfield sites on neighborhoods. We find a significant association between brownfield sites and lower neighborhood property values, increased foreclosures, and premature deaths. Furthermore, using a case study of Louisville's East Russell neighborhood, we demonstrate the possible benefits of brownfield site remediation. Based upon the findings from the regression and the case study, we offer policy prescriptions that help address the fiscal and social costs of brownfield sites.

PART III. NEIGHBORHOOD AND CITY EFFORTS TO COMBAT CLIMATE CHANGE

How to Make Our Schools Greener and Our Students Smarter

Turkey Foot Middle School, located in northern Kentucky, the heart of coal country, is one of the greenest school buildings in the United States. Given historical and economic ties to fossil fuels, it seems improbable that Turkey Foot Middle School could become one of the most energy-efficient facilities of its time. This chapter describes how it happened. The transition from energy-intensive schools to green schools can result in enormous energy savings; higher grades and test scores; less violence, and happier students (because of daylighting in classrooms); lower building costs; and reduced water costs (from water conservation techniques and rainwater collection systems).

The Impact of Air Pollution on Public School Achievement

This chapter also examines the impacts of air pollution on educational achievement in Louisville, Kentucky, one of the most polluted cities in the United States. It examines the impacts of air pollution on educational achievement and student attendance. Yet, little research has been performed on the impacts that toxic air pollution has on student achievement in elementary schools. To determine the impacts of air pollution on student education, our study's methodology included a two-group comparison coupled with a statistical analysis using key variables. We compared 10 public elementary schools located near chemical factories in the western Louisville neighborhood known as "Rubbertown" to 10 elementary schools furthest away in the wealthy east side of Louisville with the least exposure to toxic air.

Our conclusions contradict the disinformation provided by the Louisville Air Pollution Control District. Our data indicate that: (1) air pollution has gotten a lot worse, not better, in Louisville, according to the EPA's risk screening environmental indicator; and (2) the quantities of toxic air that mostly minority school children are exposed to are 35 times higher than the mostly non-minority elementary schools elsewhere in the city. Consequently, reading, mathematics, attendance, and suspensions are far worse in schools near the polluting factories and industries. Teacher retention in schools located in areas with high levels of air pollution is problematic. One out of five teachers leaves these schools every year, as opposed to 1 out of 20 leaving schools in areas with healthier air quality.

Will Planting Eight Billion More Trees Solve Climate Chaos? No!

From cleaning our air to beautifying our world, trees are important in so many ways. It is amazing how much power a single tree has. One single tree can absorb up to 48 pounds of CO_2 a year (AEA, 2008). One single tree can provide daily oxygen for up to four people and absorb 14,000 pounds of rainwater a year (USDA Forest Service, n.d.). Trees have the ability not only to filter our air but also add beauty to our cities. Trees can also help reduce the threat of flooding by constantly absorbing groundwater and growing even stronger. Trees can also reduce the energy cost of buildings by shading houses from the hot sun, and in the winter, leafless trees allow the sun to heat homes. In Louisville, we studied how trees are beneficial in reducing energy costs. Obviously, trees also promote walkability by promoting cooler and more comfortable walks. As a result of the many benefits associated with trees and their ability to improve urban areas, it is imperative that city officials and planners start incorporating trees into cities and urban areas on a much larger scale. Similarly, city inhabitants need to raise awareness and support the planting of more trees. Just one tree can make a difference.

There are some substantial benefits of trees, but they are not a miracle cure to address the issue of heavy chemical pollution. Trees cannot significantly combat or reduce the ill effects of pollution on the reduction of lifespan. In Louisville, we have compared on a map neighborhood tree canopy concentration and its interaction with highly polluted neighborhoods and found no impact on the health of residents in terms of lifespan. Similarly, in another study, we compared tree canopy percentage in 144 semi-isolated mid-size cities and again found no correlation in conducting a regression analysis. The findings showed that lifespan, pollution, income, and race are the main factors in shaping the life chances of individuals. Moreover, the eight billion trees strategy in place of reducing pollution is often overstated, and some believe the impact of planting trees on reducing global warming is minimal, by only half a degree (George, 2023; Einhorn, 2023). But too often planting eight billion trees will not significantly lower the warming of the Earth or clean up the pollution. It's a corporate decoy designed to take the focus away from pollution and pollution regulation.

Does Walkability Matter? Exploring the Relationship Between Walkability and Housing, Foreclosure, Health, and Reducing Greenhouse Gases

The world's most respected urban planner, Jane Jacobs, in *The Death and Life of Great American Cities*, argues that the ideal neighborhood is one that is walkable. In this study, we propose to test these arguments in a study of 170 neighborhoods in a medium-sized city to see whether walkability matters in terms of measures of livability. For 50 years, we never had a reliable measure of the social and economic impact of walkable neighborhoods. This changed dramatically when walkability index measures were introduced. The more people walk or bikewalk, bike, or use buses/trains the less pollution in the air. Walkability measures gauge how accessible daily living activities are by foot and the likelihood of being car-dependent in the city where one lives. In general, inner-city neighborhoods built before the mass production of cars are more walkable than sprawling "residential only" suburban neighborhoods that are isolated from the basic necessities of everyday life. We developed a hedonic price equation that controls for recognized independent variables that predict our dependent variables and adds in our test variable, walkability. We find that walkability is statistically significant in predicting an increase in neighborhood housing values and has a significant negative correlation with foreclosures in the neighborhoods of Louisville. Finally, walkability is also associated with reduced crime in several measures.

Biking is the Best Choice for Health, Safety, and Zero Emissions

Biking is one of the most viable transportation choices for healing a broken Earth. Compared to cars, it releases none of the dangerous greenhouse gases that wreak havoc on the environment. Choosing a bike over a car means better health for bikers, longer life spans, decreased medical expenditures, and an average of $8,000 per year in savings. A bike-friendly neighborhood also increases community and housing values. The chapter features a case study of the economic benefits of increased bicycle usage by auto-commuters to the University of Louisville. Our research suggests that increased urban biking means neighborhoods will become more prosperous and sustainable. The local economic benefits are significant in shifting car-centered transportation toward a more bike-friendly culture in urban neighborhoods. Fewer cars and more bikes, along with bike-oriented infrastructure, could translate into higher property values, more jobs, decreased traffic congestion, and, ultimately, more money in the consumer's pocket. We estimate that the Old Louisville neighborhood could see an infusion of up to $26 million in new community spending simply through a reduction in car dependency. This spending on housing repairs, installation of community gardens, and locally operated grocery stores and restaurants (expected results of the cost savings created by the reduction or discontinuance of car-based commuting) represents the potential green dividend of investments in urban bicycling. Another important finding is that people who ride bikes will be healthier due to increased physical activity and improved air quality. But, most importantly, the net zero impact of carbon emissions from bicycles means a reduction in greenhouse gases and, hopefully, a way to heal the Earth back to normalcy.

Designing and Building Affordable and Attractive Housing for Working People

There is little doubt that Covington HOPE VI is one of the best green housing developments for low- and moderate-income people in the country. In an era of climate chaos, this deserves high praise. River's Edge will become LEED Certified and is further pursuing certification at the Silver Level, making it one of the only affordable, green housing developments. River's Edge is a major accomplishment. It provides a compass for affordable and green housing advocates on how to get this done. Put simply, the promise of LEED is to cut energy costs by one-half of what the poor normally pay. We have a great deal of praise for the LEED model, but we also suggest ways that LEED certification guidelines can be improved to decrease energy consumption. In an era of climate change, building, siting, and designing sustainable, affordable homes works to the advantage of residents, neighborhoods, cities, and the world. The United States has only a handful of multi-family housing communities that are affordable, environmentally friendly, and energy efficient. Covington, Kentucky's HOPE VI development provides one of the best examples to date on how to design energy-efficient, affordable developments—both inside the home as well as locating the development in a place that reduces car dependency by encouraging walking, biking, and public transit. This chapter illustrates both the highlights and shortcomings of Covington's HOPE VI green development and describes how to improve the quality of life with green infrastructure by exploring the three environmental/energy programs used in the development: LEED, Enterprise, and Energy Star.

Portland, The Best Livable City in America: Equity, Health, and Safety

This chapter is about how all of us can act against climate change. Today, Portland is considered one of the most livable and sustainable mid-sized cities in the United States and is different from most other U.S. cities in one crucial way: it has reduced its per capita carbon emissions. Portland has not only become a city with a steadily shrinking environmental footprint, but one that is experiencing an economic boom based on the "green" economy, a massive influx of young and college-educated residents, and steadily rising rates of walking, bicycling, and transit use. How has the city accomplished all of this? The answer is: by embracing urbanism. Over the last 40 years, the city of Portland has instituted policies that incentivized the construction of dense, mixed-use development. Combined with generous investments in streetcars, light rail, and bicycling infrastructure, Portland has created a dense, walkable urban environment that has been attracting new residents and new investment ever since. In this chapter, we will put the city of Portland under the microscope and learn what makes it tick. The other 270 mid-sized cities in America should look to Portland as an example of how investing in alternative transit and dense, mixed-use neighborhoods can pay off handsomely; an example of how "being green" and economically thriving do not have to be competing ideas. Portland's charm is that it is a weird, unconventional, and exciting place to live, and the city is living proof that investing in the "green" lifestyle pays off in a big way.

Amsterdam: Planning and Policy in the World's Most Livable City

Amsterdam includes many components that make it an ideal city: dense and compact neighborhoods near work and school, biking and walking for over half the population, light and hard rail access, minimal personal usage of cars, historic preservation, personal freedom, and support for the arts to attract young people. Many of the social, economic, and environmental problems facing Amsterdam are considerably fewer than in cities in the United States, and in most cases, Western Europe. Amsterdam, at this moment in history, might be the world's greatest city because of its ability to ensure necessities, freedom, and creativity. In this chapter, the author compares, on a per capita basis, differences between Amsterdam/Holland and the United States. The data show that Amsterdam has lower rates of crime, murder, rape, drug usage (cocaine and marijuana), teenage pregnancy, diabetes, obesity, suicide, abortion, infant mortality, dependence on fossil fuels, and homelessness. There is also considerably less racial segregation. People live longer because of Amsterdam's walkability, bike usage, and access to parks. People living in Amsterdam seem more tolerant, secure, happier, and healthier compared to citizens in the United States. Amsterdam is by no means perfect, but in comparison to many other democratic industrial cities, it is a far better place for citizens of all races, religions, and incomes. There is much to learn from the Dutch about important solutions to policy and planning, and Amsterdam provides us with valuable lessons. The Dutch have designed a modern city that significantly reduces greenhouse gases by encouraging half of the nation to ride bikes regularly and up to 90 percent to use bikes occasionally.

Historic Preservation as a Sustainability Strategy to Foster Pro-Environmental Cultures

This chapter examines the impacts of historic preservation on jobs, property values, and reducing fossil fuel consumption. Louisville, Kentucky is a national leader in preservation, ranking first in the White House's Preserve America initiative with 73 recognized communities. Tax incentive programs have been an effective tool for creating positive changes in historic areas. Louisville has more affordable housing than any other state in the nation, and that is largely due to having the oldest housing stock in the nation. We have more shotgun housing stock than post-Katrina New Orleans. In our previous books looking at 144 medium-sized semi-isolated cities, we found over several decades that the greater amount of preserved older housing stock, the lower mean housing costs (Gilderbloom and Appelbaum, 1988; Gilderbloom, 2008). Historic preservation results in more job creation than most other public investments. In the presence of escalating gas prices and assorted environmental practices, we show that neighborhoods containing historic districts exhibit higher increases in median neighborhood housing values than undesignated neighborhoods. This chapter also shows that environmentalism and historic preservation are linked together and complement each other. Residents of historic urban neighborhoods exhibit more environmentally friendly behavior, particularly those living in single-family homes. More importantly, they recycle materials that reduce energy consumption with embedded energy.

PART IV. PUT A MASK ON POLLUTION

Climate of Hope: Cities Leading the Way

It's going to get worse. Look around, and you'll see that no place is safe from the ravages of climate chaos. Every state has experienced the pain of climate chaos—Hawaii, Florida, New York, Arizona, California, and of course, Kentucky. East to West, North to South, we are witnessing the slow destruction of the Earth as we once knew it. Or look around the world: North Pole, Africa, Middle East, China, Russia, and South America. We have a climate crisis that is destroying our way of life with record high temperatures. We are facing a catastrophic crisis that will destroy the Earth as we know it in eight years, according to the United Nations. Our best chance to reverse climate change is our cities. By 2050, 70 percent of the population will live in cities. Imagine our world in 2050, where seven out of 10 people reside in cities. The world's population is projected to be around 9.7 billion, which means an astonishing 6.6 billion people will be living in cities. The relentless urban migration is up 55 percent since 2018, which means the brunt of climate change will be experienced in metropolises. A documentary titled *Climate of Hope: Cities Saving the World* is being developed in conjunction with this book. Our mission is to inspire and inform more front-line neighborhood crusaders to advocate for city-driven climate change mitigation worldwide. We aim to create a powerful grassroots movement for change that will amplify the distribution and momentum of the book's message. Every block, every neighborhood, every family matters. Climate is about more than data; it's about human lives. Cities, at their core, are dynamic organisms. They evolve, adapt, and with the right nudge, can heal.

What We Hope to Accomplish with This Book

This book is bookended with the writings of Pope Francis and a dynamic speech at the United Nations by Nobel Prize winner Albert Gore. Our message is clear: we are not interested in merely slowing down devastating climate change (and the resulting raging fires, droughts, flooding, starvation, wars, and mass migration), but we have the audacity to reverse it with a proposed 80 percent reduction in greenhouse gases. I recently spoke with Nobel Prize winner Al Gore, and he assured me that our proposals are doable, but we need an army of big-thinking scientists to do the research and advocacy. You might have seen me featured in the film *Rubbertown* on Amazon, which upset the big polluters in West Louisville, or read my pieces in the *Harvard Medical School Primary Care Review* (Gilderbloom et al., 2021) or *The New York Times* (Gilderbloom, 2016).

Our approach is impartial scientific research using big data sets with plenty of ideas to make enough "good trouble." Too many administration leaders embrace pseudo-science that is compliant with a political agenda, not intellectual rigor. Some people don't want scientific research to happen; they want to serve greedy corporate polluters who choose millions of dollars over the needs of people. Polluted neighborhoods are marked by dramatically reduced lifespans, thousands of boarded-up or demolished houses, and 50 percent of elementary school students unable to perform functional math or reading. This culture promotes "censorship, self-censorship, and declining standards." Top scholars who embrace traditional norms of science, rigor, and impartial analysis are harassed with investigations and threats.

It's time to stand up for science and recognize how it can improve the livability of our city. Too many people are unnecessarily dying in over 1,000 places in America where there is heavy, dangerous pollution. Politicians from both parties assert that pollution regulations kill jobs and hurt the economy. They portray the EPA as a villain. Presidential candidates have run on platforms to abolish the EPA. Many Democrats have jumped on the same bandwagon. They seem to think they can only win in the South by perpetuating the idea that air, water, and soil protections kill jobs. This claim is false, and the empirical evidence does not back it up. To test the argument that clean air regulations kill the economy, we ranked states by air quality and unemployment levels using data from the United States Health Foundation and the Bureau of Labor Statistics. The unemployment rate in "clean" states (those in the top third) is just 4.2 percent. In the "dirty" states (those in the bottom third), it's 6.9 percent. More importantly, residents in states with cleaner air live longer.

Perhaps the fundamental divide is no longer between blue versus red states but rather green (clean) versus brown (dirty) states. And the green states are winning; people who live there have better job opportunities, a cleaner environment, and longer lifespans. Candidates in brown states should rethink their losing strategy of attacking environmental regulation. Most people—including Republicans—want clean air, water, and soil in their communities. Instead of perpetuating the lie that government cannot help, Democrats and moderate Republicans can win elections in Kentucky by embracing the need for thoughtful regulation to protect healthy environments and careful investments in education. Becoming a cleaner, greener state is the key to better employment opportunities, longer healthy lifespans, and more sustainable communities for everyone in Kentucky and throughout the South.

The EPA is the United States' primary protector of its air, water, and soil. Yet the agency is often written off as little more than a job killer by political opponents and those in industry. Particularly in the South, elections are won by politicians who decry the job-killing effects of environmental protection.

The environmental movement needs to get wiser because it is losing ground. Proponents of environmental protection need to change the debate from jobs to healthy life expectancy. They might begin by asking how many people want to continue living lives shortened by 10, 12, 14 years because of polluted air and environmental degradation in distressed neighborhoods in many other cities nationwide. Environmental issues should be debated as the life and death issues that they really are.

Environmental justice means that all neighborhoods, minority and impoverished, are given the same environmental protections as wealthy neighborhoods. Environmental degradation is the major reason why thousands of people die prematurely. Lower-income communities and racial and ethnic minority populations have historically been the targets of disproportionate amounts of pollution.

People live longer, up to 14 years longer, in neighborhoods that have clean air than residents of polluted places. This is an empirical reality. Inequalities are generated within and among cities that contribute to varying life chances, including the length of life itself. But instead of responding with more medical treatments, why not address the causes of these premature death rates? Why not remove the environmental toxins from our neighborhoods? We are spending billions of dollars to find cures for diseases that kill while ignoring the causes of these problems. Most of these problems are rooted in our industrialized environment of cars, industry, and fossil fuels, chemical factories (which is a large focus of this book since it is ignored by scholars).

Properly designed research studies show, in short, that remediating environmental degradation and reducing air pollution could help cities reduce worrisome neighborhood disparities in premature deaths. Across America, states with cleaner air have citizens who live longer. The CDC has estimated the causes of variations in "healthy life expectancy"—that is, living longer without major medical problems. It finds that residents in eight states (most in the South) are more likely than Americans in general to experience premature deaths linked to air pollution.

A person's neighborhood is a powerful predictor of how long he/she will live. If our residence is in close proximity to dangerous environmental contaminants, we are likely to have a shorter life. While we know from studies done outside the United States that this can be a major problem for neighborhoods close to environmental contaminants, this issue has not received the attention it deserves in the United States.

According to the Air Pollution Control District and Russ Barnett, former director of the Kentucky Institute for the Environment and Sustainable Development, Louisville produces more greenhouse gases per capita than any other city in the USA. Thirty percent of the world's greenhouse gases come from the United States, and the USA represents only 6 percent of the world's population. Foreign-owned chemical companies locate in the USA. because of lax regulations that European countries would never allow. The EPA Toxics Release Inventory identifies 1,000 places around the USA where dangerous chemical companies release toxic chemicals in small and medium-sized places. These small and medium-sized places are bullied into believing that chemical companies bring jobs and that pollution is not a threat to people, places, and the planet. This book shows they are wrong.

Frustratingly, saving lives and cleaning the air require bucking politics as usual in the South, where elections are routinely won by politicians who decry environmental protections as job killers. Ironically, this overlooks how much killing of people rather than jobs is going on in polluted states and hides the reality that the U.S. states with the cleanest air have lower unemployment rates than states with the dirtiest air. In the 10 states with the cleanest air, unemployment averages two percentage points less than in the eight southern states with the dirtiest air. Or put another way, thousands more have jobs in clean states than dirty states. Proponents of environmental protection need to shift the spotlight to healthy life expectancy, asking why so many people—especially in the South—need to die prematurely just because their cities and states refuse to clean up the air, water, and soil around them.

REFERENCES

Arbor Environmental Alliance (AEA). (2008). Neighborhood Revitalization Initiative. Retrieved April 11, 2019, from http://www.arborenvironmentalalliance.com/carbon-tree-facts.asp.

Banzhaf, S. (2012). *Superfund Taint and Market*. Stanford, CA: Stanford University Press.

Beelen, R., Raaschou-Nielsen, O., Stafoggia, M., Andersen, Z. J., Weinmayr, G., Hoffmann, B., Wolf, K., Samoli, E., Fischer, P., Nieuwenhuijsen, M., Vineis, P., Xun, W. W., Katsouyanni, K., Dimakopoulou, K., Oudin, A., Forsberg, B., Modig, L., Havulinna, A. S., Lanki, T., Turunen, A., Oftedal, B., Nystad, W., Nafstad, P., De Faire, U., Pedersen, N. L., Östenson, C. G., Fratiglioni, L., Penell, J., Korek, M., Pershagen, G., Eriksen, K. T., Overvad, K., Ellermann, T., Eeftens, M., Peeters, P. H., Meliefste, K., Wang, M., Bueno-de-Mesquita, B., Sugiri, D., Krämer, U., Heinrich, J., de Hoogh, K., Key, T., Peters, A., Hampel, R., Concin, H., Nagel, G., Ineichen, A., Schaffner, E., Probst-Hensch, N., Künzli, N., Schindler, C., Schikowski, T., Adam, M., Phuleria, H., Vilier, A., Clavel-Chapelon, F., Declercq, C., Grioni, S., Krogh, V., Tsai, M. Y., Ricceri, F., Sacerdote, C., Galassi, C., Migliore, E., Ranzi, A., Cesaroni, G., Badaloni, C., Forastiere, F., Tamayo, I., Amiano, P., Dorronsoro, M., Katsoulis, M., Trichopoulou, A., Brunekreef, B., & Hoek, G. (2014). Effects of Long-Term Exposure to Air Pollution on Natural-Cause Mortality: An Analysis of 22 European Cohorts within the Multicentre

ESCAPE Project. *The Lancet*, 383(9919), 785–795. https://doi.org/10.1016/S0140
-6736(13)62158-3.

Been, V. (1995). Analyzing Evidence of Environmental Justice. *Journal of Environmental Land Use and Environmental Law*, 11, 1–37.

Been, V., & Gupta, F. (1997). Coming to the Nuisance or Going to the Barrios-A Longitudinal Analysis of Environmental Justice Claims. *Ecology Law Quarterly*, 24, 1–56.

Bhan, G. (2009). "This Is No Longer the City I Once Knew." Evictions, the Urban
Poor and the Right to the City in Millennial Delhi. *Environment and Urbanization*,
21(1), 127–142. https://doi.org/10.1177/0956247809103009.

Bullard, R. D. (1990). *Dumping in Dixie: Race, Class, and Environmental Quality
(Vol. 3)*. Boulder, CO: Westview Press.

Bullard, R. D. (2000). *Dumping in Dixie: Race, Class, and Environmental Quality*.
3rd ed. Boulder, CO: Westview Press.

Bullard, R. D., Mohai, P., Saha, R., & Wright, B. (2008). Toxic Wastes and Race at
Twenty: Why Race Still Matters after All of These Years. *Environmental Law*, 38,
371.

Chay, K. Y., & Greenstone, M. (2003). The Impact of Air Pollution on Infant Mortality: Evidence from Geographic Variation in Pollution Shocks Induced by a Recession. *The Quarterly Journal of Economics*, 118(3), 1121–1167.

Chetty, R., Cutler, D., Stepner, M., & Abraham, S. (2016). The Association between
Income and Life Expectancy in the United States; 2001–2014. *Journal of the
American Medical Association*, 315(16), 1750–1766.

Clauretie, T., & Daneshvary, N. (2009). Estimating the House Foreclosure Discount
Corrected for Spatial Price Interdependence and Endogeneity of Marketing Time.
Real Estate Economics, 37(1), 43–67.

Einhorn, C. (2023). How Much Can Trees Fight Climate Change, Massively, but Not
Alone. *The New York Times*, November 13, 2023.

Ellis-Petersen, H. (2019). Bhopal's Tragedy Has Not Stopped: The Urban Disaster
Still Claiming Lives. *The Guardian*, December 8, 2019.

George, Z. (2023). Can Planting a Trillion New Trees Save the World? *The New York
Times Magazine*, July 13, 2023.

Gilderbloom, J. I. H. (2008). *Invisible City: Poverty, Housing and New Urbanism*.
Austin, TX: University of Texas Press.

Gilderbloom, J. I. H. (2009). Amsterdam: Planning and Policy for the Ideal City?
Local Environments, 14(5), 473–494.

Gilderbloom, J. I. H. (2016). The Causes of Shorter Life Expectancies in America.
The New York Times, April 18, 2016.

Gilderbloom, J. I. H., Ambrosius, J., Squires, G., Hanka, M., & Kenitzer, Z. (2012).
Investors: The Missing Piece in the Foreclosure Racial Gap Debate. *Journal of
Urban Affairs*, 34(5), 559–582.

Gilderbloom, J. I. H., & Applebaum, R. P. (1988). *Rethinking Rental Housing*. Philadelphia, PA: Temple University Press.

Gilderbloom, J. I. H., Appelbaum, R. P., Dolny, M., & Dreier, P. (1992). Sham Rent
Control Research: A Further Reply. *Journal of the American Planning Association*,
58(2), 220–224.

Gilderbloom, J. I. H., Kingsberry, I., & Squires, G. D. (2021). How Many More Children Must Be Hurt by Pollution? *Harvard Medical School Primary Care Review*, March 3, 2021. https://info.primarycare.hms.harvard.edu/review/children -hurt-pollution.

Gilderbloom, J. I. H., Meares, W. L., & Riggs, W. (2013). *How Superfund Sites Kill Places and People: An Examination of Neighborhood Housing Values, Fore-closures, and Lifespan* (Forthcoming). Institute of Urban Affairs. University of Louisville.

Gilderbloom, J. I. H., Meares, W. L., & Squires, G. D. (2022). Pollution, Place, and Premature Death: Evidence from a Mid-sized City. *Local Environment: The International Journal of Justice and Sustainability*, 25(6), 419–432. https://doi.org/10 .1080/13549839.2020.1754776.

Gilderbloom, J. I. H., Reed, L., Turner, D., Brazley, M., & Squires, G. D. (2021). Pollution is a Form of Racial Injustice Crippling Western Louisville. *Courier Journal*, January 28, 2021. https://www.courier-journal.com/story/opinion/2021/01/28/ louisville-air-quality-west-end-suffers-while-others-prosper/6558314002/.

Gilderbloom, J. I. H., Riggs, W., Frederick, C., Squires, G., & Quenichet, K. (2022). The Missing Link of Air Pollution: A Closer Look at the Association between Place and Life Expectancy in 146 Mid-Sized Cities. *International Journal of Strategic Energy and Environmental Planning,* 4(3), 31–51.

Gilderbloom, J. I. H., & Squires, G. (2022). Put a Mask on Pollution: Connect the Dots Between COVID-19 and Unhealthy Air. *Social Policy*, 51(4), 28.

Gilderbloom, J. I., Squires, G. D., & Meares, W. L. (2020). Mama I Can't Breathe: Louisville's Dirty Air Has Steep Medical Costs. *Local Environment: The International Journal of Justice and Sustainability*, 25(8), 619–626. https://doi.org/10 .1080/13549839.2020.1789570.

Gilderbloom, J. I. H., Washington, C. B., Quenichet, K., Manella, C., Dwenger, C., Slaten, E., Sarr, S., Altaf, S., & Frederick, C. (2020). What Cities are the Most Dangerous to Your Health? Ranking the Most Polluted Mid-Size Cities in the United States. Pre-prints with *The Lancet*, January 9, 2020. https://papers.ssrn.com/sol3/ papers.cfm?abstract_id=3506217.

Goodell, J. (2023). *The Heat Will Kill You First*. New York, NY: Little, Brown, and Company.

Goodstein, R., & Lee, Y. (2010). Do Foreclosures Increase Crime? *SSRN*, September 5, 2010. https://ssrn.com/abstract=1670842 or http://dx.doi.org/10.2139/ssrn .1670842.

Hamilton, J. (1995). Testing for Environmental Racism: Prejudice, Profits, Political Power? *Journal of Policy Analysis and Management*, 14(1), 107–132.

Hipp, J. R., & Lakon, C. M. (2010). Social Disparities in Health: Disproportionate Toxicity Proximity in Minority Communities over a Decade. *Health and Place*, 16(4), 674–683. 10.1016/j.healthplace.2010.02.005.

Lewis, S. (2019). Planting a Trillion Trees Could be the Most Effective Solution to Climate Change. *CBS News*, July 8, 2019.

McKenna, P., & Bruggers, J. (2021). Louisville's Super-Polluting Chemical Plant Emits Not 1, but 2 Potent Greenhouse Gases. *Inside Climate News,* April 5, 2021.

https://insideclimatenews.org/news/05042021/chemours-lousiville-super-polluting-chemical-plant-greenhouse-gas-emissions/.

Meares, W., Gilderbloom, J. H., Squires, G., & Jones, A. (2021). Pollution and the Pandemic: Explaining the Differences in COVID-19 Rates Across 146 U.S. Communities. *International Journal of Strategic Energy and Environmental Planning,* 4(2), 9–26.

Organization for Economic Co-operation and Development (OECD). (2012). Environmental Outlook to 2050: The Consequences of Inaction. *OECD Publishing,* March 15, 2012. https://doi.org/10.1787/9789264122246-en.

Pastor, M., Sadd, J., & Hipp, J. (2001). Which Came First? Toxic Facilities, Minority Move-In, and Environmental Justice. *Journal of Urban Affairs,* 23(1), 1–21.

Peterson, E. (2013). Rubbertown and Health: The Whole Series. *Louisville Public Media,* January 27, 2013. http://wfpl.org/post/Rubbertown-and-health-whole-series-0.

Physicians for Social Responsibility. (n.d.). Website Slogan. *Physicians for Social Responsibility.* Retrieved May 31, 2014, from http://www.psr.org/.

Pope, C. A., Burnett, R. T., Thun, M. J., Calle, E. E., Krewski, D., Ito, K., & Thurston, G. D. (2002). Lung Cancer, Cardiopulmonary Mortality, and Long-term Exposure to Fine Particulate Air Pollution. *Journal of the American Medical Association,* 287(9), 1132–1141. 10.1001/jama.287.9.1132.

Samet, J. M., Dominici, F., Curriero, F. C., Coursac, I., & Zeger, S. L. (2000). Fine Particulate Air Pollution and Mortality in 20 U.S. Cities, 1987–1994. *New England Journal of Medicine,* 343, 1742–1749. https://doi.org/10.1056/NEJM200012143432401.

Sand, B. (2024). Everyone Says Trees Are Good. This Scientist Wants to Prove It. *Washington Post,* January 29, 2024.

Smith, C. L. (2009). Economic Deprivation and Racial Segregation: Comparing Superfund Sites in Portland, Oregon and Detroit, Michigan. *Social Science Research,* 38(3), 681–692.

Swanston, S. F. (2012). An Environmental Justice Perspective on Superfund Reauthorization. *Journal of Civil Rights and Economic Development,* 9(2), 18.

The Lancet. (2013). Editorial. Action on Ambient Air Pollution. *The Lancet,* 382(9897), 1000. https://doi.org/10.1016/S0140-6736(13)61960-1.

US Forest Service. (2019). Benefits of Trees. Retrieved April 11, 2019, from https://www.fs.fed.us/learn/trees.

Wong, E. (2013). Early Deaths Linked to China's Air Pollution Totaled 1.2 Million in 2010, Data Shows. *The New York Times,* April 2, 2013.

Part I

NATIONAL STUDIES ON THE IMPACT OF POLLUTION

Chapter 2

The Missing Link of Air Pollution

A Closer Look at the Association Between Place and Life Expectancy in 146 Mid-Sized Cities

John Hans Gilderbloom, William Riggs,
Chad Frederick, Gregory Squires,
and Karrie Quenichet

Recent research has generated controversy concerning whether air pollution has an impact on life expectancy. Chetty et al. found that "variations in life expectancy were correlated with health behaviors and local area characteristics" (Chetty et al., 2016). Contradicting land use and planning literature, they could not connect air pollution to shorter life expectancies. They said that ". . . physical aspects of the local environment affect health, for example, through exposure to air pollution" were not supported by their work. While the research underscores decades of urban planning work indicating land use and public health are connected, it takes a more passive approach to the impacts of vehicle miles traveled and air quality—stopping short of connecting the compounding effects of local environmental conditions on health trajectories.

The Chetty team has encouraged further analysis of their data. In our view, one limitation of their work was that the spatial scale of the measurements taken was at the national level. The team chose not to test whether pollution impacted lifespan by city despite available air pollution test variables on 547 places in the data set of 741 urban commuting zones (CZs). Our work uses a smaller spatial scale to investigate whether air pollution is associated with variations in the median human lifespan in cities. We gathered EPA airborne emissions data on four measures: sulfur oxides (SOx), nitrogen oxides (NOx), and particulate matter (PM 2.5 μ and PM 10 μ). Much of this air pollution is caused by local stationary power sources and vehicles. This research is positioned to complement land use planning, transportation planning, and engineering research aimed at reducing the harmful impacts of air pollution emissions from a human health and climate perspective.

BACKGROUND

A vast amount of research on local air pollution illustrates its negative effects on human health and planning considerations with far-ranging consequences for people of all races, genders, ages, income levels, and occupations.

Impacts of Air Pollution on Human Health

Numerous scholarly articles argue that pollution adversely impacts human health (Frumkin et al., 2004; Pope, 2000). These effects can be exacerbated by proximity to higher amounts of air pollution. While multiple variables affect the quality of life for the groups discussed, air pollution appears to be a key component. These works span the globe and cross disciplines from public health to city planning and engineering. Worldwide, air pollution was responsible for 7 million premature deaths in 2010 (Kuehn, 2014). The Organization for Economic Co-operation and Development (OECD) states that by 2050, air pollution will be the number one cause of rising premature death rates, ahead of current primary causes including sanitation and dirty water (Organisation for Economic Co-operation and Development, 2012). *The Lancet* estimates that 1.2 million premature deaths occur annually in Chinese cities due to polluted air, shortening life expectancies by four to thirteen years [9,13] (Rydin et al., 2012; Čapek, 1993). A study by the EPA estimated that a Trump Administration rollback of pollution restrictions on coal plants (that covered requirements for cleaning coal ash and toxic heavy metals such as mercury, arsenic, and selenium) could result in up to 1,400 more premature deaths annually in the United States (Friedman, 2018).

Air pollution is potentially even more damaging for infants. Studies have verified a link between fetal exposure to pollution and higher infant mortality rates (Chay & Greenstone, 1999). Reductions in atmospheric carbon monoxide (CO) levels reduce instances of low birth weight and development issues as children grow older (Currie & Walker, 2009; Almond & Currie, 2011). Prinz et al. have shown that prenatal exposure to various types of air pollution negatively affects academic test scores and economic outcomes (Prinz et al., 2018). Lower labor force participation and lower earnings are correlated with higher exposure to pollution (Isen et al., 2017; Borgschulte et al., 2018).

Urban Planning Considerations

Related public health issues connect to the urban planning discipline. For example, many studies have considered the connection between increased walking and cycling, decreased emissions, and increased public health outcomes, encouraging planning for community walkability as a mitigation strategy (Doyle et al., 2006; Riggs, 2015). One of the most cited studies is Frank et al.'s, which found that for every 5 percent increase in walkability, there were over 5 percent fewer grams of NOx and volatile organic compounds emitted (Frank et al., 2006). Economic planners have shown health benefits in terms of jobs and productivity. For example, Hanna and Oliva found that a 19.7 percent decrease in SOx emissions results in an increase of 3.5 percent in working hours for immediate neighbors within 3.1 miles (5 km) of a heavy polluter; this includes a 6 percent increase in the probability of those residents working over 40 hours weekly and a 2.5 percent increase in the probability of them working over 10 hours weekly (Hanna & Oliva, 2015).

Planning and environmental literature shows that many of the areas with the most sources of air pollution are located in racial and ethnic minority neighborhoods (Pahl, 2015; Appelbaum et al., 1976; Pastor et al., 2007; Bullard et al., 2008; Bullard, 2000; Keeley & Short, 2018; Been, 1995). While there has been debate as to populations that were forced to self-select into these locations, recent work suggests that many air polluting sources were located in minority neighborhoods after their racial composition had been established (Hipp & Lakon, 2020; Smith, 2009; Swanston, 2012). The evidence consistently supports the premise that minority populations are disproportionately impacted by pollution (Tessum et al., 2019; Hamilton, 1995; Banzhaf, 2012; Gilderbloom et al., 2014; Riggs, 2011).

Recent work by Chetty et al. explores how *place* shapes life chances yet provides inconclusive results with regard to the impacts of air quality. While the data from Chetty provide an accurate measurement of lifespan by city and affirm that places with higher incomes have longer life expectancies using granular life expectancydata, it fails to assess highly localized impacts of air pollution (Chetty et al., 2016; Frederick et al., 2018; Frederick et al., 2016; Gilderbloom et al., 2020; Riggs & Gilderbloom, 2016). Numerous studies emphasize that from a land use and transportation perspective, local conditions matter in relation to air pollution exposure (Currie & Neidell, 2005; Dockery et al., 1993; Samet et al., 2000; Caiazzo et al., 2013). They also relate to global climate-related initiatives to reduce greenhouse gas (GHG) emissions, helping to reduce some of the burden of these issues.

Reductions in fossil fuel combustion processes correspond with reductions in CO_2 emissions, lowering levels of criteria pollutants and particulates (Cifuentes, 2001). The benefits of reducing GHGs correspond to roughly a half million fewer premature deaths due to other types of air pollution and to a slowing in the effects of global warming by 2030 (West et al., 2013). A study by Burtraw et al. estimated that the total short-term health benefits gained through ancillary NOx and SOx reductions, and taxing all GHGs could justify the initial cost of the tax (Burtraw et al., 2003).

In the realms of planning and land use, the connections between air pollution and health have focused on urban heat island effects and climate mitigation. For example, Brian Stone's work advocated for air quality management from a pollution and climate adaptation perspective; he sees much of the public health benefits sourced from urban heat reduction measures (Stone et al., 2014; Stone, 2005). While Schweitzer and Zhou provided an analysis of how more compact cities can result in fewer air emissions, they do not address how the benefits of reducing pollution emissions connect back to public health in compact cities (Schweitzer & Zhou, 2010).

Boswell, Seale, and Greve argue for climate mitigation strategies in land use and transportation planning, yet cite secondary sources in making their case for the health benefits of climate-related pollution emissions reduction strategies (Boswell et al., 2012).

In this context, we position our work to extend the planning literature, underscoring the health benefits of compact cities, and re-evaluating Chetty et al.'s work at a smaller spatial scale (Chetty et al., 2016). This is positioned in the spirit of action, akin to the response to Stone by Winkleman, that we need to "re-imagine how our communities and regions look and function" while addressing the adverse impacts of air quality (Winkleman, 2007).

Methodologies

In contrast to the work of Chetty et al., we consider an air quality analysis that focuses on a local spatial scale. Instead of the larger state and CZs, we use the much smaller and generally more consistent unit of *county* as our level of analysis. While smaller geographical units may be used, given the research question (*viz.* "Is pollution in cities associated with variations in lifespan?") and data restraints (EPA data used here is available for roughly 547 counties in the United States), we chose to minimize the reductionist fallacy rather than the ecological fallacy.[1]

When using U.S. counties as the unit of analysis, two concerns make replicating recent research difficult: spatial lag and case selection. While we recognize many local policies happen at the level of cities, we use the county geographic framework as a policy lens to consider city policies, since most of our case cities contain one or multiple counties. In this sense, a limitation of this study is our reference to city policy, and cities are reflective of county geography. Most of the cases used in extant research either border or are situated near other cases of similar or larger populations. They often share labor, transportation, housing markets, and public policies. Adjacent counties and cities are subject to policy spillovers, whereby the impacts of one city's policies can be measured in geographically adjacent jurisdictions. Airborne emissions disregard areal units and political boundaries.

Case Selection

For case selection, researchers often use all available cities for which there is data. Since there is great variety among counties, this approach is known to urban scholars as misleading. It is highly problematic to include giant metropolises like New York City in a sample set that is dominated by smaller U.S. cities. Instead of lumping together categorically different urban centers, we select cases by using a well-validated set of decision rules from the field of urban affairs that addresses both the spatial lag and case selection concerns. We applied the research protocols found in Appelbaum's *Size, Growth, and U.S. Cities* (1978) and other studies which used this sampling technique (Appelbaum et al., 1976; Frederick et al., 2018; Appelbaum, 1978; Gilderbloom & Appelbaum, 1987; Molotch, 1976; Florida, 2002; Hanka & Gilderbloom, 2008). This approach has been used to study a wide range of interurban phenomena, including rent differentials, income inequality, public health, and quality of life.

Case selection begins with the universe of all incorporated places as defined by the U.S. Census. The population is then further reduced to those cities with populations over 50,000 which are not located within 20 miles (32.2 km) of another city of a similar size. Figure 2.1 shows the location of the cities used for this study. The counties in the dataset are neither evenly distributed nor clustered, but are randomly distributed throughout the continental United States (Moran's I: 0.049, $z = 1.62$). The county as a unit of analysis allows for a more granular assessment than the larger CZs. The distance between counties decreases problems of spatial proximity.

Dependent Variable

We borrow the lifespan measure and solid life span averages for cities from Chetty, who uses Social Security mortality data (Chetty et al., 2016). We focus on the dependent variable of *life expectancy* for four demographic groups: poor men and women, and wealthy men and women. Poor and wealthy are defined as the lowest and highest income quartiles, respectively. This measure is considered the most accurate measure of life expectancy by city. Traditionally, lifespan was estimated, Chetty's numbers are based on Social Security data of when a person actually died, with a mean created for each county.

ISOLATED MIDSIZE U.S. CITIES

Figure 2.1 One Hundred Forty-Eight Semi-isolated, Mid-Sized U.S. Cities. *Source*: Center for Sustainable Urban Neighborhoods, formerly University of Louisville, transferred to Neighborhood Associates, Washington, D.C.

Control Variables

All the remaining variables used in this study are from the U.S. Census 2013 three-year estimates measured at the geographical level of the county. Researchers often use variables without regard for multicollinearity and endogeneity, namely unemployment and the Gini coefficient, a measure of statistical dispersion. We use single measures for control variables that capture broad sociological and geographical phenomena associated with life expectancy: *latitude, population, density, median household income, percent black (African American), percent college educated,* and *commute mode diversity.*

Test Variables

The National Emissions Inventory, released every three years, provides a comprehensive estimate of the air emissions of criteria pollutants and hazardous air pollutants from emissions sources. State, local, and tribal air agencies provide data that supplement EPA data [61–63] (US EPA, 2016; US EPA, 2015). Four EPA variables, described below, are used in these studies [6,64–68].

- *Particulate matter (PM10 & PM2.5):* Particulate matter refers to inhalable particles of sized 2.5 and 10 micrometers which can become deeply lodged in an individual's respiratory system or bloodstream. Studies have linked particle pollution exposure to numerous health problems (e.g., premature deaths and increased respiratory symptoms) [69,70].

• *Oxides of nitrogen (NOx):* Oxides of nitrogen are reactive, poisonous gases, commonly emitted by automobiles, power plants, industrial boilers and turbines. Breathing high concentrations of NO_2 can irritate the airways which can aggravate respiratory diseases, like asthma, causing difficulty breathing and coughing [61,71].

• *Oxides of sulphur (SOx):* Oxides of sulfur are colorless gases that can be detected by taste and smell. The main sources of SO_2 are from fossil fuel combustion in power plants, vehicles and other industrial facilities [61,72]. Short-term exposure to SO_2 can harm the human respiratory system, making breathing difficult [70,71].

Analysis

As shown in table 2.1, we use cluster analysis to group the five cleanest and dirtiest cities as reflected in those that have the least criterial pollutants and the most pollutants, respectively. To test whether pollution had a significant net impact on life expectancy across cities ($N = 148$), we constructed a series of regression models using the backward entry removal method. The basic regression equation is as follows:

$$y = \beta0 + \beta1*\text{latitude} + \beta2*\text{density} + \beta3*\text{population} + \beta4*\text{income} + \beta5*\text{percent black} + \beta6*\text{percent college} + \beta7*\text{multimodality}, + \beta8*\text{EPA measures} + \varepsilon,$$

Where: y is the dependent variable of *life expectancy* for one of our four demographic groups, $\beta0$ is the constant, $\beta1$ through $\beta8$ are coefficients to be estimated, and ε is the error term. Regarding multicollinearity, tolerances are all above the critical threshold of .2. The presence of competing dependencies was explored and determined to be negative: in no case did two variables both show problematic values on the variance proportion table. Together, our methodology provides an accurate estimation and comparison of the impacts of the control and test variables on life expectancies in U.S. cities.

Results

As shown in tables 2.2, 2.3, and 2.4, our conventional regression model predicting urban lifespan shows that the air pollution test variable using four different EPA measures is a statistically significant predictor of lifespan in some cases, having a stronger impact over the standard control variables. For each

Table 2.1 The Impacts of Air Pollution on Life Expectancy, Low vs. High Emission Cities

City, State	County	Poor Women	Rich Women	Difference	Poor Men	Rich Men	Difference
Clean cities							
Flagstaff, AZ	Coconino	83.87	87.74	3.87	79.11	87.02	7.91
Yuma, AZ	Yuma	83.63	86.25	2.62	79.12	82.71	3.59
Redding, CA	Shasta	82.07	87.49	5.42	77.14	84.75	7.61
Yakima, WA	Yakima	83.11	86.88	3.77	79.18	86.02	6.84
Santa Barbara, CA	Santa Barbara	84.86	88.7	3.84	80.56	86.35	5.79
	Group average	83.508	87.412	3.904	79.022	85.37	6.348
Dirty cities							
Roanoke, VA	Roanoke	79.2	87.58	8.38	73.23	84.22	10.99
Columbus, OH	Franklin	79.6	87.58	7.98	74.38	85.06	10.68
Richmond, VA	Richmond	79.13	87.13	8	73.62	85.14	11.52
Terre Haute, IN	Vigo	77.89	88.65	10.76	73.03	83.45	10.42
Louisville, KY	Jefferson	80.02	87.37	7.35	74.03	84.72	10.69
	Group average	79.168	87.662	8.494	73.658	84.518	10.86
Btw group average		4.34	-0.25	-4.59	5.364	0.852	-4.512

Sources: 1. EPA data; 2. Gilderbloom, J.H., Riggs, W., Frederick, C., Squires, G., & Quenichet, K. (2022). The Missing Link of Air Pollution: A Closer Look at the Association Between Place and Life Expectancy in 146 Mid-Sized Cities. International Journal of Strategic Energy and Environmental Planning, V4.4; 3. Center for Sustainable Urban Neighborhoods, formerly University of Louisville, transferred to Neighborhood Associates, Washington, D.C.

Table 2.2   The Impact of Air Pollution on Wealthy Men and Women (LE factors)

	Wealthy Women Base	Wealthy Men Base	Wealthy Men NOx
Latitude			
Density			
Population	-0.133		
HH income	0.052		
Black	0.001	0.001	0.002
College	0.001	0.001	0.001
CMD		0.014	0.019
PM 10	-0.318	-0.331	-0.249
PM 2.5	0.514	0.511	0.534
NOx		0.170	0.161
SO$_2$			-0.136
F	28.462***	45.336***	35.241***
R	0.610	.697	.705
Adj. R^2	0.359	0.475	0.482
N	148	148	148

Sources: 1. EPA data; 2. Gilderbloom, J.H., Riggs, W., Frederick, C., Squires, G., & Quenichet, K. (2022). The Missing Link of Air Pollution: A Closer Look at the Association Between Place and Life Expectancy in 146 Mid-Sized Cities. International Journal of Strategic Energy and Environmental Planning, V4.4; 3. Center for Sustainable Urban Neighborhoods, formerly University of Louisville, transferred to Neighborhood Associates, Washington, D.C.

	Base		PM10		PM2.5		NO₃		SO₂	
Latitude	-0.276	0.001	-0.386	0.001	-0.314	0.001	-0.266	0.001	-0.260	0.001
Density										
Population	0.132	0.057			0.127	0.062	0.180	0.013	0.153	0.028
HH Income										
Black	-0.555	0.001	-0.528	0.001	-0.450	0.001	-0.442	0.001	-0.485	0.001
College			0.229	0.001	0.150	0.050				
CMD	0.205	0.005			0.142	0.066	0.200	0.006	0.211	0.004
PM10			-0.285	0.001						
PM2.5					-0.253	0.003				
NO₃							-0.188	0.038		
SO₂									-0.159	0.032
F	19.993***		22.706***		15.917***		17.247***		17.352***	
R	.599		.623		.635		.615		.616	.599
Adj. R²	0.341		0.371		0.378		0.356		0.357	0.341
N	148		148		148		148		148	148

Figure 2.2 The Impact of Air Pollution on Poor Women (LE factors).*Sources: 1.* EPA data; 2. Gilderbloom, J.H., Riggs, W., Frederick, C., Squires, G., & Quenichet, K. (2022). The Missing Link of Air Pollution: A Closer Look at the Association Between Place and Life Expectancy in 146 Mid-Sized Cities. International Journal of Strategic Energy and Environmental Planning, V4.4; 3. Center for Sustainable Urban Neighborhoods, formerly University of Louisville, transferred to Neighborhood Associates, Washington, D.C.

	Base		PM10		PM2.5		NO₃		SO₂	
Latitude	-0.198	0.002	-0.206	0.001	-0.193	0.002	-0.213	0.001	-0.183	0.003
Density							0.286	0.046		
Population	0.165	0.004	0.157	0.005	0.171	0.003	0.165	0.008	0.185	0.001
HH Income										
Black	-0.670	0.001	-0.626	0.001	-0.599	0.001	-0.611	0.001	-0.604	0.001
College										
CMD	0.288	0.001	0.273	0.001	0.285	0.001	0.261	0.001	0.293	0.001
PM10			-0.113	0.069						
PM2.5					-0.122	0.071				
NO₃							-0.397	0.004		
SO₂									-0.151	0.012
F	48.508***		40.118***		40.096***		35.528***		41.599***	
R	.759		.765		.765		.776		.771	
Adj. R²	0.564		0.571		0.571		0.585		0.580	
N	148		148		148		148		148	

Figure 2.3 The Impact of Air Pollution on Poor Men (LE factors).*Sources: 1.* EPA data; 2. Gilderbloom, J.H., Riggs, W., Frederick, C., Squires, G., & Quenichet, K. (2022). The Missing Link of Air Pollution: A Closer Look at the Association Between Place and Life Expectancy in 146 Mid-Sized Cities. International Journal of Strategic Energy and Environmental Planning, V4.4; 3. Center for Sustainable Urban Neighborhoods, formerly University of Louisville, transferred to Neighborhood Associates, Washington, D.C.

group, we first provide the base model with no pollutants. Although the same factors are present for both poor men and poor women (*latitude, population, percent black,* and *multimodality*), the base model explains greater variation for poor men. Another difference is that when particulate matter pollutants are introduced, education becomes significantly positive for women's life expectancy (LE).

Aside from *percent black*, the most significant factor in both the poor men's and women's models, the pollution variables are of comparable strength to all the other significant factors: *latitude, population,* and *multimodality.* Factors comprising the poor men's models are similar. However, poor women seem more vulnerable to *particulate matter,* while men are more impacted by NOx and SOx. Another difference is that compared to poor women, in no circumstances does education play a role for poor men. *Density* only plays a role when NOx is introduced to the model. Finally, living in a wealthier county does not seem to have a measurable impact on the LE of lower-income populations. *Population* seems to have a positive impact, while colder, more northerly *latitudes* have a negative impact.

As is often the case, for the model exploring higher-income individuals, the data is less compelling. For example, there is little discernible correlation between LE and pollutants and income for women. In contrast, college education replaces *percent black* as the strongest control variable. Furthermore, unlike the case for lower-income women, *population size* has a negative association with LE, yet *latitude* has no measurable effect. Unlike nearly every other model, the built environment in terms of *multimodality* has no role in the LE of wealthy women. For wealthy men, the model differs only in that *multimodality* plays a role, *population* does not, and NOx has a measurable impact on their LE.

Our analysis illustrates that air pollution has a greater tendency to occur in the nation's poorest cities. Residents die before their time—by as much as 5 years—and this is concentrated in Black communities. As tables 2.2, 2.3, and 2.4 illustrate, NOx, SOx, and particulate matter are all are negatively correlated with life expectancies. Most acutely, PM10 is strongly associated with reduced life expectancy. This extends the work done by Chetty and others, showing pathways between urban emissions and health outcomes—a true missing link in the research dialogue.

Discussion

Chetty et al. examined associations between life expectancy and income, finding that low-income populations in bigger cities with highly educated populations and more government expenditures had higher life expectancies than low-income populations in smaller cities with less educated populations and less government expenditures (Chetty et al., 2016). Our regression analysis extends that work and confirms most of the Chetty team's findings. There are social mobility benefits correlated to health outcomes for both men and women. While median household income has little role, education does matter. We find that these benefits are attained primarily for those with higher incomes. The team chose not to test for the impacts of pollution using their data set. Other research shows that pollution results in unwalkable cities, lower housing values, greater risk of foreclosure, and reduced tax revenues to support essential services (Gilderbloom et al., 2020; Meares & Riggs, 2016; Gilderbloom et al., 2015; Riggs & Sethi, 2019).

Though other behaviors matter (e.g., smoking and income levels), our research dispels several popular notions while confirming other literature from the planning field. Our research clearly shows that people are affected by the four types of pollution. It extends other research that shows that toxic air can contribute to countless other problems (e.g., higher rates of respiratory and heart diseases, more miscarriages, and certain types of cancer). The external costs of added health care are borne by the affected residents and society as a whole. In cities like Louisville, Kentucky, these impacts result in concentrated areas of health disparity (City of Louisville, 2017). In other cities such as Yuma, Arizona, people live up to five years longer than their counterparts since the air is cleaner, despite having the lowest tree canopy levels of any U.S. city. Trees provide the benefits of reducing energy consumption by shading buildings, encouraging walking, reducing flooding, improving the mental health of children, reducing traffic speed, and capturing/sequestering pollutants from the local environment (Gilderbloom et al., 2014; Louv, 2008).

The policy implications of our findings are important for cities: reducing pollution in cities improves life expectancy. While Chetty et al. updated the conversation on differing life expectancies across geographic regions, their data does little to explain causality, which is necessary for policy change (Chetty et al., 2016). In fact, Chetty argues that there is little proof that pollution matters. We attempt to fill the gap between environmental degradation and lower life expectancies in mid-size cities in the United States.

CONCLUSIONS

Our research represents the largest sample of U.S. cities (N=148) ever studied on the impact of pollution on lifespan. We use four reliable EPA measures that are used to study mid-sized cities that are semi-independent to assess inter-city differentials in lifespan.

Our research determined that four measures of airborne pollution (*PM10, PM2.5, NOx, SOx*) show statistically significant relationships with shortened lives of the poor. The greater the environmental pollution in a city, the greater the reduction in median life expectancy—even when controlling for race, gender, and income. Environmental degradation should be a more explicit focus of future research and public policy to better understand and address the health consequences of pollution. As the Lancet Commission (Landrigan et al., 2018) notes:

> *Pollution is the largest environmental cause of disease and premature death in the world today. Diseases caused by pollution were responsible for an esti-mated 9 million premature deaths in 2015—16% of all death worldwide—three times more deaths than from AIDS, tuberculosis, and malaria combined and 15 times more than form all wars and other forms of violence. In the most severely affected countries, pollution-related diseases are responsible for more than one death in four.*

Using a world map, the Lancet Commission implies that the problem of air pollution in the United States is not as severe when compared to the consequences in some other countries (e.g., China and India). As we have documented, there are cities where pollution caused thousands of people to needlessly suffer. As we found in our research in west Louisville, 60,000 people die prematurely by an average of 10 years (Gilderbloom, Meares, and Squires, 2020; Gilderbloom, Squires, and Meares, 2020). Higher levels of air pollution are also linked to other major challenges such as climate change and COVID-19.

Our research not only confirms the Chetty argument that *place* matters but also identifies the significant roles of gender and income in shaping life chances. It further confirms research that shows when cities provide more multi-modal transportation options like walking, biking, and transit, people get more exercise, pollution levels are lower, and city resident life expec-tancy increases (Frederick et al., 2018). Our research shows that the decision makers in industry and government impose on a person's everyday personal problems. It shows how *place* is also a proxy for the level of environmental degradation in a city and how it impacts our lifespan (Gilderbloom et al., 2014).

Air quality varies significantly by nation, state, city, and neighborhood and can cause significant health problems. Dr. Donald Schwartz, director at the Robert Wood Johnson Foundation, pointed out, "Even with our investment in healthcare we (the U.S.) rank 15th out of 17 developed western countries for life expectancy." A new culture of health can only evolve from a transforma-tion of our built environment (Green, 2013). We now know what is causing significant reductions in life expectancy in cities: poverty, gender, unhealthy behaviors, and air quality. We show that pollution is a key cause of shortened lifespans by up to five years. Regulatory enforcement of clean air matters in reducing the impact of pollution on health, housing, learning, and neighbor-hoods. It is a life and death issue.

42

John Hans Gilderbloom et al.

Originally published in the *AEE International Journal of Strategic Energy and Environmental Planning (IJSEEP)*. The Association of Energy Engineers (AEE), founded in 1977, is a non-profit professional association of more than 17,000 members and over 32,000 certified professionals in more than 100 countries worldwide dedicated to serving members and other industry professionals to save energy, reduce GHGs, make buildings perform better, and help reach global goals for Net-Zero. For more about AEE, visit www.aeecenter.org. Revised and reprinted with permission from *IJSEEP* editor, Dr. Stephen Roosa.

NOTE

1. Researchers are rightfully concerned about the ecological fallacy but also seem so with the atomistic fallacy. This occurs when the characteristics of cities and neighborhoods are assumed to result from characteristics of smaller units such as Census blocks.

REFERENCES

Almond, D. and Currie, J. (2011). Killing me softly: The fetal origins hypothesis. *Journal of Economic Perspectives*, 25(3), 153–172.

Appelbaum, R. (1978). *Size, Growth, and U.S. Cities*. Praeger.

Appelbaum, R., Bigelow, J., Kramer, H., Molotch, H. and Relis, P. (1976). *The Effects of Urban Growth: A Population Impact Analysis*. Praeger.

Banzhaf, H. (2012). *The Political Economy of Environmental Justice*. Stanford University Press.

Been, V. (1995). Analyzing evidence of environmental justice. *Journal of Land Use and Environmental Law*, 11(1), 1–36.

Bernard, S., Samet, J., Grambsch, A., Ebi, K. and Romieu, I. (2001). The potential impacts of climate variability and change on air pollution-related health effects in the United States. *Environmental Health Perspective*, 109(2), 199–209.

Boman, B., Forsberg, A. and Järvholm, B. (2003). Adverse health effects from ambient air pollution in relation to residential wood combustion in modern society. *Scandinavian Journal of Work and Environmental Health*, 29(4), 251–260.

Borgschulte, M., Molitor, D. and Zou, E. (2022). Air pollution and the labor market: Evidence from wildfire smoke. *IZA Institute of Labor Economics Discussion Paper* (IZA DP No. 15373). https://docs.iza.org/dp15373.pdf.

Boswell, M., Greve, A. and Seale, T. (2012). *Local Climate Action Planning*. Island Press.

Bullard, R. (2000). *Dumping in Dixie: Race, Class, and Environmental Quality*. Westview Press.

Bullard, R., Mohai, P., Saha, R. and Wright, B. (2008). Toxic wastes and race at twenty: Why race still matters after all of these years. *Environmental Law*, 38, 371.

Burtraw, D., Krupnick, A., Palmer, K., Paul, A., Toman, M. and Bloyd, C. (2003). Ancillary benefits of reduced air pollution in the U.S. from moderate greenhouse gas mitigation policies in the electricity sector. *Journal Environmental and Economic Management*, 453, 650–673.

Caiazzo, F., Ashok, A., Waitz, I., Yim, S. and Barrett, S. (2013). Air pollution and early deaths in the United States. Part I: Quantifying the impact of major sectors in 2005. *Atmospheric Environment*, 79, 198–208.

Čapek, S. (1992). Environmental justice, regulation, and the local community. *International Journal of Health Services,* 22(4), 729–746.

Čapek, S. (1993). The 'environmental justice' frame: A conceptual discussion and application. *Social Problems*, 40(1), 5–24.

Centers for Disease Control and Prevention. (2021). Particle Pollution. *Centers for Disease Control and Prevention*, July 16, 2021. Retrieved October 8, 2021, from https://www.cdc.gov/air/particulate_matter.html.

Chay, K. and Greenstone, M. (1999). The impact of air pollution on infant mortality: Evidence from geographic variation in pollution shocks induced by a recession. *National Bureau of Economic Research*, working paper 7442. doi: 10.3386/w7442.

Chetty, R., et al. (2016). The association between income and life expectancy in the United States, 2001–2014. *Journal of the American Medical Association*, 315(16), 1750–1766.

Cifuentes, L., Borja-Aburto, V., Gouveia, N., Thurston, G. and Davis, D. (2001). Climate change: Hidden health benefits of greenhouse gas mitigation. *Science*, 293(5533), 1257–1259. doi: 10.1126/science.1063357.

City of Louisville. (2017). Health equity report. Retrieved October 8, 2021, from https://louisvilleky.gov/government/center-health-equity/health-equity-report.

Currie, J. and Neidell, M. (2005). Air pollution and infant health: What can we learn from California's recent experience? *The Quarterly Journal of Economics*, 120(3), 1003–1030.

Currie, J. and Walker, W. (2009). Traffic congestion and infant health: Evidence from E-Z Pass. *National Bureau of Economic Research*, working paper 15413. https://www.nber.org/system/files/working_papers/w15413/w15413.pdf.

Dockery, D., et al. (1993). An association between air pollution and mortality in six U.S. cities. *New England Journal of Medicine*, 329(24), 1753–1759. doi: 10.1056/NEJM199312093292401.

DOEE. (2021). Department of agriculture, water and the environment. *Department of Climate Change, Energy, the Environment and Water*. Retrieved October 8, 2021, from https://www.dcceew.gov.au/.

Doyle, S., Kelly-Schwartz, A., Schlossberg, M. and Stockard, J. (2006). Active community environments and health: The relationship of walkable and safe communities to individual health. *Journal of the American Planning Association*, 72(1), 19–31.

Florida, R. (2002). *The Rise of the Creative Class: And How It's Transforming Work, Leisure, Community and Everyday Life*. Basic Books.

Frank, L., Sallis, J., Conway, T., Chapman, J., Saelens, B. and Bachman, W. (2006). Many pathways from land use to health: Associations between neighborhood walkability and active transportation, body mass index, and air quality. *Journal of the American Planning Association*, 72(1), 75–87.

Frederick, C., Riggs, W. and Gilderbloom, J. (2016). *Multi-modality and Public Health: A Multivariate Analysis of 148 U.S. Cities.* [Conference presentation]. Urban Affairs Association, San Diego, CA, United States.

Frederick, C., Riggs, W. and Gilderbloom, J. (2018). Commute mode diversity and public health: A multivariate analysis of 148 U.S. cities. *International Journal of Sustainable Transportation*, 12(1), 1–11. doi: 10.1080/15568318.2017.1321705.

Friedman, L. (2018, August 21). Cost of new EPA coal rules: Up to 1,400 more deaths a year. *The New York Times.* Retrieved October 8, 2021, from https://www.nytimes.com/2018/08/21/climate/epa-coal-pollution-deaths.html.

Friedrich, R. and Bickel, P., eds. (2001). *Environmental External Costs of Transport.* Springer.

Frumkin, H., Frank, L. and Jackson, R. (2004). *Urban Sprawl and Public Health: Designing, Planning, and Building for Healthy Communities.* Island Press.

Gilderbloom, J. and Appelbaum, R. (1987). *Rethinking Rental Housing.* Temple University Press.

Gilderbloom, J., Meares, W. and Riggs, W. (2014). How brownfield sites kill places and people: An examination of neighborhood housing values, foreclosures, and lifespan. *Journal of Urban International Residential Placemaking and Urban Sustainability*, 9(1), 1–18. doi: 10.1080/17549175.2014.905488.

Gilderbloom, J., Meares, W. and Squires, G. (2020). Pollution, place, and premature death: Evidence from a mid-sized city. *Local Environment,* 25(6), 419–432.

Gilderbloom, J., Riggs, W. and Meares, W. (2015). Does walkability matter? An examination of walkability's impact on housing values, foreclosures and crime. *Cities*, 42(part A), 13–24. doi: 10.1016/j.cities.2014.08.001.

Gilderbloom, J., Squires, G. and Meares, W. (2020). Mama, I can't breathe. Louisville's dirty air has steep medical and economic costs. *Local Environment*, 25(8), 619–626.

Gilderbloom, J., Squires, G., Riggs, W. and Čapek, S. (2017). Think globally, act locally: Neighborhood pollution and the future of the earth. *Local Environment*, 22(7), 894–899. doi: 10.1080/13549839.2017.1278751.

Green, J. (2013). The dirt: Uniting the built and natural environments, preserving Nigeria's forests. *THE DIRT.* https://dirt.asla.org.

Hamilton, J. (1995). Testing for environmental racism: Prejudice, profits, political power? *Journal of Policy Analysis and Management,* 14(1), 107–132.

Hanka, M. and Gilderbloom, J. (2008). How one-way thinking is hurting historic downtown neighborhoods. *Invisible City Poverty Housing and New Urbanism.* Retrieved January 26, 2014, from http://sun.louisville.edu/preservation/one-way-streetver12-012908-5B1-5D%20.pdf.

Hanna, R. and Oliva, P. (2015). The effect of pollution on labor supply: Evidence from a natural experiment in Mexico City. *Journal of Public Economy*, 122, 68–79.

Hipp, J. and Lakon, C. (2020). Social disparities in health: Disproportionate toxicity proximity in minority communities over a decade. *Health Place*, 16(4), 674–683. doi: 10.1016/j.healthplace.2010.02.005.

Isen, A., Rossin-Slater, M. and Walker, W. (2017). Every breath you take—every dollar you'll make: The long-term consequences of the clean air act of 1970. *Journal of Political Economy*, 125(3), 848–902.

Keeley, M. and Short, L. (2018). *Urban Sustainability in the U.S: Cities Take Action.* Springer.

Kuehn, B. (2014). WHO: More than 7 million air pollution deaths each year. *Journal of the American Medical Association*, 311(15), 1486–1486.

Landrigan, P., et al. (2018). The Lancet Commission on pollution and health. *The Lancet*, 391(10119), 462–512.

Louv, R. (2008). *Last Child in the Woods: Saving Our Children from Nature-deficit Disorder.* Algonquin books.

Meares, W. and Riggs, W. (2016). Walkability and the benefits of place-based housing: An examination of Louisville's HOPE VI neighborhoods. *Housing Society*, 43(2), 103–125. doi: 10.1080/08882746.2017.1293428.

Miller, K., et al. (2007). Long-term exposure to air pollution and incidence of cardiovascular events in women. *New England Journal of Medicine*, 356(5), 447–458. doi: 10.1056/NEJMoa054409.

Molotch, H. (1976). The city as a growth machine: Toward a political economy of place. *American Journal of Sociology*, 82(2), 309–332. doi: 10.2307/2777096.

Organisation for Economic Co-operation and Development (OECD). (2012). OECD Environmental Outlook to 2050: The Consequences of Inaction. *OECD Publishing*, March 15, 2012. doi: 10.1787/9789264122246-en.

Pahl, R. (2015). *Whose City? And Further Essays on Urban Society*. Penguin.

Pastor, M., Sadd, J. and Morello-Frosch, R. (2007). Still toxic after all these years: Air quality and environmental justice in the San Francisco Bay Area. [Semiar Presentation]. U.S. Environmental Protection Agency, Region 9, San Francisco, CA, United States.

Pope, C. (2000). Epidemiology of fine particulate air pollution and human health: Biologic mechanisms and who's at risk? *Environmental Health Perspective*, 108, 713–723. doi: 10.2307/3454408.

Pope, C., Bates, D. and Raizenne, M. (1995). Health effects of particulate air pollution: Time for reassessment? *Environmental Health Perspective*, 103(5), 472–480. doi: 10.2307/3432586.

Prinz, D., Chernew, M., Cutler, D. and Frakt, A. (2018). Health and economic activity over the lifecycle: Literature review. *National Bureau of Economic Research*, working paper 24865. doi: 10.3.386/w24865.

Riggs, W. (2011). Walkability and housing: A comparative study of income, neighborhood change and socio-cultural dynamics in the San Francisco Bay area. [PhD. Thesis]. *UC Berkeley*. doi: <http://escholarship.org/uc/item/ 5970164x>.

Riggs, W. (2014). Inclusively walkable: Exploring the equity of walkable neighborhoods in the San Francisco Bay area. *Local Environment, 5*, 527–554. doi: 10.1080/13549839.2014.982080.

Riggs, W. (2015). Walkability: To quantify or not to quantify. *Journal of Urban International Residential Placemaking and Urban Sustainability*, 10(1), 125–127. doi: 10.1080/17549175.2015.1111926.

Riggs, W. and Gilderbloom, J. (2016). The connection between neighborhood walkability and life longevity in a midsized city. *Focus*, 13(1). http://digitalcommons .calpoly.edu/focus/vol13/iss1/11.

Riggs, W. and Sethi, S. (2019). Multimodal travel behavior, walkability indices, and social mobility: How neighborhood walkability, income and household characteristics guide walking, biking and transit decisions. *Local Environment*, 25, 57–68.

Rydin, Y., et al. (2012). Shaping cities for health: Complexity and the planning of urban environments in the 21st century. *The Lancet*, 379(9831), 2079–2108. doi: 10.1016/S0140- 6736(12)60435-8.

Samet, J., Dominici, F., Curriero, F., Coursac, I. and Zeger, S. (2000). Fine particulate air pollution and mortality in 20 U.S. cities, 1987–1994. *New England Journal of Medicine*, 343(24), 1742–1749. doi: 10.1056/NEJM200012143432401.

Schweitzer, L. and Zhou, J. (2010). Neighborhood air quality, respiratory health, and vulnerable populations in compact and sprawled regions. *Journal of the American Planning Association*, 76(3), 363–371. doi: 10.1080/01944363.2010.486623.

Smith, C. (2009). Economic deprivation and racial segregation: Comparing superfund sites in Portland, Oregon and Detroit, Michigan. *Social Science Research*, 38(3), 681–692.

Stone, B. (2005). Urban heat and air pollution: An emerging role for planners in the climate change debate. *Journal of the American Planning Association*, 71(1), 13–25.

Stone, B., et al. (2014). Avoided heat-related mortality through climate adaptation strategies in three U.S. cities. *PLOS One*, 9(6), e100852. doi: 10.1371/journal.pone .0100852.

Swanston, S. (2012). An environmental justice perspective on superfund Reauthorization. *Journal of Civil Rights and Economic Development*, 9(2), 18.

Tessum, C., et al. (2019). Inequity in consumption of goods and services adds to racial–ethnic disparities in air pollution exposure. *Proceedings of the National Academy of Science*, 116(13), 6001–6006. doi: 10.10-3/pnas.1818859116.

US EPA. (2014). NAAQS table. *United States Environmental Protection Agency*, April 10, 2014. Retrieved October 8, 2021, from https://www.epa.gov/criteria-air -pollutants/naaqs-table.

US EPA. (2015a). 2011 national emissions inventory data. *United States Environmental Protection Agency*, June 3, 2015. Retrieved October 8, 2021, from https:// www.epa.gov/air-emissions-inventories/previous-nei-reports.

US EPA. (2015b). Overview of greenhouse gases. *United States Environmental Protection Agency*, December 23, 2015. Retrieved October 8, 2021, from https://www .epa.gov/ghgemissions/overview-greenhouse-gases.

US EPA. (2016a). Basic information about air emissions monitoring. *United States Environmental Protection Agency*, May 9, 2016. Retrieved October 8, 2021, from https://www.epa.gov/air-emissions-monitoring-knowledge-base/basic-information -about-air-emissions-monitoring.

US EPA. (2016b). Health and environmental effects of particulate matter. *United States Environmental Protection Agency*, April 26, 2016. Retrieved October 8,

2021, from https://www.epa.gov/pm-pollution/health-and-environmental-effects
-particulate-matter-pm.

US EPA. (2018). Regulatory impact analysis for the proposed emission guidelines for greenhouse gas emissions from existing electric utility generating units. *United States Environmental Protection Agency*, August, 2018. https://www.epa.gov/sites /default/files/2018-08/documents/utilities_ria_proposed_ace_2018-08.pdf.

West, J., et al. (2013). Co-benefits of global greenhouse gas mitigation for future air quality and human health. *National Climate Change*, 3(10), 885–889. doi: 10.1038/ NCLIMATE2009.

Winkleman, S. (2007). Comment on stone: Could the worst of times for the planet be the best of times for planning? *Journal of the American Planning Association*, 73(4), 418–420. Retrieved August 29, 2016, from https://www.researchgate.net /profile/Steve_Winkelman/publication/233068391_Comment_on_Stone_Could_ the_Worst_of_Times_for_the_Planet_Be_the_Best_of_Times_for_Planning/links /543548 af0cf2bf1f1f2868e0.pdf.

Wong, E. (2013). Air pollution linked to 1.2 million premature deaths in China. *The New York Times*, April 1, 2013. Retrieved October 8, 2021, from https://www .nytimes.com/2013/04/02/world/asia/air-pollution-linked-to-1-2-million-deaths-in -china.html#:~:text=BEIJING%20%E2%80%94%20Outdoor%20air%20pollution %20contributed,leading%20causes%20of%20death%20worldwide.

World Health Organization. (2018). Climate change and health. *World Health Orga-nization*, October 12, 2023. Retrieved October 8, 2021, from https://www.who.int/ news-room/fact-sheets/detail/climate-change-and-health.

Chapter 3

What Cities Are Most Dangerous to Your Life Expectancy? Toward a Methodology of Livability

John Hans Gilderbloom, Christopher Bird,
Gregory Squires, Chad Frederick,
Ellen Slaten, Karrie Ann Quinchet,
Carla J. Snyder, and William Riggs

The Clean Air Act is under assault. While there is some resistance in green cities, most health departments and universities are silent about the ill effects of air pollution. During the Trump administration, the enforcement of clean air regulations was lax. Data on the levels of poor air quality that was once easily obtainable became less accessible. Previously available EPA air quality data is no longer being collected, due to the false belief that disclosing locations with high toxic air hurts economic development.

Air pollution comprises one of the most detrimental environmental human health risks in the world. In 2012, on the authority of the World Health Organization (WHO), recurring exposure to household and ambient air pollutants, such as particulate matter (PM), contributed to the deaths of 7 million people worldwide (Sosa et al., 2017). Their study considered PM and associated polycyclic aromatic hydrocarbon (PAH) levels in outdoor air, identifying their possible emission sources and analyzing health risks in the city of Tandil, Argentina (Sosa et al., 2017). Sixteen priority PAHs (categorized as PM2.5) were considered to be probable human carcinogens and were listed by the International Agency for Research on Cancer as priority pollutants. The study showed that out of 8,855 Tandil children under five years of age, 4 cases of death were attributable to short-term outdoor PM10 exposure (Sosa et al., 2017).

They also found 21 lethal cases out of the total population of 123,871 people in Tandil. Another study established that exposure to nitrogen dioxide

(NO_2), a gaseous air pollutant associated with motor vehicle emissions, is attributed to mortality in a number of time series studies (Costa et al., 2017). Their study noted associations between NO_2 and non-accidental, circulatory and respiratory deaths in single-pollutant models (Costa et al., 2017).

PM10 and PM2.5 lethality is associated with cardiac and respiratory causes, decreased lung capacity in children and asthmatic adults, and increased chronic obstructive pulmonary diseases (Leiva et al., 2013). The evidence indicates that the relative estimated risks (based on the EPA criteria document's review of PM10 studies) include an increased all-age mortality of 2.5 percent to 5.0 percent for each 1 mg/m^3 increase in the PM10 concentrations or 25 mg/m^3 increase in the PM2.5 concentrations. A study by Leiva et al. ruled out tobacco smoking, passive tobacco smoke exposure, occupational exposure to fine particles, temperature and alcohol use as causes of cerebrovascular hospital admissions (Leiva et al., 2013). The authors found that fine-particle pollution was related to at least a 15 percent difference in death rates between the least and most polluted cities. The information also determined that when the PM2.5 concentrations increase by 10 mg/m^3, the risk of emergency hospital admissions for cerebrovascular causes increases by 1.29 percent. The goal of our research is to understand how mid-sized U.S. cities compare to one another in their pollution emissions.

LITERATURE REVIEW

A vast amount of academic research regarding air pollution illustrates its negative effects on quality of life with far-ranging consequences for people of all races, genders, ages, income levels and occupations. These effects can be exacerbated by proximity to higher levels of air pollution. While multiple variables affect the quality of life for the groups discussed, air pollution is a key component. Worldwide, air pollution was responsible for 7 million premature deaths in 2010 (Kuehn, 2014). The Organization for Economic Co-operation and Development (OECD) states that by 2050, poor air quality will be the number one cause of premature death rates rising ahead of current primary causes from sanitation and dirty water (Organization for Economic Co-operation and Development, 2012). A study published in *The Lancet* estimates that 1.2 million premature deaths occur annually in Chinese cities due to air pollution (The Lancet, 2013; Capek, 1992; Capek, 1993; Wong, 2013; Tessum et al., 2019). Furthermore, an EPA study showed that a Trump administration proposed rollback of pollution restrictions on coal-fired plants could result in up to 1,400 more premature deaths annually in the United States (U.S. Environmental Protection Agency, 2018).

Air pollution is potentially more damaging to the health of infant children. While much research on the fetal origins hypothesis centers on how undernutrition, disease, and maternal health habits create lasting effects in newborns, a growing base of literature links the proximity to air pollution to adverse health effects in utero (Almond & Currie, 2011). Studies have shown the link between fetal exposure to pollution and higher infant mortality rates as well as the link between reductions in environmental carbon monoxide (CO) and reductions in instances of low birth weight (Chay and Greenstone, 1999; Currie et al., 2009; Currie and Walker, 2011). Infants weighing less than 5.5 lbs. (2.5 kg) at birth often suffer from early sickness or infections and may have development issues as they age (Centers for Disease Control and Prevention, 2014). Each of these studies concur that the proximity of a pregnant mother to higher levels of air pollution has a key role in the health of the newborn child. These effects are found in seemingly healthy babies later in life. Prenatal exposure to a variety of types of air pollution negatively affects academic test scores and economic outcomes later in life (Prinz et al., 2018). Further work illustrates that adult non-health endpoints are affected as well. Both lower labor force participation and lower earnings are correlated with higher exposure to pollution at birth and later in life (Isen et al., 2017; Borgschulte et al., 2019).

Another study determined that a 19.7 percent decrease in sulfur oxide (SOx) emissions results in an increase of 3.5 percent in working hours for immediate neighbors within 3.1 miles (5 km) of a heavy polluter (Hanna and Oliva, 2015). The finding noted a 6 percent increase in the probability of those residents working over 40 hours per week and 2.5 percent increase in the probability of them working over 10 hours per week (Hanna and Oliva, 2015). Motti's study of Santiago, Chile, found that women's working hours decreased significantly during those weeks of the year with the city's highest levels of pollution though total work hours remained consistent (Montt, 2018). Montt posits that this discrepancy was due to the likelihood of women assuming the care of children and the elderly, groups who are most likely to be affected by poor air quality (Montt, 2018).

Research literature on environmental justice issues has long established that hazardous sites, including sources of air pollution, are disproportionately located in racial and ethnic minority neighborhoods (Capek, 1993; Pahl, 1975; Appelbaum et al., 1976; Capek and Gilderbloom, 1992; Been, 1995; Bullard, 1990; Bullard, 2000; Pastor et al., 2001; Keeley and Benton-Short, 2019; University of Wisconsin Population Health Institute, 2014). There is debate as to whether hazardous industries were relocated to minority neighborhoods or minority population moved to such sites after the hazards were present (Bullard, 2000; Pastor et al., 2001). Recent studies indicate that the hazardous sites were located in minority neighborhoods after racial compositions are established (Smith, 2009; Hipp and Lakon, 2010; Swanston, 2012). This can be due to the neighborhood's lack of political power or weak real estate markets (Hamilton, 1995; Banzhaf, 2012). Consistent evidence supporting the disproportionate impact of environmental contaminants on racial and ethnic minority neighborhoods raises concerns about the possible links to shortened lifespans. Evidence shows that living in countries, states, cities, or neighborhoods with high levels of air pollution shortens life expectancies by four to thirteen years (Kuehn, 2014; The Lancet, 2013; Wong, 2013).

The world's recent focus on reducing dependency on greenhouse gas (GHG) emitting compounds also helps to reduce other types of air pollution. A study by Burtraw et al. showed that the total short-term health benefits gained through ancillary NOx and SOx reductions by taxing all GHGs justifies the initial costs of the tax (Burtraw et al., 2003). Reductions in fossil fuel combustion corresponds with reductions in particulate air pollution (Cifuentes et al., 2001).

Greenhouse gases trap heat in the Earth's atmosphere, altering the planet's climate. Major sources of GHGs include electric generation and industrial process plants fired with coal, oil and natural gas and vehicles that consume fossil fuels. The co-benefits of reducing GHGs correspond to roughly a half million fewer premature deaths due to other types of air pollution plus a slowing of the effects of global warming by 2030 (West et al., 2013). The EPA classifies four different types of emissions as GHGs: carbon dioxide (CO_2), methane (CH_4), nitrous oxide (NO) and fluorinated gas (F2) (U. S. Environmental Protection Agency, 2019). The WHO suggests that increases in GHG emissions affect human health by contaminating air and drinking water, spoiling crops, and destroying man-made shelter (World Health Organization, 2018). This selection of work is only part of the research dedicated to understanding how air pollution affects human health, environmental justice and productivity. While there is much work to be done to fully understand how, why and where these effects are most costly, it appears that there are significant benefits to living in places with lower air pollution.

Methodology

Our study methodology involves establishing case selection criterion, identifying a sample of
U.S. counties to study, developing a process to exclude certain cases, and selecting and assessing test variables.

Case Selection Criteria

We first established an appropriate case selection criterion. Urban researchers tend to use all available cities for which there is data. Since there is great variety among cities, this approach is known to urban scholars to be misleading. It is problematic to include large metropolises like New York City in a sample set that is dominated by cities with much lower populations. Overly large geographies of metropolitan statistical areas (MSAs) or commuting zones (CZs) are more appropriate for regional research. Most CZs contain multiple counties, such as San Francisco and Chicago which include both generally suburban counties with generally cleaner air and urban counties with more polluted air. Instead, we use the much smaller and generally more consistent unit of county as our level of analysis. The county as the unit of consideration allows for a more granular analysis than the much larger MSAs or CZs. The county-level has the additional benefit of representing entities which are politically governed. The MSAs are too large and the Census tracts too small to provide transferable findings to municipal leaders; findings uncovered at the level of county are more actionable.

The problem of spatial lag occurs in situations where cities are either immediately adjacent to or in close proximity to another city of similar size or larger. As such, they likely share labor, transportation and housing markets. Airborne pollution emissions disregard political boundaries. Neighboring cities are also subject to policy spillovers, in which the impacts of one city's policies can be measured in adjacent jurisdictions. Such factors tend to obscure empirical findings.

To identify an appropriate set of municipalities to study, we draw on Appelbaum's *Size, Growth, and U.S. Cities* and Molotch's *City as a Growth Machine* (Appelbaum, 1978; Molotch 1976). What we call the *Molotch/Appelbaum Method* has been replicated by other researchers to predict urban riots, rents, health and happiness (Friedland, 1982; Gilderbloom and Appelbaum, 1987; Gilderbloom, 2008; Frederick and Gilderbloom, 2018).

Our case selection methodology yields a subset of U.S. cities that we call *semi-isolated, mid-size cities*. Selection using this approach begins with the universe of all incorporated places defined by the U.S. Census. The number of cities is further reduced to those cities with populations over 50,000 which are not located within another 20 miles (32.2 km) of another city of a similar size. This results in a subset of ($N = 142$) semi-isolated, mid-size cities.

Excluded Cases

In two instances, one county contained two semi-isolated, mid-size cities. In these cases, we kept the larger of the two cities in the list. For example,

Bakersfield and Delano are both in Kern County, California. We retained Bakersfield and omitted Delano to avoid double counting the emissions data of Kern County. In Santa Barbara County, California, we kept Santa Barbara and omitted Santa Maria. Four semi-isolated, mid-size cities in Virginia were excluded from our comparison: Lynchburg, Richmond, Roanoke and Harrisonburg. These *independent cities* are not politically part of a county, though they may be surrounded by one. We omitted these cities from our analysis to preserve our methodology that uses the county as the unit of comparison.

Test Variables

The U.S. Clean Air Act (CAA) is a federal law, enacted in 1970 and last amended in 1990. It authorized the existence of the EPA and establishing the National Ambient Air Quality Standards (NAAQS). The EPA regulates air emissions from stationary and mobile sources in an attempt to protect public health. The National Emissions Inventory (NEI) is a comprehensive and detailed estimate of air emissions of criteria pollutants and hazardous air pollutants from air emissions sources which is released every three years. State, local, and tribal air agencies provide data which supplements EPA data. The NEI is considered the guiding source for environmental emissions in the United States (U.S. Environmental Protection, 2017).

We gathered the 2011 NEI data for four air pollutants for the selected 142 semi-isolated, mid-size cities: PM10, PM2.5, NOx and SOx. The data file contains the emissions by sector for all U.S. counties. While the EPA has publicly available online tools to query the 2011 NEI data, this tool only presents one location at a time. The data presented is, to our knowledge, the first of its kind in comparing county-level pollution. Moreover, a University of Louisville graduate student working for the EPA was instructed to destroy these four variables. At the time, scientists only had access to PM10 and PM2.5, which created mismeasurement error because the other two variables could not be found: Nitrogen Oxide and Sulfur Oxide. The student feared that destroying government data could land him in jail, so he put the data through a "back door," and made it hard to find. He later told a graduate seminar class at the University of Louisville how to get access to all four variables. Combining the four variables together resulted in different rankings, with Louisville ranked number two.

POLLUTANTS

Next, we discuss the EPA criteria air pollutants selected for our study of U.S. counties.

Particulate Matter

Particulate matter (PM) is an air pollutant classification based on particle diameters. The EPA provides data on two types. Particulate matter 10 (PM10) is the category for a mixture of solid particles and liquid droplets found in the air that are 10 micrometers or less in diameter. Similarly, PM2.5 are particles 2.5 micrometers or less in diameter. These particles are small enough to be inhaled deep within the lungs where they may be deposited and result in adverse health effects.

PM10 and PM2.5 may be different sizes and shapes and can be composed of hundreds of different chemicals. The chemical properties of PM depend on the source of emission. PM is both a primary and secondary pollutant (Frumkim et al., 2004). Some are emitted from a point source, such as construction sites, unpaved roads, fields, smokestacks or fires. Most particles form in the atmosphere as a result of complex reactions of chemicals such as SOx and NOx, which are pollutants emitted from power plants, industries and motor vehicles with internal combustion engines (U.S. Environmental Protection Agency, 2018).

PM aerosols affect climate directly by scattering and absorbing solar radiation. This contributes to global warming while reducing the radiation flux at the Earth's surface. Aerosols are also an indirect radiative because the presence of particles in the atmosphere has a role in cloud formation (Fuzzi et al., 2015). Adverse health effects related to PM exposure concern mainly respiratory and cardiovascular systems (Centers for Disease Control and Prevention, 2014; Pope et al., 1995; Boman et al., 2003). There is evidence that PM affects atherosclerosis and leads to adverse birth outcomes. Children and elderly populations are most sensitive to the impacts of exposure to PM (Fuzzi et al., 2015). Epidemiological and human exposure studies show that both long- and short-term exposure to PM correlate with cardiovascular and respiratory morbidity and mortality (Anderson et al., 2012). PM in outdoor air pollution is designated as a Group I carcinogen by the International Agency for Research on Cancer (Straif et al., 2013). A study involving 312,944 people in nine European countries revealed that there was no safe level of particulates in the airstream. For every increase of 10 $\mu g/m^3$ in PM10 and PM2.5, the lung cancer rate rose 22 percent and 36 percent respectively (Hamra et al., 2014).

Nitrogen Oxides

Nitrogen oxides are a group of highly reactive and poisonous gases, the most common being nitrogen dioxide (NO_2). Most airborne NOx comes from combustion-related emissions sources, primarily fossil fuel combustion by electric utilities, high-temperature operations at other industrial sources, and operation of motor vehicles (U.S. Environmental Protection Agency, 2016).

Nitrogen oxides react with other airborne chemicals to form PM. Exposure to particulate matter aggravates chronic respiratory and cardiovascular diseases, alters host defenses, damages lung tissue, leads to premature deaths, and possibly contributes to cancer (Bernard et al., 2001). When exposed to the ultraviolet (UV) rays in sunlight, NOx molecules break apart and form ozone (O_3), a GHG which contributes to smog. Ground-level ozone exacerbates chronic respiratory diseases and causes short-term reductions in lung function (Bernard et al., 2001). Breathing air with high concentrations of NOx irritates airways in the human respiratory system. Such exposures over short periods aggravates respiratory diseases, particularly asthma, leading to respiratory symptoms (such as coughing, wheezing or difficulty breathing), hospital admissions, and visits to emergency rooms. Longer exposures to elevated concentrations of NO_2 contributes to the development of asthma and potentially increases susceptibility to respiratory infections (U.S. Environmental Protection Agency, 2016).

Sulfur Oxides

Sulfur oxides (SOx) are a group of reactive and toxic gases, the most common being sulfur dioxide (SO_2). These are colorless gases that can be detected by taste and smell. SOx are emitted primarily through fossil fuel combustion (Rankin cycle) processes by power plants and industrial facilities. Other sources of SOx emissions include: industrial processes such as extracting metal from ore, natural sources such as volcanoes, locomotives, ships, other vehicles and heavy equipment that combust fossil fuels with a high sulfur content. Short-term exposure to SOx causes harm to the human respiratory system, making breathing difficult (Boman et al., 2003; Bernard et al., 2001). Atmospheric SOx react with water, oxygen, and other chemicals to form damaging PM, and contribute to acid rain by forming sulfuric acid (H_2SO_4).

Analysis: Which Mid-Size Cities Have the Dirtiest Air?

Our research is designed to empower citizens with local information regarding air pollution. A total of 142 mid-size cities met our criterion of having populations greater than 50,000, not being located within 20 miles (32.2 km) of another city of 50,000, and with EPA data providing accurate measures of the four types of air pollutants. The public has the right to know which cities have dangerous levels of pollution that can cause health problems and reduce life expectancies. To determine how to retrieve this data for your locale, follow these steps:

Step 1: Download the NEI 2011 data for *All Sectors* from the following link: https://www.epa.gov/air-emissions-inventories/2011-national-emissions -inventory-nei-data. This file is quite large (over 90MB) when unzipped. The download may take a few minutes.

Step 2: Extract the comma-separated values (CSV) file from the zipped folder and open in Microsoft Excel or an alternative spreadsheet. The following steps may differ when using other types of software.

Step 3: Highlight each column containing data and click the *Filter* button in the top ribbon. The *Filter* button can be found under the *Data* heading in the top ribbon. After completing this step, the top heading of each row (under row A) should appear with a dropdown button.

Step 4: Click the dropdown button in Column D, *St_usps_cd*. This allows you to filter the large amount of data in this file by state. Check only the box for the state in which your city resides. For example, someone searching for emissions data for Indianapolis, Indiana would check only the box next to "IN".

Step 5: Click the dropdown button in Column F, *County Name*. Check only the box for the county in which your city is located. For example, someone searching for emissions data for Indianapolis, Indiana would check only the box next to "Marion" for "Marion County, Indiana."

Step 6: Click the dropdown button in Column H, *Pollutant cd*. Check only the box for the pollutant you wish to retrieve data for. For example, someone searching for PM10 emissions data (still for Indianapolis, Indiana) would check the box next to *PM10-PRI*.

Step 7: Scroll to the first empty cell under Column K, *Totalemissions*. Highlight the cell and click *Autosum (Σ)* in the top ribbon. Record the calculated number. The *Autosum* ribbon can be found under the home tab and is shown under the Σ. This number is the total emissions for the selected pollutant as estimated by the NEI.

Step 8: Repeat step 6 for each pollutant desired. Record the last number in Column K after completing Step 6 each time. These numbers are the total emissions for each selected pollutant as estimated by the NEI. The last cell with a value in Column K will remain auto-summed even after different pollutants are selected.

Step 9: Compare each value recorded for total emissions by selected pollution in Table 3.1A of this article. Steps 4–6 can be repeated for any county, city (using county as proxy), state, and tribal sector as desired. In all cases, higher numbers reflect higher pollution levels.

Ranking Mid-Size Cities by Air Pollution Levels

Tables were developed to show the wide variations in levels of air pollution. They also show which cities have dangerously high levels of pollution as measured by the EPA. Table 3.1A (see Appendix) lists the 142 cities and their pollution levels for each of the four considered pollutants. Table 3.2 shows the 10 cities with the highest and lowest levels of PM10. Table 3.3 shows the 10 cities with highest and lowest PM2.5 levels. Table 3.4 shows the 10 cities with highest and lowest NOx levels. Table 3.5 shows the 10 cities with highest and lowest SOx levels.

Table 3.2   PM10 Emissions (tons)

City	State	County	Rank	PM10
Las Cruces	NM	Doña Ana	1	67,065
Santa Fe	NM	Santa Fe	2	65,890
Wichita	KS	Sedgwick	3	39,029
Cheyenne	WY	Laramie	4	35,765
Long View	TX	Gregg	5	28,564
Lubbock	TX	Lubbock	6	28,118
Gulfport	MS	Harrison	7	25,598
Bend	OR	Deschutes	8	24,099
Fargo	ND	Cass	9	23,986
Bakersfield	CA	Kern County	10	23,553
Clarksville	TN	Montgomery	132	3,463
Toms River	NJ	Ocean County	133	3,451
La Crosse	WI	La Crosse	134	3,223
Rocky Mount	NC	Nash	135	3,179
Scranton	PA	Lackawanna	136	3,130
Athens	GA	Clarke	137	2,473
Vineland	NJ	Cumberland	138	2,270
Tyler	TX	Smith	139	2,211
Wilmington	NC	New Hanover	140	1,974
Waco	TX	McLennan	141	1,759

Sources: 1. 2011 NEI data; 2. Gilderbloom, J.H., Bird, C., Squires, G., Frederick, C., Slaten, E., Quenichet, K.A., Taskar, A., Snyder, C.J., and Riggs, W. (2022). What Cities are Most Dangerous to your Life Expectancy? Toward a Methodology of Livability. International Journal of Strategic Energy and Environmental Planning, V4.4; 3. Center for Sustainable Urban Neighborhoods, formerly University of Louisville, transferred to Neighborhood Associates, Washington D.C.

Table 3.3   PM2.5 Emissions (tons)

City	State	County	Rank	PM2.5
Bend	OR	Deschutes	1	13,802
Bakersfield	CA	Kern County	2	12,208
Santa Fe	NM	Santa Fe	3	10,673
Duluth	MN	St. Louis	4	9,391
Flagstaff	AZ	Coconino	5	8,975
Lake Charles	LA	Calcasieu	6	8,684
Las Cruces	NM	Doña Ana	7	8,346
Mobile	AL	Mobile	8	7,684
Louisville	KY	Jefferson	9	7,672
Wichita	KS	Sedgwick	10	7,182
Vineland	NJ	Cumberland	132	1,168
Clarksville	TN	Montgomery	133	1,156
Evansville	IN	Vanderburgh	134	1,066
Greenville	NC	Pitt	135	1,045
La Crosse	WI	La Crosse	136	1,028
Bloomington	IN	Monroe	137	1,028
Wilmington	NC	New Hanover	138	933
Columbus	GA	Muscogee	139	914
Rocky Mount	NC	Nash	140	906
Athens	GA	Clarke	141	489

Sources: 1. 2011 NEI data; 2. Gilderbloom, J.H., Bird, C., Squires, G., Frederick, C., Slaten, E., Quenichet, K.A., Taskar, A., Snyder, C.J., and Riggs, W. (2022). What Cities are Most Dangerous to your Life Expectancy? Toward a Methodology of Livability. International Journal of Strategic Energy and Environmental Planning, V4.4; 3. Center for Sustainable Urban Neighborhoods, formerly University of Louisville, transferred to Neighborhood Associates, Washington D.C.

Table 3.4   Oxides of Nitrogen Emissions (tons)

City	State	County	Rank	NOx
Bakersfield	CA	Kern County	1	46,852
Louisville	KY	Jefferson	2	37,796
Jacksonville	FL	Duval	3	36,981
Columbus	OH	Franklin	4	36,846
Pittsburgh	PA	Allegheny	5	35,455
Lake Charles	LA	Calcasieu	6	35,343
Duluth	MN	St. Louis	7	34,183
Mobile	AL	Mobile	8	32,303
Long View	TX	Gregg	9	23,724
Toledo	OH	Lucas	10	23,624
Idaho Falls	ID	Bonneville	132	4,055
Greenville	NC	Pitt	133	3,831
Dubuque	IA	Dubuque	134	3,827
Dothan	AL	Houston	135	3,651
San Angelo	TX	Tom Green	136	3,553
Bloomington	IN	Monroe	137	3,462
Eau Claire	WI	Chippewa	138	3,409
Athens	GA	Clarke	139	3,049
Kokomo	IN	Howard	140	2,685
Manhattan	KS	Riley	141	2,548

Sources: 1. 2011 NEI data; 2. Gilderbloom, J.H., Bird, C., Squires, G., Frederick, C., Slaten, E., Quenichet, K.A., Taskar, A., Snyder, C.J., and Riggs, W. (2022). What Cities are Most Dangerous to Your Life Expectancy? Toward a Methodology of Livability. International Journal of Strategic Energy and Environmental Planning, V4.4; 3. Center for Sustainable Urban Neighborhoods, formerly University of Louisville, transferred to Neighborhood Associates, Washington D.C.

Table 3.5   Oxides of Sulfur Emissions (tons)

City	State	County	Rank	SOx
Terre Haute	IN	Vigo	1	55,945
Lake Charles	LA	Calcasieu	2	41,135
Louisville	KY	Jefferson	3	39,231
Gulfport	MS	Harrison	4	32,925
Sioux City	IA	Woodbury	5	29,693
Jacksonville	FL	Duval	6	20,852
Mobile	AL	Mobile	7	20,673
Green Bay	WI	Brown	8	18,307
Amarillo	TX	Potter	9	15,388
Columbus	GA	Muscogee	132	84
St. George	UT	Washington	133	82
Conway	AR	Faulkner	134	77
Yuba City	CA	Sutter	135	75
Abilene	TX	Taylor	136	72
Laredo	TX	Webb	137	71
Bismarck	ND	Burleigh	138	68
Bowling Green	KY	Warren	139	61
Evansville	IN	Vanderburgh	140	43
Kokomo	IN	Howard	141	35

Sources: 1. 2011 NEI data; 2. Gilderbloom, J.H., Bird, C., Squires, G., Frederick, C., Slaten, E., Quenichet, K.A., Taskar, A., Snyder, C.J., and Riggs, W. (2022). What Cities are Most Dangerous to your Life Expectancy? Toward a Methodology of Livability. International Journal of Strategic Energy and Environmental Planning, V4.4; 3. Center for Sustainable Urban Neighborhoods, formerly University of Louisville, transferred to Neighborhood Associates, Washington D.C.

Table 3.6 Composite Emissions Rank

City	State	County	Rank
Mobile	AL	Mobile	1
Louisville	KY	Jefferson	2
Lake Charles	LA	Calcasieu	3
Duluth	MN	St. Louis	4
Bakersfield	CA	Kern County	5
Long View	TX	Gregg	6
Gulfport	MS	Harrison	7
Jacksonville	FL	Duval	8
Baton Rouge	LA	East Baton Rouge Parrish	9
Colorado Springs	CO	El Paso	10
Bowling Green	KY	Warren	132
Evansville	IN	Vanderburgh	133
Eau Claire	WI	Chippewa	134
La Crosse	WI	La Crosse	135
Yuba City	CA	Sutter	136
Greenville	NC	Pitt	137
Kokomo	IN	Howard	138
Athens	GA	Clarke	139
Rocky Mount	NC	Nash	140
Columbus	GA	Muscogee	141

Sources: 1. 2011 NEI data; 2. Gilderbloom, J.H., Bird, C., Squires, G., Frederick, C., Slaten, E., Quenichet, K.A., Taskar, A., Snyder, C.J., and Riggs, W. (2022). What Cities are Most Dangerous to Your Life Expectancy? Toward a Methodology of Livability. International Journal of Strategic Energy and Environmental Planning, V4.4; 3. Center for Sustainable Urban Neighborhoods, formerly University of Louisville, transferred to Neighborhood Associates, Washington D.C.

Table 3.6 combines the measures of air pollution with one total ranking. This was developed by averaging the rank of each city as determined from tables 3.1A and tables 3.2–3.5 using the following equation:

$$CR = (R10 + R25 + RNOX + RSOX) \div 4$$

Where: CR = Composite ranking of cleanest and dirtiest cities considered. R10 = City's rank among all considered in PM10 emissions. R25 = City's rank among all considered in PM2.5 emissions. RNOX = City's rank among all considered in NOX emissions. RSOX = City's rank among all considered in SOX emissions.

The Implications of Pollution

In April 2016, an issue of *Journal of the American Medical Association* (*JAMA*) listed Louisville, Kentucky was one of the U.S. cities with the shortest life spans among poor men, about 1,643 days shorter than those in Santa

Barbara, California (Chetty et al., 2016). How do you explain this 4.5-year gap in lifespan? The lead author, Raj Chetty, believes that *place* shapes lifespans but fails to identify the causes of such differences (Detroit versus San Francisco also has a 4.5-year gap). The *JAMA* article failed to mention the causes of reduced life expectancy in Louisville and other cities: environmental toxins in the air, water, and soil (Gilderbloom, 2016). His own data proved this with a simple observation: Detroit, Gary, Indianapolis, Tulsa, and Louisville have significantly higher air pollution levels compared with coastal cities like Los Angeles, Santa Barbara, Santa Rosa, San Francisco, and New York. This better explains the nearly five-year difference in life spans among these cities in his tables. The overwhelming evidence shows that if you live in countries, states, cities, or neighborhoods with more air pollution, your life will be shortened by 4 to 13 years (Kuehn, 2014; The *Lancet*, 2013; Wong, 2013).

Chetty et al.'s methodological approach of using supersized metropolitan regions with one or more counties diminishes the measure of air pollution's impact on humans (Chetty et al., 2016). The influence of exposures at the street and neighborhood levels becomes averaged out. Louisville is a good example of why smaller geographic areas should be used: they are where the effects of pollution are most severe. For example, a study released by the city of Metro Louisville found that life expectancies can be as much as 13 years longer in several east-end neighborhoods compared to several neighborhoods on the less affluent west side of town. Our published research shows conclusively that shorter life spans are attributable to: 1) the toxic air pollution emitted from chemical and other industrial plants; 2) the brownfields that dot the communities where these less people live; and 3) the loss of the tree canopy caused by air pollution and other causes (Gilderbloom, 2016).

Past research that considered 148 mid-size cities revealed that improved health (using four different measures including life span) was correlated with reduced pollution by promoting walking, biking, car sharing and transit instead of single-car occupancy use (Frederick et al., 2018). Though race and income may shape where you live and work, the major cause of reduced life expectancy is living near unhealthy air, water and soil. Shorter life spans are found in other parts of the southern United States where pollution levels are the highest. According to the Centers for Disease Control and Prevention, 73 percent of U.S. citizens self-report that they are healthy at age 65. This percentage decreases to 62 percent in Kentucky, Mississippi and Alabama, 63 percent in West Virginia, percent in Tennessee, 67 percent in Arkansas and Louisiana, and 68 percent in Oklahoma (Centers for Disease Control and Prevention, 2014). In Oregon, a state with stronger environmental regulations, including clean air protections ranking in the top third for U.S. states, 78 percent of those aged 65 rate themselves as healthy. Roughly 55,000 more of those who are 65 in Kentucky feel unhealthy compared to the national average. When Kentucky is compared to the less polluted state of Oregon, one with the country's cleanest water, air and soil, nearly 100,000 Oregonians feel healthier at age 65.

Worldwide, air pollution was responsible for 7 million premature deaths in 2010 according to a study sponsored by the WHO (Kuehn, 2014). By 2050, according to the OECD, poor air quality will be the number one cause of premature death rates rising ahead of sanitation and dirty water, which are currently the primary causes (Organization for Economic Co-operation and Development, 2012).

Facts Over Politics

One of the problems is American exceptionalism: that it might be a problem elsewhere, but in the United States, scientists are often pressured not to attack polluting industries. This problem has been real for the lead author and his colleagues who have been pressured by public officials, industry leaders, and university administrators not to openly criticize polluting industries. Louisville's mayor acknowledged the wide disparity of life spans between different neighborhoods. Nevertheless, the healthy neighborhoods' task force representing foundations, industry, educators and the medical community ignored the connections between urban environmental degradation and health (Greater Louisville Project, 2013).

Instead, those victimized were often blamed. Claims are made that air pollution was not the problem, but rather health issues were caused by lifestyle choices of diet, exercise, smoking, lack of education and income. Such claims cite the *County Health Rankings and Roadmaps* program enacted by the Robert Wood Johnson Foundation (RWJF) in collaboration with the University of Wisconsin Population Health Institute (University of Wisconsin Population Health Institute, 2014). Through this program, communities can identify and implement solutions that make it easier for people to be healthy in their schools, workplaces and neighborhoods. Tracking nearly every county in the nation, these rankings illustrate a more subjective view as to what impacts the health of those living in the nation's counties (University of Wisconsin Population Health Institute, 2014). The program supports the argument that the impact of air, water, housing and transit pollution has a meager impact (only 10 percent) on a person's health.

Among the most serious problems with the County Health Rankings and Roadmaps is that they exclude three of the four EPA primary measures.

Findings are based solely on levels of PM2.5 which misrepresents each county's air quality and shows only negligible differences among counties in air quality. Our analysis shows that when the impacts of the four EPA measures are combined, there are wide differences in air quality and the impacts of toxic air measurements are minimized. Efforts to correct these methodological errors have been rejected. Moreover, it contributes to arguments published in the world's leading medical journals (e.g., *Lancet* and *JAMA*) that inaccurately suggest that local air pollution in the United States is not a serious medical concern (Chetty et al., 2016; University of Wisconsin Population Health Institute, 2014).

The RWJF endorsed and helped fund the Louisville study. We obtained a copy of the data set, which allowed us to test it and insert additional control variables that measure environmental toxins (e.g., location of industrial polluters with high EPA negative scores and identified toxic brownfield sites). In Louisville, our regression analysis teases out the net impacts of each factor, showing that proximity to polluted areas is as important as race and income in explaining the shortened lives in certain Louisville neighborhoods (Gilderbloom, 2016).

Limitations of the Methodology Applied

The utility of this methodology in predicting shortened lifespans in mid-size, semi-isolated cities is limited to the availability of the EPA data and using all four measures of pollution together as one. In mid-size, semi-isolated cities for which the EPA data is available, this methodology provides a means of determining how pollution impacts people's lives. By utilizing all four measures of pollution, it provides a more accurate picture than the RWJF study or the County Health Rankings and Roadmaps program, which uses only one air quality measure (University of Wisconsin Population Health Institute, 2014). The same methodology has been used for predicting high COVID-19 rates in cities and confirming the relationship between high levels of air pollution and the number of virus cases (Meares et al., 2022; Gilderbloom et al., 2022). This methodological tool is not appropriate for large cities or cities that lack EPA guidelines or measurements.

CONCLUSION

Livability indexes are often flawed because they do not accurately measure quality of life or account for the significant impacts of pollution on the livability of neighborhoods and cities. The problem is that pollution is either not included in the livability index or it is minimized or mismeasured. Perhaps pollution is often ignored because it is sometimes invisible to the naked eye.

Published research has tested whether high pollution levels impact health, neighborhoods and cities. First, one study found that pollution has an independent effect of reducing life expectancies (i.e., with differences in groups of poor people, of five years for men, and of four years for women). Another study found a similar impact when controlling for levels of smoking, drinking, and walkability. Both regression models yielded similar results. Instead of confronting environmental pollution of air and soil and their adverse effects on life spans, billions of dollars are expended in search of treatments and cures, which often means more cutting, gadgets, implants, radiation and pills. In the field of urban sustainability, these external costs are substantial.

Our research offers a more comprehensive assessment. There were 142 mid-size cities that met our research criterion of having populations greater than 50,000 and were not located near another other city of similar size. We obtained EPA data that provided accurate measures for four types of criteria air pollutants (i.e., PM10, PM2.5, NOx and SOx). Using this data set, we proved that pollution has a negative impact on life expectancy. Focusing on one highly polluted city which we ranked number two among 142 cities, we confirmed that high levels of pollution cause a reduced life expectancy. A study found that Louisville lost five years in life expectancy (Chetty et al., 2016). The Louisville Public Health Department found that citizens near high levels of pollution in Louisville's West End lost anywhere from 10 to 12 years of life expectancy whether in nearly all black or white neighborhoods (Gilderbloom et al., 2020).

Part of the cure for many of our cities is to monitor and reduce pollution in the air, water, soil and streets, and plant more trees. We recommend that all cities have EPA professional pollution monitoring stations. The external costs of PM2.5 and PM10 emissions are substantial. It is notable that there are few studies showing the separate, individual effects of unique pollutants encompassed within PM2.5 and PM10 emissions. These airborne pollutants travel together, so it is very difficult to isolate the effects of each constituent found in PM2.5 and PM10. It could be that combined pollutants amplify and intensify the ruinous effects of others. Most past studies only measured the environmental levels of PM2.5 and PM10 and correlated them with rises in certain adverse health conditions in the surrounding affected area without a control. In addition, the investigations into the pollutant effects were limited to shorter lag times, most likely due to budget and time constraints.

We need to adjust our thinking to be more proactive instead of reactive in resolving public health issues that are associated with pollution. As the Physicians for Social Responsibility point out, we should be "preventing what we cannot cure" (Physicians for Social Responsibility, 1985). One such preventive measure is ensuring that our communities, including our poor inner-city neighborhoods, enjoy a healthy environment. A cleaner environment also addresses the challenge of climate change, since reductions in pollution from the use of fossil fuels also reduces GHG emissions. It is a win-win for everyone and improves our stewardship of fragile Earth. You might believe that Flint, Michigan is an exception, but there are many cities similar to Flint in the United States and Louisville is among them. Proponents of environmental protection need to shift the focus on healthy life expectancies, asking why so many people—especially in the southern states—needlessly die too soon.

ACKNOWLEDGMENTS

We thank the many students who worked to gather data in graduate sustainability courses. We appreciate the insights and research of Debby Warren, the support of Alexander Bain and Afnan Rashid, and the financial support provided by Bobby Austin of Neighborhood Associates in Washington, D.C. and Marilyn Melkonian, CEO of Telesis Corporation. We thank Christopher Bird, who risked his professional career by providing EPA data that was initially hidden from public access.

Originally published in the *AEE International Journal of Strategic Energy and Environmental Planning (IJSEEP)*. The Association of Energy Engineers (AEE), founded in 1977, is a non-profit professional association of more than 17,000 members and over 32,000 certified professionals in more than 100 countries worldwide dedicated to serving members and other industry professionals to save energy, reduce GHGs, make buildings perform better, and help reach global goals for Net-Zero. For more about AEE, visit www.aeecenter.org. Revised and reprinted with permission from *IJSEEP* editor, Dr. Stephen Roosa.

A preliminary version of this article was previously presented in prepublication format in *The Lancet: Ranking the Most Polluted Mid-Sized Cities in the United States* (2020) by Gilderbloom, J., Bird, C., Quenichet, K., Manella, C. Dwenger, C., Rose, L., Sarr, S. Altaf, S. and Frederick, C.

APPENDIX

Table 3.1A   2011 NEI Data for Semi-Isolated, Mid-Size U.S. Cities (emissions in tons)

U.S. City	State	County	PM10	PM2.5	NOx	SOx
Anchorage	AK	Anchorage	8,134	1,916	12,298	429
Auburn	AL	Lee	9,088	3,229	5,193	743
Dothan	AL	Houston	7,492	2,140	3,651	543
Mobile	AL	Mobile	19,680	7,684	32,303	20,673
Montgomery	AL	Montgomery	9,367	4,298	10,725	2,042
Tuscaloosa	AL	Tuscaloosa	14,534	5,254	16,958	2,072
Conway	AR	Faulkner	9,772	1,958	4,314	77
Fort Smith	AR	Sebastian	7,584	2,317	4,487	272
Jonesboro	AR	Craighead	13,140	2,632	4,599	166
Flagstaff	AZ	Coconino	17,829	8,975	17,341	670
Lake Havasu City	AZ	Mohave	12,761	1,547	15,846	134
Yuma	AZ	Yuma	9,223	1,733	8,366	86
Bakersfield	CA	Kern County	23,553	12,208	46,852	3,072
Salinas	CA	Monterey	8,217	2,965	13,398	470
Santa Barbara	CA	Santa Barbara	5,524	1,578	10,285	441
Yuba City	CA	Sutter	3,698	1,189	5,340	75
Chico	CA	Butte	7,248	3,038	7,771	181
Redding	CA	Shasta	5,948	2,858	9,435	205
Colorado Springs	CO	El Paso	15,106	4,499	21,605	9,599
Grand Junction	CO	Mesa	4,352	1,416	7,412	109
Pueblo	CO	Pueblo	7,703	1,915	12,670	3,241
Gainesville	FL	Alachua	8,740	3,050	9,247	2,329
Jacksonville	FL	Duval	12,427	5,542	36,981	20,852
Pensacola	FL	Escambia	9,416	3,731	16,620	3,038
Tallahassee	FL	Leon	8,418	3,565	7,469	388
Albany	GA	Dougherty	5,591	2,367	5,145	1,706
Athens	GA	Clarke	2,473	489	3,049	250
Augusta	GA	Richmond	7,008	2,489	10,092	4,295
Columbus	GA	Muscogee	3,729	914	4,292	84
Savannah	GA	Chatham	7,135	2,281	16,308	10,944
Valdosta	GA	Lowndes	8,780	2,353	5,755	784
Ames	IA	Story	7,672	1,823	5,791	3,536
Davenport	IA	Scott	9,318	2,326	8,111	5,295
Dubuque	IA	Dubuque	9,764	1,962	3,827	1,068
Sioux City	IA	Woodbury	10,854	3,512	16,738	29,693
Waterloo	IA	Black Hawk	9,499	2,089	4,684	387
Idaho Falls	ID	Bonneville	12,900	2,198	4,055	143
Pocatello	ID	Bannock	9,494	1,545	4,985	93
Champaign	IL	Champaign	17,812	3,434	7,775	630
Decatur	IL	Macon	13,599	2,976	8,306	12,928
Peoria	IL	Peoria	11,023	2,565	11,308	14,213
Rockford	IL	Winnebago	9,309	2,225	7,319	439
Springfield	IL	Sangamon	14,756	3,428	8,424	3,385

(Continued)

Table 3.1A (Continued)

U.S. City	State	County	PM10	PM2.5	NOx	SOx
Bloomington	IN	Monroe	5,701	1,028	3,462	1,511
Evansville	IN	Vanderburgh	5,659	1,066	6,271	43
Fort Wayne	IN	Allen	13,759	2,736	14,789	272
Kokomo	IN	Howard	6,288	1,341	2,685	35
Lafayette	IN	Tippecanoe	10,478	2,213	8,074	7,522
Terre Haute	IN	Vigo	12,547	4,570	12,046	55,945
Manhattan	KS	Riley	9,440	3,427	2,548	316
Wichita	KS	Sedgwick	39,029	7,182	18,153	197
Bowling Green	KY	Warren	5,489	1,524	5,735	61
Lexington	KY	Fayette	8,341	1,772	7,845	582
Louisville	KY	Jefferson	15,898	7,672	37,796	39,231
Owensboro	KY	Daviess	6,096	1,583	7,656	9,014
Baton Rouge	LA	E. Baton Rouge Parish	12,872	5,832	22,052	11,812
Lafayette	LA	Lafayette Parish	5,556	1,953	8,954	1,324
Lake Charles	LA	Calcasieu	14,668	8,684	35,343	41,135
Shreveport	LA	Caddo Parish	8,618	3,593	19,620	1,859
Frederick	MD	Frederick	4,683	1,482	6,816	439
Portland	ME	Cumberland	6,652	2,807	11,792	3,429
Ann Arbor	MI	Washtenaw	9,072	2,786	11,616	189
Flint	MI	Genesee	10,732	3,119	12,293	176
Saginaw	MI	Saginaw	12,099	3,216	6,469	524
Lansing	MI	Ingham	8,076	2,340	9,906	7,625
Duluth	MN	St. Louis	21,556	9,391	34,183	7,032
Rochester	MN	Olmsted	8,860	3,138	6,362	651
St. Cloud	MN	Stearns	11,372	3,916	7,607	280
Columbia	MO	Boone	12,699	2,261	7,703	7,024
Joplin	MO	Jasper	13,312	2,697	6,189	9,068
Springfield	MO	Greene	19,321	3,773	13,432	8,840
St. Joseph	MO	Buchanan	6,199	1,251	6,658	2,014
Gulfport	MS	Harrison	25,598	4,892	16,468	32,925
Jackson	MS	Hinds	20,162	2,887	9,550	87
Billings	MT	Yellowstone	18,823	3,927	10,656	7,515
Great Falls	MT	Cascade	11,310	1,957	4,331	104
Missoula	MT	Missoula	17,843	5,981	5,425	388
Asheville	NC	Buncombe	4,295	1,536	8,663	2,679
Fayetteville	NC	Cumberland	4,600	1,365	8,318	287
Greenville	NC	Pitt	3,513	1,045	3,831	184
Jacksonville	NC	Onslow	3,746	1,596	4,434	827
Rocky Mount	NC	Nash	3,179	906	4,412	100
Wilmington	NC	New Hanover	1,974	933	9,946	13,844
Bismarck	ND	Burleigh	8,057	1,422	4,338	68

(Continued)

Table 3.1A   (Continued)

U.S. City	State	County	PM10	PM2.5	NOx	SOx
Fargo	ND	Cass	23,986	5,616	10,234	892
Grand Forks	ND	Grand Forks	13,535	3,257	4,776	816
Grand Island	NE	Hall	10,442	2,187	7,378	2,378
Lincoln	NE	Lancaster	19,193	3,402	16,990	4,254
Manchester	NH	Hillsborough	6,574	3,165	7,602	1,464
Toms River	NJ	Ocean County	3,451	1,711	7,970	493
Vineland	NJ	Cumberland	2,270	1,168	4,132	696
Las Cruces	NM	Doña Ana	67,065	8,346	11,506	209
Sante Fe	NM	Santa Fe	65,890	10,673	6,936	382
Rochester	NY	Monroe	10,599	3,652	16,726	6,959
Syracuse	NY	Onondaga	9,483	2,856	12,533	3,574
Utica	NY	Oneida	8,305	2,763	5,378	1,115
Akron	OH	Summit	6,237	2,128	17,500	4,311
Canton	OH	Stark	7,858	2,712	15,163	567
Columbus	OH	Franklin	14,690	4,802	36,846	441
Toledo	OH	Lucas	6,650	2,986	23,624	12,715
Youngstown	OH	Mahoning	4,664	1,361	9,362	1,481
Lawton	OK	Comanche	16,193	4,320	6,607	385
Bend	OR	Deschutes	24,099	13,802	6,486	858
Medford	OR	Jackson	14,086	5,976	6,844	612
Salem	OR	Marion	16,848	6,615	10,074	676
Erie	PA	Erie	5,825	2,444	11,373	1,659
Lancaster	PA	Lancaster	17,405	5,402	13,810	1,209
Pittsburgh	PA	Allegheny	8,892	5,740	35,455	15,080
Reading	PA	Berks	10,868	4,486	14,404	5,669
Scranton	PA	Lackawanna	3,130	1,567	5,840	459
Columbia	SC	Richland	9,225	3,536	16,035	8,343
Greenville	SC	Greenville	11,060	3,035	12,860	657
Rapid City	SD	Pennington	10,801	4,660	6,813	1,047
Sioux Falls	SD	Minnehaha	12,158	2,453	5,918	425
Chattanooga	TN	Hamilton	3,469	1,514	14,444	833
Clarksville	TN	Montgomery	3,463	1,156	5,103	479
Jackson	TN	Madison	5,209	2,014	5,938	238
Knoxville	TN	Knox	7,315	1,974	17,137	567
Abilene	TX	Taylor	12,659	1,954	5,053	72
Amarillo	TX	Potter	10,952	2,541	13,531	15,388
Corpus Christi	TX	Nueces	16,912	5,547	18,528	1,518
El Paso	TX	El Paso	16,915	3,025	19,152	588
Laredo	TX	Webb	6,368	1,312	16,967	71
Long View	TX	Gregg	28,564	5,507	23,724	8,489
Lubbock	TX	Lubbock	28,118	4,694	8,854	143
San Angelo	TX	Tom Green	12,970	2,856	3,553	204
Tyler	TX	Smith	2,211	1,569	8,076	634
Victoria	TX	Victoria	13,612	2,315	7,104	106
Waco	TX	McLennan	1,759	1,240	11,661	1,305
Wichita Falls	TX	Wichita	9,694	2,156	9,398	617
St. George	UT	Washington	10,955	1,560	4,937	82

(Continued)

Table 3.1A (Continued)

U.S. City	State	County	PM10	PM2.5	NOx	SOx
Bellingham	WA	Whatcom	5,683	3,078	10,781	8,011
Yakima	WA	Yakima	9,923	3,613	8,904	193
Eau Claire	WI	Chippewa	5,280	1,575	3,409	118
Green Bay	WI	Brown	7,120	2,383	14,161	18,307
Janesville	WI	Rock	8,587	2,454	6,524	85
La Crosse	WI	La Crosse	3,223	1,028	4,500	220
Madison	WI	Dane	18,012	5,277	18,777	1,769
Charleston	WV	Kanawha	8,530	3,090	15,449	13,365
Casper	WY	Natrona	21,919	3,583	5,047	394
Cheyenne	WY	Laramie	35,765	4,924	11,922	348

Sources: 1. 2011 NEI data; 2. Gilderbloom, J.H., Bird, C., Squires, G., Frederick, C., Slaten, E., Quenichet, K.A., Taskar, A., Snyder, C.J., and Riggs, W. (2022). What Cities are Most Dangerous to Your Life Expectancy? Toward a Methodology of Livability. International Journal of Strategic Energy and Environmental Planning, V4.4; 3. Center for Sustainable Urban Neighborhoods, formerly University of Louisville, transferred to Neighborhood Associates, Washington D.C.

REFERENCES

Almond, D. and Currie, J. (2011). Killing me softly: The fetal origins hypothesis. *Journal of Economic Perspectives*, 25(3), pages 153–172. doi:10.1257/jep.25.3.153.

Anderson, G., Krall, J., Peng, R. and Bell, M. (2012). Is the relation between ozone and mortality confounded by chemical components of particulate matter? Analysis of 7 components in 57 US communities. *American Journal of Epidemiology*, 176, pages 726–732. doi:10.1093/aje/kws188.

Appelbaum, R. (1978). *Size, Growth, and U.S. Cities*. New York, New York: Praeger Publishers.

Appelbaum, R., Bigelow, J., Krammer, H., Molotch, H. and Relis, P. (1976). *The Effects of Urban Growth: A Population Impact Analysis*. New York, New York: Praeger Publishers.

Banzhaf, S. (2012). *The Political Economy of Environmental Justice*. Stanford, California: Stanford University Press.

Been, V. (1995). Analyzing evidence of environmental justice. *Journal of Environmental Land Use and Environmental Law*, 11, pages 1–37.

Bernard, S., Samet, J., Grambsch, A., Ebi, K. and Romieu, I. (2001). The potential impacts of climate variability and change on air pollution-related health effects in the United States. *Environmental Health Perspectives*, 109(2), pages 199–209. doi:10.2307/3435010.

Boman, B., Christoffer, A., Forsberg, B. and Järvholm, B. (2003). Adverse health effects from ambient air pollution in relation to residential wood combustion in modern society. *Scandinavian Journal of Work, Environment and Health*, 29(4), pages 251–260. doi:10.5271/sjweh.729.

Borgschulte, M., Molitor, D. and Zou, E. (2019). Air pollution and the labor market: Evidence from wildfire smoke. *NBER*, working paper 29952. Cambridge, Massachusetts. National Bureau of Economic Research, Inc.

Bullard, R. (1990). *Dumping in Dixie: Race, Class, and Environmental Quality.* Boulder, Colorado: Westview Press.

Bullard, R. (2000). *Dumping in Dixie: Race, Class, and Environmental Quality* (3rd ed.). Boulder, Colorado: Westview Press.

Burtraw, D., Krupnick, A., Palmer, K., Paul, A., Toman, M. and Bloyd, C. (2003). Ancillary benefits of reduced air pollution in the U.S. from moderate greenhouse gas mitigation policies in the electricity sector. *Journal of Environmental Economics and Management*, 45(3), pages 650–673. doi:10.1016/s0095-0696(02)00022-0.

Capek, S. (1992). Environmental justice, regulations, and the local community. *International Journal of Health Services*, 22(4), pages 729–746.

Capek, S. (1993). The environmental justice frame: A conceptual discussion and an application. *Social Problems*, 40(1), pages 5–24. doi:10.2307/3097023.

Capek, S. and Gilderbloom, J. (1992). *Community Versus Commodity: Tenants and the American City.* Albany, New York: State University of New York Press.

Centers for Disease Control and Prevention. (2014). Particle pollution. https://www.cdc.gov/air/par-ticulate_matter.html.

Chay, K. and Greenstone, M. (1999). The impact of air pollution on infant mortality: Evidence from geographic variation in pollution shocks induced by a recession. NBER working paper 7442. Cambridge, Massachusetts. *National Bureau of Economic Research, Inc.* doi:10.3386/w7442.

Chetty, R., Stepner, M., Abraham, S., Lin, S., Scuderi, B., Turner, N., Bergeron, A. and Cutler, D. (2016). The association between income and life expectancy in the United States, 2001–2014. *Journal of the American Medical Association*, 315(16), pages 1750–1766. doi:10.1001/jama.2016.4226.

Cifuentes, L., Borja-Aburto, V., Gouveia, N., Thurston, G. and Davis, D. (2001). Climate change: Hidden health benefits of greenhouse gas mitigation. *Science*, 293(5533), pages 1257–1259. doi:10.1126/science.1063357.

Costa, A., Hoek, G., Brunekreef, B. and Ponce de Leon, A. (2017). Effects of NO2 exposure on daily mortality in São Paulo, Brazil. *Environmental Research*, 159, pages 539–544. https://doi.org/10.1016/j.envres.2017.08.041.

Currie, J., Neidell, M. and Schmieder, J. (2009). Air pollution and infant health: Lessons from New Jersey. *Journal of Health Economics*, 28(3), pages 688–703. doi:10.3386/w14196.

Currie, J. and Walker, W. (2011). Traffic congestion and infant health: Evidence from E-Z Pass. *American Economic Journal*, 3(1), pages 65–90. doi: 10.3386/w15413.

Frederick, C. and Gilderbloom, J. (2018). Commute mode diversity and income inequality: An inter-urban analysis of 148 midsize U.S. cities. *Local Environment: The International Journal of Justice and Sustainability*, 23(1), pages 54–76. doi: 10.1080/13549839.2017.1385001.

Frederick, C., Riggs, W. and Gilderbloom, J. (2018). Commute mode diversity and public health: A multivariate analysis of 148 U.S. cities. *International Journal of Sustainable. Transportation*, 12(1), 1–11. doi: 10.1080/15568318.2017.1321705.

Friedland, R. (1982). *Power and Crisis in the City*. London: MacMillan.

Frumkin, F., Frank, L. and Jackson, R. (2004). *Urban Sprawl and Public Health: Designing, Planning, and Building for Healthy Communities*. Washington, DC: Island Press.

Fuzzi, S., Baltensperger, U., Carslaw, K., Decesari, S., Denier van der Gon, H., Facchini, M. and Gilardoni, S. (2015). Particulate matter, air quality and climate: Lessons learned and future needs. *Atmospheric Chemistry and Physics*, 15(14), pages 8217–8299. doi: 10.5194/acp-15-8217-2015.

Gilderbloom, J. (2008). *Invisible City: Poverty, Housing, and New Urbanism*. Austin, Texas: University of Texas Press.

Gilderbloom, J. (2016). The Causes of Shorter Lifespans. *The New York Times*, April 18, 2016.

Gilderbloom, J. and Appelbaum, R. (1987). *Rethinking Rental Housing*. Philadelphia, Pennsylvania: Temple University Press.

Gilderbloom, J., Meares, W. and Squires, G. (2020). Pollution, place and premature death: Evidence from a midsized city. *Local Environment: International Journal of Sustainability and Justice*, 25(6), pages 419–432. https://doi.org/10.1080/13549839.2020.1754776.

Gilderbloom, J., Riggs, W., Frederick, C., Squires, G. and Quenichet, K. (2022). The missing link of air pollution: A closer look at the association between place and life expectancy in 146 mid-sized cities. *International Journal of Strategic Energy and Environmental Planning*, 4(3), 31–51.

Greater Louisville Project. (2013). Building a healthier Louisville. Louisville, Kentucky: Community Foundation of Louisville.

Hamilton, J. (1995). Testing for environmental racism: Prejudice, profits, political power? *Journal of Policy Analysis and Management*, 14(1), pages 107–132.

Hamra, G., Guha, N., Cohen, A., Laden, F., Raaschou-Nielsen, O., Samet, J., Vineis, P., Forastiere, F., Saldiva, P., Yorifuji, T. and Loomis, D. (2014). Outdoor particulate matter exposure and lung cancer: A systematic review and meta-analysis. *Environmental Health Perspectives*, 122(9), pages 906–911. doi:10.1289/ehp.1408092.

Hanna, R. and Oliva, P. (2015). The effect of pollution on labor supply: Evidence from a natural experiment in Mexico City. *Journal of Public Economics*, 122, pages 68–79. doi: 10.3386/w17302.

Hipp, J. and Lakon, C. (2010). Social disparities in health: Disproportionate toxicity proximity in minority communities over a decade. *Health and Place*, 16(4), pages 674–683. doi: 10.1016/j.healthplace.2010.02.005.

Isen, A., Rossin-Slater, M. and Walker, W. (2017). Every breath you take–every dollar you'll make: The long-term consequences of the Clean Air Act of 1970. *Journal of Political Economy*, 125(3), pages 848–902. doi: 10.3386/w19858.

Keeley, M. and Benton-Short, L. (2019). *Urban Sustainability in the U.S.: Cities Take Action*. Cham: Palgrave Macmillan. doi: 10.1007/978-3-319-93296-5.

Kuehn, B. (2014). WHO: More than 7 million air pollution deaths each year. *Journal of the American Medical Association*, 311(15), page 1486. doi :10.1001/jama.2014.4031.

Leiva, G., Santibañez, D., Ibarra, E., Matus, C. and Seguel, R. (2013). A five-year study of particulate matter (PM2.5) and cerebrovascular diseases. *Environmental Pollution*, 181, pages 1–6. doi: 10.1016/j.envpol.2013.05.057.

Meares, W., Gilderbloom, J., Squires, G. and Jones, A. (2022). Pollution and the pandemic: Explaining differences in COVID-19 rates across 146 U.S. communities. *International Journal of Strategic Energy and Environmental Planning*, 4(2), 9–26.

Molotch, H. (1976). The city as a growth machine. *American Journal of Sociology*, 82(2), pages 309–332.

Montt, G. (2018). The gendered effects of air pollution on labor supply. *International Labor Office* working paper 27. https://www.ilo.org/global/research/publications/workingpapers/WCMS_625863/langen/index.htm.

Organization for Economic Co-operation and Development (OECD). (2012). Environmental outlook to 2050: The consequences of inaction. doi: 10.1787/9789264122246-en.

Pahl, R. (1975). *Whose City?* London: Penguin.

Pastor, M., Sadd, J. and Hipp, J. (2001). Which came first? Toxic facilities, minority move-in, and environmental justice. *Journal of Urban Affairs*, 23(1), pages 1–21.

Physicians for Social Responsibility. (1985). "We must prevent what we cannot cure: Mobilizing health professionals on issues that represent the gravest dangers to human health." Nobel Peace Prize recipients. https://www.psr.org.

Pope, C., Bates, D. and Raizenne, M. (1995). Health effects of particulate air pollution: Time for reassessment? *Environmental Health Perspectives*, 103(5), pages 472–480. doi: 10.1289/ehp.95103472.

Prinz, D., Chernew, M., Cutler, D. and Frakt, A. (2018). Health and economic activity over the lifecycle: Literature review. NBER working paper 24865. Cambridge, Massachusetts. *National Bureau of Economic Research, Inc.* doi: 10.3386/w24865.

Smith, C. (2009). Economic deprivation and racial segregation: Comparing Superfund sites in Portland, Oregon and Detroit, Michigan. *Social Science Research*, 38(3), pages 681–692.

Sosa, B., Porta, A., Colman-Lerner, J., Noriega, R. and Massolo, L. (2017). Human health risk due to variations in PM10-PM2.5 and associated PAHs levels. *Atmospheric Environment*, 160, pages 27–35. doi: 10.1016/j.atmosenv.2017.04.004.

Straif, K., Chen, A. and Samet, J., eds. (2013). *Air Pollution and Cancer*. International Agency for Research on Cancer Scientific Publication no. 161. www.iarc.fr/wpcontent/uploads/2018/07/AirPollu-tionandCancer161.pdf.

Swanston, S. (2012). An environmental justice perspective on superfund reauthorization. *Journal of Civil Rights and Economic Development*, 9(2), pages 565–572.

Tessum, C. et al. (2019). Inequity in consumption of goods and services adds to racial–ethnic disparities in air pollution exposure. *Proceedings of the National Academy of the Sciences*, 166(13), pages 6001–6006. doi: 10.1073/pnas.1818859116.

The Lancet. (2013). Editorial. Action on ambient air pollution. *The Lancet*, 382(9897), page 1000. doi: 10.1016/S0140-6736(13)61960-1.

University of Wisconsin Population Health Institute. (2014). County health rankings key Findings. https://www.countyhealthrankings.org/sites/default/files/2014%20County%20Health%20 Rankings%20Key%20Findings.pdf.

U.S. Environmental Protection Agency. (2016). Basic information about NO2. https://www.epa.gov/no2-pollution/basic-information-about-no2#What%20is%20NO2.

U.S. Environmental Protection Agency. (2017). Basic information about air emissions monitoring. https://www.epa.gov/air-emissions-monitoringknowledge-base/basic-infor-mation-about-air-emissions-monitoring.

U.S. Environmental Protection Agency. (2018). Regulatory impact analysis for the proposed emission guidelines for greenhouse gas emissions from existing electric utility generating units; revisions to emission guideline implementing regulations; revisions to new source review program (Report no. 452/R-18-006). https://www.epa.gov/sites/production/files/201808/documents/utilities_ria_pro-posed_ace_2018-08.pdf.

U.S. Environmental Protection Agency. (2019). Overview of greenhouse gases. https://www.epa.gov/ghgemissions/overview-greenhouse-gases.

West, J. et al. (2013). Co-benefits of global greenhouse gas mitigation for future air quality and human health. *Nature Climate Change*, 3(10), pages 885–889. doi:10.1038/NCLIMATE2009.

Wong, E. (2013). Air pollution linked to 1.2 million premature deaths in China. *The New York Times*, April 1, 2016. https://www.nytimes.com/2013/04/02/world/asia/air-pollution-linked-to-1-2-million-deaths-in-china.html.

World Health Organization. (2018). Climate change and health. https://www.who.int/en/news-room/fact-sheets/detail/climate-change-and-health.

Chapter 4

Pollution and the Pandemic

Explaining Differences in COVID-19 Rates Across 146 U.S. Communities

Wesley L. Meares, John Hans Gilderbloom,
Gregory D. Squires, and Antwan Jones

Over 31 million cases of SARS-COV-2 and over 950,000 deaths from COVID-19 have been reported globally since the first incident of the disease was reported in China on November 17, 2019 (Huang & Wang, 2019; Johns Hopkins University, 2020). The U.S. represents 4.23 percent of the world's population but has the largest share of global cases and deaths, with over 6.8 million individuals with COVID-19 and nearly 200,000 deaths attributed to COVID since the first laboratory-confirmed case on January 21, 2020 (Johns Hopkins University, 2020; Jorden et al., 2020). While the pandemic has affected the lives of individuals, it has also substantially affected social institutions such as the economy, politics and government, and the health care system, and these effects have had unprecedented regional consequences in the United States. The subsequent quarantine and its politicization of the quarantine affected COVID-19 rates of infection as well as air pollutants that were quelled from the reduction in transportation emissions (Berman & Ebisu, 2020).

The quality of air varies significantly by state and city in the United States., and that variation could contribute to the variation in COVID-positive cases and COVID-related deaths in the United States. This study makes a unique contribution in that it looks at 146 semi-isolated counties in the United States to better understand the spatial effect that air quality and sociodemographic and political characteristics have on confirmed COVID-19 case counts and deaths. This study has broad significance. No peer-reviewed articles have yet been published on inter-community

COVID-19 cases and deaths. Preliminary research from Harvard University suggests that COVID-19 is higher in cities with high pollution levels (Wu, Nethery, Sabath, Braun, & Dominici, 2020). Pollution is the largest environmental cause of premature deaths, with some studies estimating over 9 million deaths (Landrigan, Fuller, Hu, et al., 2018). Some research suggests that in highly polluted countries such as China, which were also hit hard with COVID-19, the reduction in air pollution translated to a decrease in all-cause mortality during the pandemic (Chen, Wang, Huang, Kinney, & Anastas, 2020).

METHODS

The data used for this study were obtained from the 2018 American Community Survey (ACS), Johns Hopkins University, *The New York Times*, Social Explorer, County Business Patterns, Local Government Websites, and the University of Washington in St. Louis' Atmospheric Composition Analysis Group. From ACS data, we used the 2018 three-year estimates measured at the U.S. Bureau of the Census's geographic level of place (U.S. Bureau of the Census, 2020), although the unit of analysis, described below, is the county.

Case Selection

The units of analysis for the current study are the counties in 146 mid-size cities, defined by the U.S. Bureau of the Census as cities with at least 50,000 residents and that are not located within 20 miles of another city of similar size. Figure 4.1 shows the location of the counties in this study. The counties of the dataset are neither evenly distributed nor clustered, but randomly distributed throughout the continental United States (Moran's I: 0.049, z = 1.62). The county as a unit of analysis allows for a finer grain of analysis than the much larger commuting zone. The distance between counties decreases problems of spatial proximity.

   The 146 cities that emerge from the 700 total places are significant for the current study. It is highly problematic to include giant metropolises like New York City or Los Angeles in a sample that has 700 mid-size communities that are 50,000 or more. Louisville, Yuma, or Erie are more typical of American cities than the six giant metropolises. For comparative purposes, it's unworkable to combine both kinds of cities together; it is like comparing a giant outlier the size of an elephant with a more typically sized mouse. Instead of combining categorically different urban centers and then adjusting for urbanicity, this study selects cases using a well-validated set of decision rules from social science literatures, which sufficiently address spatial lag and case selection concerns (Appelbaum, 1978; Molotch, 1976).

Measures

Two outcome measures are used to measure COVID at the place level: the number and rate of COVID cases (per 100,000) as of August 19, 2020, and the number and rate of COVID-19 deaths (per 100,000) in the county. The focal independent measure, air pollution exposure, is captured by PM2.5 or inhalable particles of size 2.5 microns or less in width. These are estimates that were calculated by the Atmospheric Composition Analysis Group for the year 2016 (Van Donkelaar et al., 2019).

Additionally, this study controls for several measures in the analyses. The percentage of the labor force that is considered essential is derived from the Brookings Institution's definition of essential workers based on 4-digit NAICS industries (Tomer & Kane, 2020). This list of essential workers was derived from the guidance provided by the U.S. Department of Homeland Security in 2020 for jurisdictions to identify a broad array of what is considered the essential workforce during the COVID-19 pandemic. Given the data limitations, it is impossible to isolate the number of essential workers in these industries. For example, for hospitals, doctors and nurses are essential, but the administrative staff and janitorial staff are included in the number of employed. While they may be essential to keep the hospital functioning, they are not necessarily in what might be considered essential workers for treating COVID-19 patients. Despite this concern, this broad measure is seen as a more accurate measure since several will continue working to keep the organization functioning.

For the political and policy environment of these cities, this study includes the political affiliation of the highest-ranking member of the legislative body. This is a binary measure where Republican is coded as 1 and other affiliations (Democrat/Independent/Nonpartisan) are coded as 0. Also included is a measure of the number of days that a stay-at-home order, or equivalent, was in place at the state level. Third, there is a measure indicating whether a state has had a mask mandate in place for at least 15 days before August 19, 2020. This measure is captured at the state level due to the controversies surrounding state power over cities (Herman, 1971).

To account for inter-city differentials, measures for the total population of the county, population density per square mile, the percentage of non-white residents, median household income, the median value of owner-housing units, the percentage of the workforce 16 years or older that commutes by means other than driving alone, and a region variable identifying a southern location are included in the analyses. To address potential confounding, the study also includes the percentage of the population that is older than 65 and the percentage of residents considered to have three or more high-risk factors.

Statistical Analyses

Descriptive statistics were generated for each variable (see table 4.1) and bivariate correlations were run between the dependent and independent variables. To illustrate potential key relationships, the 10 counties with the highest and lowest rates of COVID-19 infection and deaths were identified, and descriptive statistics on all the independent measures were conducted. Finally, to assess the independent effects of pollution, a series of ordinary least squares regression models were run.

Regarding multicollinearity, tolerances are all well above the critical threshold of 0.2 and the variance inflation factor does not exceed the critical threshold of 4. Furthermore, the presence of competing dependencies was also explored and determined to be negative: in no case did two variables both show problematic values on the variance proportion table. Together, this methodology provides an accurate estimation and comparison of the impact of the control and test variables on COVID-19 cases and deaths across mid-sized cities.

Results

Figure 4.1 compares the number of COVID-19 cases for the top and bottom 10 counties in our sample. It shows that COVID-19 cases are much higher in highly polluted counties. Pollution is generally twice as high in counties with high COVID-19 cases. Also, COVID-19 cases are much lower in counties that are not located in the South. In our top 10 counties with the fewest COVID-19 cases, none are in the South, but in our bottom 10 counties with highest COVID-19 cases, eight are in the South. We also found evidence that states with Republican Governors have higher rates of COVID-19, and some evidence that Republican ayors tend to have higher COVID-19 cases. A larger percentage of non-whites tend to have higher cases of COVID-19. Higher population and density levels also show higher rates of COVID-19. Those counties with higher pre-existing conditions have a higher number of COVID-19 cases. We also found that there was no difference in COVID-19 cases associated with the share of the population over the age of 65.

Figure 4.2 compares the number of COVID-19 deaths for the top 10 and bottom 10 counties. Interestingly, several top and bottom cases in terms of deaths were different from those included in the top and bottom cases, with five dropping out of the top 10 in deaths and eight in the bottom 10. As we showed earlier, the number of COVID-19 cases and deaths was consistent with pollution playing a big role. It shows that COVID-19 deaths are much higher in highly polluted cities. Pollution is generally twice as high in cities with high COVID-19 deaths. Also, COVID-19 deaths are not necessarily found in all Southern cities like we found with the COVID-19 cases. There is no strong evidence that Republican mayors tend to have higher COVID-19 deaths. Counties with a larger percentage of non-whites tend to have higher death rates. Higher population and density levels also show higher levels of COVID-19 deaths. Those counties with higher pre-existing conditions have higher numbers of COVID-19 deaths. We were surprised that front-line workers, use of shared transit with others, and income had no difference. The percentage of elderly is slightly higher in cities with lots of COVID-19 deaths.

In tables 4.2 and 4.3, the regression model predicting COVID-19 cases and deaths shows that the air pollution test variable using PM 2.5 is a statistically significant predictor. In fact, it has an equal if not stronger impact than some standard control variables (95 percent CI Model 1: 138.43, 654.5; Model 2: 68.008, 230.69; Model 3: 1.9, 23.56; Model 4: 1.19, 7.71). Table 4.2 is a regression analysis showing the net impact of key variables in our analysis of variations in COVID-19 cases. We also found some surprising results which were not apparent with the bivariate analysis. Across the board, pollution (95 percent CI Model 1: 138.43, 654.5; Model 2: 68.008, 230.69), percent seniors (95 percent CI Model 1: −428.47, −138.55; Model 2: −120.39, −29.00), percent with pre-existing conditions (95 percent CI Model 1: 62.22, 316.81; Model 2: 67.67, 147.93), mask mandate (95 percent CI Model 1: −2399.78, −291.2; Model 2: −770.34, −105.65), density (95 percent CI Model 1: −4.96, −1.77; Model 2: −1.061, −0.06) and ayor's political party (95 percent CI Model 1: 385.54, 2135.8; although in Model 2 it is only approaching significance, $p = .56$) are significant predictors. Also, emerging significance is found when looking at just the raw number of cases, such as population (95 percent CI Model 1: 0.02, 0.02) and percent who commute with others (95 percent CI Model 1: −264.5, −36.88). The amount of explained variation is strong, with 75 percent of the variance in raw counts of cases accounted for by these variables.

The regressions in table 4.3 were less impressive in terms of explaining COVID-19 deaths, but given the exploratory nature of the study, they still have value. Once again, the most consistent predictor was pollution levels (95 percent CI Model 3: 1.9, 23.56; Model 4: 1.19, 7.71). The only other consistent predictors were the percentage of people with pre-existing conditions (95 percent CI Model 3: 1.82, 12.48; Model 4: 1.76, 4.98) and the percentage of seniors (95 percent CI Model 3:). The impact of race (95 percent CI Model 4: .237, 1.10) and median household income (95 percent CI Model 4: .00, .00) fell for the raw number of deaths and then became significant when we measured the per capita death rate. The impact of the mask mandate did not impact the raw number of deaths, but it did approach significance in terms of the death rate (p = .07, 95 percent CI Model 4: −25.631, 1.01). Moreover, the impact of living in the South and the mayor being a Republican was lost in predicting COVID-19 deaths. The amount of explained variations was at half to one-third of that explained in the last two equations found in table 4.3.

Discussion

Conventional wisdom suggests that until a safe and effective vaccine is discovered and made widely available, the most important steps that individuals can take, and public officials can encourage, are to wear masks when in public, maintain social distance when in public, and wash hands frequently. Moreover, even if a vaccine is found, the distribution will not be fully implemented until the summer of 2021 or later. And like a flu shot, many will still get COVID-19 with a vaccine and many others will refuse to get a vaccine. In other words, the best health practices even with a vaccine will be government ordering masks on in public, social distancing, and regularly washing hands. Unfortunately, some public officials have long given mixed signals, which reduce the effectiveness of these practices (Santucci, 2020). But there are other steps that can be taken.

The findings of this study highlight the importance of air pollution as a contributor to the number of COVID-19 cases and deaths. This is consistent with a wealth of research on the social determinants of health, including a recent research paper on the impact of pollution on COVID-19 (Wu, Nethery, Sabath, Braun, & Dominici, 2020). A key conclusion of that study was that "our results underscore the importance of continuing to enforce existing air pollution regulations to protect human health both during and after the COVID-19 crisis." Unfortunately, the U.S. Environmental Protection Agency appears to be moving in the opposite direction, as it has weakened or eliminated many rules and regulations designed to protect the environment (Friedman, 2020; Landrigan et al., 2018).

Our research found that rates of COVID-19 and deaths from the virus vary widely across mid-size communities in the United States. Several factors are statistically significantly associated with either the number or rates of cases or deaths, or both including the presence of mask mandates, population density, percentage of seniors, share with pre-existing conditions, racial composition, and median income—all expected. The political party of local mayors was also a factor, with jurisdictions having Democratic mayors experiencing fewer incidents. Surprisingly, the share of essential workers was not a significant predictor.

These findings suggest the importance of learning more about the dynamics of local communities. What actions are local elected officials, business leaders, and non-profit organizations taking? Has grassroots organizing been a factor? Are local foundations supporting COVID-19 education and relief efforts? Hopefully, there will be a vaccine or treatment of some kind before all the potentially informative research and action is taken. But even if remedies for this pandemic are found in the relatively near future, such work can inform and ameliorate other social determinants of health generally.

There is a limit, of course, to what can be done in response to the impact of several of these variables. The share of the population that is elderly cannot be quickly altered, though the care of the elderly might be. Population density is not likely to change at the city or county level, though individual neighborhoods could see changes as properties deteriorate or new developments grow. The political party of local officials generally must wait until the next election.

Pollution levels, which are the focus here, may not appear at first glance to be something that can be quickly changed. But with appropriate leadership (and perhaps local organizing to pressure leaders), older industrial sites can be remediated in many cases. The EPA could change course and re-introduce what are in fact life-saving environmental protection rules, though such action is unlikely to take place until at least January of 2021. Local and state leaders could act on their own, and many around the world are doing so (Emanuel, 2020; Keeley & Benton-Short, 2019). Everyone hopes for a quick resolution of the current pandemic. But in the meantime, we might follow the guidance of the Physicians for Social Responsibility, who asserted in reference to the costs of pollution, "We must prevent what we cannot cure" (Physicians for Social Responsibility, 2020). See Figure 43.

Originally published in the *AEE International Journal of Strategic Energy and Environmental Planning* (*IJSEEP*). The Association of Energy Engineers (AEE), founded in 1977, is a non-profit professional association of more than 17,000 members and over 32,000 certified professionals in more than 100 countries worldwide dedicated to serving members and other industry professionals to save energy, reduce GHGs, make buildings perform better, and help reach global goals for Net-Zero. For more about AEE, visit www.aeecenter.org. Revised and reprinted with permission from *IJSEEP* editor, Dr. Stephen Roosa.

Table 4.1 Descriptive Statistics

Variable Name	Description	Min.	Max.	Mean	Std. Deviation	Source
Cumulative confirmed COVID 19 cases as of Aug 19 2020	Cumulative confirmed COVID 19 cases as of Aug 19 2020	185	27,184	4079.73	4430.41	Johns Hopkins
Cumulative COVID 19 confirmed deaths as of Aug 19 2020	Cumulative COVID 19 confirmed deaths as of Aug 19 2020	0	1023	96.88	131.02	Johns Hopkins
Cumulative confirmed COVID 19 cases Rate per 100,000 as of Aug 19 2020	Rate of cumulative cases of COVID 19 per 100,000 as of August 19, 2020	226	5742	1531.19	994.98	Johns Hopkins
Cumulative COVID 19 confirmed death rate per 100,000 as of Aug 19 2020	Rate of cumulative COVID 19 deaths per 100,000 as of August 19, 2020	0	190	33.84	33.67	Johns Hopkins
Total population	Total population of the county	53,332	127,533	262353.87	213230.55	ACS
Population density (Per Sq. Mile)	Population density	8	2395	383.28	400.59	ACS
PM 2.5	0.01° × 0.01° grid resolution PM2.5 prediction 2016	2.2	12.5	6.804	2.01	Atmospheric Composition Analysis Group
Percent of nonwhite residents	Percentage of residents that identified as not white	4.57	78.99	22.9038	14.83	ACS
Median household income	Median household income (In 2018 inflation adjusted dollars)	$35,516	$91,999	$53,870.66	$8,998.17	ACS
Median value	Median housing value of owner units	$85,200	$649,800	$176,187.67	$76,921.27	ACS
Percent of Commuters That Do Not Drive Alone	Percentage of Workers, 16 and over, that commute to work by not driving alone.	11.6	35.99	19.58	4.33	ACS

Variable Name	Description	Min.	Max.	Mean	Std. Deviation	Source
Southern states	Whether or not a state is in the Southern United States	0	1	0.43	0.50	ACS
Percentage of residents considered to have high risk factors	Percentage of residents with three or more health risk factors	16.91	40.65	25.18	3.83	Social Explorer
Percentage of the population older than 65	Percentage of the population that is 65 or older	8.6	28.8	15	3.01	ACS
Mask mandate at least 15 days	Binary variable indicating that the state the city is in has had a mask mandate for at least 15 days from 8/19/2020	0	1	0.69	0.46	*The New York Times* Corona Virus Data
Number of days with stay at home order	The number of days that a stay at home order, or equivalent, was in place	0	80	33.02	17.33	*The New York Times* Corona Virus Data
Percentage of essential workers	Department of Homeland Security definition of essential jobs based on industry	29.12	62.46	47.16	6.07	County Business Patterns
Mayor's political party	Binary variable indicating whether or not a mayor is Republican	0	1	0.25	0.43647	Local Government Websites

Sources: 1. Data sources in table; 2. Meares, W., Gilderbloom, J.H., Squires, G., & Jones, A. (2021). Pollution and the Pandemic: Explaining the Differences in COVID-19 Rates Across 146 U.S. Communities. International Journal of Strategic Energy and Environmental Planning, V.4.2; 3. Center for Sustainable Urban Neighborhoods, formerly University of Louisville, transferred to Neighborhood Associates, Washington, D.C.

TABLE 2 TOP TEN AND BOTTOM TEN CASES BASED ON CUMLATIVE CASES PER 100,000 AS OF 8/19/2020

Rank	County Name	Cumulative COVID 19 Cases Rate per 100,000	Total Population	Population Density	PM2.5	Percent of Nonwhite Residents	Median Household Income	Median Value	Percent of Commuters That Do not Drive Alone	Southern	Percentage of Residents with High Risk Factors	Percentage of Residents More than 65 Years	Mask or Mandate at Least 15 days	Number of Days Stay-at-Home Orders was in effect	Percentage of Essential Workers	Mayor's Political Party
1	Cascade County, Montana	226	81,746	30	3.4	11.93	48,160	171700	20.13	No	26.45	17.86	Yes	30	46.13	Nonpartisan
2	Mesa County, Colorado	247	149,998	45	3.8	5.95	53,683	214400	21.62	No	25.5	17.93	Yes	32	49.57	Republican
3	Shasta County, California	288	179,085	47	4.2	12.98	50,905	242500	18.19	No	23.14	19.89	Yes	50	48.22	Nonpartisan
4	Jackson County, Oregon	288	214,267	77	2.6	8.63	50,851	261700	23.94	No	25.95	21.04	Yes	53	44.68	Republican
5	Natrona County, Wyoming	311	80,610	15	3.1	5.86	60,550	202600	15.73	No	23.32	13.75	No	0	48.93	Nonpartisan
6	Missoula County, Montana	337	115,983	45	5.2	8.31	51,370	271400	27.56	No	19.32	14.55	Yes	30	38.53	Democrat
7	St. Louis County, Minnesota	348	200,680	32	3.3	7.74	53,344	152000	20.84	No	24.36	18.26	Yes	31	52.16	Democrat
8	Deschutes County, Oregon	362	180,640	60	2.2	6.51	63,680	316400	25.56	No	19.76	19.1	Yes	53	32.11	Democrat
9	Chippewa County, Wisconsin	446	63,635	63	4.4	5.03	57,883	159000	11.86	No	19.72	16.92	Yes	47	46.57	Nonpartisan
10	Erie County, Pennsylvania	451	275,972	345	4.4	13.23	49,016	130000	20.93	No	27.59	16.79	Yes	30	54.84	Democrat
	Sum	3304	1542016	759	39.1	86.17	539,147	2141700	212.36		235.11	176.09		356	455.74	
	Average	330.4	154201.6	75.9	3.92	8.617	539,447	214170	21.236		23.511	17.609		35.6	45.574	
137	Montgomery County, Alabama	3268	226,941	289	9.5	63.49	47,960	157000	15.44	Yes	30.91	14.18	Yes	26	48.57	Independent
138	Lafayette Parish, Louisiana	3348	240,091	893	9.2	30.3	54,726	177500	15.86	Yes	22.68	12.25	Yes	39	50.96	Republican
139	Escambia County, Florida	3370	311,522	474	8.3	31.87	49,086	173600	23.94	Yes	22.72	16.2	No	32	49.48	Republican
140	Calcasieu Parish, Louisiana	3574	200,182	188	8.6	29.78	49,452	149400	14.44	Yes	22.93	14.26	Yes	39	50.83	Republican
141	Webb County, Texas	3660	272,053	81	8.1	4.57	42,293	119900	19.06	Yes	40.65	8.96	Yes	29	58.82	Democrat
142	Woodbury County, Iowa	3768	102,398	117	7.7	13.83	55,483	117700	16.32	No	23.8	14.43	No	0	45.28	Nonpartisan
143	Victoria County, Texas	3977	91,970	104	6.8	12.44	55,631	136900	19.81	Yes	33.16	15.41	Yes	29	49.77	Nonpartisan
144	Yakima County, Washington	4551	249,325	53	4.1	21.87	49,871	167700	20.74	No	29.66	13.19	Yes	42	55.77	Nonpartisan
145	Nueces County, Texas	4863	360,486	430	7.6	10.07	55,648	130700	16.52	Yes	33.26	13.69	Yes	29	50.24	Republican
146	Yuma County, Arizona	5742	207,829	38	12.5	25.01	44,058	122000	18.41	Yes	34.21	18.01	No	31	36.93	Republican
	Sum	40111	2262797	2672	81.7	243.23	503,838	1382900	180.21		291.98	140.58		296	496.05	
	Average	40111	226279.7	267.2	8.17	24.323	500,838	138290	18.024		29.198	14.058		29.6	49.605	

Figure 4.1 **Top 10 and Bottom 10 Cases Based on Cumlative Cases per 100,000 as of 8/19/2020.** *Sources:* 1. Data sources in table 4.1; 2. Meares, W., Gilderbloom, J.H., Squires, G., & Jones, A. (2021). Pollution and the Pandemic: Explaining the Differences in COVID-19 Rates Across 146 U.S. Communities. International Journal of Strategic Energy and Environmental Planning, V.4.2; 3. Center for Sustainable Urban Neighborhoods, formerly.

TABLE 3 TOP TEN AND BOTTOM TEN CASES BASED ON CUMULATIVE DEATHS PER 100,000 AS OF 8/29/2020

Rank	County	Cumulative Confirmed COVID 19 Cases Rate	Total Population	Population Density	PM2.5	Percent of Nonwhite Residents	Median Household Income	Median Value	Percent of Commuters That Do not Drive Alone	Southern	Percentage of Residents with High Risk Factors	Percentage of Residents More than 65 Years	Mask Mandate at Least 15 days	Number of Days Stay in Order was in effect	Percentage of Essential Workers	Mayor's Political Party
1	Chippewa County, Wisconsin	0	63,685	63	4.4	5.03	57,288	159000	17.86	No	19.72	16.92	Yes	47	40.57	Nonpartisan
2	Jackson County, Oregon	0.93	214,267	77	2.6	8.63	50851	261700	23.94	No	25.95	21.04	Yes	53	44.68	Republican
3	La Crosse County, Wisconsin	1.1	118,016	254	5.4	8.7	55479	167100	17.7	No	21.81	16.2	Yes	47	48.57	Democrat
4	Natrona County, Wyoming	1.24	80,610	15	3.1	5.86	60550	202600	15.73	No	23.32	13.75	No	0	48.93	Nonpartisan
5	Missoula County, Montana	1.72	115,983	45	5.2	8.31	51270	271400	27.56	No	19.32	14.55	Yes	30	38.53	Democrat
6	Santa Fe County, New Mexico	2.01	146,917	78	3.4	16.48	59192	282600	21.17	Yes	30.27	22	Yes	38	35.67	Democrat
7	Bannock County, Idaho	2.35	85,065	76	5	10.18	49739	152500	23.16	No	21.94	13.58	No	37	45.19	Republican
8	Boone County, Missouri	2.83	176,515	257	5.9	18.91	54043	179800	21.51	No	17.61	11.42	No	28	47.73	Independent
9	Laramie County, Wyoming	3.07	97,682	96	2.8	11.06	64220	214300	18.11	No	22.69	15.07	No	0	47.97	Republican
10	Mesa County, Colorado	3.33	149,998	45	3.8	5.95	50683	214400	21.62	No	25.5	17.93	Yes	32	49.57	Republican
	Sum	18.58	1250698	946	41.6	99.11	556915	2105100	208.36		228.13	162.26		31.2	447.41	
	Average	1.858	125069.8	94.6	4.16	9.911	55691.5	210510	20.836		22.813	16.226		31.2	44.741	
137	Berks County	90.25	416,642	487	9.2	17.35	61522	174200	20.16	No	25.67	16.58	Yes	30	47.76	Democrat
138	Mohave County	91.72	206,064	15	7.8	9.49	43,266	151100	19.99	Yes	31.98	28.8	No	31	45.95	Nonpartisan
139	Yakima County	97.46	249,325	58	4.1	21.87	49871	167700	20.74	No	29.66	13.19	Yes	42	55.77	Nonpartisan
140	Lackawanna County	100.26	211,454	461	6.8	8.39	50875	149700	19.99	No	28.37	19.28	Yes	30	52.07	Independent
141	Cumberland County	103	153,400	317	7.8	32.79	52593	162500	19.17	No	27.85	14.55	Yes	80	52.52	Republican
142	Mahoning County	112.96	231,064	562	7.8	20.04	44682	103400	14.83	No	23.83	20.01	Yes	99	51.74	Democrat
143	Caddo Parish	122.4	248,361	283	7.6	53.61	40666	144400	14.06	Yes	30.52	15.97	Yes	39	53.46	Democrat
144	Yuma County	145.31	207,829	38	12.5	25.01	44058	122000	18.41	Yes	34.21	18.01	No	31	36.93	Republican
145	Ocean County	172.82	591,939	942	8.2	8.83	68021	272900	17.49	No	24.6	22.27	Yes	80	45.71	Republican
146	Doughtry County	190.01	91,049	277	8.6	73.03	37633	103900	21.65	Yes	28.27	14.63	No	22	39.3	Democrat
	Sum	1226.19	2607127	3440	80.4	270.41	493387	1551800	186.49		285.96	183.29		424	481.21	
	Average	122.619	260712.7	344	8.04	27.041	49338.7	155180	18.649		28.596	18.329		42.4	48.121	

Figure 4.2 Top 10 and Bottom 10 Cases Based on Cumlative Deaths pre 100,000 as of 8/19/2020. *Sources:* 1. Data sources in table 4.1; 2. Meares, W., Gilderbloom, J.H., Squires, G., & Jones, A. (2021). Pollution and the Pandemic: Explaining the Differences in COVID-19 Rates Across 146 U.S. Communities. International Journal of Strategic Energy and Environmental Planning, V.4.2; 3. Center for Sustainable Urban Neighborhoods, formerly.
Note: nst. = Unstandardized coefficients. Beta = standardized. p < 0.1+. *p < 0.05. **p < 0.01. ***p < 0.001.

Table 4.2 OLS Regression Confirmed COVID 19 Cases

	Model 1: Cumulative Number of Cases			Model 2: Cumulative Case Rate per 100 K		
	Unst.	95% CI	P-Value	Unst.	95% CI	P-Value
(Constant)	−3329.02	(−10383.17, 3725.14)	.352	−1858.43	(−4082.09, 365.22)	.101
Total population	.02	(.02, .02)	.000	.000	(−.001, .001)	.647
Population density (Per Sq. Mile)	−3.36	(−4.96, −1.77)	.000	−0.56	(−1.061, −.06)	.029
PM 2.5	396.47	(138.43, 654.51)	.003	149.35	(68.01, 230.69)	.000
Percent of nonwhite residents	−17.97	(−51.97, 16.02)	.298	4.98	(−5.736, 15.7)	.36
Median household income	.00	(−.07, .07)	.989	.02	(−.003, .04)	.083
Median value	.01	(−.004, .01)	.231	.000	(−.002, .003)	.778
Percent of commuters that do Not drive alone	−150.69	(−264.5, −36.88)	.01	−26.47	(−62.35, 9.40)	.147
Southern states	1638.60	(620.39, 2656.81)	.002	386.55	(65.583, 707.52)	.019
Percentage of residents consider to have high risk factors	189.51	(62.22, 316.81)	.004	107.8	(67.67, 147.93)	.000
Percentage of the population older than 65	−283.51	(−428.47, −138.55)	.000	−74.7	(−120.39, −29)	.002
Mask mandate at Least 15 days	−1345.49	(−2399.78, −291.2)	.013	−438	(−770.34, −105.65)	.01
Number of days with stay at home order	11.44	(−17.8, 40.69)	.44	−1.57	(−10.78, 7.65)	.738
Percentage of essential workers	52.2	(−24.97, 129.37)	.183	7.58	(−16.75, 31.9)	.539
Mayor's political party	1260.67	(385.54, 2135.8)	.005	268.83	(−7.04, 544.69)	.056
F	29.95			10.6		
R^2	.76			.53		
Adj R^2	.74			.48		

Sources: 1. Data sources in table 4.1; *2.* Meares, W., Gilderbloom, J.H., Squires, G., & Jones, A. (2021). Pollution and the Pandemic: Explaining the Differences in COVID-19 Rates Across 146 U.S. Communities. International Journal of Strategic Energy and Environmental Planning, V.4.2; *3.* Center for Sustainable Urban Neighborhoods, formerly University of Louisville, transferred to Neighborhood Associates, Washington, D.C.

Note: nst. = Unstandardized coefficients. Beta = standardized. $p < 0.1+$. $*p < 0.05$. $**p < 0.01$. $***p < 0.001$.

Table 4.3 OLS Regression Confirmed COVID 19 Deaths

	Model 3: Cumulative Deaths			Module 4: Cumulative Deaths Rate per 100 K		
	Unst.	95% CI	P-Value	Unst.	95% CI	P-Value
(Constant)	−280.08	(−575.58, 15.42)	.063	−122.03	(−211.14, −32.91)	.008
Total population	.000	(.00, .00)	.000	−8.65E-06	(.000, .000)	.661
Population density (Per Sq. Mile)	.01	(−.5, .08)	.697	.000	(−.024, .02)	.711
PM 2.5	12.75	(1.9, 23.56)	.021	4.45	(1.19, 7.71)	.008
Percent of nonwhite residents	.29	(−1.13, 1.71)	.69	.67	(.237, 1.10)	.003
Median household Income	.00	(.00,.01)	.081	.00	(.00, .00)	.031
Median value	.00	(−.001, .00)	.155	.00	(.00, .00)	.05
Percent of commuters that do Not drive Alone	−4.04	(−8.81, .72)	.096	−.42	(−1.86, 1.02)	.562
Southern states	−20.24	(−62.9,22.4)	.35	−9.8	(−22.66, 3.06)	.134
Percentage of residents consider to have high risk factors	7.15	(1.82, 12.48)	.009	3.37	(1.76, 4.98)	.000
Percentage of the population older than 65	5.3	(−.77, 11.38)	.086	2.23	(.4, 4.06)	.017
Mask mandate at least 15 days	−28.57	(−72.73, 15.6)	.203	−12.31	(−25.631, 1.01)	.07
Number of days with stay at home order	1.29	(.06, 2.51)	.04	.31	(−.06, .68)	.101
Percentage of essential workers	−1.83	(−5.07, 1.4)	.264	−.54	(−1.51, .44)	.279
Mayor's political party	16.48	(−20.18, 53.14)	.376	.55	(−10.51, 11.61)	.922
F	10.33			4.9		
R^2	.52			.34		
Adj R^2	.47			.27		

Sources: 1. Data sources in table 4.1; 2. Meares, W., Gilderbloom, J.H., Squires, G., & Jones, A. (2021). Pollution and the Pandemic: Explaining the Differences in COVID-19 Rates Across 146 U.S. Communities. International Journal of Strategic Energy and Environmental Planning, V.4.2; 3. Center for Sustainable Urban Neighborhoods, formerly University of Louisville, transferred to Neighborhood Associates, Washington, D.C.

Note: nst. = Unstandardized coefficients. Beta = standardized. $p < 0.1+$. *$p < 0.05$. **$p < 0.01$. ***$p < 0.001$.

ISOLATED MIDSIZE U.S. CITIES

1 inch ≈ 508 km

0   125 250   500 Miles

• Isolated Cities

Figure 4.3 One Hundred Forty-eight Semi-Isolated, Mid-Sized U.S. Cities. *Source*: Center for Sustainable Urban Neighborhoods, formerly University of Louisville, transferred to Neighborhood Associates, Washington, D.C.

REFERENCES

Appelbaum, R.P. (2019). *Size, Growth, and US Cities*. Praeger Publishers.

Berman, J.D., & Ebisu, K. (2020). Changes in US air pollution during the COVID-19 pandemic. *Science of the Total Environment*, 739, 139864.

Chen, K., Wang, M., Huang, C., Kinney, P.L., & Anastas, P.T. (2020). Air pollution reduction and mortality benefit during the COVID-19 outbreak in China. *The Lancet Planetary Health*, 4(6), e210–e212.

Emanuel, R. (2020). *The Nation City: Why Mayors are Now Running the World*. Alfred A. Knopf.

Freidman, L. (2020). New research links air pollution to higher Coronavirus death rates. *The New York Times*, April 17, 2020. https://www.nytimes.com/2020/04/07/climate/air-pollution-coronavirus-covid.html.

Herman, P.W. (1971). A historical examination of the City-State controversy over control of the New York City subways. *The Urban Lawyer*, 78–98.

Huang, C., Wang, Y., & Li, X., et al. (2020). Clinical features of patients infected with 2019 novel coronavirus in Wuhan, China. *The Lancet*, 395, 497–506.

Johns Hopkins University. (2020). COVID-19 in the USA. *Johns Hopkins University and Medicine*, September 18, 2020. Retrieved September 20, 2020, from https://coronavirus.jhu.edu/.

Jorden, M.A., & Rudman, S.L., CDC COVID Response Team, et al. (2020). Evidence for limited early spread of COVID-19 within the United States, January–February 2020. *Morbidity and Mortality Weekly Report*, 69, 680–4.

Kavanagh, M.M., & Singh, R. (2020). Democracy, capacity, and coercion in pandemic response—COVID 19 in comparative political perspective. *Journal of Health Politics, Policy and Law*, 45(6), 997–1012. doi: 10.1215/03616878-8641530.

Keeley, M., & Benton-Short, L. (2019). *Urban Sustainability in the US*. Springer.

Landrigan, P.J., Fuller, R., Acosta, N.J., Adeyi, O., Arnold, R., Baldé, A.B., Bertollini, R., Bose-O'Reilly, S., Boufford, J.I., Breysse, P.N., & Chiles, T. (2018). The Lancet Commission on pollution and health. *The Lancet*, 391, 462–512.

Landrigan, P.J., Fuller, R., Hu, H., et al. (2018). Pollution and global health–an agenda for prevention. *Environmental Health Perspectives*, 126(8).

Molotch, H. (1976). The city as a growth machine: Toward a political economy of place. *American Journal of Sociology*, 82, 309–32.

Physicians for Social Responsibility. (2020). Retrieved September 21, 2020, from https://www.psr.org/.

Santucci, J. (2020). The White House has sent conflicting messages on wearing masks and the new coronavirus cases. *USA Today*, July 6, 2020. Retrieved September 20, 2020, from https://www.usatoday.com/story/news/politics/2020/07/05/trump-white-house-give-mixed-messages-masks-coronavirus-spread/5368455002/.

Tomer, A., & Kane, J.W. (2020). How to protect essential workers during COVID-19. *Brookings Report*, March 31, 2020. Retrieved September 20, 2020, from https://www.brookings.edu/research/how-to-protect-essential-workers-during-covid-19/.

US Bureau of the Census. (2020). Geographies. *United States Census Bureau*, September 20, 2020. Retrieved September 20, 2020, from https://www.census.gov/programs-surveys/geography/geographies.html.

Van Donkelaar, A., Martin, R.V., Li, C., & Burnett, R.T. (2019). Regional estimates of chemical composition of fine particulate matter using a combined geoscience-statistical method with information from satellites, models, and monitors. *Environmental Science & Technology*, 53, 2595–611.

Wu, X., Nethery, R.C., Sabath, B.M., Braun, D., & Dominici, F. (2020). Exposure to air pollution and COVID-19 mortality in the United States. *Science Advances*, 6(45), eabd4049. https://doi.org/10.1101/2020.04.05.20054502.

Chapter 5

Automobile Addiction Kills the Earth

The Need for Multimodality

Chad Frederick, William Riggs,
and John Hans Gilderbloom

Achieving positive public health in urban areas has been a challenge for the last two centuries, from the days of tenements and settlement houses to suburban growth and the rise of non-contagious illnesses like cancer and heart disease (Corburn, 2007). Complicating matters, the central culprit has changed from coal-fired smokestacks of industry to the single-occupant vehicle (SOV). Since the city is the common denominator, it has carried much of the blame, as if poor public health outcomes were intrinsic to compact built environments. However, a large body of research has illuminated the positive relationships between good health outcomes and compact urban form. Indeed, researchers have connected the lack of walkability, open space, and multimodal commutes to accidents, obesity, and related diseases (Dannenberg, Frumkin, & Jackson, 2011; Doyle, Kelly-Schwartz, Schlossberg, & Stockard, 2006; Frank & Pivo, 1994; Frank et al., 2006; Frumkin, 2002; Sallis, Frank, Saelens, & Kraft, 2004; Sturm & Cohen, 2004).

A significant component of that research has demonstrated the connection between development type and forms of active transportation such as walking, biking, or transit. What we now know is that compared to other modes of transportation, car use promotes a greater amount of sedentary behavior than walking, biking, bus, or rail. Furthermore, the research literature indicates that cars create more pollution than any other form of transportation, causing health problems such as asthma and heart disease (Frank et al., 2006). Dependency on cars compounds negative health impacts from environmental injustices such as neighborhood degradation and segregation (Day, 2006).

While the research in this area is useful and informative, it also has limitations that hamper the ability of planners and policymakers to advance better transportation policies. For one, much of the existing research has focused on the negative impacts of sprawl. As a result, many policymakers consider sprawl the opposite of urban density. And yet, dense cities can also be automobile dependent (AD) (Eidlin, 2005), and "sprawling" cities like Tokyo can be multimodal (MM); or, the ability to access the city using a variety of modes of transportation. Thus, the opposite of sprawl is not merely density, but rather rests at the intersection of many other characteristics, including accessibility (Galster et al., 2001; Torrens & Alberti, 2000). The focus on the negative impacts of sprawl comes at the expense of identifying the positive impacts of multimodality (McLoughlin, 1991).

Furthermore, quantitative research on transportation and the built environment tends to focus on output variables (e.g., Vehicle Miles Traveled, or VMT), and then works back to the individual behaviors that produce them. While some research singles out the benefits of travel behavior such as walking or biking, fewer researchers broadly analyze the impact of multimodal commuting, which may better reflect urban outcomes than research on the complex modal choice sets at the individual level. Overall, planners and policymakers start the policy process with vague ideas of *what not to do*, and with less empirical support for *what should be done*, particularly in reinforcing the connectivity between sustainable modes of travel. This *ad hoc* approach to policy results in serious contradictions. A common example is the development of places meant to be walkable and bikeable, but which also encourage automobile use through Euclidean zoning and parking requirements.

This work responds to the call for more quantitative research across multiple communities in the areas of transportation, the built environment, and public health (Dannenberg et al., 2011; Johnson, 2001; Litman, 2012), including studies focused on how "travel activity affect public health" (Litman, 2013). Ewing and colleagues (2003) suggested future research "refine measures of urban form, improve measures of physical activity, and control for other individual and environmental influences on physical activity, obesity, and related health outcomes." In that light, this work extends the dialogue away from the harmfulness of sprawl and toward the health and sustainability benefits of supporting multimodal travel.

LITERATURE

Since WWII, the automobile has reshaped cities. While there were many economic and mobility-related benefits of auto-related *travel*, the focus on auto-centric *environments* has also exacerbated issues related to segregation, pollution, health, and social isolation/loss of community (Blokland & Rae, 2016; Putnam, 2001). Today, many Americans travel in cars for hours each day while commuting to work; longer commute times are associated with increased levels of obesity (Zhang et al., 2014). Gilderbloom et al. (2009) found, when comparing American cities to Amsterdam, with its compact design, green transportation system, and high density that residents in Amsterdam experienced longer life spans and enjoyed improved health. This has long been one of the critiques of sprawling development patterns: they center on automobile transportation and lead to less walking, higher pollution, and poorer health (Foster, 1979; Jackson, 2003; Wachs, 1984). For example, Ewing and colleagues (2003) illustrate a significant association between urban form, physical fitness, and other related health outcomes. Using a sprawl index generated through principal components analysis of U.S. Census data, they found positive health outcomes associated with less sprawl at both the county and MSA levels, including healthier behaviors. They found that, "Those living in sprawling counties were likely to walk less, weigh more, and have greater prevalence of hypertension than those living in compact counties" (ibid. pg. 55).

Ewing's findings support the earlier work of Frank and colleagues (2004), who identified a correlation between land use and physical fitness. Using data from the 13-county Metro Atlanta region, Frank studied the probability that one would become obese based on density, connectivity, physical activity, and mix of land uses (e.g., residential, commercial, office, and institutional). Across the board for age and gender, there was a decrease in the likelihood of obesity with incremental increases in the mix of land use. The study found that

> Each additional hour spent in a car per day was associated with a 6% increase in the likelihood of obesity. Conversely, each additional kilometer walked per day was associated with a 4.8% reduction in the likelihood of obesity. (*ibid*, pg. 90)

They also suggested,

> there must be more to land use mix affecting obesity than car time or walk distance. If it is not just car time or walk distance, then what else is it about land use mix that is so strongly associated with obesity? (pg. 94)

Research published since then suggests that this is a complex equation, but points toward the idea that multimodal travel can be healthier and more sustainable (McLoughlin, 1991). The physical design of the built environment can promote or restrict multimodality in complex ways. For example, the absence of a tree canopy can reduce walkability (Owen, 2009). Wide roads, as found in many subdivisions in the United States, encourage rapid vehicular travel that diminishes the safety of bikers and walkers. Swift, Painter, and Goldstein (1998) found that vehicle speed increases as road width increases. Conversely, a decrease in road width constriction speeds increases the safety of neighborhoods for walk and play. And yet, standard road design for two-lane streets is 38 feet across, a distance that encourages speed and increases risk. This is likely due to the fact that engineers think of the *capacity* of the street before its utility. They imagine the street as a conduit for moving *through* places, not getting people *to* places (Schwartz, 2015). Even with buffers between the street and sidewalk, the wide-lane platform takes longer to traverse and is less safe for pedestrians.

Race plays another critical role at the intersection of public health, transportation, and the built environment, particularly concerning the ongoing situations facing many African Americans, Hispanics, and Native Americans. Barriers to physical activity include infrastructural divestment in terms of environmental justice, poor air quality, urban heat islands, and access to amenities such as parks (Bedimo-Rung, Mowen, & Cohen, 2005; Day, 2006; Harlan & Ruddell, 2011; Ruddell, Harlan, Grossman-Clarke, & Buyantuyev, 2009; Wen, Zhang, Harris, Holt, & Croft, 2013; Zwickl, Ash, & Boyce, 2014).

While there is a strong relationship between the built environment and health, there are many connections between the environment and health independent of physical activity. The stress of the daily commute, the social isolation, the limited access to healthy food by transit-using residents, noise, pollution, and other issues affect both physical and mental health outcomes (Evans, 2003; Frumkin, 2002; Newman & Kenworthy, 2006). Hypertension, for example, is associated with sprawl (Ewing et al., 2003). Still, the mental health aspects of sprawl, as well as any benefit of walkable or compact environments, are unclear (Sallis et al., 2009).

Comprehensive indices of sprawl and compactness produce useful insights, but since they aggregate various physical properties and measures of behavior, they are less actionable than single measures. One exception is Cervero and Murakami (2010), who, after studying 370 urban areas, found that per capita *vehicle miles traveled* (VMT) is strongly predicted by the percentage of commuters using a single-occupant vehicle (SOV). Using a structural equation model, they found that the percentage of commuters using a SOV is the single most important factor in the total coefficient for VMT, surpassing income, road density, job access, and several other variables. Indeed, they found the importance of the proportion of automobile commuters for predicting VMT to be almost precisely the opposite of population density.

Although less than 20 percent of all travel in cities is related to work, the commute is a good representation of urban travel patterns, capacities, and habits (Cervero & Murakami, 2010; Santos, McGuckin, Nakamoto, Gray, & Liss, 2011). If one can use alternatives to the car for the work commute, then it is possible that non-automobile travel is available for other purposes as well (Frank & Pivo, 1994). As it turns out, many people visit places where other people work. Thus, the work commute is a useful proxy for the built environment's influence on physical activity.

Hypotheses

If automobile dependent environments harm public health, then multimodal built environments should be associated with better health outcomes. However, cities suffer from a host of issues that negatively influence health, making density a poor proxy for mobility. What are the public health differences between automobile dependent and multimodal cities? We hypothesize that multimodal cities, defined as cities supporting higher commute mode diversity and therefore a more varied travel behavior choice set, should have measurably better health outcomes after controlling for density and other built environment and public health factors.

METHODOLOGY

Our framework is an adaptation of Harvey Molotch and Richard Appelbaum's methodology used to study the impact of population growth on quality of life in the 1970s. We are unaware of other studies that have used a similar methodology to study the impact of transportation on health. We take a unique approach not used in previous studies:

1. Instead of analyzing an arbitrary sample of non-representative cities, we identify a large population of mid-size cities;
2. Instead of using cases that contain multiple core cities, we use counties with isolated urban cores having fewer policy spillovers and tighter labor and housing markets;
3. We do not look at a few related variables measured at a single geographic scale, but at 15 broad and distinct measures of health and activity at both the county and metropolitan scales using bivariate and multivariate analyses;
4. Instead of using a characteristic of the built environment or an output variable as the test variable, we use travel behavior to study multimodality's association with health outcomes.

Level and Unit of Analysis

To more accurately analyze the relationships under study, we adapted the well-reviewed decision rules which have been used in the academic works of Appelbaum (1976), Appelbaum, Bigelow, Kramer, Molotch, and Relis (1976), Appelbaum and Follett (1978), and Gilderbloom and Appelbaum (1987). This recognizes the variety of typologies in the United States and the numerous methodological and interpretive challenges this creates, particularly related to scale and boundaries (Appelbaum, 1976; Gilderbloom, 2009). While the methodologies of the aforementioned studies constituted a major advance in urban scholarship, this research recognizes and improves upon some of their shortcomings, namely: earlier work used an arbitrary population cap of 400,000, which raised statistical and methodological concerns; their bivariate analyses compared subjective classes such as the "top 25" vs. the "bottom 25"; finally, they did not use multiple regression analysis.

We evaluate all "places" defined by the U.S. Census that had a population of 50,000 or more (792 cities in 2013), were at least 20 miles from the nearest city of 50,000 or more, and had municipal government (classifying it as a "city," not just a "Census-designated place"). These restrictions reduce the number of cities to 148. This unit—the mid-sized, semi-isolated city—has unique merits, including more unitary markets and fewer "spillover effects" from large, influential neighboring jurisdictions. These cities should be considered as "not simply a sample; they constitute all such places in the United States" (Appelbaum, 1976, pg. 13). In short, the unit of analysis is the county housing a comparatively isolated U.S. city with a population of over 50,000.[1]

Two research assistants working independently of one another, with funding provided by the University of Louisville's Center for Sustainable Urban Neighborhoods, verified the dataset. The assistants measured distances in Google Maps to check the inclusion of cities fitting the decision rules, and they verified the data using geoID codes, ensuring the valid transfer from the various data sources. Figure 5.1 shows these cities as neither evenly

ISOLATED MIDSIZE U.S. CITIES

• Isolated Cities

1 inch = 508 km

0 125 250   500 Miles

Figure 5.1 Cities of the Dataset. *Source*: Center for Sustainable Urban Neighborhoods, formerly University of Louisville, transferred to Neighborhood Associates, Washington D.C.

distributed nor clustered, but randomly distributed around the United States (Moran's I = .048, z = 1.62, $p \leq .011$), reducing the impact of regional biases.

Data

From a data standpoint, we use 21 variables that measure travel health, leisure, and quality of life alongside our key test variable of *Commute Mode Diversity* (CMD). CMD is the "percent of people commuting by means other than a single-occupant vehicle." Our goal is to use a variety of sources, measurements, and constructs related to regional health. Table 5.1 reports on four basic characteristics: the years represented, the unit of measurement, the data source, and the geographic scale of the measurement. Summary statistics are found in table 5.2. Data is explained as follows.

US Census Data and Tiger Line Files—*Commute Mode Diversity* (CMD). Our CMD variable stems from U.S. Census 2013, three-year estimates and

Table 5.1 Variable Descriptions

Dependent Variables	Years	Values	Source	Scale
Overall	2014	Rank	Healthways	MSA
Emotional health	2014	Rank	Healthways	MSA
Physical health	2014	Rank	Healthways	MSA
Healthy behaviors	2014	Rank	Healthways	MSA
Overall	2007	Score	Sperling's	MSA
Leisure	2007	Score	Sperling's	MSA
Quality of life	2007	Score	Sperling's	MSA
Percent below average health	2006–12	Percent	BRFSS	County
Poor physical health days	2006–12	Count	BRFSS	County
Poor mental health days	2006–12	Count	BRFSS	County
Access to exercise space	2010–13	Percent	ESRI	County
Percent physically inactive	2011	Percent	CDC	County
Percent obese	2011	Percent	CDC	County
Years potential life lost	2010–12	Count	NCHS	County
Percent low birth weight	2006–12	Percent	NCHS	County
Independent variables				
Latitude	2010	Degrees	U.S. Census	Both
Density	2013	Count	U.S. Census	Both
Population	2013	Count	U.S. Census	Both
Median household income	2013	Count	U.S. Census	Both
Percent population Black	2013	Percent	U.S. Census	Both
Percent population college educated	2013	Percent	U.S. Census	Both
Commute mode diversity	2013	Percent	U.S. Census	Both

Sources: 1. Data sources in table; 2. Kentucky Institute for the Environment and Sustainable Development; 3. Center for Sustainable Urban Neighborhoods, formerly University of Louisville, transferred to Neighborhood Associates, Washington D.C.

Table 5.2 Descriptive Statistics of Metro Areas for Mid-size U.S. Cities

MSA Dependent Variables	N	Minimum	Maximum	Mean	Std. Dev.
Gallup/healthways overall	68	4	188	104	56
Emotional health	68	1	188	94	59
Physical health	68	5	188	107	60
Healthy behaviors	68	8	313	122	67
Sperling's overall	143	11	100	59	18
Leisure	143	0	95	39	24
Quality of life	143	0	98	40	30
MSA independent variables					
Latitude	148	21.33	61.18	38.16	5.43
Density	148	7	2,967	221	303
Population	148	78,671	19,839,700	528,315	1,692,956
Median HH income	148	35,149	89,605	48,171	7,587
Percent population Black	148	0.3	53.0	11.2	11.9
Percent college educated	148	11.4	51.4	26.2	7.4
Commute mode diversity	148	12.0	49.5	20.0	5.1
County dependent variables					
Below average health	147	5.8	26	15.8	4.24
Poor physical health days	148	2.2	5.3	3.6	0.64
Poor mental health days	147	1.9	5.2	3.5	0.66
Access to exercise space	148	52.4	97.2	81.2	9.94
Percent physically inactive	148	11.9	33.6	23.7	4.55
Percent obese	148	14.2	38.6	28.9	4.15
Years potential life lost	148	3,945	10,897	7,232	1,487
Percent low birth weight	148	4.9	15.8	8.3	1.81
County independent variables					
Latitude	148	21.33	61.18	38.16	5.43
Density	148	7.22	3,414.65	422.57	522.81
Population	148	50,759	1,229,582	261,187	212,233
Median HH income	148	31,024	82,061	47,186	7,713
Percent population Black	148	0.3	70.1	12.8	14.0
Percent college educated	148	11.4	51.5	27.6	8.0
Commute mode diversity	148	11.1	35.6	20.2	4.76
*- Gallup values are rankings, from best (1) to worst (400)					

Sources: 1. Data sources in table 5.1; 2. Kentucky Institute for the Environment and Sustainable Development; 3. Center for Sustainable Urban Neighborhoods, formerly University of Louisville, transferred to Neighborhood Associates, Washington D.C.

Tiger line files. Three-year data is a compromise between the timeliness of the 1-year estimates and the accuracy of the 5-year estimates. It is constructed based on theory from Cervero and Murakami (2010), who found the proportion of commuters using an single-occupant vehicle (SOV) highly predictive of VMT and reflects the percentage of workers who use some mode of travel other than the SOV commute. This includes modes of transportation such as car-pooling, bicycling, walking, cabs, and mass transit. We further frame our *Commute Mode Diversity* variable in theories of networked travel, complexity, and mode substitution, and theories of planned behavior which recognize the increasing multimodal nature of travel decisions (Arentze, Borgers, Timmermans, & DelMistro, 2003; Batty, 2007; Dugundji & Walker, 2005; Godin & Kok, 1996; Lin & Wang, 2014; Piatkowski, Krizek, & Handy, 2015; Riggs, 2016, 2017). We argue that focusing on multiple diverse modes in aggregate is an important development and best represents the complex choice sets that individuals face on a daily basis in making travel decisions. While we recognize that some commuters have complex commutes that involve multiple modes, our data is limited in our ability to assess these linked trips. That said, the Census travel data is appropriate for our analysis, since our main purpose in using *Commute Mode Diversity* is to further uncover the relationship between multimodal built environments and public health outcomes. The CMD for cities ranges from a low of 11 percent in Houston County, Alabama to a high of 36 percent in Honolulu County, Hawaii.

Gallup/Healthways Well-Being Index. Three MSA-level indicators come from the Gallup/Healthways Well-Being Index (WBI). The WBI has been used in urban sustainability research (Cloutier, Jambeck, & Scott, 2014; Cloutier, Larson, & Jambeck, 2014) as well as mental health research (Kahneman & Deaton, 2010). Gallup/Healthways ranks MSAs from the best (1) to the worst of over 300 MSAs. While the sample size is low (61), this is sufficient for a complementary analysis. These are rankings: lower scores indicate better performance.

Sperling's Cities Ranked and Rated. Sperling's Best Places collects data on diverse aspects of urban life in over 400 metropolitan areas. The *Overall* score includes measures of crime, employment, healthcare costs and access, leisure, quality of life, and several other subscores. In contrast to the Gallup/Healthways variables, these are scores: higher numbers indicate better performance. Of the 148 cases, 143 MSAs provide data for these measures.

Behavioral Risk Factor Surveillance System. Data on *Below Average Health*, *Poor Physical* and *Mental Health Days* are for the years 2006–2012. The survey question determining the percent of residents with *Below Average Health* asks if "your health is excellent, very good, good, fair, or poor?"

Center for Disease Control's Diabetes Interactive Atlas data for percent *Physically Inactive* and *Obesity* come from the year 2011 and reflect the percentage of adults aged 20 and over who report no leisure-time physical activity and adults who report a BMI of 30 or more, respectively.

National Center for Health Statistics. *Years of Potential Life Lost per 100,000* (YPLL) population is a measure of premature death for the years 2010–2012. The statistic is the total difference in the ages of residents at the time of death from the reference age of 75 years. Thus, someone who died at age 25 would add 50 years to the YPLL; someone who died at 78 would add zero to the total.

Methods

Urban researchers often compare the "top 25" and "bottom 25" cities for a given measurement or compare those cities that lie on either side of the median of the chosen test variable. Such groups are intuitive but arbitrary. We instead use *k*-means cluster analysis of the standardized *z*-scores of our cities' commute mode diversity to identify similarity for three groups. This gives us a county grouping of high *Commute Mode Diversity* (n = 25) and low CMD counties (n = 68) that are separated by a large (n = 55) unexamined third group of "medium modality" counties. Similarly, there are 23 MM MSAs and 55 AD MSAs, with 70 MSAs in the medium modality group. We designate cities performing best on the metric of *Commute Mode Diversity* as multi-modal (MM), while the worst performing as automobile dependent (AD).

Following this, we conduct multivariate Ordinary Least Squares (OLS) modeling consistent with Tabachnick, Fidell, & Osterlind (2001). We use six critical controls that capture some fundamental sociological and geographical influences over urban health outcomes: latitude, density, population, median household income, percent African American, and percent population with a bachelor's degree. The basic regression equation is as follows:

$$y\lambda = \beta_0 + \beta_1 * cmd\lambda + \beta_2 * latitude + \beta_3 * density\lambda + \beta_4 * population\lambda + \beta_5 * income\lambda + \beta_6 * percent\ black\lambda + \beta_7 * percent\ college\lambda + \varepsilon,$$

where *y* is one of 15 dependent variables, λ is a power transformation, β_0 is the constant, *mm* is multimodality, β_1 through β_7 are coefficients to be estimated, and ε is the error term.

Box-Cox transformation was used to estimate the best value for λ. Independent variables based on percentages were transformed using the natural *log*, while counts and measurements employed *log*10. Variables enter into the model normally (at once). The small number of outliers was treated by Winsorization: outlier values are given a replacement value of 1 plus the closest non-outlier value (Ghosh & Vogt, 2012). The inspection of residual plots raised no concerns.

Tolerances are all well above the critical threshold of .10; the lowest being .56. Correspondingly, VIFs never exceed 10; in fact, no VIF exceeded a value of 2, indicating a comfortable distance from unacceptable limits. Furthermore, the presence of competing dependencies was negative: in no case did two variables both show values above .40 on the variance proportion table.

FINDINGS

Bivariate Analysis: Differences in Multimodality by Exemplars

We conducted a series of *t*-tests on cities grouped into high and low car-dependent cities to uncover substantive differences between the means of the dependent variables. High *Commute Mode Diversity* cities are referred to as multimodal (MM), while low CMD cities are considered automobile dependent (AD).

Metropolitan Statistical Areas. The bivariate analysis for these variables contains between 37 and 77 cases. As expected, the cluster analysis of the 148 cities identified more cases associated with automobile dependency: the dominant commuter paradigm in the United States is the SOV commute (see table 5.3).

Table 5.3 Difference of Means: Metropolitan Exemplars

Variable	Type	N	Mean	Std. Dev.	Sig.	Mean Diff.
G/HW overall	AD	28	121.5	48.5	.001	67.6
	MM	9	53.9	53.7		
Emotional health	AD	28	107.0	59.6	.102	37.8
	MM	9	69.2	56.1		
Physical health	AD	28	121.7	57.0	.006	64.8
	MM	9	56.9	59.0		
Healthy behaviors	AD	28	148.3	32.8	.001	98.3
	MM	9	50.0	29.6		
Sperling's overall	AD	54	52.2	17.2	.001	−17.0
	MM	21	69.2	22.0		
Leisure	AD	54	30.2	22.3	.001	−25.0
	MM	21	55.2	22.5		
Quality of life	AD	54	27.5	20.9	.001	−40.5
	MM	21	68.0	34.1		
Latitude	AD	54	36.73	4.30	.059	.259
	MM	23	39.32	7.47		
Density	AD	54	225	152	.306	−149
	MM	23	374	673		
Population	AD	54	390,686	321,068	.267	−997,269
	MM	23	1,387,955	4,190,858		
Median HH income	AD	54	45,512	4,626	.002	−8,952
	MM	23	54,464	12,180		
Percent pop black	AD	54	18.1	12.9	.001	11.3
	MM	23	6.8	7.3		
Percent bachelor	AD	54	23.9	4.6	.001	−9.6
	MM	23	33.5	9.6		
CMD	AD	54	15.7	1.3	.001	−13.5
	MM	23	29.2	5.4		

Sources: 1. Data sources in table 5.1; 2. Kentucky Institute for the Environment and Sustainable Development; 3. Center for Sustainable Urban Neighborhoods, formerly University of Louisville, transferred to Neighborhood Associates, Washington D.C.

Gallup/Healthways Well-Being Index. The means for three of the four tests are significantly different: the dissimilarity between AD and MM cities on the *Physical Health* measure is considerable (64.8), with the *Overall* score next at 67.6. *Healthy Behaviors* show the greatest mean difference between groups at 98.3.

For *Emotional Health*, the 28 AD cities (mean = 107) score below the overall mean of 94, while the nine MM cities with a mean score of 69 perform much better. Still, it presents the smallest average difference (37.8) with broad variation among the MM cities. As such, it is not a significant difference.

Sperling's Cities Ranked and Rated. We were able to analyze data on 75 MSAs. All three variables report significantly different means between the 54 AD and 21 MM cities. The *Overall* score—a combination of several measures—shows the lowest mean difference at 17. The MM cities' mean of 62 is closer to the total mean score of 59 than is the AD cities' mean of 52, indicating the difference largely in AD cities' poor performance.

Leisure shows a larger mean difference of 25, while *Quality of Life* shows the largest difference of over 40, approaching half of the total range of the measurement. These results indicate that AD MSAs have slightly fewer leisure opportunities than average, while MM MSAs perform well above average for quality of life.

Counties. The bivariate analysis for these variables contained over 93 cases (see table 5.4). Again, as expected, there are more AD cities than MM cities.

Behavioral Risk Factor Surveillance System: While a higher percentage of residents report having *Below Average Health* in AD counties, there is a nonsignificant mean difference of only 1.63 percent. This is not surprising considering the myriad ways one can be unhealthy that do not depend on physical activity. However, for the number of *Poor Physical Health Days*, data suggests a significant difference of one-third of a day per month, or one fewer day every three months. *Poor Mental Health Days* show no significant difference.

US Census Tigerline Files. *Access to Exercise Space* shows a significant difference favoring MM cities, with AD and MM cities' means nearly equidistant from the overall mean of 148 cities. Over 85 percent of residents in MM cities are deemed to have adequate access to parks and other amenities, compared to only 75 percent in AD cities, and 81 percent overall.

Center for Disease Control's Diabetes Interactive Atlas. The average percentage of physically inactive residents for all 148 cities is 24 percent. Worse are AD cities at 26 percent. Multimodal counties are considerably farther from the mean at only 20 percent, a significant difference. Similarly, the overall percentage obese is 29 percent, with 31 percent obesity in AD counties

Table 5.4 Difference of Means: County Exemplars

Variable	Type	N	Mean	Std. Dev.	Sig.	Mean Diff.
Perc. below aver. health	AD	69	16.28	3.51	.081	1.63
	MM	24	14.65	4.91		
Poor physical health days	AD	69	3.73	0.63	.033	0.33
	MM	25	3.40	0.69		
Poor mental health days	AD	69	3.54	0.66	.538	0.09
	MM	24	3.45	0.61		
Access to exercise space	AD	69	78.36	10.77	.004	−7.24
	MM	25	85.60	9.32		
Percent physically inactive	AD	69	25.91	3.62	.001	6.57
	MM	25	19.35	4.66		
Percent obese	AD	69	30.77	3.21	.001	5.62
	MM	25	25.15	4.58		
Years potential life lost	AD	69	7,736	1,292	.001	1,439
	MM	25	6,297	1,552		
Percent low birth weight	AD	69	8.90	1.83	.001	1.64
	MM	25	7.26	1.58		
Independent variables						
Latitude	AD	69	37.74	4.42	.287	−1.32
	MM	25	39.06	7.21		
Density	AD	69	424	406	.525	−117.65
	MM	25	542	881		
Population	AD	69	253,625	191,692	.397	−43,008
	MM	25	296,633	274,379		
Median household income	AD	69	46,166	5,894	.087	−3,900
	MM	25	50,066	10,444		
Percent african american	AD	69	17	15	.002	0.09
	MM	25	8	1		
Percent college educated	AD	69	26	6	.001	−0.08
	MM	25	34	10		
Commute mode diversity	AD	69	16.24	1.69	.001	−12.12
	MM	25	28.36	2.72		

1 Gallup/Healthways rankings' best score is 1; lower values reflect better outcomes.
AD = automobile dependent cluster; MM = multimodal cluster

Sources: 1. Data sources in table 5.1; 2. Kentucky Institute for the Environment and Sustainable Development; 3. Center for Sustainable Urban Neighborhoods, formerly University of Louisville, transferred to Neighborhood Associates, Washington D.C.

vs. a significantly smaller 25 percent in MM core counties; such a difference corresponds to several thousand people in an average-sized county.

National Center for Health Statistics. For *Years Potential Life Lost*, the difference between the means of the two groups is significant: counties with multimodal commute patterns suffer 1,400 fewer years of life lost per 100,000 people than do AD counties. The percent of children born with low birthweight averages 8 percent for the 148 counties. Those with AD commute patterns are worse than average at 9 percent vs. a significantly lower 7 percent for MM cities.

Table 5.5 OLS Regression; Associations between Commute Mode Diversity and Public Health Outcomes

	Gallup/HW Healthy Beh.		Sperling's Overall		Sperling's Leisure		Sperling's QoL		Perc. Below Aver. Health		Poor Physical Health Days	
	Beta	Sig.	Beta	Sig.	Beta	Sig.	Beta	Sig.	Beta	Sig.	Beta	Sig.
Constant		.987		.075		.071		.592		.112		.007
Latitude	.047	.742	-.040	.625	.242	.001	.012	.863	-.358	.001	-.094	.247
Density	-.198	.260	-.132	.157	.034	.663	-.020	.795	-.091	.258	-.026	.787
Population	-.114	.541	.192	.039	.661	.001	.083	.284	.151	.059	.233	.017
Median HH income	.032	.839	-.097	.286	.017	.822	.088	.249	-.080	.283	-.247	.007
Percent population Black	.291	.065	-.138	.138	-.088	.266	-.192	.015	.082	.281	.014	.882
Percent college graduates	-.035	.813	.614	.001	-.022	.749	.573	.001	-.565	.001	-.477	.001
Commute mode diversity	-.389	.009	.097	.254	.305	.001	.150	.036	.079	.247	.035	.672
F	4.441***		14.786***		27.864***		28.871***		30.905***		14.334***	
R	.584		.659		.769		.774		.780		.646	
Adjusted R^2	.264		.405		.570		.579		.589		.388	
N	68		143		143		143		147		148	

Sources: 1. Data sources in table 5.1.; 2. Kentucky Institute for the Environment and Sustainable Development; 3. Center for Sustainable Urban Neighborhoods, formerly University of Louisville, transferred to Neighborhood Associates, Washington D.C.

Note: nst. = Unstandardized coefficients. Beta = standardized. $p < 0.1+$. $*p < 0.05$. $**p < 0.01$. $***p < 0.001$.

Regression Analyses

The regression analysis makes a powerful case for how transportation choices can impact the health of residents. In table 5.5, four health measures out of six show statistical association with transportation choices. Our major finding is that commute mode affects the years of projected life lost—if your primary mode of travel is single-occupancy automobiles, your life will be shortened because of car accidents or lack of exercise. The beta weight indicates that it is nearly as important as having a college education and more important than income. If your city encourages multimodal transportation, you are less likely to be obese and will exercise more. Moreover, with a more active transportation system, the average weight of newborn babies will be significantly higher. The amount of explained variation in the model is strong as well. The results of 12 OLS regression models are provided on tables 5.5 and 5.6.

We did not provide the regression model for Gallup/Healthways' *Emotional Health* since it was not significant in the bivariate analysis and only income and education contribute to the model (both positive). While significant in the bivariate analysis, Gallup/Healthways' *Overall* and *Physical Health* scores did not show an association with commute mode diversity when introducing six controls: only the percent of the population with a bachelor's degree or higher explained variation in either model.

Gallup/Healthways Well-Being Index. The Gallup/Healthways data suffer somewhat from a lower sample size. However, we note that while the adjusted R-squared for *Emotional Health* (.310) and *Healthy Behaviors* (.264) models are similar, only the Healthy Behaviors model is significant for *Commute Mode Diversity*. These results mirror the bivariate analyses. Thus, we believe the model for Gallup/Healthways data is accurate for the purposes of identifying basic associations.

Lower scores on this variable are desirable; the negative coefficient for multimodality indicates a better healthy behavior ranking. As expected, CMD is a strong explanatory factor in *Healthy Behaviors*, which include walking and biking. In fact, perhaps surprisingly, it is the *only* one: none of the other control variables explain variation in healthy behavior across metro areas.

Sperling's Cities Ranked and Rated. Sperling's *Overall* score is not a significant model. We include the results here first because it was significant in the bivariate analysis and because it provides further insights. First, density neither helps nor harms this score, an important consideration. Second, a higher population indicates a better score: more people mean more variety in nearly every aspect of urban life. Although often taken to be proxies for one another, education contributes to the model while income does not. Education contributes the most to the model with a large standardized beta of .614.

Table 5.6 OLS Regression; Associations between Commute Mode Diversity and Public Health Outcomes

	Poor Mental Health Days		Perc. Access to Exercise		Perc. Phys. Inactive		Percent Obese		Years Pot. Life Lost		Percent Low Birth Weight	
	Beta	Sig.	Beta	Sig.	Beta	Sig.	Beta	Sig.	Beta	Sig.	Beta	Sig.
Constant		.001		.142		.038		.079		.001		.001
Latitude	-.034	.723	-.025	.759	.004	.943	.209	.001	-.136	.034	-.255	.001
Density	.069	.549	.070	.478	-.061	.421	.023	.766	-.416	.001	-.167	.024
Population	.113	.325	.325	.001	.058	.438	-.061	.425	.337	.001	.153	.037
Median HH income	-.307	.005	.181	.050	-.121	.088	-.080	.271	-.199	.006	-.037	.594
Percent population Black	.103	.350	-.273	.004	.443	.001	.488	.000	.354	.001	.511	.001
Percent college graduates	-.177	.067	.300	.001	-.390	.001	-.432	.000	-.277	.001	.066	.279
Commute mode diversity	.087	.375	-.028	.736	-.232	.001	-.241	.000	-.249	.001	-.331	.001
F	4.528***		13.902***		37.181***		33.864***		35.538***		40.418***	
R	.431		.640		.806		.793		.800		.818	
Adjusted R^2	.145		.381		.633		.610		.622		.652	
N	147		148		148		148		148		148	

Sources: 1. Data sources in table 5.1; 2. Kentucky Institute for the Environment and Sustainable Development; 3. Center for Sustainable Urban Neighborhoods, formerly University of Louisville, transferred to Neighborhood Associates, Washington D.C.

Note: nst. = Unstandardized coefficients. Beta = standardized. $p < 0.1+$. *$p < 0.05$. **$p < 0.01$. ***$p < 0.001$.

For *Leisure, Commute Mode Diversity* is a powerful contributor to the model (beta = .305). Population is the most remarkable with a standardized beta of .661. Again, larger populations imply greater diversity in interests and more variety in leisure-related outlets and amenities. Density again neither harms nor helps the score. Higher latitudes indicate better scores; this may reflect more recreational opportunities associated with seasonal changes, such as skiing and ice skating.

Commute mode diversity makes a modest contribution to the *Quality of Life* model (beta = .150); only the percent African American (−.192) and educational attainment (.573) contribute.

Behavioral Risk Factor Surveillance System. While the percent *Below Average Health* variable was significantly different between AD and MM cities in the bivariate analysis, only latitude and education make significant contributions to the model.

Similarly, the control variables clarified the relationship between *Poor Physical Health Days* and *Commute Mode Diversity*. Higher populations indicate more poor health days; more variety in lifestyles implies more variety in health risks. Curiously, higher density does not. One possibility is that the dataset does not include cities with the higher densities that may be necessary to facilitate aggressive disease vectors. As expected, higher median incomes and education militate against poor health.

Like Gallup/Healthways' *Emotional Health* score, *Poor Mental Health* is not associated with CMD. We find it important to note, however, that neither population size nor density is associated with variation in mental health. Again, this may be a function of the smaller, less dense cities that can be found in the United States as a whole. Similar to Gallup/Healthways' *Emotional Health* score, income makes a significant contribution, and education comes close to significance.

US Census Tigerline Files. Despite the bivariate findings, *Access to Exercise* is not significantly associated with CMD after controlling for other factors. Higher populations, incomes, and education levels are associated with more access, possibly reflecting a combination of diverse interests, market provision, and demand for amenities (or selection into counties with higher amenities).

The percentage of African Americans also contributes to the model (−.273) similarly in strength to population (.325) and education (.300), albeit in the opposite direction. We see this as a strong indication of the ongoing situations facing many African Americans, including infrastructural divestment in terms of environmental justice and access to amenities such as parks.

Center for Disease Control's Diabetes Interactive Atlas. The model for percent *Physically Inactive* has the second highest adjusted R-squared of the

12 at .633. The largest contributor to the model is the percent African Americans, followed by education and *Commute Mode Diversity*. Unexpectedly, density is not significant. The significance of African Americans echoes the previous model: Blacks and other minorities in the United States lack access to recreational amenities.

The *Obesity* variable also responds well to the model, explaining well over half of the variation in obesity among the 148 counties (R-squared of .610). Similar to the previous model for physical inactivity, *Commute Mode Diversity*'s contribution is third behind the percent of population that is African American and college education. Again, surprisingly, density has no measurable impact on obesity when controlling for other factors.

National Center for Health Statistics. When a county has higher *Years of Potential Life Lost*, the residents of the county are less likely to achieve the national average life expectancy. Therefore, a low YPLL is desirable. This model is robust, explaining close to half of the variation for this variable; the third highest adjusted R-squared of the 12 at 0.622.

The model shows that an increase in multimodality is related to less premature death. *Commute Mode Diversity*'s strength is on par with other significant factors and is more impactful than median household income. Every factor contributes to YPLL. Strongest of all is density, which reduces premature death; this may reflect more access to better healthcare and emergency services.

The *Low Birthweight* model is the strongest of the 12 with an adjusted R-squared of .652. However, the percent population that is Black carries the bulk of its explanatory value with a standardized beta of .511. CMD takes second place with a positive contribution (−.331), followed by latitude (−.255). Density contributes to healthier infants, possibly related to accessibility of more and better healthcare, as well as support for active lifestyles. Perhaps most surprisingly of all, neither education nor income is significant.

DISCUSSION

We take these findings to mean that having a high *Commute Mode Diversity*, and therefore having a more multimodal opportunity for healthy travel, is of benefit to public health. It is likely that there are more healthy travel options for day-to-day living. They also strongly suggest that the benefit of density has been overstated. For density to make a positive impact, it may need a diverse travel environment and multimodal context. Density might be *necessary* for a compact multimodal environment but is not *sufficient*. Future research efforts will focus on interactions between density, multimodality, and other aspects of the built and social environments.

We found strong associations with *Commute Mode Diversity* and benefi-cial physical behavior; focusing on walking and bicycling would likely make the differences between walkable cities and automobile dependent cities more dramatic. CMD is thus a conservative variable. However, the focus of *Commute Mode Diversity* is not merely increased exercise *per se*, but rather a more active lifestyle in general, including the sociability that results from increased face-to-face interactions. Still, future research should continue to examine the impact of individual commute modalities.

Finally, there are more travel choices in high CMD geographies: more ways to travel than in geographies where transportation is dominated by SOV. This is an important finding as the key policy suggestion of urban sprawl researchers in the past has been for cities to pursue infill develop-ments that will enable people to become more active. However, there was no statistical difference between exemplar MSAs or counties regarding density. This research provides evidence that while density is not a sufficient condi-tion for multimodality—and perhaps only marginally necessary—the ability to traverse environments using diverse modes is critical, and a focus on travel behavior policy may produce more effective results when paired with efforts to improve density. Additional quantitative and qualitative analyses, includ-ing more work on mode substitution behavior paired with land use policy, may inform this issue more than the data presented here.

CONCLUSION

Using both bivariate and multivariate analysis, we find powerful statistical evi-dence that high modal diversity and providing a multimodal travel environment are just as important a concern for public health as many other sociological, geographical, and economic concerns, including race, density, latitude, educa-tion, and income. Auto-dependency in American cities can have harmful health effects compared to cities with greater transportation choices. People living in multimodal, green cities are more likely to live longer and better; their quality of life is significantly higher, and they live life in better physical condition.

Commute Mode Diversity's impact can hardly be overstated. The power of CMD as a component of the built environment is impressive, especially when you consider its explanatory value compared to race, income, and edu-cation. While much of the benefit of multimodality is a function of exercise, an additional component is the kind of social environment it supports. This is why we believe multimodality is not merely a commute choice; it is type of living arrangement.

Still, multimodality, density, and other facets of the built environment are not cure-alls. Urban policymakers would do well to stop looking for the built environment to solve the problems of social inequality in general, and instead address where the built environment itself produces inequality. Increased walking and bicycling will not compensate for social health inequalities that result from poverty, lack of education, environmental racism, and poor social policy.

On the other hand, we should not underestimate the importance of increased multimodality in our cities and the decreased use of the single-occupant vehicle as a means for traversing urban environments. Since the automobile is an icon of the middle class, developing countries are looking to the United States for inspiration: Chinese and African cities are planning for increased automobile use. A major shift in U.S. urban transportation policy would have vast benefits far beyond our borders. In an era of climate change, a reduction in greenhouse gas means a healthier planet; a healthier planet means healthier people.

NOTE

1. This approach has the additional benefit of reducing the contribution from multiple counties. Of the 148 cities, all but 18 are completely enveloped in a single county. These exceptions contain only minor proportions of the cities' population: close inspection of the boundaries reveals that these additional areas result from annexations of infrastructure assets, such as an airport or a reservoir.

REFERENCES

Appelbaum, R. P. (1976). *The Effects of Urban Growth: A Population Impact Analysis*. New York: Praeger Publishers.

Appelbaum, R. P., Bigelow, J., Kramer, H. P., Molotch, H. L., & Relis, P. M. (1976). *The Impacts of Growth: A Population Impact Analysis*. New York: Praeger Publishers.

Appelbaum, R. P., & Follett, R. (1978). Size, growth, and urban life a study of medium-sized American cities. *Urban Affairs Review*, 14(2), 139–168.

Arentze, T., Borgers, A., Timmermans, H., & DelMistro, R. (2003). Transport stated choice responses: Effects of task complexity, presentation format and literacy. *Transportation Research Part E: Logistics and Transportation Review*, 39(3), 229–244.

Batty, M. (2007). *Cities and Complexity: Understanding Cities with Cellular Automata, Agent-Based Models, and Fractals*. The MIT Press. Retrieved from http://dl.acm.org/citation.cfm?id=1543541.

Bedimo-Rung, A. L., Mowen, A. J., & Cohen, D. A. (2005). The significance of parks to physical activity and public health: A conceptual model. *American Journal of Preventive Medicine*, 28(2), 159–168.

Blokland, T., & Rae, D. (2016). The end to urbanism: How the changing spatial structure of cities affected its social capital potentials. In T. Blokland & M. Savage (Eds.), *Networked Urbanism: Social Capital in the City*, 23–40. Burlington, Vermont: Ashgate Publishing.

Cervero, R., & Murakami, J. (2010). Effects of built environments on vehicle miles traveled: Evidence from 370 US urbanized areas. *Environment and Planning A*, 42(2), 400–418.

Cloutier, S., Jambeck, J., & Scott, N. (2014). The Sustainable Neighborhoods for Happiness Index (SNHI): A metric for assessing a community's sustainability and potential influence on happiness. *Ecological Indicators*, 40, 147–152.

Cloutier, S., Larson, L., & Jambeck, J. (2014). Are sustainable cities "happy" cities? Associations between sustainable development and human well-being in urban areas of the United States. *Environment, Development and Sustainability*, 16(3), 633–647.

Corburn, J. (2007). Reconnecting with our roots American urban planning and public health in the twenty-first century. *Urban Affairs Review*, 42(5), 688–713.

Dannenberg, A., Frumkin, H., & Jackson, R. (2011). *Making Healthy Places: Designing and Building for Health, Well-Being, and Sustainability*. Washington, DC: Island Press.

Day, K. (2006). Active living and social justice: Planning for physical activity in low-income, Black, and Latino communities. *Journal of the American Planning Association*, 72(1), 88–99.

Doyle, S., Kelly-Schwartz, A., Schlossberg, M., & Stockard, J. (2006). Active community environments and health: The relationship of walkable and safe communities to individual health. *Journal of the American Planning Association*, 72(1), 19–31.

Dugundji, E., & Walker, J. (2005). Discrete choice with social and spatial network interdependencies: An empirical example using mixed generalized extreme value models with field and panel effects. *Transportation Research Record: Journal of the Transportation Research Board*, 1921(1), 70–78.

Eidlin, E. (2005). The worst of all world: Los Angeles, California, and the emerging eeality of dense sprawl. *Transportation Research Record: Journal of the Transportation Research Board*, 1902(1), 1–9.

Evans, G. W. (2003). The built environment and mental health. *Journal of Urban Health*, 80(4), 536–555.

Ewing, R., Schmid, T., Killingsworth, R., Zlot, A., & Raudenbush, S. (2003). Relationship between urban sprawl and physical activity, obesity, and morbidity. *American Journal of Health Promotion*, 18(1), 47–57.

Foster, M. (1979). City planners and urban transportation: "The American response, 1900–1940". *Journal of Urban History*, 5(3), 365.

Frank, L. D., Andresen, M. A., & Schmid, T. L. (2004). Obesity relationships with community design, physical activity, and time spent in cars. *American Journal of Preventive Medicine*, 27(2), 87–96.

Frank, L. D., & Pivo, G. (1994). Impacts of mixed use and density on utilization of three modes of travel: Single-occupant vehicle, transit, and walking. *Transportation Research Record*, 1466, 44–52.

Frank, L. D., Sallis, J. F., Conway, T. L., Chapman, J. E., Saelens, B. E., & Bachman, W. (2006). Many pathways from land use to health: Associations between neighborhood walkability and active transportation, body mass index, and air quality. *Journal of the American Planning Association*, 72(1), 75–87.

Frumkin, H. (2002). Urban sprawl and public health. *Public Health Reports*, 117(3), 201.

Galster, G., Hanson, R., Ratcliffe, M. R., Wolman, H., Coleman, S., & Freihage, J. (2001). Wrestling sprawl to the ground: defining and measuring an elusive concept. *Housing Policy Debate*, 12(4), 681–717.

Ghosh, D., & Vogt, A. (2012). *Outliers: An Evaluation of Methodologies*. Paper presented at the Joint Statistical Meetings.

Gilderbloom, J., & Appelbaum, R. (1987). *Rethinking Rental Housing*. Philadelphia: Temple University Press.

Gilderbloom, J. I., Hanka, M. J., & Lasley, C. B. (2009). Amsterdam: Planning and policy for the ideal city? *Local Environment*, 14(6), 473–493.

Godin, G., & Kok, G. (1996). The theory of planned behavior: A review of its applications to health-related behaviors. *American Journal of Health Promotion*, 11(2), 87–98. https://doi.org/10.4278/0890-1171-11.2.87.

Harlan, S. L., & Ruddell, D. M. (2011). Climate change and health in cities: Impacts of heat and air pollution and potential co-benefits from mitigation and adaptation. *Current Opinion in Environmental Sustainability*, 3(3), 126–134.

Jackson, L. E. (2003). The relationship of urban design to human health and condition. *Landscape and Urban Planning*, 64(4), 191–200.

Johnson, M. P. (2001). Environmental impacts of urban sprawl: A survey of the literature and proposed research agenda. *Environment and Planning A*, 33(4), 717–735.

Kahneman, D., & Deaton, A. (2010). High income improves evaluation of life but not emotional well-being. *Proceedings of the National Academy of Sciences*, 107(38), 16489–16493.

Leiferman, J. A., & Evenson, K. R. (2003). The effect of regular leisure physical activity on birth outcomes. *Maternal and Child Health Journal*, 7(1), 59–64.

Lin, T., & Wang, D. (2014). Social networks and joint/solo activity–travel behavior. *Transportation Research Part A: Policy and Practice*, 68(October), 18–31. https://doi.org/10.1016/j.tra.2014.04.011.

Litman, T. (2012). *If Health Matters: Integrating Public Health Objectives in Transportation Planning*. Retrieved from Victoria, British Columbia.

Litman, T. (2013). Transportation and public health. *Annual Review of Public Health*, 34, 217–233.

McLoughlin, B. (1991). Urban consolidation and urban sprawl: A question of density. *Urban Policy and Research*, 9(3), 148–156.

Newman, P., & Kenworthy, J. (2006). Urban design to reduce automobile dependence. *Opolis*, 2(1), 35–52.

Owen, D. (2009). *Green Metropolis: Why Living Smaller, Living Closer, and Driving Less Are the Keys to Sustainability*. New York: Riverhead Books.

Piatkowski, D. P., Krizek, K. J., & Handy, S. L. (2015). Accounting for the short term substitution effects of walking and cycling in sustainable transportation. *Travel Behaviour and Society*, 2(1), 32–41. https://doi.org/10.1016/j.tbs.2014.07.004.

Putnam, R. D. (2001). *Bowling Alone: The Collapse and Revival of American Community*. New York: Simon and Schuster.

Riggs, W. (2016). Cargo bikes as a growth area for bicycle vs. auto trips: Exploring the potential for mode substitution behavior. *Transportation Research Part F: Traffic Psychology and Behaviour*, 43, 48–55. https://doi.org/10.1016/j.trf.2016.09.017.

Riggs, W. (2017). Painting the fence: Social norms as economic incentives to non-automotive travel behavior. *Travel Behaviour and Society*, 7, 26–33. https://doi.org/10.1016/j.tbs.2016.11.004.

Ruddell, D. M., Harlan, S. L., Grossman-Clarke, S., & Buyantuyev, A. (2009). Risk and exposure to extreme heat in microclimates of Phoenix, AZ. In P. S. Showalter & Y. Lu (Eds.), *Geospatial Techniques in Urban Hazard and Disaster Analysis*, 179–202. Dordrecht: Springer.

Sallis, J. F., Frank, L. D., Saelens, B. E., & Kraft, M. K. (2004). Active transportation and physical activity: Opportunities for collaboration on transportation and public health research. *Transportation Research Part A: Policy and Practice*, 38(4), 249–268.

Sallis, J. F., Saelens, B. E., Frank, L. D., Conway, T. L., Slymen, D. J., Cain, K. L., . . . Kerr, J. (2009). Neighborhood built environment and income: Examining multiple health outcomes. *Social Science & Medicine*, 68(7), 1285–1293.

Santos, A., McGuckin, N., Nakamoto, H. Y., Gray, D., & Liss, S. (2011). *Summary of Travel Trends: 2009 National Household Travel Survey*.

Schwartz, S. (2015). *Street Smart: The Rise of Cities and the Fall of Cars*. New York: PublicAffairs.

Sturm, R., & Cohen, D. A. (2004). Suburban sprawl and physical and mental health. *Public Health*, 118(7), 488–496.

Swift, P., Painter, D., & Goldstein, M. (1998). *Residential Street Typology and Injury Accident Frequency*. Longmont: Swift and Associates.

Tabachnick, B. G., Fidell, L. S., & Osterlind, S. J. (2001). *Using Multivariate Statistics*. 4th ed. Boston: Allyn and Bacon.

Torrens, P. M., & Alberti, M. (2000). Measuring sprawl. *Centre for Advanced Spatial Analysis*, working paper no. 27. University College London.

Wachs, M. (1984). Autos, transit, and the sprawl of Los Angeles: The 1920s. *Journal of the American Planning Association*, 50(3), 297–310.

Wen, M., Zhang, X., Harris, C. D., Holt, J. B., & Croft, J. B. (2013). Spatial disparities in the distribution of parks and green spaces in the USA. *Annals of Behavioral Medicine*, 45(1), 18–27.

Zhang, X., Holt, J. B., Lu, H., Onufrak, S., Yang, J., French, S. P., & Sui, D. Z. (2014). Neighborhood commuting environment and obesity in the United States: An urban–rural stratified multilevel analysis. *Preventive Medicine*, 59, 31–36.

Zwickl, K., Ash, M., & Boyce, J. K. (2014). Regional variation in environmental inequality: Industrial air toxics exposure in US cities. *Ecological Economics*, 107, 494–509.

Chapter 6

Reducing Greenhouse Gas Emissions and the Need for Renewable Energies

Stephen A. Roosa

Environmental problems have emphasized the ultimate fragility of the Earth's ecosystems. Due to our unresponsive choices, climate changes are occurring at an accelerated pace. Our common future is becoming grim. No matter what actions we take today, the consequences of increasing carbon levels will continue to plague the Earth for generations to come. Never before have human-induced environmental changes had such a global impact. Environmental changes caused by climate change can be locally devastating. At this juncture, nothing less than the sustainability of life on our planet is at stake. This chapter outlines the sustainability issues regarding greenhouse gases (GHGs) and explores approaches to decarbonization. There are tough decisions ahead on the path to global decarbonization.

The environmental changes associated with greenhouse gases have been festering for a long time. Our patterns of energy use since the early 1900s, which have focused on the use of fossil fuels, have contributed to our increasingly dire situation. The results of our past insensitivity to the environment are becoming more noticeable, the consequences less dismissible. With the acceptance that an over-abundance of atmospheric carbon dioxide (CO_2) is impacting Earth's ecosystems in detrimental ways, a new age is dawning. The economies of the Hydrocarbon Age are at an impasse. Indeed, the availability of certain fossil fuels is reaching a tipping point in some regions of the world. We are changing our landscapes because of economic development and technological advancements—terra-forming them to meet the needs of agriculture, industry, urban expansion, and regional transportation networks. Many of our cities have evolved to become "post-industrial" and are unlike anything that our forebearers might have imagined. Cities in Brazil, Mexico, China, and India are reeling from population growth, poor development choices, and environmental damage. Access to energy is a key to sustainability. Yet over a billion people lack electricity, a prime mover of the information age. How can people learn to use computers and access the internet when they lack the electricity needed to light a single light bulb?

Countries seeking economic and political control over carbon-based energy resources are at times exerting their military muscle, threatening, and waging regional conflicts. Examples include the rise of OPEC to control oil supplies in the late 1970s and the Russian Federation's cut-off of natural gas supplies to the European Union in 2022. The impacts of each country's policies and economic systems are intensifying the problems associated with carbon emissions rather than resolving them. However, the political dialogue is shifting from outright rejection of the existence of carbon-induced climate problems to the development of strategies and legislation to mitigate them.

Sustainability requires foresight. At this point in history—when creative solutions are needed—the backdrop of conflicting influences, political stalemates, and economic turmoil makes long-term solutions difficult to implement. Energy resources, environments, and economic systems are intimately linked. For example, energy-consuming systems account for 95 percent of man-made CO_2 emissions. How can we cost-effectively reduce energy-related emissions of GHGs to stabilize the climate while meeting the endless demand for energy? The seminal inconvenient truth is that a consensus regarding how to proceed remains elusive. The stakes are evident. All that is required is an increase in atmospheric temperatures of 2–3°C and the climate will evolve to being like it was in the Pliocene period, about 3 million years ago, when sea levels were estimated to be about 25 meters (80 feet) higher than today. Already we are seeing ice shelves melt away, warmer oceans, increased coastal flooding, and more intense storms.

Mankind's continued misuse of carbon-based fuels has created uncertainty and increased risks. While there are both natural and man-made sources of atmospheric carbon, natural endowments that reduce atmospheric carbon levels are under attack by our activities. Formed over hundreds of millions of years, carbon compounds were safely stored by natural processes that worked very slowly. These processes gradually reduced atmospheric concentrations by storing carbon in natural sinks, such as soils, vegetation, underground, and in the oceans. Over time, they transformed the Earth and created conditions to enable human evolution. There are currently no viable substitutes for these natural processes that have effectively sequestered massive quantities of carbon for eons.

Hope for solutions can be found in policies, programs, and technologies. Governments are creating policies, corporations are reconstructing strategic plans, and institutions are redefining their missions. Agendas are in flux, and new programs are being launched and implemented. There is an evolving consensus on the horizon, one that will change how we prioritize our efforts to become more sustainable. Innovative technologies, such as renewable energy systems, are proving to have the potential to transform our energy supply systems into ones that are carbon neutral. These technologies focus on the use of solar power, wind energy, geothermal energy, tidal energy, biofuels, and creative energy storage resources. The time has come to refocus our resources toward finding solutions that reduce or eliminate carbon emissions.

INCREASED USE OF CARBON-BASED FUELS

Today, the carbon compounds that are being released into the atmosphere come from many sources. Some occur naturally. Others are sourced from the primary combustion of fossil fuels such as coal, natural gas, and oil. The use of coal represents a large portion of the CO_2 being released into the atmosphere. While coal use is decreasing in the United States, its use continues unabated in many developing countries. Despite its high levels of CO_2 emissions and its low thermal energy conversion efficiency (approximately 37 percent), coal usage plays an important role in the global energy supply. Regardless, ways must be found to reduce the carbon emissions generated from combustion processes.

What is happening is that we are increasing our use of carbon-based fuels—and the atmospheric concentrations of CO_2, methane (CH_4) and other greenhouse gases are increasing. A consensus has evolved that global climate change is a result. The debate is over. We know that carbon emissions are directly linked to the use of fossil fuels, and they are responsible for global climate change. Mankind's activities are exacerbating this problem and creating climatic disruption, loss of biodiversity, and economic uncertainty. Despite increased awareness, efforts to address the impacts of climate change are in their infancy. They have yielded little impact on global emissions. In fact, carbon emissions worldwide continue to increase. This is due to the growing quantities of fossil fuels being extracted and how they are used in combustion processes.

In the face of mounting evidence to the contrary, some believe that the management of carbon emissions is unnecessary as the Earth's environment is perpetually self-correcting. While this is arguable for the long term (one that exceeds man's presence on the planet), the impacts to our environment remain unaddressed and the costs of future mitigation continue to increase. One of the biggest obstacles in managing atmospheric carbon emissions is that many environmental issues are seen as international in scope rather than local; local efforts to reduce emissions are trumped by international policies. Existing institutions and national bureaucratic structures often cannot handle and were not created to deal with such problems effectively. Without the institutional capacity necessary to manage and resolve such problems on an international scale, inertia and inaction often result.

While local and regional responses can have powerful impacts, supranational governmental organizations (e.g., the United Nations and the European Union) have also proved effective. International action to prevent additional damage to the ozone layer required cooperative and concerted efforts and proved to be effective, evolving into the Montreal Protocol. Treaties such as the Kyoto Protocol and the Paris Agreement were designed to reduce greenhouse gases. The Kyoto Protocol is an international treaty which became active in February 2005; its goal was to stabilize atmospheric concentrations of GHG at levels that balance anthropogenic (man-made) interference with the climate system.

The Paris Agreement, signed by representatives of 193 countries in April 2016, was an important step toward GHG policy on an international scale. Laurent Fabius, France's foreign minister, said this agreement was an "ambitious and balanced" approach and a "historic turning point" in our goal of reducing global warming. Perhaps its most important near-term objective is to quickly achieve the global peaking of GHG emissions. Many of the countries participating in this agreement have made substantial gains that illustrate the achievements accompanying their progressive measures. Ultimately, we need to create a pathway for the Earth's climate resilience. After all, it is the only habitable planet we have. For the Paris Agreement to be effective, all countries must cooperatively participate. While the administrative branch of the U.S. federal government abstains, many state and local governments have established GHG reduction programs, upending federal leadership. Despite efforts to reduce GHG emissions in the United States, as of April 2016, the United States has the dubious distinction of being one of the world's largest emitters of greenhouse gases on a per capita basis. It is also responsible for well over a quarter of all GHGs presently found in the atmosphere.

From a technical perspective, there are numerous ways to manage atmospheric levels of carbon. Emissions can be reduced or carbon can be extracted from the atmosphere. Reducing emissions is the least costly and disruptive option. Like other atmospheric pollutants, once emitted into the atmosphere, carbon is diffused and becomes difficult and expensive to extract. The technologies available to remove carbon from the atmosphere are commercially unproven and many are untested. Thus, reducing the quantities of carbon compounds prior to their release into the atmosphere holds far greater promise.

Many of the adverse environmental consequences we face are caused by increased atmospheric carbon concentrations, and their impact has only recently become understood. Why is this? As with serious environmental problems encountered in the past (e.g., natural resource depletion or water pollution), there is a time lag before the environmental impacts of human actions become apparent. Environmental changes must be meticulously observed before the causes and effects are reliably assessed. Theories then evolve as to why the changes are occurring. Potential impacts must be weighed. To do this, the scientific community must make comparative measurements of before and after conditions, link the environmental changes to the causes, and consider an assortment of remedial actions. Prior to suggesting approaches to mitigation, a wide array of intervention methodologies and technologies must be developed, tested, and deployed. After a period of societal denial and disbelief, a political consensus must emerge before public resources can be used. This is time-intensive and laborious. Another alternative is to adapt to the new and developing circumstances. As the climate warms and sea levels rise, we can move inland seeking higher ground. Yet this means the abandonment of many low-lying coastal communities and cities and the loss of tens trillions of dollars of infrastructure. This is beginning to happen today. A village in Alaska was moved inland by the U.S. Army Corps of Engineers at a cost exceeding $250 million, justified by rising seas caused by climate change.

When management and adaptation fail, mitigation often proves costly. A recent example is the ironic request of oil refining industries in southwest Texas. Their representatives have consistently denied the effects of climate change while their products have contributed to atmospheric carbon levels. Yet they have recently requested the use of state and federal taxpayer funds, several hundreds of millions of dollars, to build levees to protect them from increasing sea levels which they say are being caused by rising water levels in the Gulf of Mexico.

To complicate matters, there is no single solution to our problem. The "smoking gun" is often more readily identified than the elusive "silver bullet"—and none may exist. In fact, scientific investigations often conclude that a set of customized mitigation approaches is necessary. Such contextual responses vary in their identification of the source of the problems, local circumstances, economic costs, and the scope of the solution. In some cases, deploying recommended technologies has unintended results that are not necessarily favorable. For example, roughly 40 percent of the corn grown in the United States is harvested to manufacture ethanol, which is used as a renewable biofuel. The use of corn for fuel production is increasing pressure on corn prices and stressing the use of corn as feed for livestock.

A scientific consensus regarding the environmental damage caused by carbon emissions has only recently become available. Increasingly, the deleterious consequences of carbon-induced climate change have become more apparent. The consensus is that carbon emissions are contributing to possibly irreversible changes in the world's climate. The time of inaction has passed, and the time of initiative is upon us; we must act now or our common future is in peril.

Our patterns of energy consumption dictate the success of our mitigation efforts. Sustainable energy sources are an important component of our solutions. Renewable energy sources can be categorized as sustainable and inexhaustible, while most nonrenewable energy sources are potentially unsustainable and likely exhaustible. The keys to reducing carbon emissions include reducing the use of carbon-based fuels by improving efficiency and fuel substitution. The concentrations of carbon in fossil fuels vary, as do the amounts of carbon released into the atmosphere during combustion processes. New energy sources must produce fewer carbon emissions. One way to do this is to use renewable forms of energy. These are forms of energy that are derived from and replaced rapidly by natural processes. They include solar thermal and photovoltaic, wind power, hydropower, tidal power, biomass, and geothermal energy.

Carbon Emissions

Global carbon emissions from man-made sources more than tripled from 1950 to 2000, and they continue to increase. From 1990 to 1999, global emissions increased at a rate of 1.1 percent annually, jumping to 3 percent annually despite the creation of the Kyoto Protocol (Friedman, 2008). Projections indicate that CO_2 emissions will increase to 42.9 billion metric tons by 2030 (Energy Information Administration, 2007). Increases of such magnitude are unprecedented. The potential impacts to the Earth's ecosystems are unknown. It is likely that further damage to the Earth's climate will occur. This damage will have unforeseen consequences and impact our economies, our health, our resources, and our settlements. These changes are likely to occur so rapidly that we may be unable to adjust quickly enough to avoid a global catastrophe.

Atmospheric carbon emissions in the United States in 2016 were estimated to be 5.2 gigatons. In addition, 0.5 gigatons were emitted as methane, a potent GHG that is at least 25 times more damaging in the short term than CO_2. Methane gas is used as a combustion fuel for electrical production and escapes into the atmosphere from fermentation, landfills, coal mining operations, manure, natural gas systems, and other sources. Since methane combustion releases less CO_2 per unit of heat generated than other hydrocarbon fuels, it is seldom the fuel of choice. With a half-life of seven years, atmospheric methane can oxidize, producing CO_2 and water (Wikipedia, 2008). In comparison to CO_2, each methane molecule has a relatively large global warming impact that diminishes in a comparatively short period of time. Methane tends to concentrate in the stratosphere and in tropical regions. For these reasons, efforts to mitigate atmospheric carbon concentrations must include preventing the release of methane gas. The amount of methane present in the atmosphere has increased from 700 parts per billion (ppb) in 1750 to 1,745 ppb in 1998, with more than half due to human activity (Wikipedia, 2008).

There is wide variability in national emissions rates. The carbon emissions of the world's countries are unevenly distributed. The United States emits CO_2 at more than twice the per capita rates of Germany, Japan, or Russia. In more populous, industrializing countries, the rates of increase are becoming unmanageable. In 2004, China and India combined for 22 percent of world emissions, yet their share is anticipated to further increase to 31 percent by 2030 (Energy Information Administration, 2007). China surpassed the United States in 2016 as the world's No. 1 carbon dioxide polluter. The United States ranks among the highest on a per capita basis with an index of roughly 514.4 metric tons emitted per person annually in 2022.

Within countries, GHG emissions vary widely across economic sectors. Greenhouse gases from power generation, industrial processes, and transportation account for over 52 percent of the world's total greenhouse gas emissions. CO_2 emissions tend to be concentrated in urban areas. People in the Netherlands produce roughly 8.1 tons of CO_2 per capita per year, while residents of India produce only 1.4 tons per capita annually.

Strategies to reduce carbon emissions will impact all sectors of the economy, and costs are likely to be unevenly distributed regardless of the strategies employed. Therefore, policies that focus on reducing carbon emissions must be broadly based. A consensus must be reached, and programs must be implemented to achieve the goals established by the policies that are adopted. While policies and programs are replicable, they must be overarching in their principles, consistent in their goals, and locally adaptable. The technologies employed must be appropriate, deployable, and economically viable.

The Costs of Reducing U.S. Carbon Emissions

The cost of reducing U.S. greenhouse gases is difficult to estimate. The projections range from almost nothing to hundreds of billions, if not trillions, of dollars. According to a report entitled *Reducing U.S. Greenhouse Emissions: How Much at What Cost?* by McKinsey and Company, annual greenhouse gas emissions in the United States are projected to increase to 9.7 gigatons— one gigaton is equal to one billion (10^6) tons—in 2030 if no remedial actions are undertaken (McKinsey and Company, 2007). This report identifies the primary causes for the projected growth of U.S. carbon emissions:

- The anticipated long-term expansion of the economy.
- Growth in the use of energy by buildings, appliances, and transportation is due to a projected population growth increase of 70 million.

- The continued reliance on carbon-based electrical power generation from the construction of new coal-fired power plants that lack carbon capture and storage (CCS) technology.
- A gradual decline in the ability of U.S. forests and agricultural lands to absorb carbon is forecasted to decrease from 1.1 gigatons in 2005 to 1.0 gigatons in 2030.

The McKinsey and Company report provides case projections that consider abatement opportunities such as the greater use of coal as an energy source using capture and storage (CCS), expanded use of nuclear power, renewable energy, and biofuels, along with vehicle efficiency improvements and energy efficiency upgrades for buildings. These carbon abatement projections establish a cost of $50 per ton or less. However, motor vehicle efficiency improvements typically require investments greater than $50 per ton.

Potential for Reducing Carbon Emissions

There remain opportunities to reduce carbon emissions. The potential in the United States can be measured by how policies, programs, and technologies are effectively synchronized to implement this goal. There must be a focus on managing broadly-based initiatives that successfully reduce carbon emissions. However, the pressures of population growth and development temper the belief that we can reduce carbon emissions simply by initiating additional energy conservation and efficiency improvements. We find that most technological solutions, such as pre- and post-combustion carbon capture are expensive. Finding terrestrial carbon sinks can be difficult since storage of carbon for long periods of time is not guaranteed.

The 2007 study by McKinsey and Company evaluates five categories of technologies that can be employed to reduce carbon emissions and estimates their potential impact. They are:

- Reducing the carbon intensity from electric power production by using alternative energy and CCS technologies (800–1,570 megatons).
- Improved energy efficiency in buildings and appliances (710–870 megatons).
- Implementing carbon reduction opportunities in the industrial sector (620–770 megatons).
- Expanding natural carbon sinks to capture and store more carbon (440–590 megatons).
- Increasing vehicular efficiency and using less carbon-intensive fuels (340–660 megatons).

These approaches, if considered additive, offer an estimated total reduction in carbon emissions ranging from 2,910 to 4,460 megatons, or a reduction from the base year (2005) emissions ranging from 40 percent to 62 percent. Substantial infrastructure efficiency improvements will be necessary. If implemented, such initiatives would place the United States at a per capita carbon emission level close to that of Germany or Japan.

The U.S. is improving the efficiency of vehicles and new buildings. It is clear that the country needs to direct more resources and efforts toward reducing emissions from electrical energy production and improving the efficiency of existing buildings and appliances—areas where the greatest reductions in carbon emissions are possible. The need to expand the U.S. economy and support a growing population makes the implementation of carbon emission mitigation programs challenging. However, there are ways to reduce the impact of carbon emissions without adversely impacting the economy. Reducing electrical demand (kW) is a prime example. Projections indicate that improvements in building and appliance energy efficiency combined with industrial sector initiatives could offset 85 percent of the incremental demand for electricity through 2030 (FOXNews.com, 2009). This scenario is certainly possible, yet the increased need for electricity through 2030 must be met while decreasing the number of operating coal-fired plants.

Nuclear power generation is a potentially viable option as it does not generate CO_2 emissions. However, the realities of deploying utility-scale nuclear electrical generation are difficult to ascertain. Primary concerns include the mining process for nuclear fuels, operational safety, water consumption, development costs, and waste storage. Indeed, the high development costs, long planning horizons, and nuclear waste disposal issues associated with nuclear power stations remain largely unresolved.

It has become increasingly difficult to estimate the costs of new utility-scale nuclear plants. A project in South Carolina, one of the country's poorest states, failed in 2017 after $9 billion had been expended on construction; this boondoggle produced zero electricity and left ratepayers with the tab. A new nuclear power plant, the first in theUnited States in over 30 years, is now operating in Waynesboro, Georgia. Bechtel assumed responsibility for the Plant Vogtle project (to add reactors 3 and 4) in 2017. This project was subsidized by more than $12 billion in governmental loan guarantees. With an estimated initial cost of $14 billion, delays, labor issues, cost overruns, and inflation during the latter years of construction increased the project cost to over $31 billion. Add in $3.7 billion that the original contractor, Westinghouse, paid to the Vogtle owners to walk away from constructing the reactors, and the total approaches $35 billion. Seven years behind schedule, the total capital costs for the two new reactors are an astounding $15.8 million per MW—making new nuclear power in the United States the world's most expensive way to generate electricity. China's Three Gorges Dam hydro-power project, the world's largest electrical generating station of any type, cost the equivalent of $32 billion and has a generating capacity of 22,500 MW, about 10 times that of Vogtle Units 3 and 4. Until development cost issues are resolved, utility-scale nuclear energy is not economically viable.

Impact of CO_2 Regulations and Economics

U.S. greenhouse gas regulations and policies are in flux. In April 2009, the U.S. EPA released a proposed finding on CO_2 and five additional greenhouse gases (methane, nitrous oxide, hydrofluorocarbons, perfluorocarbons, and sulfur hexafluoride). The agency stated that "in both magnitude and prob-ability, climate change is an enormous problem" and that greenhouse gases "endanger public health and welfare within the meaning of the Clean Air Act" (FOXNews.com, 2009). It cited man-made pollution as a "compelling and overwhelming" cause of global warming. According to EPA Adminis-trator Lisa Jackson, the finding "confirms that greenhouse gas pollution is a serious problem now and for future generations" (U.S. Supreme Court ruling, 2007). This finding was prompted by a Supreme Court decision in April 2007 ruling that greenhouse gases are pollutants as classified by the Clean Air Act and regulation is required if human health is threatened (FOXNews.com, 2009; Shin, 2009).

This redirection of U.S. policy is already impacting the utility industry's plans to decrease coal-fired electrical power generation. AES Corporation withdrew an application to construct a 600 MW coal-fired power plant in Oklahoma as part of a "broadened strategy to reevaluate" their growth plans (Shin, 2009). Louisiana Generating was recently sued by the U.S. Envi-ronmental Protection Agency (EPA) for failing to install modern pollution control equipment when its Big Cajun Plant underwent major modifications (Shin, 2009). Michigan has placed several power plants on hold while Gover-nor Jennifer Granholm attempts to shift her state to power from cleaner, more sustainable energy sources (Shin, 2009). In Kansas, a state with a success-ful wind power development program, former Governor Kathleen Sebelius vetoed three legislative attempts to approve two large coal plants proposed by Sunflower Electric. While U.S. states are making impressive policy gains, many federal government programs languish as a result of administrative policy changes beginning in 2016.

Perhaps one of the strongest developing arguments against the develop-ment of coal generation concerns its long-term economics. The startling fact is that coal-fired power plants are fast becoming non-competitive. Most states are using less coal. In Connecticut, New Hampshire, and Massachu-setts, coal consumption has declined by over 75 percent. In others, coal has effectively become an intermittent power source. The Energy Information Agency determined that "natural gas-fired power plants are cheaper to build with capital costs ranging from $676 to $2,095 per kilowatt, depending on the technology." This is the primary reason why natural gas now generates 34 percent of the electricity in the United States. The math shows that for the coal industry to compete with natural gas, it would require additional subsidies of $1.2 trillion and need to modify construction practices and combustion processes; its comparatively high rates of source energy use, adverse environmental impacts, and higher ownership and lifecycle costs would all require resolution.

Sustainable Buildings

The goal of sustainable buildings is to provide higher quality construction with healthy and comfortable indoor environments. They address issues such as site development, recycling, energy use, water consumption, renewable energy, and indoor air quality.

There are many opportunities in practice to include green design features and components in buildings that address their sustainability. Developing a green building project often requires trade-offs. It involves considering how buildings are designed and constructed at each stage of the project development and delivery process. Green buildings are designed to use less energy. The energy they require for operations is often harvested on-site by using renewable energy, such as solar photovoltaics, wind power, or geothermal energy.

The standards for green construction are evolving. The International Energy Conservation Code (IECC) requires that certain energy-efficient design methodologies be used in construction practices. Developed by the International Code Council, it addresses the design of energy-efficient building envelopes and the installation of energy-efficient systems while emphasizing system performance. It is comprehensive and provides regional guidelines with specific requirements based on climate zones. New construction materials and products are available that offer improved design solutions. Many state and local governments in the United States use the IECC or have developed their own codes that incorporate sustainable design requirements.

Numerous assessment systems for sustainable buildings are used throughout the world. The Building Research Establishment Environmental Assessment Method (BREEAM), the world's longest established method of assessing, rating, and certifying the sustainability of buildings, has been used in over 50 countries. Leadership in Energy and Environmental Design (LEED) offers an interrelated family of green building certification programs for interiors, new construction, existing buildings, schools, hospitals, core and shell projects, neighborhood development, among others. LEED considers a wide range of sustainable construction features, components, and project types. It uses both prescriptive and proscriptive criteria. LEED projects are credited for design attributes including site sustainability, energy and water consumption, indoor air quality, environmental criteria, transportation, and use of green building standards. According to Richard Fedrizzi of the United States Green Building Council, LEED's developer, "this movement has created a whole new system of economic development."

When required by regulation, IECC requirements are included in BREEAM or LEED projects in the United States. This holds the designers of buildings being certified accountable by mandating higher energy efficiency standards. Many green building technologies, such as high efficiency windows, solar arrays, and day-lighting applications are readily observable. Regardless, it is discouraging to owners that many important engineered features of green buildings go unnoticed as they are hidden from view in mechanical rooms and spaces not accessible to visitors. Examples of these technologies include computer control systems to manage energy and water use, rainwater collection systems, lighting control systems, under-floor airflow systems, geothermal heating and air conditioning systems, among others.

Validating green building performance is challenging due to the subjective concepts involved, the evolving nature of standards, and the local variability of construction practices. Measurement and verification have an important role as they provide procedures that verify the energy use, water consumption, and emissions reductions green buildings offer.

Many building owners recognize the importance of sustainable buildings, the features of sustainable buildings, and the importance of resource and energy use in green construction. They are often the leaders in promoting green design features in lighting, electrical, and mechanical systems that are found in a growing number of buildings. Many newer structures are designed to be zero net energy (ZNE), net positive energy (NPE), or carbon neutral buildings. ZNE buildings have been completed in Japan, Malaysia, the United States, and elsewhere. The use of green construction techniques and technologies will continue to expand and more examples of green buildings will become available. Future research will respond to concerns about increased construction costs and more fully assess actual energy and environmental impacts.

The Role of Transportation Systems

In 2022, the U.S. transportation industry consumed the equivalent of 100 quadrillion Btus of source energy, or 28 percent of its total primary energy. Plans to reduce GHGs must include ways to reduce impacts from the combustion of transportation fuels. Most of our mobility fuels are derived from fossil fuels, primarily oil and, to a lesser extent, natural gas. Products include diesel fuels and gasoline. Combustion of these fuels creates environmental problems, adding greenhouse gases, pollutants, ozone, and soot to the atmosphere. While mobility fuel use continues to increase, we have other options. Solutions that decrease these pollutants include reducing the use of vehicles by planning solutions, manufacturing vehicles that are more efficient, using alternative means of transportation, and using fuels that are less polluting. Fuel substitution is another. These can lead us to a more sustainable energy future.

Planning solutions include infrastructure improvements and densification. We can design our cities to minimize the impact of automobiles by discouraging use of vehicles, providing green wedges from the city center to its outskirts, and brownfield redevelopment. Well-designed urban planning approaches encourage people to walk, bicycle, and skateboard by locating them in proximity to urban amenities and creating transit centers. Infrastructure improvements such as bikeways, bus rapid transit, and light rail have a place as planning solutions.

We can provide more efficient vehicles. This has been a focus of successful legislative efforts to improve fleet vehicle mileage requirements. It has been accomplished by improving the efficiency of the combustion engines used in automobiles and trucks and by reducing vehicle weights and sizes.

Alternative means of transportation enable consumer choice by providing multiple transportation options. Arriving at the central train station in Amsterdam by bicycle, canal boat, bus, tram, or taxi, you can connect to rail service that will get you anywhere in Europe. While you are not allowed to park a vehicle at the station, there is a three-story garage for bicycles.

The railways in the Netherlands operate on wind power. Yet renewable fuels in the United States comprise only a small portion of its transportation fuels. Examples include biogas-derived compressed natural gas, ethanol, hydrogen vehicles using fuel cells, electric vehicles that are charged using renewable energy, and many others. In Atlanta, one of the largest landfills produces gas to operate their garbage trucks. The main commercially produced biofuel in the United States is ethanol, created using corn. In most processes, we use the corn kernels for this purpose as the sugar content is higher than in other bulky materials. Producing cellulosic ethanol from these waste materials is more costly but often provides a higher energy production ratio, like using sugar cane.

Providing alternatives to fossil fuels for transportation systems is challenging. Gasoline and diesel fuels have remarkably high energy production ratios. However, greenhouse gas emissions are substantial. Not so with biofuels, which recycle the carbon dioxide they extract from the atmosphere. Many

researchers believe algae may be a future solution. When grown under specific conditions, they produce oil plus byproducts that can be used to make
ethanol. Solar PV-powered vehicle charging stations are becoming more
common in the United States. Some are designed for fleet mobility uses. One
example is the City of Asheville, North Carolina which has electric-powered
police cars and its own vehicle charging stations. Las Vegas, Nevada, has its
own hydrogen fueling station for hydrogen-powered vehicles.

Tough Decisions on the Path to Decarbonization

The term *decarbonization* literally means the reduction of carbon. It is often
included in lists that identify future trends, but we are seeing it become policy
today. The term refers to the transition to an economic system that sustain-
ably reduces and compensates for atmospheric emissions of carbon dioxide,
a potent greenhouse gas (GHG). Decarbonization has been identified as one
of the top three trends in the energy sector (Marr, 2022). In a broader sense,
it refers to the transition away from the use of fuels and compounds that
increase GHG emissions. The concept is indeed transformative because it
challenges us to transition from using fossil fuels to clean energy resources.

Gaining traction in the 1950s, the world's historical CO_2 trend shows a
trajectory of emissions that increases unabated since the 1950s (see figure
6.1). Though total emissions declined in 2020 due to the economic impact
of COVID-19, they increased to 36.3 billion tones in 2021 as the world's
economic situation improved.

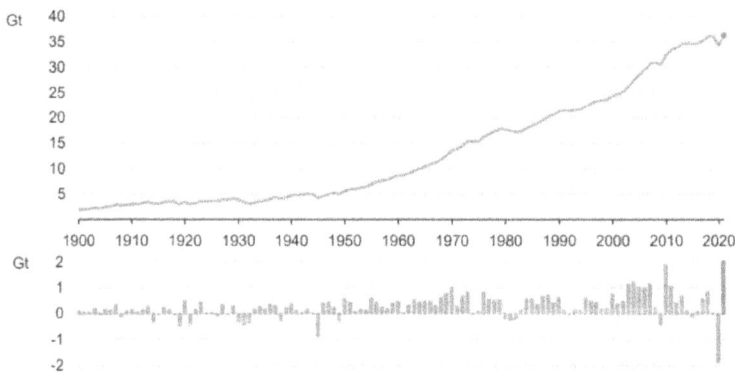

**Figure 6.1 Total CO_2 Emissions from Energy Combustion and Industrial Processes (top
graph) and Their Annual Change (bottom graph).** *Source*: International Energy Agency.
Global energy review: CO_2 emissions in 2021. https://iea.blob.core.windows.net/assets
/c3086240-732b-4f6a-89d7-db01be018f5e/GlobalEnergyReviewCO2Emissionsin2021
.pdf, accessed August 3, 2022.

The goal of decarbonization is to create a CO_2 emissions-free global economy (Volkswagen AG, 2022). This is a daunting, if not seemingly impossible, objective to achieve in the short term. Long-term reductions in GHG emissions will require substantial increases in infrastructure expenditures by industrial and governmental entities. Many advanced economies have emphasized decarbonization measures in their recent economic improvement plans (International Energy Agency, 2021). Given the past trajectory of greenhouse gas emissions, getting to zero emissions is a questionable goal. Regardless, we are not without tools that aid in this effort.

How can GHG emissions be reduced? Numerous ways have been proposed to achieve decarbonization. One strategy is to lower or eliminate the use of fossil fuels in combustion processes, ultimately lowering total GHG emissions. As the world's population increases, per capita fossil fuel consumption would show incremental declines. Some argue that this is beginning to happen. Per capita fossil fuel consumption began to level off in 2011, peaked in 2018, and has since declined. This has been attributed to energy conservation and efficiency improvements, fuel-switching (e.g., substituting renewables and natural gas for coal), and economic conditions. An increase in this measure was recorded in 2021 and attributed primarily to economic growth (BP, 2022).

Another solution is to expedite the transition to using energy that does not emit GHGs during combustion processes or otherwise lacks a combustion process. Realistic approaches include creating pathways to incrementally transition to clean energy. *Clean energy* usually refers to renewable energy plus nuclear energy since these energy resources produce negligible quantities of GHGs (Roosa, 2022a). The vision of decarbonization is to evolve to clean energy electric grids using local generation from renewable energy sources (e.g., wind, geothermal, hydropower, and solar power). Excess power can be stored for later use by using batteries (e.g., lithium-ion, sodium-sulfur, vanadium-redox, etc.), compressed air, pumped hydropower storage, or directed to a process storage solution such as using water for electrolysis to create hydrogen and oxygen (Roosa, 2022). Another alternative is to generate electricity with package nuclear power systems (Energy Information Administration, 2007). In 2021, renewable energy sources and nuclear power combined provided a larger share of global electricity generation than coal (International Energy Agency, 2021). The use of renewable energy, led by wind and solar power, has continued to increase and now accounts for 13 percent of total power generation. Renewable electrical generation increased by about 17 percent in 2021 (bp, 2022). It has accounted for over half of the increase in global power generation over the past two years (bp, 2022). The United States is making progress. The country's CO_2 emissions have declined from 5,672 million tons in 2002 to 4,826 million tons in 2022, a 14 percent decline over the 20-year period while its gross domestic product has grown. Zero carbon electrical power reached an all-time high of a 41 percent share of total production in 2023.

Tracking the Bottom Line

Pathways to decarbonization are often sector-specific. Instituting building decarbonization measures has been proposed (International Energy Agency, 2021; Roosa, 2022). Communities have created pathways to institute clean energy programs that use energy-efficient buildings and support electrification (Roosa, 2022b; Annan, 2022). Technological approaches include developing net zero buildings and processes using green hydrogen for mobility applications (Wiegand-Jackson, 2022). Another approach is to convert all appliances and heating systems to operate using electricity and generate all the electricity required using renewable resources. To make this happen, industries would need to make major infrastructure changes. Key elements of an industrial decarbonization strategy would include gathering energy data, establishing emissions reporting methodologies, establishing decarbonization project generation and drivers, creating a marginal abatement cost curve, expanding electrification, and enforcing decarbonization checklists (Miller, 2022). For industrial concerns, making changes to functional processes often face formidable internal resistance. We need synergistic, multi-faceted approaches to resolving our energy-related transportation emissions. It begins with rethinking the design of our transportation infrastructure, offering more efficient vehicles, offering more options, and greater use of renewable energy systems.

Most years since the 1950s have yielded increased CO_2 emissions (see Figure 1). This trend must be reversed. As energy engineers and managers, the world looks to us to identify the pathways that lead to successful decarbonization. Not all efforts toward decarbonization will be effective. There will be unforeseen risks. Implementing our solutions will require rigorous analysis, compromise, and ultimately trade-offs. Thankfully, central governments (i.e., those in the United States, China, and the European Union) are starting to provide more funding for decarbonization efforts that will enable more of our projects to go forward. Regardless of the approaches we choose to achieve decarbonization, now is the time to act.

A stabilization triangle approach to developing the potential of wedges has been adopted by individual countries and the International Governmental Panel on Climate Change (IPCC) to address climate change solutions. It supported international agreements including the Kyoto Protocol (in force in 2005), the Copenhagen Accord (2009), and the Paris Agreement (legally binding in November 2016). Emissions reduction pathways and the predicted changes in atmospheric temperatures stem from implementation of strategic GHG reduction approaches in the Paris Agreement. It provides reference cases and focuses on the role of the Paris Agreement. It supports the international consensus to implement efforts to prevent further increases in atmospheric temperatures.

Has there been progress toward our GHG reduction goals? Perhaps. From 2000 to 2009, global emissions increased at a rate of 2.9 percent annually; from 2010 to 2019, global emissions increased at a rate of 1.0 percent annually, the same rate of increase as from 1990 to 1999. The lower rate of emissions in the last decade is due to the implementation of climate mitigation policies and technologies, economic issues, and other factors. Total emissions are now 37.5 $GtCO_2$/year, about 10 $GtCO_2$/year more than in 1990. The progress so far has resulted from more energy-efficient buildings, changes to transportation systems, energy efficiency improvements, fuel substitution, energy storage systems, and carbon recycling and sequestration.

The hope is that the more dire implications of climate change are preventable. Strategic approaches to reducing GHG emissions have already been identified by researchers, and newer technological solutions will be available to mitigate climate change. Many of the technologies needed to implement the solutions have been tested and proven. The key goal is to transition our economies away from using primarily fossil fuels toward clean energy resources. Knowing what solutions are workable now leads us to the next phase, which is to implement the most workable solutions at larger scales.

ACKNOWLEDGMENTS

Segments of this chapter were initially published in *The Sustainable Development Handbook, Carbon Reduction: Policies, Strategies and Technologies,* and select issues of *Strategic Planning for Energy and the Environment.* The author extends thanks and credit to the Fairmont Press, Inc. for the use of previously published material.

REFERENCES

Annan, G. (2022). Building decarbonization transformation. *International Journal of Strategic Energy and Environmental Policy*, 4(5), pages 61–68.

bp. (2022). Statistical review of world energy, 71st edition. *BP*, 2022. https://www.bp .com/content/dam/bp/business-sites/en/global/corporate/pdfs/energy-economics/ statistical-review/bp-stats-review-2022-full-report.pdf, accessed August 29, 2022.

Energy Information Administration. (2007). International energy outlook 2007. *U.S. Energy Information Administration*, May 2007. http://www.eia.doe.gov/oiaf/ieo/emissions.html, accessed November 3, 2007.

FOXNews. (2009). EPA takes first step toward regulating pollution linked to climate change. *Fox News*, April 17, 2009. http://www.foxnews.com/politics/first100days/2009/04/17/epa-takes-steps-regulating-pollution-linked-climate-change/, accessed April 29, 2009.

Friedman, T. (2008). *Hot, Flat and Crowded.* New York: Farrar, Straus and Giroux, p. 214.

International Energy Agency. (2022). Global energy review: CO_2 emissions in 2021. *IEA*, March 2022. https://iea.blob.core.windows.net/assets/c3086240-732b-4f6a-89d7-db01be018f5e/GlobalEnergyReviewCO2Emissionsin2021.pdf, accessed August 3, 2022.

Marr, B. (2022). The 3 biggest future trends (and challenges) in the energy sector. *Forbes*, February 11, 2022. https://www.forbes.com/sites/bernardmarr/2022/02/11/the-3-biggest-future-trends-and-challenges-in-the-energy-sector/?sh=71f4a87127b7, accessed August 2, 2022.

McKinsey and Company. (2007). *Reducing U.S. greenhouse gas emissions: How much at what cost?* U.S. greenhouse gas abatement mapping initiative, executive report. *McKinsey & Company*, December 2007. http://www.mckinsey.com/client-service/ccsi/greenhousegas.asp, accessed February 3, 2008.

Miller, R. (2022). Decarbonization strategies for industrials. *International Journal of Strategic Energy and Environmental Policy*, 4(2), pages 26–37.

Roosa, S. (2022a). What to do about climate change? *International Journal of Strategic Energy and Environmental Policy*, 4(3), pages 8–30.

Roosa, S. (2022b). Creating pathways to clean energy. *International Journal of Strategic Energy and Environmental Policy*, 4(5), pages 69–76.

Shin, L. (2009). EPA review reverberates through U.S. energy industry. *Inside Climate News*, February 18, 2009. https://insideclimatenews.org/news/18022009/epa-review-reverberates-through-us-energy-industry/, accessed April 29, 2009.

U.S. Supreme Court. (April 2, 2007). ruled in *Massachusetts v. EPA* that carbon dioxide could be regulated as a pollutant under the Clean Air Act and that the government had a responsibility to issue a determination based on science. The Bush administration failed to act on this ruling.

Volkswagen A. G. (2022). Decarbonization–what is it? *Volkswagen Group*, 2022. https://www.volkswagenag.com/en/news/stories/2019/03/decarbonization-what-is-it.html, accessed August 2, 2022.

Wiegand-Jackson, L. (2022, May). On the path to net zero: green hydrogen as a decarbonization solution. *International Journal of Strategic Energy and Environmental Policy*, 4(5), pages 8–23.

Wikipedia. (2008). Methane. *Wikipedia* http://en.wikipedia.org/wiki/Methane, accessed March 8, 2008. The chemical formula for this process is $CH_4 + 2O_2 -> CO_2 + 2H_2O$.

Part II

LOCAL STUDIES ON THE NEGATIVE IMPACT OF POLLUTION

Chapter 7

How to Do a Pollution Audit in Your City

Russell Barnett, John Hans
Gilderbloom, and Bunny Hayes

Environmental justice has been defined as the pursuit of equal justice and equal protection under the law for all environmental statutes and regulations without discrimination based on race, ethnicity, and/or socioeconomic status (Bullard, 1990; Bullard, 2008; Capek, 1992; Capek, 1993; Capek & Gilderbloom, 1992; Hipp & Lakon, 2010). This concept applies to governmental actions at all levels—local, state, and federal—as well as private industry activities. Lower-income communities and minority populations have historically been the target of many sources of pollution: air pollution from industrial sites, toxic contamination from incinerators and brownfields, contamination of ground and source water, and lead exposure in low-income communities.

The Environmental Justice movement got its start in 1982 when North Carolina conducted a study to identify a site for a hazardous waste/PCB landfill. Although several physically suitable sites were identified, the state selected one in Warren County, a poor, predominantly black community in the southern portion of the state. Civil and state-rights activists staged demonstrations where 500 people were arrested. National studies (Toxic Waste and Race) found that race was a significant factor in siting hazardous waste facilities and that 60 percent of African Americans and Hispanics live in community housing near unregulated toxic waste sites. Race was found to be the most potent variable in predicting where these facilities were located—more powerful than poverty, land values, and home ownership (Gilderbloom & Wright, 1993; Bullard, 2000a; Bullard, 1990). The study found that:

(1) Three out of five African Americans live in communities with abandoned toxic waste sites;

(2) Sixty percent (15 million) of African Americans live in communities with one or more abandoned toxic waste sites;

(3) Three of the five largest commercial hazardous waste landfills are in predominantly African American or Latino communities and account for 40 percent of the nation's total estimated landfill capacity; and

(4) African Americans are heavily overrepresented in the population of cities with the largest number of abandoned toxic waste sites.

In 1994, President Clinton signed EO 12898 requiring federal agencies to make environmental justice part of their mission (Atkinson et al., 1996). Understanding environmental justice issues requires one to understand the history of how a site or area evolved to the point where public health risks increased to unacceptable levels, while environmental degradation incrementally deteriorated. To address environmental justice issues, the public must have an awareness of the cause and effect of the public health and environmental issues. The University of Louisville Kentucky Institute for the Environment and Sustainable Development and the Center for Sustainable Urban Neighborhoods prepared the following document to explain environmental justice issues in Louisville. This document is used as a reference for tours conducted by the Institute for university classes, public interest groups, and government officials.

ENVIRONMENTAL ISSUES: LOUISVILLE, KENTUCKY (KLEBER, 2000)

1. Bourbon Stockyard
2. Beargrass Creek and Urban Streams
3. Flood Control
4. Combined Sewers
5. Weinberg Co.—Trolley Barn Site
6. Distler's Warehouse
7. Bourbon Distilleries
8. Ohio River and MacAlpine Lock and Dam
9. Gallagher Power Plant
10. Chickasaw Park Lake
11. Ford Motor Assembly Plant
12. Morris Forman Wastewater Treatment
13. Liquor distilleries polluting and scarring the neighborhood
14. Rubbertown
15. Carbide Industries
16. Lubrizol, Zeon, and PolyOne

17. DuPont
18. Dow Chemical
19. American Synthetic Rubber
20. Momentive Specialty Chemical
21. Lake Dreamland
22. Lees Lane Superfund Site
23. Railroad Yards

Bourbon Stockyard

The Bourbon Stockyards was the oldest continuously operating stockyard in the United States (1864–1999). The surrounding area known as Butchertown has been associated with the meat industry since the 1830s. The force behind the development of Butchertown was the need of farmers in the Bluegrass to sell their pork and beef, and the desire of residents of the Deep South to buy it. Louisville was the obvious spot to slaughter animals and ship their meat south on the Ohio River. Because herds entered Louisville on the old Frankfort Pike (now Frankfort Avenue), the butchers opened shops, usually in their backyards, in the surrounding area. In 1834, the Bourbon House, a hotel for farmers located between Washington Street and Story Avenue in Louisville, became the nucleus of stockyards that were located in the neighborhood. In 1864, a stockyard was built at Main and Johnson streets. It was incorporated as Bourbon Stock.

Yard Company in 1875. By the late 1800s, it included a modern public market with docks, offices, and other services, allowing the company to dominate the Kentucky cattle market for the next century. In the first half of the twentieth century, the plant was expanded to correspond with the extension of the Louisville cattle market, but by mid-century, the market declined due to a change from the railroad to trucking as the major mode of transportation. Early in its operation, all undesired parts of animals (hides, blood, guts, and hooves) were disposed of in the adjacent Beargrass Creek. Many butchers opened shops on Story Avenue because a fork of Beargrass Creek just south of there was convenient for dumping waste. The butchers were soon joined by tanners, coopers who made barrels used in shipping meat, and others. Soap makers and candle makers could scrape fat off the banks of the creek. At that time, the creek discharged into the Ohio River at 2nd Street. The entrails would be flushed to the downtown area where they would be caught on rocks and rot. The stench was overwhelming, and the creek was diverted at the Stockyard due north to the Ohio River (Bourbon Stock Yards, 2024a; Bourbon Stock Yards, 2024b).

Beargrass Creek and Urban Streams

The Beargrass Creek Watershed is probably the most diverse watershed in terms of geographic area and land usage. The creek was an early source of drinking water, fed by eight major springs. It was also the site for several grist mills and one mill used in the manufacture of paper. The entire area covers approximately 61 square miles and is divided by three major sub-basins—the Muddy Fork, the Middle Fork, and the South Fork. Land use percentages are shown in table 7.1.

All three sub-basins flow from suburban areas that were developed since the early 1960s and continue to develop. They flow through many older neighborhoods, such as the Highlands, Germantown, Phoenix Hill, and Butchertown before they come together and flow into the Ohio River just above the downtown Louisville area.

The Beargrass Creek Watershed represents a sampling of practically every water quality problem that can be imagined, including combined and sanitary sewer overflows; package wastewater treatment plants; septic tank seepage; urban stormwater runoff (nonpoint source pollution from streets, lawns, and parking lots); erosion and sedimentation problems; and flood management (Rubbertown Community Advisory Council, 2020).

The entire watershed, including the piped sections in the City of Louisville, receives stormwater and wastewater flow from residential, commercial, and industrial customers. Approximately three miles of the open sections of the creek were piped or channelized from the 1930s to the 1960s.

Based on federal water quality standards, the Commonwealth of Kentucky's Division of Water sets the uses designated for Beargrass Creek and monitors its progress in meeting specific water quality standards. Based on these standards, the Kentucky Division of Water (KDOW) classifies all three forks of Beargrass Creek as not meeting the designated-use criteria for primary contact recreation (such as swimming and wading) or for aquatic life to live. Also contributing to the creek's decline, and harmful to human health, are vehicle emissions such as oil, antifreeze, benzene from exhaust, and heavy metals including zinc, copper, and lead. Other pollutants found in

Table 7.1 Beargrass Creek—Land Use Percentages (1998)

Sub-Watershed	Total Impervious	Undeveloped	Commercial	Parks	Public	Industrial	Residential
Middle Fork	39.00	11.87	13.28	3.91	15.94	2.81	52.19
Muddy Fork	35.00	9.36	19.10	3.37	16.48	4.12	47.57
South Fork	42.00	12.78	22.38	3.39	14.39	6.27	40.79
Averages	38.67	12.18	19.67	3.52	15.01	5.14	44.48

Source: Kentucky Institute for the Environment and Sustainable Development

measurable quantities in Beargrass Creek are household chemicals, including solvents, paints, pesticides, and herbicides. Stormwater runoff from poorly managed construction sites contributes tons of silt and sediment that suffocate aquatic habitat. During the past 20 years, most of the package wastewater treatment plants along Beargrass Creek have been removed by MSD trunk sewer construction. The same trunk sewers have also eliminated almost all the septic tanks in the watershed, although a few still exist.

Future water quality initiatives in the watershed will focus on combined sewer and sanitary sewer overflows, flooding, nonpoint source pollution, and natural stream corridor restoration. There are currently 63 combined sewer overflows and approximately 32 sanitary sewer overflows in the Beargrass Creek Watershed. With more than 80 miles of open drainage channels, creeks, and smaller tributaries, the access to nonpoint source pollution in the form of storm runoff is almost endless (Atkinson, 1996).

These issues will be harder to address, will take longer to correct, and will cost more. For instance, petroleum products that are deposited on roadways and parking lots will be much more difficult to collect and treat than the flow from a single pipe to a treatment plant. In addition, the use of herbicides and fertilizers is growing, increasing the negative impact on stream quality and aquatic life. Correcting these problems must begin with public education and participation in identifying the most cost-effective solutions (Rubbertown Community Advisory Council, 2020).

Flood Control

The Ohio River periodically floods (Kleber, 2000). The worst flood on record was in January 1937 when 75 percent of the city was underwater (150 sq miles) and 230,000 people were forced out of their homes. Ninety people died, with $54.3 million in financial losses attributed to the flood. The Ohio River crested at a record 57.15 feet (27.15 feet above flood stage) after record rains of 19.17 inches during the month. After this flood, the U.S. Army Corps of Engineers was authorized to construct a levee to protect Louisville, which was completed in 1957. A levee is an earthen dam that runs parallel to the river to prevent the river from expanding outward. Over the years, the flood protection works are now 29 miles long. In the downtown area, to minimize the amount of land taken by the levee, a floodwall was constructed. A floodwall is constructed of concrete and has water-tight doors that are normally open to allow traffic to travel to the river but are closed during flood events. There are 52 street openings in the floodwall and over 150 valves and gates that must be closed when the Ohio River goes into flood stage. When these openings are closed, the city relies on a system of 16 flood-pumping stations to pump rainwater inside the levee system up over the levees into the river. Buildings constructed between the river and levee must be designed for periodic flooding. Recent floods in the Mississippi River and other rivers around the United States have demonstrated that levees are not a guarantee that flood damage will not occur. Levee failures or flood heights that overtop levees can result in astronomical monetary damage and loss of life. The current approach to flood protection focuses more on preventing the construction of homes and buildings in the floodplain.

Combined Sewer

The sewer system owned, operated, and maintained by MSD has evolved for almost a century and a half into an extensive network of both sanitary and combined sewers, diversion structures, mechanical regulators and other flow control devices, wastewater treatment plants, and pump stations (Kleber, 2000). Before 1850, sewers in Louisville were built mainly of cut stone and brick. The sewers appear to have been intended mainly for the purpose of storm water disposal and usually flowed toward the Ohio River or nearest surface drain. It was not until sometime in the 1860s, with the introduction of a public water system, that the idea of constructing laterals to convey wastes directly from the houses into the sewers became adopted as common practice.

The Metropolitan Sewer District was established in 1946 to provide sewer service and limited flood protection for the old City of Louisville and Jefferson County. At that time, Louisville had about 750 miles of combined sewer pipes that discharged both sanitary and storm runoff directly to the Ohio River and Beargrass Creek. It also had an Ohio River floodwall system that was still on the drawing board; no wastewater treatment plants; and a post-World War II baby boom that was driving suburban growth into rural Jefferson County.

That massive suburban growth, framed by the 1950s through the early 1990s, created, quite possibly, the most serious water pollution problems in our community . . . the proliferation of over 300 "package" (or portable) wastewater treatment plants and over 40,000 individual septic tank systems. To eliminate these constant sources of water pollution, MSD built and acquired over 1,000 miles of sewer lines; constructed two new regional wastewater treatment plants; and expanded four others. Today, the Louisville Metro sewer system includes over 3,000 miles of combined and sanitary sewers; over 300 sewerage pumping stations; and six regional wastewater treatment plants that process over 160 million gallons of wastewater each day. Almost all these advancements were made as an investment by the local community and its citizens. MSD spent its last federal grant monies in 1987.

Since then, it has received no funding from federal or state governments for sanitary sewer improvements. Consequently, MSD and its customers have invested over $1 billion during the past 15 years to improve local water quality.

Of our 3,000-mile sewer system, over 700 miles of the system is old, combined storm and sanitary sewer, with over 400 miles of the system at least a century old. Some sewers predate the American Civil War. There are 63 combined sewer overflows and about 32 sanitary sewer overflows in the Beargrass Creek Watershed. On average, the Louisville area experiences about 30 days each year that are wet enough to cause discharges from CSOs and SSOs. On the remaining 300 or so dry days, almost all wastewater is channeled to a treatment plant. On the wettest days, millions of gallons of diluted sewage are discharged into our streams. Only by relieving these overloaded sewers can we prevent sewage from backing up into homes and basements (Louisville Metropolitan Sewer District, 2020).

Weinberg Company—Trolley Barn Site

The former Louisville Street Railway Complex, or "Trolley Barn," is located at 1701 West Muhammad Ali Blvd. (Kleber, 2000). It is an example of a brownfield site, long abandoned due to environmental contamination. Brownfields are defined as "abandoned, idled or underutilized industrial and commercial facilities where expansion or redevelopment is complicated by real or perceived contamination." Within Louisville's downtown area, 25 percent of the area can be classified as a brownfield. This site is the second pilot brownfield site prioritized by the City of Louisville for assessment and remediation under the city's National Brownfield Pilot Program. The Louisville Metro government has won three Phoenix Awards for brownfield cleanups: University of Louisville football stadium (old rail yard), the Home of the Innocents (150-year-old stockyard), and the Louisville Waterfront (200-year-old shipping port).The Trolley Barn building was built in 1879 and used to store and maintain the city's mule-drawn trolley cars and stable the mules. The first mule-drawn trolleys were introduced in Louisville in 1864, and the first electric trolleys appeared in 1889 (Green Street line). By 1901, all mules had been replaced by electric cars. The building was later used as a warehouse, most recently by the Weinberg Co., which was in the business of blending and distributing janitorial cleaning supplies, including some pesticides and herbicides. During these operations, the soil floor was contaminated with a variety of chemicals and pesticides. The property was acquired by the Louisville Housing and Urban Development (HUD) Department as part of its urban renewal program for the Russell neighborhood in February 1997. The property was placed on HUD's acquisition list in 1995, but purchase negotiations were delayed due to concerns over the potential cost of environmental cleanup. The Trolley Barn site lies within one of Louisville's old industrial corridors, part of which is littered with abandoned and underutilized properties that cost the city $8.7 million annually in lost property tax revenues.

The property was thought to have unacceptable levels of hazardous substances in the soil based on a Level I environmental assessment and limited soil sampling conducted by a consultant to HUD in 1995. The consultant estimated the cleanup costs to be as high as $30 million to remove soil down to groundwater and for groundwater remediation. In early 1996, the city received $200,000 in federal funding for a pilot Brownfield Remediation program and placed the site on its list for further investigation and action. A Level II environmental assessment was conducted to further define the extent of contamination. Groundwater sampling revealed that the site was not contributing to groundwater contamination. Further soil samples showed that soil contamination was limited. This new data led to the conclusion that the extensive soil removal and groundwater remediation requirements suggested by the Level I assessment would not be necessary. The estimated cleanup costs were revised to $80,000. HUD proceeded with the property acquisition, and soil cleanup was initiated.

The property is now a $25 million, 50,000 square foot African American Heritage Center to celebrate African American culture and highlight the many struggles and major accomplishments of black Kentuckians from the earliest days to the present. The centerpiece is the huge Trolley Barn, which, with its vaulted ceiling and wide-open spaces, serves as a "Great Hall" for special events. Other buildings feature a genealogical research center, a permanent exhibit on black history, an auditorium, traveling exhibits, studio space for artists, and some retail space. The city has expended over $800,000 to buy the building and to study alternative uses for the property. The city received federal transportation funds ($1.5 million) in 1999 to renovate the building for a museum, and an additional $12 million has been raised from private and local sources. The first phase to refurbish the buildings was completed in May 2001. During this phase, the building's structure was shored up and lead paint was removed at a cost of $2.5 million. The Second Phase ($7.8 million) of constructing the Great Hall took 18 months to complete. This was done by renovating over 550 livable, safe, and beautiful homes in a civil housing development called City View Park; the project was led by Washington D.C. Telesis with help from the University of Louisville Center for Sustainable Urban Neighborhoods (Gilderbloom et al., 2020a; Gilderbloom, 2020b; Gilderbloom, 1993). It also anchored the Louisville Central Development Corporation, which built 100 new urbanist homes, and the Louisville Urban League, which built another 85 homes. This is a feel-good story, but along with this victory, the gigantic problem of air, water, and soil pollution persists.

Distler's Warehouse

A tobacco warehouse once stood in this mostly residential Portland neighborhood. A public middle school is across the street from the site. In 1977, the warehouse was raided by the FBI investigating the illegal storage and disposal of hazardous waste by Donald Distler. They were investigating the illegal disposal of hazardous waste (hexachloropentadiene and octachlorocyclopentene used in making pesticides) down a manhole into the city sewer system, which had shut down the Morris Forman Treatment plant. For 90 days, the plant was shut down to remove the chemicals while 100 million gallons of untreated wastewater were discharged directly into the Ohio River. It took another two years to clean the sewer lines. Mr. Distler was the owner and operator of a waste disposal company, Kentucky Liquid Recycling. He operated two illegal hazardous waste sites in Bullitt County, south of Louisville, and was being investigated for dumping hazardous waste into Louisville's sewer system. Aware of the ongoing investigation, Mr. Distler rented the warehouse and stored 2,500 barrels of hazardous waste at this site, many of them rusted and leaking. As a result of the FBI's investigation, Mr. Distler was convicted of criminal environmental crimes and in 1982 sentenced to two years in prison and fined $50,000 (at the time, the most severe environmental sentence in the nation). The two sites in Bullitt County were eventually listed on the federal Superfund for cleanup. Distler's Brickyard, located in West Point, was placed on the federal Superfund list in 1982. Over $7.4 million was used to clean up the site. Groundwater at the site is still being pumped and treated by the state. The other site, called Distler's Farm, was also placed on the Superfund list in 1982. Located on the Jefferson County/Bullitt County border, over $1.2 million were expended to clean up the site. Groundwater is also still being pumped and treated by the state.

After the raid at this warehouse, the state and federal governments conducted an emergency cleanup to remove and properly dispose of the barrels of hazardous waste. The site did not qualify as a federal Superfund site, and the state had inadequate resources to complete the cleanup at the site. The warehouse owner, who was unaware of Mr. Distler's activities and was not involved in the waste business, was unable to rent the warehouse due to the contamination. It was used briefly as a building supply warehouse, which went bankrupt in 1994. On April 23, 1997, two young boys started a fire that burned the building to the ground.

The property owner, unable to sell or rent the property, stopped paying property taxes. The cost of cleaning up the site has fallen on the state and federal government. However, it does not rank high on the federal list, and the state has over 700 other sites statewide with a budget of less than $2 million a year for site cleanups (Gilderbloom et al., 2013; Gilderbloom & Wright, 1993). A contractor for the City of Louisville has cleared the remains of the warehouse and other debris from the site. The city initiated foreclosure action on the property for delinquent property taxes. The property was purchased by an individual at the courthouse for nonpayment of taxes, and a nursing home was built on the site. While the property deed has a declaration that this site contains hazardous waste, there is no obligation for the property owner to inform residents of this fact.

Liquor Distilleries

Louisville has long been associated with bourbon. For over 200 years, distilleries have been a key economic driver in the city. At one time, over 30 distilleries were located on a short section of Main Street. Louisville serves as a hub for the Bourbon Trail. The trail includes 18 distilleries and attracted 1.2 million visitors in 2022. There are about a dozen distilleries in Louisville in 2024, but there is a new trend in building micro-distilleries. Ninety-eight percent of the world's bourbon is produced in Kentucky, and Louisville is the largest producer within Kentucky. The concentration of bourbon has neighborhood consequences, both financial and possibly health-related.

Bourbon is produced through a process that has not changed much over centuries. The process includes: 1) grinding grains (>51 percent corn and wheat, barley, and rye); 2) distilling the fermented mixture to create alcohol vapors which are condensed by cooling into a liquid; 3) maturing the liquid (alcohol content of 125 proof) for up to 20 years in new, charred oak barrels; and 4) bottling. It is in the maturing process that some of the alcohol vapors escape from the oak barrels. On average, 2–4 percent of the liquid is lost each year through evaporation. The exact amount varies by the size of the oak barrels used, the climate, and the location of the barrel in the storage warehouse (a higher location in the warehouse is warmer and the evaporation rate is higher). The alcohol released into the atmosphere is known as the "Angel's Share" (Marshall, 2019). This airborne alcohol (ethanol) accelerates the growth and stimulates spore germination of a sac fungus named *Baudoinia compniacensis*. This fungus has been observed on a variety of substrates in the vicinity of distilleries, spirits maturation facilities, bonded warehouses, and bakeries. The fungus is a habitat colonist with a preference for airborne alcohol, earning it the nickname whiskey fungus.

Baudoinia species have the ability to withstand high temperatures, allowing them to colonize roof habitats. The fungus is black in color and physically resembles a thick (up to 0.6 inches) crust. It can form on a wide range of substrates, including trees, concrete, PVC plastic, masonry, steel, stone, galvanized and shingle roofing, and outdoor furniture. *Baudoinia compniacensis* was first investigated in 1872 at the request of Antonin Baudoin, the Director of the French Distillers Association. It has since then been reported worldwide in areas adjacent to distilleries. Periodic studies of the fungus were conducted throughout the twentieth century, but a definitive study was not conducted until 2007, when the warehouse staining fungus was again rediscovered by Scott et al. during a survey commissioned by a Canadian distiller of fungal colonists of outdoor surfaces near spirit maturing warehouses.

The impact of the fungus on vegetation and public health is still not clear. Conflicting evidence on its impact on vegetation has been reported. The impact of *Baudoinia compniacensis* on public health is mostly nonexistent. Nevertheless, Indiana public health officials recommend that the removal of the black fungus requires an N95 mask, goggles, and gloves, which is telling. Moreover, there are no peer-reviewed articles on the health impacts of *Baudoinia compniacensis,* which is also concerning. The impact on buildings is predominantly a cosmetic problem and does not seem to impact building integrity. Unlike other fungi which feed on the decay of building materials, *Baudoinia compniacensis*' mode of nutrition is from the atmosphere. The fungus anchors itself firmly to surfaces, making it difficult to remove. Power washing buildings with fungicide or a disinfectant costs up to $1,000 for a residence, and the fungus will quickly reappear (Marshall, 2019). A group of Kentucky homeowners in Frankfort filed a class action lawsuit in 2012, but it was dismissed. Distillers have argued that it would be too costly to control emissions of ethanol and could adversely affect the flavor of the bourbon. Property owners of homes adversely impacted by the fungus are subject to fines from code enforcement officers or neighborhood associations. Airborne ethanol has chronic non-cancer health effects at concentrations of 2,200 $\mu g/m^3$ or higher, as determined by the American Conference of Governmental Industrial Hygienists (2019). California legislation mandates for its wine and liquor industry that the black mold problem be significantly reduced by a machinery process that takes out the black mold that goes into the air. This has been done successfully in California for years and it has not affected the quality of wine, which is considered the best in the world. Throughout West Louisville you can see the damage of black fungus scarring buildings and lowering their valuation, many of which get abandoned. Mixed with black fungus and other pollutants, 5,000 housing units have been abandoned (Marshall, 2019) (see figures 7.1, 7.2, 7.3).

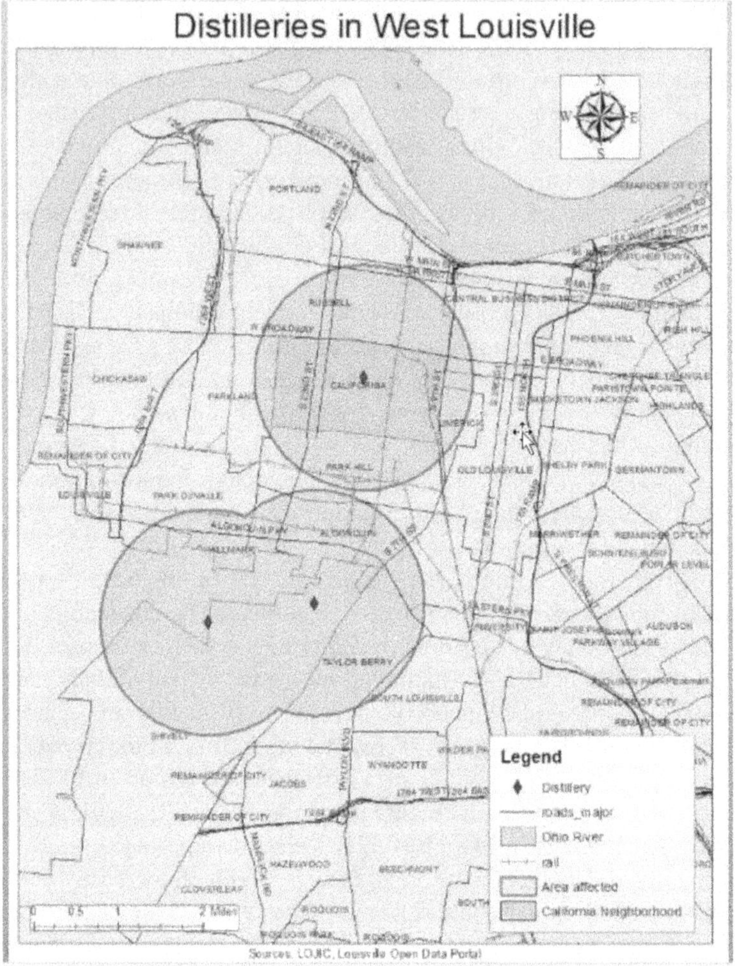

Figure 7.1 Location of Liquor Distilleries. *Source*: LOJIC, Louisville Open Data Portal

Ohio River and McAlpine Lock and Dam

Louisville owes its existence to the Ohio River. It was founded as people were forced to get off the river and portage around the Falls of the Ohio, the only falls in the river from Pittsburgh to the Gulf of Mexico. Over this 3-mile stretch of the river, the river falls approximately 26 feet. The site was not an ideal location for a city. The land adjacent to the gate in the levee providing access to the lock was once a slough that was filled in with municipal waste. The downtown area was a pond-dotted bottomland with poor drainage.

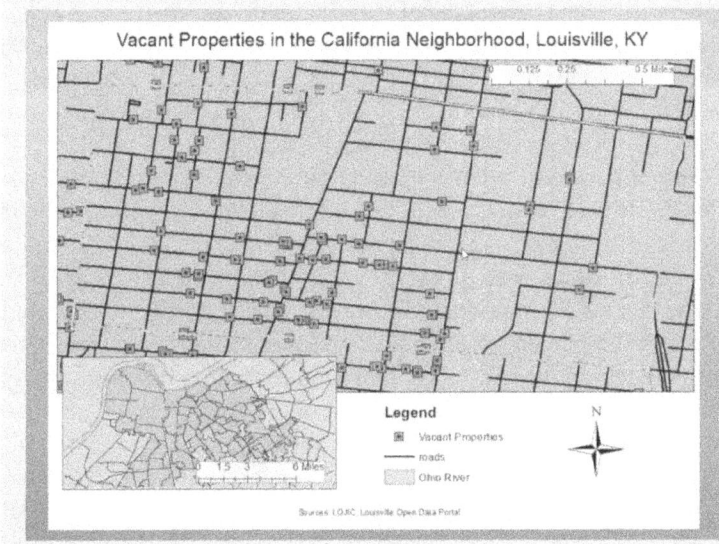

Figure 7.2 How Many Houses have been Abandoned? *Source*: LOJIC, Louisville Open Data Portal

Figure 7.3 A Once Beautiful Shotgun Home Ruined by Nearby Liquor Distilleries. *Source*: Image by John Hans Gilderbloom, Center for Sustainable Urban Neighborhoods, formerly University of Louisville, transferred to Neighborhood Associates, Washington D.C.

As the city grew, poor drainage led to health problems. Its streets were muddy and full of stagnant, stinking potholes. In 1822, a malaria epidemic resulted in the death to 25 percent of the population. Ditches were dug to drain the ponds and create land that was suitable for building. The same drainage ditches were used as sewers and to dispose of waste. To improve drainage, in the late 1850s, a canal was dug to divert Beargrass Creek to the Ohio River above downtown. The old stream bed was lined with bricks, covered with dirt, and used as one of Louisville's first sewers, emptying directly into the Ohio River. It would not be until 1958 that sewage from the city was treated prior to discharge into the Ohio River.

The Ohio River is 981 miles long and flows from Pittsburgh, Pennsylvania, to the Mississippi River at Cairo, Illinois. Louisville is located approximately 600 miles downstream from Pittsburgh (Parrish, 2024). More than 25 million people, or 10 percent of the U.S. population, live in the Ohio River Basin. The river is an important highway for barge traffic. Over 250 million tons of cargo (35 percent of the country's inland waterborne commerce) are transported on the river each year. This is equivalent to the cargo transported through the Panama Canal annually. Coal, oil, gas, and other energy products make up approximately 70 percent of the commerce traveling by barge (Parrish, 2024).

Almost 25 percent of the cargo on the Ohio River passes through Louisville (20 tows a day, 57 million tons, value = $11.7 billion). In 1830, the Louisville and Portland Canal Company completed a 1.9-mile-long canal on the Kentucky side of the river to bypass the Falls of the Ohio. This allowed boats to float around the falls, avoiding the need to unload and reload in Louisville. In 1927, the McAlpine Dam was constructed at Louisville (Parrish, 2024). The dam is 8,627 feet long, the longest on the Ohio River. Its primary function is to keep the water level in the river a minimum of nine feet throughout the year, to allow the larger boats to use the river. The backwater from the dam stretches upriver for 75 miles to the next dam at Markland, Indiana (there are 20 locks and dams between Pittsburgh and the Mississippi River). Various locks were constructed to allow boats to pass the falls dating back to the early nineteenth century through 1921. The current lock, constructed in 1961, is 1,200 feet long and 110 feet wide and can raise or lower a boat 37 feet (Parrish, 2024; Charles, 2024).

The walls of the lock are 80 feet high, and the entrance and exit are controlled through two 320-foot steel gates. Culverts 16 by 18 feet allow water to flow into the lock to raise boats and to drain the lock to lower boats. Each foot raised or lowered requires 1,000,000 gallons of water, and the lock can be filled or emptied in about 13 minutes. Barges provide an economic means of transporting material. The number of miles one ton of cargo can be transported per gallon of fuel is 514 miles by barge, 202 miles by rail, and 59 miles by truck. One 15-barge tow can transport 22,500 tons of coal. It would take 225 railroad cars or 900 trucks to transport the same amount. The construction of a second 1,200-foot lock was a $420 million project.

From a geological standpoint, the Ohio River is young. The river formed on a piecemeal basis beginning between 2.5 and 3 million years ago. The earliest ice ages occurred at this time and dammed portions of north-flowing rivers. The Teays River was the largest of these rivers.

Approximately 300,000 years ago, a glacier of the Pleistocene Ice Age advanced southward and blocked the Teays River, forming a large lake. The lake eventually overflowed, carving through the separating hills and beginning a new route for the river. The large new river subsequently drained glacial lakes and melted glaciers at the end of the Ice Age. The valley grew during and following the Ice Age. The middle Ohio River formed in a manner similar to the formation of the upper Ohio River. A north-flowing river was temporarily dammed southwest of present-day Louisville, creating a large lake until the dam burst. A new route was carved to the Mississippi. The cycle of damming and overflowing occurred repeatedly as the glaciers pressed south, ultimately forming what is now called the Deep Stage Ohio River. The process was repeated all over again as subsequent glaciers covered the region, most notably the Wisconsin Glacier, which gradually changed the river's path about 100,000 years ago to what we see today.

The Ohio River has long been both Louisville's source of drinking water and the means of disposing of its sewage and waste. As cities grew in the basin (Pittsburgh, Cincinnati, Huntington, Ashland, and Evansville), its waters became very polluted. Aquatic life in the river was destroyed. In 1948, the eight states in the basin entered a compact to work together to clean up the river. Water quality standards were established jointly by the states and federal government through the Ohio River Valley Water Sanitation Commission, commonly referred to as ORSANCO. One of the first priorities was to have the cities treat their waste prior to discharging it into the river. Louisville was required to construct a plant that treats its sewage in 1958. Today all the cities and industries along the river treat their sewage and waste, and water quality has dramatically improved. In 1989, no portion of the main stem of the Ohio River met water quality standards. Today almost 30 percent of the river does meet water quality standards. But the river is far from clean. Bacteria levels are still extremely high. The entire river is posted for "No Swimming" due to health concerns about the level of bacteria from sewers, septic tanks, and animal wastes. Kentucky has posted the river for "No Fishing" for certain types of fish (carp, catfish, white bass, and paddlefish), due to concern about potential pollutants. Pollutants may bioaccumulate in fish as they feed on smaller fish, invertebrates, algae, and aquatic plants. Each of these consumes pollutants from the water and sediment (Parrish, 2024).

Why? What is the source of these pollutants? The major source now is from nonpoint sources. These are sources that do not flow through a pipe but rather are from areas where water flows across the surface directly into a stream or river (think of rainwater running off a large parking lot). Groundwater will eventually move into a tributary or into the river. Any contamination that is picked up on the surface or in the soil will eventually be deposited into the river. Researchers are now demonstrating that air pollutants also are a major nonpoint source. Air pollutants washed out of the sky by rain or deposited on the ground and washed into a stream can add to the pollution load. Some of the major pollutants in the Ohio River today that are of concern are pesticides, heavy metals, bacteria, PCBs, and nutrients (nitrogen, phosphorus).

Gallagher Power Plant

The Gallagher Station, located in Floyd County, Indiana, was a two-unit coal-fired generating facility operated by Duke Power Company. The plant's four 140 MW units came online between 1958–1961. The total aggregate capacity of the plant's four identical units was 560 megawatts (enough energy for 200,000 homes). The plant was retired from service in June 2021. At one time, it was planned that the plant would be phased out to be replaced with a nuclear power plant proposed by Indiana Power and Light at Marble Hill, Indiana. The nuclear plant was not built, and the plant remained online despite the fact that it had no scrubber system to reduce air emissions. A bag-house was added in 2008. The prevailing wind is from the southwest, blowing most of the air pollutants from the plant across state lines into Kentucky. The plant is ranked as the 62nd highest emitter of SO_2 in the United States with almost 50,000 tons per year in emissions. New plants are permitted to release 1.2 pounds of sulfur dioxide per million Btu's. Gallagher's emission rate is 3.22 pounds per million Btu's, or more than three times the rate of LG&E plants. The plant releases approximately 7,000 tons of nitrogen oxides. The Jefferson County Air Pollution Control District estimates that 8–9 percent of the nitrogen oxides (a precursor to ozone) in the Louisville metropolitan area are attributable to the plant. New plants are permitted to release 0.15 pounds per million Btu's. Gallagher releases 0.44 pounds. Although closed, Duke Energy projects that final closure (removing structures and closing coal ash piles) will not be completed until 2050. The plant overlooks Shawnee Park, one of the original Olmsted Parks in Louisville (there are a total of 19 parks designed by Olmsted; he also designed the wide parkways linking the parks, which were constructed in 1915).

There are abundant experimental and epidemiological data showing that air toxics from coal-fired facilities, including aldehydes, butadiene, vinyl chloride, and fine particles, have pronounced effects on cardiovascular function and disease. Kentucky's death rate per 100,000 from cardiovascular disease is the eighth highest in the nation at 217.5 deaths from cardiovascular disease per 100,000 population (2021). Cardiovascular diseases affect a large proportion of the human population; consequently, even a small increase in risk could translate into a larger number of deaths than are caused by other diseases such as cancer or asthma. The susceptibility of the heart and car-diovascular tissue to environmental pollutants is underscored by the spate of recent studies showing an association between air particulates and cardio-vascular deaths. Although the mechanisms by which particulates affect heart disease are not entirely known, it is likely that long-term mutagenic changes, which are key steps in carcinogenesis, are also relevant to the development and progression of cardiovascular disease. In addition, pulmonary effects of environmental pollutants could indirectly impair cardiovascular health.

Kentucky appealed to the U.S. EPA in the early 1980s under Section 126 of the Clean Air Act about the interstate pollution allowed by Indiana. The appeal was rejected. In 1999, eight northeastern states complained about interstate pollution from coal-fired power plants in the midwest that pre-vented them from complying with national ambient air quality standards. EPA concurred and has ordered all plants in Indiana and Kentucky to reduce their NOx emissions by 65 percent by 2003. LG&E has proposed installing catalytic converters (total cost $700 million) to meet NOx reduction require-ments. Cynergy plans to meet these reduction requirements at their Gibson Power Plant (3,000 MW) in Indiana and take no action at Gallagher. In March 2000, Gallagher was one of 28 plants in the midwest cited for violations of the Clean Air Act for making improvements without installing required air pollution-control equipment. Cynergy challenged this violation.

Chickasaw Park Lake

This small pond is used by residents for recreational fishing. The lake covers 1.5 acres and is less than six feet in depth. In 1995, the State Environmental Protection Agency tested the fish for dioxins (polychlorinated dibenzo-p-dioxin), a group of chemicals that are some of the most toxic man-made pollutants. Testing showed that the concentration in some fish exceeded standards. All the fish were removed from the pond in 1996. The source of the dioxins is unknown. The water in the pond is from the public water sup-ply system. Potential sources in the area, such as the old Ashland Oil refinery across the road, were removed decades ago. Another potential source could be from fish that were caught in the Ohio River by fishermen and released into the pond (a state violation).

Sediment samples showed that low concentrations of dioxin could be detected (124 to 358 ppt, background is 329 ppt). The pond was restocked in the fall of 1997, and the fish retested in 1998. The retested fish also showed levels above the one-in-a-million risk standard adopted by the state. Funds from the American Rescue Plan are being used for the restoration of the lake. The Olmsted Parks Conservancy is coordinating with Louisville Parks and

Recreation to install a large "butterfly and bee" pollinator meadow in the park as a tribute to Muhammad Ali.

Ford Automobile Assembly Plant

In 1925, Henry Ford constructed this plant on a 22.5-acre site at a cost of $1.5 million. The building was designed by Albert Kahn, who was responsible for almost all the major industrial plants of the Big Three and other auto manufacturers in the United States. His designs provided efficient and practical solutions to a growing industrial environment, and he was one of the first users of reinforced concrete. The plant was designed for 1,000 employees and the production of 400 cars a day. The roof is made of glass to provide natural lighting throughout the facility. Ford Motor Co. also had a plant on the corner of 3rd Ave. and Eastern Parkway (1916). The plant was initially built to make Model T Fords, and then switched in 1927 to make the replacement Model A's. In 1937, the plant was flooded, and after the waters receded, the plant disassembled 325 water-damaged vehicles. During WWII, the plant was converted to military production and assembled 93,389 military jeeps. From 1925 to 1955, the plant produced 1,608,710 vehicles. Ford Motor has three manufacturing facilities in Jefferson County, and more Ford trucks are manufactured in Louisville than any other place in the world.

Morris Forman Wastewater Treatment Plant

The Morris Forman Wastewater Treatment plant was constructed in 1958 and is Metropolitan Sewer District's (MSD) largest treatment plant with a design capacity of 105 million gallons per day. The plant treats an average of 114 million gallons per day. Sewer lines in Louisville date back to the 1850s. The sewer lines carried stormwater and household and industrial waste directly into the Ohio River and Beargrass Creek. In 1946, with Louisville's rapid growth, the Kentucky General Assembly and Louisville Board of Alderman created MSD. In 1948, the eight states in the Ohio River Valley (IL, IN, OH, WV, PA, NY, TN, KY) entered an interstate compact to clean up the river, which had become a 981-mile open sewer. Under the compact, Louisville was required to build a wastewater treatment plant. It was not until 1958 that any treatment was conducted prior to discharge into the river. The cost of the original construction was $12 million ($129 million in 2024). Initially, the plant only provided primary treatment capable of removing 65 percent of the solids in the wastewater, much more than the 45 percent required by state and federal laws at the time. Today, the plant can remove 97 percent of the solids with primary and secondary treatment processes. The secondary treatment process was constructed in 1976 but, due to problems was not able to meet the secondary treatment requirements until 1985. The present connected population to the plant is 495,000, with a daily average flow of 114 million gallons per day and a daily stormflow of 338 million gallons. The plant is located on the lowest point of land in Jefferson County.

The plant uses activated sludge to treat wastewater. Wastewater passes through two complete stages of treatment and is then disinfected with sodium hypochlorite and de-chlorinated prior to discharge into the river. The major operating units of the plant are:

- Preliminary treatment—mechanical screens remove solids, and grit is settled out;
- Primary treatment—4 primary settling tanks (1 million cu. ft.) allow solids to settle to the bottom and floatable materials to rise to the surface;
- BioRoughing towers—wastewater flows over plastic media where biological slime growth is attached; the slime breaks down organic material in the wastewater;
- Secondary aeration batteries—partially treated wastewater is mixed mechanically with pure oxygen (produced on-site) to further decompose organic material;
- Secondary clarification—wastewater flows into 20 clarifiers where solids settle to the bottom, and clear liquid flows to the next process;
- Disinfection—sodium hypochlorite is added to kill pathogens remaining in the water, with a 30-minute contact time.

The plant had a history of creating odors; however, in October 2002, the initial phase of odor controls was installed. The solids processing system turns the sludge into dried biosolid pellets suitable as a fertilizer. Solids removed from wastewater are broken down in four giant oxygen-free anaerobic digesters. The solids from the digesters are dewatered in five centrifuges. Methane gas produced in the digesters is used to generate heat to dry the dewatered sludge. The four dryers each can remove 18,000 pounds of water per hour and are the largest drying trains in use in the United States. Any off-gases are treated in a scrubber to remove particulates and by oxidizers to eliminate odors. The solids processing facility cost $82 million. MSD is currently installing aluminum covers over open areas of the treatment process to capture any odors.

Rubbertown

The petrochemical industrialization in west Louisville began in 1918 with the construction of the 2000 barrel/day Standard Oil of Kentucky Refinery (now the Chevron Terminal and Tank Farm). Over the next two decades, two additional refineries were constructed (Aetna Oil and Louisville Refining), and both were eventually purchased by Ashland Oil. With the outbreak of World War II, the demand for rubber became the impetus for the development of the Complex. A modern nation could not hope to defend itself without rubber (Pastor et al., 2001).

The construction of a military airplane used one-half ton of rubber; a tank needed about one ton, and a battleship required 75 tons. Each person in the military required 32 pounds of rubber for footwear, clothing, and equipment. Tires were needed for all kinds of vehicles and aircraft. Recognizing the critical need for rubber, in June 1940 President Roosevelt formed the Rubber Reserve Company (RRC). The RRC set objectives for stockpiling rubber, conserving the use of rubber in tires by setting speed limits, and collecting scrap rubber for reclamation. Major world sources (90 percent) of natural rubber in Southeast Asia were under Japanese control. The U.S. Office of Production Management eventually built 15 synthetic rubber plants nationwide using German technology. Ultimately, the nation spent as much on its rubber program as it did on the atomic bomb. All the initial plants were constructed under the supervision of the U.S. Office of Production Management (who also considered Sheffield, Alabama, as a potential site). The government either built the plants or purchased them from their original owners, investing $92.4 million in Louisville. The first plant to be built was National Carbide in 1941. The plant manufactured acetylene gas. LG&E constructed the Paddy's Run power plant to supply the electrical needs of National Carbide. Although Paddy's Run is now closed, National Carbide is still the largest single user of LG&E produced power.

The acetylene gas was used as feedstock for a neoprene synthetic rubber plant built by E.I. DuPont de Nemours & Co. that same year. The acetylene was piped over to the DuPont plant. DuPont manufactured vinyl acetylene, which was then chlorinated to produce chloroprene. When polymerized, chloroprene is turned into neoprene, which is a type of synthetic rubber.

B. F. Goodrich also began construction in 1941. It produced a synthetic rubber product called Koroseal, which is made from vinyl chloride. It also piped acetylene gas from National Carbide as feedstock for the manufacture of vinyl chloride monomer. B. F. Goodrich was joined by Phillips Petroleum Corp. and Hycar Chemical Corp. to make various types of "nitrile" rubber from acrylonitrile, butadiene, and styrene (see figures 7.4, 7.5).

With the outbreak of war in December 1941, the federal government confiscated the DuPont plant. DuPont was retained to manage the facility. The federal government constructed a plant to make butadiene from grain alcohol (the plant is now owned by Dow). Grain alcohol was piped in from nearby distilleries on Dixie Highway. In 1943, the federal government opened what is now the American Synthetic Rubber plant to make styrene-butadiene rubber for tires. The federal government had selected this type of rubber to be

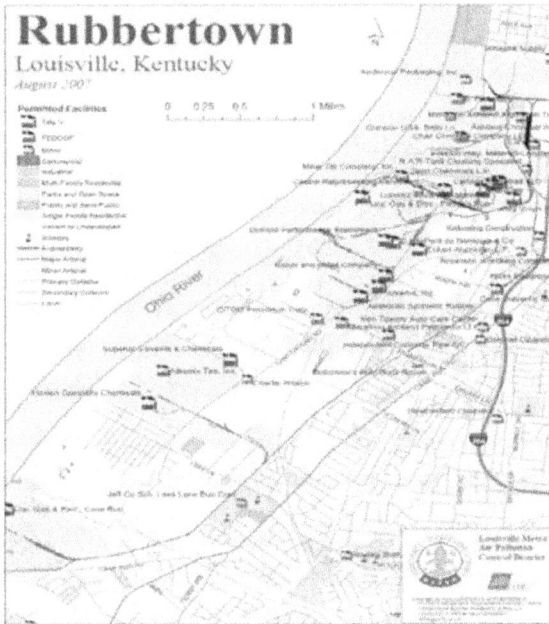

Figure 7.4 Rubbertown. *Source*: Louisville Metro Air Pollution Control District

Figure 7.5 Picture of Rubbertown Next to the Ohio River, including the Water, Soil, and Air. *Source*: Kentucky Institute for the Environment and Sustainable Development

the standard for the Department of Defense, and it is still the principal type of rubber used in tires today. National Synthetic Rubber, a consortium of five tire companies, managed the plant.

In 1941, there were an estimated 38,000 workers employed in Louisville by defense industries. In one year, industrial employment jumped 18 percent over 1940. Housing, schools, hospitals, and public infrastructure were in critical shortage as rural Kentuckians moved to Louisville. Other plants in the metropolitan area included Curtis Wright (airplanes), Naval Ordnance (gun mountings), Westinghouse (naval guns), Howard Shipyards in Jeffersonville (L.S.T. crafts), and the Hoosier Ordnance Plant (munitions). By 1944, the number of industrial employees reached a peak of 80,000. The area around Rubbertown was primarily truck farms in 1940. By the end of the war, the area had changed to residential land uses. Wartime wages and inflation are reflected in bank deposits in Louisville, which soared from $160 million in 1938 to $368 million in 1943. Peak synthetic rubber production was reached in 1944 with 195,000 tons of rubber produced, employing 4,000 workers, and Rubbertown became the world's largest producer of synthetic rubber (21 percent of national production of 920,000 tons) (Rubbertown Community Advisory Council, 2020; Peterson, 2013).

After the war, the federally owned plants were sold back to private operators. DuPont purchased its old plant and land in 1949. American Synthetic Rubber, a consortium of 20 companies, purchased the National Synthetic Rubber plant from the government in 1955 (Rubbertown Community Advisory Council, 2020).

The chief concern from the Rubbertown plants currently focuses on their release of air toxics. The concern over air toxics is not new (Table 7.2). Early public concern resulted in the Rubbertown industries commissioning a survey of the dust problem in West Louisville in 1952. Louisville was studied by the U.S. Public Health Service in 1957–1958 to determine the health impacts from air pollutants. The study showed air releases as high as 22 million pounds per month, much greater than the 5 million pounds annually today. Air toxic monitoring is being conducted through a collaborative effort of the Jefferson County Air Pollution Control District, the West Jefferson County Community Task Force, the University of Louisville, Kentucky Division for Air Quality, Rubbertown Industries, and the U.S. EPA. The partners above selected monitoring station locations, and monitoring was initiated in May 1999. A total of 12 monitoring stations were established to monitor 78 volatile organic compounds (VOCs), 63 semivolatile organic compounds, 20 metals, and 2 reactive aerosols (hydrogen fluoride and hydrochloric acid). The EPA provided monitors for six of the stations and conducted all the analyses from these six stations. After May 2000, only seven monitoring stations were maintained by the University of Louisville. Twenty-four-hour samples are taken once every 12 days (Rubbertown Community Advisory Council, 2020).

The data was analyzed by a contractor (Science International) to develop a risk assessment. The assessment showed that for the 250 chemicals analyzed, none were at concentrations high enough to have acute (immediate) health effects. However, all the monitors showed ambient concentrations that are at levels where cancer risks exceed a probability of 1 in 10,000. For other health impacts, the data showed a slightly elevated risk. The most significant chemicals posing potential health risks are:

Table 7.2 Chemicals Posing Potential Health Risks

Chemical of Concern	Health Effect	10^{-6} Risk
1,3 Butadiene	carcinogen	500
Acrylonitrile	carcinogen	130
Chloroprene	kidney, dermal	110
Chloroform	liver, kidney	77
Chromium	neurotoxin	66
Formaldehyde	carcinogen	46
Perchloroethylene	carcinogen	39
Ethyl acrylate	carcinogen	33
Benzene	carcinogen	32

Source: Center for Sustainable Urban Neighborhoods, formerly University of Louisville, transferred to Neighborhood Associates, Washington D.C.

Carbide Industries

Carbide Industries is a privately owned company that also owns a plant in Calvert City, KY. Originally, the plant had seven electric-arc furnaces and was built for $1 million. It currently operates one 50-megawatt furnace. Carbide Industries manufactures three products: calcium carbide, acetylene, and calcium hydroxide. Calcium carbide is used as an alternative energy source, providing improved furnace efficiency, increased furnace productivity, reduced costs, and a lower carbon footprint for the steel-making process. They manufacture about two rail cars daily of calcium carbide.

The process of producing acetylene is relatively simple and has remained unchanged since 1892: crushed limestone (abundant in central Kentucky) is mixed with petroleum coke (provided from refineries in the Ohio River basin) and heated to 3800°F (using large electric-arc furnaces). This produces calcium carbide ($3C + CaO \rightarrow CaC_2 + CO$) which, when mixed with water, produces acetylene gas. The same chemical process fueled old miners' lamps. Acetylene gas is used as a feedstock for vinyl chloride and as a fuel in metal cutting and welding and for iron and steel desulfurization. Calcium hydroxide is a waste byproduct. It is piped under Bells Lane and disposed of in the 10-acre waste pile on the north side of the road. In 1963, the pile was significantly higher (100 ft). On February 25 of that year, the pile sloughed off, covering Bells Lane and most of the parking lots on the south side of the road. The material was sold to LG&E for use in their air pollution control scrubbers. LG&E sells the material after it has been used to wall board manufacturers as a raw material. A fire in June 2009 damaged the plant, and a fire and explosion in March 2011 killed two people and destroyed the building. The plant has an annual revenue of $5–10 million and has 140 employees (Rubbertown Community Advisory Council, 2020).

Lubrizol, Zeon, and PolyOne

These Louisville plants produce chlorinated polyvinyl chloride (CPVC) resins and compounds, which are used in the manufacture of residential and industrial plumbing systems. The plant was originally owned by B. F. Goodrich (up until 2001). Formulated first in 1835, from 1912 to 1950 vinyl chloride was produced from acetylene and hydrogen chloride using mercuric chloride as a catalyst. Currently, the production of vinyl chloride consists of a series of well-defined steps. 1,2-dichloroethane is prepared by reacting ethylene and chlorine. In the very first study about the dangers of vinyl chloride, published in 1930, it was disclosed that exposure of test animals to just a single short-term high dose of VC caused liver damage. In 1970, Dr. P. L. Viola reported that test animals exposed to 30,000 ppm of VC developed a rare sarcoma of the liver. In 1972, Dr. Cesare Maltoni, another Italian researcher for the European VC industry, found liver tumors (including angiosarcoma) from VC exposures as low as 250 ppm for four hours a day. Dr. John Creech from B. F. Goodrich discovered angiosarcoma in the liver of three workers at the B. F. Goodrich plant in Louisville, Kentucky. To date, 26 former B. F. Goodrich workers have died from this disease. In May of 1974, the Occupational Safety and Health Administration (OSHA) proposed a maximum exposure level for vinyl chloride at no detectable level, using equipment with an accuracy of one part per million.

In 1972, the federal government imposed tighter emission standards and worker safety rules for vinyl chloride after it was linked to a fatal liver cancer associated with workers at the plant. By 1997, the Center for Disease Control and Prevention found that worker exposure was "completely eliminated." The EPA's 2001 updated Toxicological Profile and Summary Health Assessment for VC in its Integrated Risk Information System (IRIS) database lowers the EPA's previous risk factor estimate by a factor of 20 and concludes that

> because of the consistent evidence for liver cancer in all the studies . . . and the weaker association for other sites, it is concluded that the liver is the most sensitive site, and protection against liver cancer will protect against possible cancer induction in other tissues.

The plant also produces several vinyl and acrylic latex emulsions, which are used as coatings in industrial and consumer products. The vinyl chloride monomer used in the plants was formerly was produced in Louisville. The companies now purchase vinyl chloride from Westlake Monomers in Calvert City, bringing it to Louisville by rail. One of the waste products of Lubrizol is hydrochloric acid. Prior to 1998, this waste byproduct was disposed of as hazardous waste. The company is currently transporting this waste to Carbide/Graphite Group, Inc., who uses the acid to treat wastewater from their calcium hydroxide storage piles. This beneficial use of waste reduces costs for both companies while improving the environmental quality within the community.

The Lubrizol Corporation (Cleveland) entered a partnership with two separate companies, Zeon Chemicals (Japan) and PolyOne (Cleveland), who now operate portions of the original plant independently. Zeon Chemicals is engaged in the manufacture of nitrile rubber. The rubber is used in the manufacture of automotive parts, adhesives, plastic modification, wire, and cable parts. The EPA has ruled they must reduce the toxins that are cancer-causing (Giffin, 2023).

Butadiene, acrylonitrile, styrene, and ethyl acrylate are feedstocks used in the manufacture of specialty rubber. PolyOne produces polyvinyl chloride in a powder or pellet form, which is used in the manufacture of vinyl house siding, PVC pipe, vinyl windows, wire, and cable insulation.

DuPont

In 1955, the Louisville plant started manufacturing Freon-22® refrigerant and aerosol propellant. Freon-22 is used as a refrigerant in freezers and air conditioners. The plant is the nation's largest emitter of a climate super-pollutant known as hydrofluorocarbon-23 (HFC-23). As a greenhouse gas, this chemical is 12,400 times more potent than carbon dioxide, the primary chemical compound responsible for warming the planet. The company has been pledging to reduce and eliminate the production of HFC-123 since 1992 but has not reached this goal. In 2015, DuPont spun off Chemours, and the new company took over the Louisville plant's fluorochemical production. Louisville Mayor Greg Fischer declared a climate emergency in 2019, telling a youth climate strike gathering that "we must take action now." However, the Jefferson County Air Pollution Control District has not pressed the company to eliminate the HFC-22 emissions, deferring instead to the state or federal government. The production and use of HFC-23 were banned in the United States and other developed countries on January 1, 2020, under an international agreement known as the Montreal Protocol. However, Chemours is exempt from the ban because the HCFC-22 produced in Louisville is used as a feedstock to manufacture Teflon and other fluoropolymers that do not damage the earth's protective ozone layer.

Chemours vented 251 tons of HFC-23 in 2019. The emissions are equal to the annual greenhouse gas emissions of 671,000 automobiles based on a national average for annual vehicle miles driven (https://www.epa.gov/energy/greenhouse-gas-equivalencies-calculator). That eclipses the 519,000 cars and light-duty trucks currently registered in Louisville. Giffin (2023) reported that the EPA found Chemours to be one of Louisville's worst polluters.

In addition to refrigerants, DuPont Fluoroproducts still produces Freon 23, DFE, vinyl fluoride, and hydrochloric acid. Freon 23 is used as an electronic gas and a fire-extinguishing agent. DFE is used in aerosols and as a blowing agent for foams. Vinyl fluoride is used in the manufacture of a plastic (Tedlar) whose properties include excellent resistance to weathering, outstanding mechanical properties, and inertness toward a wide variety of chemicals, solvents, and staining agents. It is used in the manufacture of aircraft, cars, graphic signs, and a variety of other items. It is also used in the PV Solar industry. DuPont estimates that this market will expand 50 percent in each of the next 5 years and has announced it will invest up to $178 million to double production. One of the byproducts in the manufacture of Freon is hydrochloric acid. Prior to 1992, the plant disposed of this "waste" byproduct in two underground injection wells. In 1992, almost 30 million pounds of acid were disposed of in this manner. Since then, the company has found markets to sell the acid for beneficial reuse. McKenna and Bruggers (2021) found that this company creates greenhouse emissions that are equal to 650,000 cars circulating around Louisville in one year. It is for this reason that when you total the amount of greenhouse gas emissions from the 44 chemical companies, Louisville produces more greenhouse gases per capita than any other city in the United States. This fact came from the head of the Air Pollution Control District in a 2024 Spring lecture seminar on Environmental Management at the University of Louisville.

The DuPont plant in the 1960s employed 2,400 workers and was one of Louisville's largest manufacturers. That number bottomed out in 2009 to 180 workers. The plant today is not even in the top 40 largest manufacturers in Louisville.

Dow Chemical, Arkema

This site is now owned by Dow Chemical. They make PARALOID™ plastic additives to increase the high impact strength and natural weathering resistance of PVC (used in window frames); PARALOID™ B thermoplastic acrylic resins for coating applications on metal and concrete; and crude methyl methacrylate primarily used in the manufacture of acrylic plastics (plexiglass), toner inks, printing inks, non-yellowing finishes, and general finishes. Although the plant in the 1990s employed over 800 employees, currently the company has 150 employees.

The main chemical feedstock used at the plant is methyl methacrylate, which is brought by barge from their plant in Texas. Arkema purchased the patent for Plexiglass in July 1998 from Rohm and Haas, who owned this plant since 1960. The plant manufactures acrylic resin. The resin is used in taillight lenses, backlit signage, lighting applications, dashboard panels, medical equipment, videodisks, and in the manufacture of sheets of Plexiglass®. The company has 50 employees.

American Synthetic Rubber Corp.

Consortiums of rubber, chemical, and tire companies have run this plant for most of its history. It is now wholly owned by Michelin Tire Co. (France). Its management and operation have improved dramatically as a result. The plant produces synthetic rubber that is sold in two forms: in a 75-pound bale and in a liquid form that is shipped out in tank trucks. The bales (styrene-butadiene rubber) are shipped to tire manufacturing plants where they are mixed with other raw materials such as carbon black. The liquid rubber (polybutadiene-acrylonitrile-acrylic acid polymer) is shipped to Thiokol Space Operations, where it is used as the binder for the solid fuel in the booster rockets for the Space Shuttle program. In December 2005, the company installed a $3 million thermal oxidizer to destroy 99.99 percent of the 1,3 Butadiene emissions from the plant. Data for 2006 shows an 80 percent reduction in ambient 1,3 Butadiene levels in west Louisville. The EPA has identified American Synthetic Rubber as one of the deadliest polluters causing cancer. (Giffin, 2023; Kleber, 2000)

Lake Dreamland

Originally a dairy farm until 1931, lots for summer cottages were leased by the owner, Ed Hartlage, to wealthy Louisvillians. Located on the Ohio River and a lake created by a dam on Bramer's Run, the area was a pleasant retreat from the city. Visitors would swim, fish, and boat on the lake or ride horses along the shore. The dairy barn was converted into a dancehall for entertainment. The Ohio River flood of 1937 devastated the community. A floodwall cut through the development, and the remaining cottages were soon in poor repair. The neighborhood lacked public water, electricity, and paved roads. The construction of nearby chemical plants and their air emissions ended the era as a resort.

After World War II, surrounded by chemical plants, the abandoned cottages were offered for sale or lease. Of the 120 lots, only 5 were sold. The rest were leased for a nominal fee. Since the residents did not own the property, they were unable to obtain loans or convince the government to provide utilities or paved roads. The homes could deteriorate. The dancehall was converted in the 1950s into a nightclub, El Rancho, which was a popular club. It featured a new type of music (rock and roll), but by the 1960s was taken over by motorcycle gangs and burned down in 1967. Mr. Hartlage died in 1980, and his will specified that the property be sold. In 1987, the city purchased the land from his estate. With year-round residents, the city decided to sell individual lots for $1. Although the county will deed the land over to the residents, it has also imposed a death sentence of sorts on the community. Under county plans, current tenants will own their land, but the deeds will stipulate that when the current owner or his family leaves, the land will revert to the county. The county plans to convert the community into a land trust.

Residents still have infrastructure problems in that they have no access to wastewater treatment since they are outside the floodwall. Sewage is disposed of in "constructed pits." The pits are filled with rock, and sewage is "filtered" through the rock before flowing into groundwater and eventually into the lake or Ohio River. Visible on the lake is a thick layer of green algae from nutrient loading. The city, in 1997–1998 initiated a program to install aerated septic tanks to manage wastewater. The cost per household was estimated at $6,500, but costs, overruns, and loss of federal funding ended the program after only a few homes received new systems.

Stauffer Chemical

From 1953 to 1983, this plant produced chlorinated solvents, chloroform, and hydrochloric acid. When it was built, the $3 million plant was viewed as an economic boom for the community. The plant manufactured chloroform, which is manufactured by the chlorination of methane in a process that can be made to yield varying proportions of methyl chloride, methylene chloride, chloroform, and carbon tetrachloride. Chloroform is a probable human carcinogen. The site has extensive soil and groundwater contamination. The contaminant's half-life is estimated to be 100–200 years. It is now owned by ICI Americas and its affiliate Atkemix Ten. In 1961, a cloud of acrid gas floated over Lake Dreamland, resulting in the evacuation of 1,000 residents.

Momentive Specialty Chemicals

Momentive Specialty Chemicals (Columbus, Ohio) is the world's largest producer of binder, adhesive, coating, and ink resins for industrial applications. The plant was constructed to produce formaldehyde and phenolic resins in 1979. These products are used in a wide variety of automotive, foundry, adhesive, and wood manufacturing plants. Formaldehyde is used in the manufacture of herbicides and fungicides; fabric softeners; oil and gas applications; particleboard, plywood, and shingles; slow-release fertilizers, melamine formaldehyde (MF), and phenol formaldehyde (PF) resins; and spandex fiber. The primary feedstock is methanol. The facility is the largest (2.6 million pounds of formaldehyde annually) and most modern foundry resin production facility in the world. The facility agreed in February 2007, in response to a class action suit filed by residents in the community, to a $52 million settlement. The settlement includes payment to adjacent landowners for property devaluation, construction of a berm between loading docks and adjacent neighbors, and internal controls to reduce emissions. Momentive has a total of 223 employees on site and an annual payroll of $17 million per year.

Lee's Lane Superfund Site

Lee's Lane Landfill is a 112-acre landfill and junkyard that lies in the Ohio River floodplain (U.S. Environmental Protection Agency, 2018). Portions of the site flood every year. The site was originally a sand and gravel quarry, with a pit over 120 feet deep. From the 1940s to 1975, the site was operated as a landfill by Joseph Hofgesang and received over 2 million cubic yards of domestic, commercial, and industrial waste (estimated to be 212,400 tons). The landfill is located on 100 feet of porous alluvium. Waste is in direct contact with groundwater, and pollutants flow downgradient to the Ohio River. In 1975, residents living next to the site reported "blue sheets of flame" around their hot water heaters. Explosive levels of methane gas from the landfill were detected. Seven homes were evacuated and purchased by local authorities. The state closed the landfill the same year. In 1980, the state discovered 400 exposed drums of hazardous materials on the riverbank. They identified more than 50 chemicals, including phenolic resins, benzene, and a variety of heavy metals. Groundwater, soil, and surface water were contaminated with benzene, heavy metals including lead and arsenic, and inorganic chemicals. In 1982, the site was listed on the National Priority List for Superfund (Gilderbloom et al., 2013; Swanston, 2012).

Over $2.2 million was expended to clean up the surface of the site, and it has been delisted for any further Superfund action. The property is owned by the Hofgesang Foundation (Joseph Hofgesang died in 1972), which settled with U.S. EPA to pay $2.6 million to clean up the site. The presence of methane and other toxic gases was detected in the residential neighborhood east of the site.

An underground collection system was designed to vent the volatile gas safely so it would no longer seep into neighboring homes and yards. A cap was placed on the landfills and monitoring wells around the site were installed. A 2004 engineering consultant firm said the monitoring and gas collection system was beyond the typical useful life and needed to be updated.

MSD has since 1991 maintained the site and conducted monitoring under a 29-year agreement (to 2020) with the U.S. EPA, which alleged at the time that the sewer district was one of the parties dumping toxic waste there. The cost of fixing the methane gas collection systems was estimated to be more than $300,000. MSD took the case to court in 2007 but agreed to budget $350,000 to fix the system (U.S. Environmental Protection Agency, 2018). The site is largely abandoned although the city and other groups have studied proposals to reclaim the site into a solar farm, recreational park, or other uses. To date, no firm proposal has been submitted.

Railroad Transportation Loading Yard

The most dangerous and deadly area in Louisville, and perhaps in the United States, is the railroad transportation loading yard where highly toxic chemicals are stored in the Rubbertown area for days or weeks. If not properly guarded, industrial sabotage, terrorism, a drone attack, or a train derailment could be deadly. If someone with a hammer or brick released the toxins, it could potentially cause thousands of fatalities immediately and even more deaths long term, according to Russ Barnett (the lead author who headed up the Kentucky Institute for the Environment and Sustainable Development and before that was a top officer in the Kentucky Natural Resources—Energy and Environmental Cabinet), and extensively written about by James Bruggers in the Courier Journal. Bruggers is now with Inside Climate News (www .insideclimatenews.org). An incident of sabotage would also significantly reduce our war readiness.

The death toll could potentially surpass that of the 1984 Bhopal, India disaster in which at least 10,000 people died instantly from a chemical accident. The East Palestine train derailment caused the instant death of 40,000 animals, from fish to birds. When the second co-author (Gilderbloom) went to the West Wing for a visit with former Mayor Jerry Abramson and then later Director of Intergovernmental Affairs at the Obama White House in the Spring of 2015, I complained that anybody could walk the unguarded rail yard and cause mayhem with a hammer. Abramson responded that they were aware of this concern and hoped it didn't happen. Since then, we have raised these concerns with two other Louisville mayors (Fischer and Greenberg), and no action has been taken to put in security measures such as high fences, security cameras, or a police presence. I have talked to numerous security officials who claim this could be America's next 9/11. As the map of the United Sstates in chapter 19 shows, there are 1,000 other neighborhoods throughout the United States. that are vulnerable to these devastating threats that could end up killing people, places, and the planet. Hopefully, citizens who read this book or see the documentary will demand that our elected leaders take action.

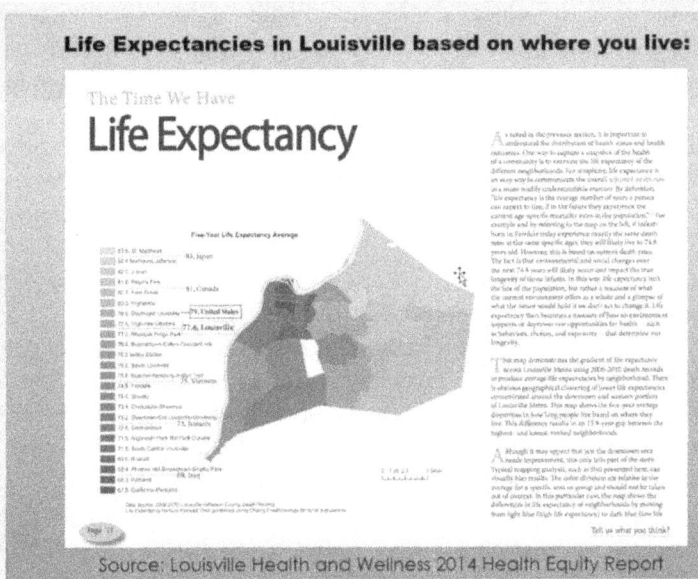

Figure 7.6 Life Expectancies in Louisville. *Source*: Louisville Health and Wellness 2014 Health Equity Report

CONCLUSION

Every city should have an environmental audit by neighborhood to expose the inequalities in the air, water, and soil. The most affected are those living near industrial pollution in poor and minority neighborhoods with shortened lifespans, lower proficiency scores, abandoned housing, significantly reduced housing equity, and higher crime (Gilderbloom, 2016; Samet et al., 2000; Gilderbloom & Squires, 2022; Wong, 2013). Chetty et al. (2016) dismiss the idea that pollution has any negative impacts and argue instead that lifespan is mostly associated with geography and income. However, when you use his own data and insert pollution measures, our data analysis of his data proves a strong correlation between pollution and lifespan (Figure 7.6). The proof is even in Chetty's maps published in *The New York Times* and *Journal of the American Medical Association*. Lifespan is much shorter in areas with high rates of pollution. Another great source of data is from the EPA, which we use throughout this book (US EPA, 2015a, 2015b).

This environmental audit was published on our University of Louisville website and created an uproar. In the words of Rachel Carson, who wrote *Silent Spring* in 1961 and was posthumously awarded the Presidential Medal of Freedom, "The people have the right to know." The EPA responded by telling the city to reduce the 75,000 pounds of toxic pollution by 6,500 pounds. This was one small step in recognizing the problem of chemical toxins as the biggest problem facing low-income/minority neighborhoods (Giffin, 2023). The U.S. Department of Justice cited the toxic neighborhoods in West Louisville as an underlying cause in the urban uprising that took place in 2020–2021 after the death of Breonna Taylor (U.S. Department of Justice, 2023). The uprising resulted in the devastating destruction of the downtown area and several other deaths.

There is growing evidence that lawsuits might be a way to prevent chemical toxins from harming people or depriving them of a full life. Louisville-based lawyer Jennifer Moore led a successful fight against Roundup weed killer on behalf of people sickened by it. Similarly, legal action was taken on behalf of those who became ill after being exposed to contaminated water at Camp Lejeune. Some environmental officials believe that a similar kind of lawsuit might force the government to take tough environmental actions in Louisville.

With 44 chemical companies, numerous liquor distilleries, and a coal-fired plant, Louisville is like a toxic wasteland. This is why companies like to relocate there because with numerous companies in the area, it is difficult to pinpoint which chemical is responsible for causing a shortened life or poor grade school performance. It cannot be considered healthy for 60,000 West Louisville workers, students, and residents to be exposed to 75,000 pounds of environmental toxins. People who live in East Louisville, where the air is cleaner, live 12 years longer (Gilderbloom et al., 2022a; Gilderbloom, 2016; Gilderbloom et al., 2021a, 2021b; City of Louisville, 2017). Moreover, in a study of 144 mid-size cities, Louisville ranked number two for having the most toxic air (Gilderbloom et al., 2022c; Gilderbloom, 2022b). This toxic air not only reduced lifespan but had a major impact on the health of public schoolchildren's proficiency scores and their health, causing them to have high absenteeism. According to CDC data, the toxic air is also correlated with higher rates of COVID-19, cancer, liver disease, and asthma, which we document throughout this book (Gilderbloom et al., 2021a, 2021b; McKenna & Bruggers, 2021).

REFERENCES

Atkinson, M. C., director. (1996). *Beargrass—The Creek in Our Backyard.* Documentary Kentucky Educational Television. Beargrass—The Creek in Our Backyard | KET.

Bourbon Stock Yards Company. (2024a). "Pigs in a stock yard." *The Filson Historical Society Digital Projects*, accessed July 10, 2024. https://filsonhistorical.omeka.net/items/show/897.

Bourbon Stock Yards Company. (2024b). "Records, 1880–1995." *The Filson Histori-cal Society Digital Projects.* University of Louisville Archives. Created by Bour-bon Stockyards Company. 1887–1962. Retrieved January 2018, from Kentucky Institute for the Environment & Sustainable Development.

Bullard, R. D. (1990). *Dumping in Dixie: Race, Class, and Environmental Quality, Vol. 3.* Boulder, CO: Westview Press.

Bullard, R. D., ed. (2000a) [1990]. *Dumping in Dixie: Race, Class, and Environmen-tal Quality*, 3rd ed. Boulder, CO: Westview Press.

Bullard, R. D. (2000b). *Dumping in Dixie: Race, Class, and Environmental Quality*, rev. ed. Boulder, CO: Westview Press.

Bullard, R. D., Mohai, P., Saha, R., & Wright, B. (2008). Toxic wastes and race at twenty: Why race still matters after all of these years. *Environmental Law*, 38(2), 371–411.

Čapek, S. (1992). Environmental justice, regulation, and the local community. *Inter-national Journal of Health Services*, 22(4), 729–746.

Čapek, S. (1993). The "environmental justice" frame: A conceptual discussion and application. *Social Problems*, 40(1), 5–24.

Capek, S., & Gilderbloom, J. (1992). *Community Versus Commodity: Tenants and the American City.* Albany, NY: State University of New York Press.

Chetty, R., Stepner, M., Abraham, S., Lin, S., Scuderi, B., Turner, N., Bergeron, A., & Cutler, D. (2016). The association between income and life expectancy in the United States, 2001–2014. *Journal of the American Medical Association*, 315(16), 1750–1766. https://doi.org/10.1001/jama.2016.4226.

City of Louisville. (2017). Health equity report. *LouisvillKY*, 2017. Retrieved Octo-ber 8, 2021, from https://louisvilleky.gov/government/center-health-equity/health -equity-report.

Giffin, C. (2023). Louisville chemical plants provide air monitoring in West End. *Courier Journal*, April 25, 2023.

Gilderbloom, J. I. H. (2016). The causes of shorter life expectancies in America. *The New York Times*, April 18, 2016.

Gilderbloom, J. I. H., Kingsberry, I., & Squires, G. D. (2021a). How many more children must be hurt by pollution? *HARVARD MEDICAL SCHOOL PRIMARY CARE REVIEW*, March 3, 2021. https://info.primarycare.hms.harvard.edu/review/ children-hurt-pollution.

Gilderbloom, J. I. H., Meares, W. L., & Riggs, W. (2013). *How Superfund Sites Kill Places and People: An Examination of Neighborhood Housing values, Foreclo-sures, and Lifespan.* Institute of Urban Affairs, University of Louisville.

Gilderbloom, J. I. H., Meares, W. L., & Riggs, W. (2013). *How Superfund Sites Kill Places and People: An Examination of Neighborhood Housing values, Foreclo-sures, and Lifespan.* Institute of Urban Affairs, University of Louisville.

Gilderbloom, J. I. H., Meares, W. L., & Squires, G. D. (2022a). Pollution, place, and premature death: Evidence from a mid-sized city. *Local Environment: The Inter-national Journal of Justice and Sustainability*, 25(6), 419–432. https://doi.org/10 .1080/13549839.2020.1754776.

Gilderbloom, J. I. H., Reed, L., Turner, D., Brazley, M., & Squires, G. D. (2021b). Pollution is a form of racial injustice crippling western Louisville. *Courier Journal*, January 28, 2021. https://www.courier-journal.com/story/opinion/2021/01/28/louisville-air-quality-west-end-suffers-while-others-prosper/6558314002/.

Gilderbloom, J. I. H., Riggs, W., Frederick, C., Squires, G., & Quenichet, K. (2022b). The missing link of air pollution: A closer look at the association between place and life expectancy in 146 mid-sized cities. *International Journal of Strategic Energy and Environmental Planning*, 4(4), 31–51.

Gilderbloom, J. I. H., & Squires, G. (2022). Put a mask on pollution: Connect the dots between COVID-19 and unhealthy air. *Social Policy*, 51(4), 15–18.

Gilderbloom, J. I. H., Squires, G., & Meares, W. (2020a). Mama I can't breathe: Louisville's dirty air has steep medical costs. *Local Environment: The International Journal of Justice and Sustainability*, 619–626. https://doi.org/10.1080/13549839.2020.1789570.

Gilderbloom, J. I. H., Washington, C. B., Quenichet, K., Manella, C., Dwenger, C., Slaten, E., Sarr, S., Altaf, S., & Frederick, C. (2020b). What cities are the most dangerous to your health? Ranking the most polluted mid-size cities in the United States. Pre-prints with *The Lancet*, January 9, 2020. https://papers.ssrn.com/sol3/papers.cfm?abstract_id=3506217.

Gilderbloom, J. I. H., & Wright, M. (1993). Empowerment strategies for low-income African American neighborhoods. *Harvard Journal of African American Public Policy*, 2, 77–95.

Hamilton, J. (1995). Testing for environmental racism: Prejudice, profits, political power? *Journal of Policy Analysis and Management,* 14(1), 107–132.

Hipp, J. R., & Lakon, C. M. (2010). Social disparities in health: Disproportionate toxicity proximity in minority communities over a decade. *Health and Place*, 16(4), 674–683. https://doi.org/10.1016/j.healthplace.2010.02.005.

Kleber, J. E., editor. (2000). *The Encyclopedia of Louisville*. Lexington, KY: University of Kentucky Press.

Lancet. (2013). Editorial. Action on ambient air pollution. *The Lancet*, 382(9897), 1000. https://doi.org/10.1016/S0140-6736(13)61960-1.

Louisville Metropolitan Sewer District. (2020). History of the organization, MSD history. *Louisville MSD*. Retrieved January 2018, from Kentucky Institute for the Environment & Sustainable Development.

Marshall, A. (2019). The dark side of the "angel's" share. *Gastro Obscura,* October 30, 2019. www.atlasobscura.com/articles/what-is-whiskey-fungus.

McKenna, P., & Bruggers, J. (2021). Louisville's super-polluting chemical plant emits not 1, but 2 potent greenhouse gases. *Inside Climate News,* April 5, 2021. https://insideclimatenews.org/news/05042021/chemours-lousiville-super-polluting-chemical-plant-greenhouse-gas-emissions/.

Meares, W., Gilderbloom, J. H., Squires, G., & Jones, A. (2021). "Pollution and the pandemic: Explaining the differences in COVID-19 rates across 146 U.S. communities. *International Journal of Strategic Energy and Environmental Planning,* 4(2), 9–25.

Parrish, C. E. (2024). McAlpine locks and dams at the falls of the Ohio. *U.S. Army Corps of Engineers*, January 2018. www.lrl.usace.army/Portals/64/dpcs/Ops/Navigation/Booklet_McAlpine.pdf.

Pastor, M., Sadd, J., & Hipp, J. (2001). Which came first? Toxic facilities, minority move-in, and environmental justice. *Journal of Urban Affairs*, 23(1), 1–21.

Peterson, E. (2013). Rubbertown and health: The whole series. *Louisville Public Media*, January 27, 2013. Retrieved from http://wfpl.org/post/Rubbertown-and-health-whole-series-0.

Rubbertown Community Advisory Council. (2020). Rubbertown history. *RCAC*, January 2020. https://www.rubbertowncac.org/rubbertown-history.

Samet, J. M., Dominici, F., Curriero, F. C., Coursac, I., & Zeger, S. L. (2000). Fine particulate air pollution and mortality in 20 U.S. cities, 1987–1994. *New England Journal of Medicine, 343*, 1742–1749. https://doi.org/10.1056/NEJM200012143432401.

Swanston, S. F. (2012). An environmental justice perspective on superfund reauthorization. *Journal of Civil Rights and Economic Development*, 9(2), 18.

US EPA. (2015a). 2011 national emissions inventory data. *U.S. Environmental Protection Agency*, June 3, 2015. Retrieved October 8, 2021, from https://www.epa.gov/air-emissions-inventories/previous-nei-reports.

US EPA. (2015b). Overview of greenhouse gases. *U.S. Environmental Protection Agency*, December 23, 2015. Retrieved October 8, 2021, from https://www.epa.gov/ghgemissions/overview-greenhouse-gases.

US EPA. (2018a). Regulatory impact analysis for the proposed emission guidelines for greenhouse gas emissions from existing electric utility generating units; revisions to emission guideline implementing regulations; revisions to new source review program (Report no. 452/R-18-006), January 2018. https://www.epa.gov/sites/production/files/201808/documents/utilities_ria_pro-posed_ace_2018-08.pdf.

US EPA. (2018b). Sixth five year review report for lees lane landfill superfund site, Jefferson County Kentucky. *U.S. Environmental Protection Agency*, August 8, 2018.

U.S. Department of Justice Civil Rights Division and United States Office Western District of Kentucky Civil Rights Division. (2023). Investigation of the Louisville Metro Police Department and Louisville Metro Government. *United States Department of Justice*, March 8, 2023.

Wong, E. (2013). Air pollution linked to 1.2 million premature deaths in China. *The New York Times*, April 1, 2013. Retrieved October 8, 2021, from https://www.nytimes.com/2013/04/02/world/asia/air-pollution-linked-to-1-2-million-deaths-in-china.html#:~:text=BEIJING%20E2%80%94%20Outdoor%20air%20pollution%20contributed,leading%20causes%20of%20death%20worldwide.

Chapter 8

"Mama, I Can't Breathe" Louisville's Dirty Air Has Steep Medical and Economic Costs

John Hans Gilderbloom, Gregory D. Squires, Bunny Hayes, and Wesley L. Meares

"Mama, I can't breathe." These were the last words of George Floyd. There are 60,120 people living and working in West Louisville who have trouble breathing. And like George Floyd, their lives are cut short by an average of ten years. On the evening of May 28, five people were shot, and tear gas was used to disperse a crowd protesting the shooting death of Breonna Taylor, an EMT worker killed by police on March 13 in a botched drug raid. They marched, chanting "Whose streets? Our streets." Thousands of Black and White protestors experienced the agony of being gassed by the police. It was painful, sickening, and scary. However, citizens in West Louisville are regularly "gassed" by pollution, causing long-term health problems. A disproportionate share of the community's folks living in West Louisville are gassed regularly, and dying prematurely, and most of them are African Americans. As "systemic racism" is addressed in Louisville, we must look at pollution in Black neighborhoods and the institutions that profit from it. West Louisville's air, water, and soil are toxic and rank among the worst in American cities. In the words of Robert Bullard, the founder of the environmental justice movement:

> The communities where police are shooting Blacks are the same communities with high asthma rates, greater incidences of diabetes and strokes, more poverty, and more deaths from COVID-19. Forty years ago, people were saying "we can't breathe, we are choking, you are killing us." We are seeing across the board people are talking about dismantling this violent system of racism not just

when a police officer kneels and chokes a person to death. It's violence when you have all this pollution pumped into a neighborhood and people are choking. (Bruggers, 2020)

West Louisville is unlivable. No wonder Louisville now has over 5,000 abandoned housing units, more than any other mid-size city in the United States. Most of the abandoned homes are in West Louisville, where Muhammad Ali grew up. In an article in the *Journal of Urban Affairs*, Gilderbloom and Meares (2021) found that cities with high levels of pollution have lower rents. And in Louisville, the average price of a home is $37,500 for a typical bungalow in West Louisville. Block after block, West Louisville looks like a war zone of bombed-out buildings and empty lots where houses once stood.

The warnings had been stated; it is not a matter of if the next pandemic will come, but rather when. Public health officials have been warning us about the next big health crisis for years (Luthra, 2020). Then it came. COVID-19 has been a global pandemic the likes of which we have not seen in almost a century. However, some areas have been hit harder than others. Minority and low-income neighborhoods, such as the neighborhoods in West Louisville, have been hit harder (Brooks, 2020; Hauck et al., 2020).

The damage of unregulated pollution has been great; worldwide, according to Lancet, nine million people—mostly people of color—die each year (Landrigan et al., 2018). In the United States, death by pollution occurs primarily in poor minority neighborhoods. Louisville might be the deadliest city in the United States.

However, what if this is not the only health crisis facing West Louisville? Another large public health crisis has been brewing in the background for years. We have heard about it and even read about it, but the explanations for the crisis have focused almost exclusively on the choices of individuals (e.g., smoking, diet, exercise, or lack thereof) rather than the primary direct contributing factors (Ungar, 2014). We are, of course talking about health disparities attributable to environmental degradation among neighborhoods in cities. In Louisville, there is significant disparity in life expectancy, where residents of West Louisville have a life expectancy roughly 10 years shorter than residents in other areas of the city (Gilderbloom, Meares, and Squires, 2020; Louisville Metro Department of Public Health and Wellness, 2014). The differences can be as high as 13 years. One City of Louisville-sponsored study found that the life expectancy in two West Louisville neighborhoods was lower than war-torn Iraq (Greater Louisville Project, 2013; Smith et al., 2011).

The impact of pollution is colorblind in West Louisville. While it is true that the majority of people who die prematurely in West Louisville are Black, one nearly all-White neighborhood near the pollution plants has a comparably reduced lifespan. In fact, the mostly all-White neighborhood in Portland is tied with the all-Black California neighborhood for having a reduction in lifespan of 13 years.

While we do not argue against the importance of genetics and personal choices in terms of the life expectancy gap, we have found that the environment (the soil, water, and air) has a significant impact on health inequality, a fact denied by local leaders. Whether or not there is a political will to admit that proximity to pollutants impacts our health and life outcomes, the environment in which we live matters.

How do we know that Louisville has the worst air quality of any mid-size city? The EPA provides the most valid and reliable measures of poor air quality. We found that Louisville ranks number one when you average out the four EPA measures of pollution, as we did in our analysis of air quality in 146 similar size cities.

A major reason for having the worst air pollution is the inordinate amount of toxic air coming out of West Louisville industries, particularly the chemical plants in an East Side neighborhood known as Rubbertown. For example, the Centers for Disease Control found that residents in neighborhoods near the chemical industries, compared to those in the East Side clean air neighborhoods 17 miles away, are twice as likely to have asthma and high blood pressure, four times more likely to have chronic obstructive pulmonary disease (COPD), seven times more likely to have heart disease, and four times more likely to have poor physical health.

Another stark measure is the comparison between West Louisville neighborhoods near Rubbertown with United States averages and the five cities with the cleanest air in the United States. Like figure 8.1A, it appears that those who live nearest chemical plants have the worst medical problems compared to U.S. national averages or America's cleanest cities like Santa Barbara, Evansville, IN, or Columbus, GA. These differences are stark and vivid.

Scientists widely agree that pollution is a source of harm to the heart, lungs, and brain and a major preexisting condition for contracting COVID-19. Maybe, "sheltering in place" is not such a good idea if you live in West Louisville, especially when doctors are now telling patients to get out of West Louisville if they live, go to school, or work there.

However, the problems were not just chemical plants spewing out toxic emissions. We also found brownfields, wherever they are located, are also a danger to the health of residents (Gilderbloom et al., 2020). We found that one of the major causes of premature deaths was proximity to Rubbertown's toxic contaminants, an effect that ranked fourth after the risk

John Hans Gilderbloom et al.

WEST VS. EAST LOUISVILLE

Health Outcomes	West	East
Asthma	16.2 2148006-21111003000	8.6 2148006-21111010309
HBP	58.2 2148006-21111012801	27.5 2148006-21111010314
Kidney	6.8 2148006-21111005000	2.1 2148006-21111010314
COPD	17.2 2148006-21111002100	4.2 2148006-211110701
Coronary Heart Disease	15.4 2148006-21111005000	2.6 2148006-211110705
Diabetes	23.7 2148006-21111012801	5.1 2148006-211110701
Mental Health	27.4 2148006-21111003000	8.6 2148006-21111010312
Physical Health	28.2 2148006-21111003000	7.9 2148006-211110701
Tooth Loss	55.1 2148006-21111003000	6.1 2148006-21111010317
Stroke	8.9 2148006-21111005000	1.7 2148006-21111010314
Unhealthy Behaviors		
Smoking	42.2 2148006-21111003000	10.5 2148006-21111007501
Physical Inactivity	55.9 2148006-21111003000	20.9 2148006-211110701
Obesity	55.5 2148006-21111003000	22 2148006-211110701
Sleep <7 Hours	52.5 2148006-21111003000	28.2 2148006-21111007501

2148006-21111003000: North of Broadway, West of Roy Wilkins Ave., East of 15th St.

2148006-21111012801: East of Rubbertown, in the neighboring census tract
2148006-21111005000: West of I-65, south of Broadway
2148006-21111002100: North of Market, West of 24th st.

All data was compiled by Jeremy Chesler using information from

Figure 8.1A   Health Issues in West Louisville (Dirty Air) Neighborhoods vs. East Louisville (Clean Air) Neighborhoods. *Source:* Centers for Disease Control and Prevention https://www.cdc.gov/500cities/index.htm

HEALTH OUTCOMES: USA VS. WEST LOUISVILLE VS. CLEAN AIR CITIES

Health Outcomes	US	West Louisville	Clean cities coverage	Columbus, GA	Evansville, IN	Greenville, NC	Lincoln, NE	Santa Barbara, CA
Asthma	9	14.94	9.8	10	10.8	10.6	8.8	8.8
HBP	32.4	57.16	29.98	36	36.8	27.3	24.3	25.5
COPD	6.6	16.84	6.60	7.6	10.1	5.9	4.9	4.8
Heart Disease	6.4	12.76	5.64	6.3	7.8	4.4	4.9	4.8
Diabetes	10.8	23.18	9.76	12.9	12.1	8	7.8	8.3
Poor Mental Health	12.4	24.4	14.1	14.8	17.6	15.6	11.3	11.3
Poor Physical Health	12.3	26	12.24	13.3	16.2	11.7	9.6	10.5
Tooth Loss	14.5	46	15.86	19.6	21	18.9	10.4	9.4
Smoking	16.4	38.96	18.58	21.6	26	18.1	15.3	11.9
No phys. Activity	26.6	50.72	27.24	34.9	34.6	26.6	21.9	18.2
Obesity	30.1	52.2	32.24	37.7	35.7	32	30.1	23.7

All data was compiled by Jeremy Chesler using information from

Figure 8.2A   Health Outcomes for USA vs. West Louisville (Dirty Air) Neighborhoods vs. Clean Air Cities. *Source:* Centers for Disease Control and Prevention https://www.cdc .gov/500cities/index.htm

factors of race, income, and crime. Brownfields also affected neighborhood life expectancy, although ranking below proximity to Rubbertown. Finally, we also found that when taken together, living near Rubbertown and near Brownfield sites had an even bigger effect on life expectancy, explaining as much as 75 percent of the 10-to-13-year variance among Louisville neighborhoods (Gilderbloom et al., 2020). Put another way, the odds that

Rubbertown is not a contributing factor to premature neighborhood death rates are less than one out of a thousand, with brownfields one out of one hundred.

Why does poor air quality affect Louisville's residents? When Mayor Fischer convened an elite task force to study disparities in life expectancy, they again blamed it on the lifestyle of the poor: smoking, drinking, diet, obesity, and education. Environmental degradation was not listed as a cause for reduced life expectancy; industrial polluters were likely pleased to get a "pass" on pollution.

Scientists around the world have debunked this "blame the victim" explanation. The cause is pollution. This is evident in comparing cities with different levels of pollution. Examining the bottom income quartile of Louisville citizens at the age at which they passed compared with similar residents of the five cleanest cities that heavily regulated air pollution, poor men in Louisville died five years earlier and poor women died four years earlier than their counterparts in the cleaner cities. More interesting is that if you look at the top income quartile of the rich in dirty cities, they live the same number of years as their counterparts in clean cities. Why? Because the rich move far away from these poisonous polluters. Poor air quality is not an issue for the rich and powerful to fight—they live, work, and go to school many miles away from these polluters (Greater Louisville Project, 2013; Smith et al., 2011).

Pollution does not only result in significantly lower life expectancy, but also results in other challenges found in dozens of scientific journal articles, including (Gilderbloom et al., 2020; Yue, 2019) higher risks of fetal damage, vascularization of the placenta, miscarriages, dementia, cancer, asthma, depression, and lower birth rates. The Louisville Health Equity Report admits these problems but assigns no fault to the deadly pollution and instead blames humans.

Pollution also results in a host of community ills such as lower school achievement scores, reduced chances of college admission, higher chances of home foreclosure, higher crime rates, greater greenhouse gases/climate change, significant reduction in homeowner equity and appreciation, reduced taxes to support essential neighborhood services, and unwalkable neighborhoods. Add to this list: COVID-19. Scientists tell us that neighborhoods with high pollution levels will cause more illness and death from this pandemic.

Louisville is considered to have one of the worst cases of polluted air in America's hundreds of mid-sized cities, but not all who live there share the burden equally. This takes on increasing significance in light of the current pandemic. We found that Rubbertown and the surrounding neighborhoods of West Louisville are uniquely vulnerable to COVID-19, which hits those with certain preexisting conditions hardest (Chow et al., 2020). A CDC study found that about 75 percent of victims who needed hospitalization had at least one such condition, and their outcomes tend to be worse (Wu et al., 2020; Chow et al., 2020). West Louisville's high rates of respiratory diseases (such as asthma and COPD) are particularly worrisome since COVID-19 can cause serious damage to the lungs (Chow et al., 2020). What's more, scientists believe exposure to air pollution is itself a risk factor for COVID-19: In a recent study of 3,080 U.S. counties, Harvard University researchers found that those with dirtier air had higher death rates from the disease (Wu et al., 2020).

In the aftermath of the pandemic, when a review is conducted of those who survived and who died because of COVID-19, the results will show that the adverse consequences were not have been equally distributed among races and places. Polluting industries are more likely to be sited in minority and low-income communities, just as Bullard (1990) found in his classic book *Dumping in Dixie* and several subsequent publications. Deaths have already been higher in poor neighborhoods in cities like the Bronx, New Orleans, Washington D.C., Memphis, and Detroit (Brooks, 2020).

Of course, air pollution is a problem in virtually all cities. President Trump's administration has struck down 95 environmental regulations since taking office, making air, water, and soil across the U.S. more vulnerable to pollution. In the wake of COVID-19, he has exempted industries from even a modicum of environmental oversight (Friedman, 2020).

Locally, the fight to do environmental justice work has met with unprecedented attacks on scientists, including threats of violence, defunding of Centers, blocking email access of scientists, removal of websites from the University of Louisville (UofL) that contained $3.5 million worth of federally funded research reports, and refusing to fund grants to continue this avenue of research.

Instead, UofL created a new center that excluded scientists doing pollution research and pivoted to blaming the lack of trees as the cause for reduced lifespan. Millions of dollars flowed in from donors connected to polluting industries (Coffman, 2019). Similarly, President Trump has also endorsed a trillion trees as the best way to combat pollution (Irfan, 2020). The lack of trees in neighborhoods is blamed for high pollution levels. As a *USA Today* (2020) editorial declared, "Tree planting to mop up carbon pollution is important. But more important is cutting emissions, and this White House has a shady environmental record."

Trees do provide positive benefits such as shade that reduces energy costs of homes and makes walking more pleasant, and they reduce the threat of flooding, ponding, and sewage overflow by soaking up rainwater. Trees on properties also provide an economic benefit by increasing the value of homes and neighborhoods according to traditional economists who run hedonic housing equations (Gilderbloom and Appelbaum, 1988). Trees are somewhat helpful, but they are hardly the answer. Instead of spending much-needed money on proven methods of reducing pollution, these funds are being diverted to an ill-advised campaign to plant 8,000 trees, hoping trees alone will fix the problem. However, this silver bullet is a blank because it ignores the real reason for dirty and dangerous air.

Unfortunately, many of these tree planting efforts are going into middle-class neighborhoods because they require neighborhood participation, and homeowners must pay a fee for each tree planted. Moreover, according to Russ Barnett, who once headed up Kentucky's Department of Natural Resources and was the former Director of the UofL Kentucky Institute for the Environment and Sustainable Development, you would need a full-grown forest, which would take 100 years to grow. However, even a full forest taking over the street and yards would only have a small impact on cleaning up life-threatening pollution. These initiatives still do not address dangerous carbon released in the air by industries, which is the biggest threat to neighborhoods, the city, and to our small planet. Gilderbloom and Dwenger (2019) found no solid correlation between tree cover and lifespan in a regression equation. Gilderbloom et al. (2020) found some of the cleanest air in the nation is in places like Yuma, Arizona, which has the lowest level of tree canopy of any city in their study. A tree on every lawn is not going to reduce rates of cancer, asthma, or COVID-19. However, reducing pollution and implementing policies to create cleaner air will.

While the Louisville powerbrokers were trying to stop research on how pollution destroys lives and communities, covering up these truths in the process, most universities think such research is normal and welcomed around the world. The World Health Organization says pollution is the second leading cause of noncommunicable diseases. Sadly, we will see even more deaths caused by the toxic mixture of pollution and COVID-19 (Landrigan et al., 2017). However, with many leaders at all levels focusing on public health, this could be the opportunity to gain the political traction to address this health inequality and promote environmental justice.

Back in Louisville, the time has come for the city to look for real solutions to public health crises. West Louisville industries get a political pardon because they produce jobs with middle-class incomes, even though they pose significant and deadly health risks. Gone should be the days of bad science, such as the fraudulent smoking studies of the 1960s (see Brenner (2004) and the 1999 film and lobbyist-controlled policy. Science and public health officials should be setting the course for public health policy. Without science, data, and facts, the poor are truly powerless to fight back. They have no valid or reliable information to fight for environmental justice.

Science and public health officials can show Louisville how to solve some of its most pressing problems, and other cities can learn from its example. If Louisville finds the will to address the pollution of West Louisville, it could prove to be a case study in best practices on how cities can confront environmental and health injustice.

The call for greater racial equity also means cleaning up the air, water, and soil. Poor people needlessly suffer more in Louisville than the same low-income people in West Coast cities. If we adopted the same tough environmental regulations as our West Coast counterparts, West Louisville would surely bloom instead of slowly dying. The unfairness between Black and White neighborhoods is stark and vivid. As the great urbanist, Jane Jacobs, once said: "everyone hungers for a first class neighborhood for both pride and dignity . . . nobody wants a second class neighborhood." First class neighborhoods are safe, healthy, sustainable, and prosperous. It is a human right; a worldwide right.

RUBBERTOWN: COMPANIES AND
SURROUNDING AREAS

The website www.Rubbertown.sunnyways2.com is a comprehensive collection of information surrounding the Rubbertown chemical industry complex. The website provides general information about Rubbertown and why all should be concerned. The chemical industry complex is introduced with location addresses and addresses of company headquarters and goes on to discuss the products and chemicals that are produced in the area. The health impacts felt by West Louisville residents are discussed through an analysis of the 2017 Health Equity Report. The last page of information concerns community advocates working toward social and environmental justice for West Louisville residents. For more information, contact Bunny Hayes at bunny .hayes@louisville.edu.

Page 2 of www.rubbertown.sunnyways2.com addresses the companies in West Louisville, Kentucky, and some of the surrounding areas. First are the actual chemical companies, their location addresses, and the address of the chemical company headquarters.

Table 8.1 Chemical Industry Complex

Company	Address	Headquarters
Chemours Louisville Works	4200 Camp Ground Road,Louisville, KY 40216	Chemours 1007 Market Street Wilmington, DE 19899
DuPont Performance Elastomers	4242 Camp Ground Roa,Louisville, KY 40216	4242 Camp Ground Road 40216
DuPont Louisville Works	4250 Camp Ground Road, Louisville, KY 40216	DuPont de Nemours 974 Centre Road Willmington DE 19805
Dow Chemical	4300 Camp Ground Road, Louisville, KY 40216	Rohm & Haas Chemicals Midland, MI 48674
Arkema Inc.	4350 Camp Ground Road, Louisville, KY 40216	Arkema Inc. 4350 Camp Ground Rd, Louisville, KY 40216
American Synthetic Rubber	4500 Camp Ground Road, Louisville, KY 40216	Michelin North America Headquarters 1 Parkway S, Greenville, SC 29615
Borden Chemical	6200 Camp Ground Road, Louisville, KY 40216	Hexion 180 East Broad Street, Columbus, Ohio 43215
Lee's Lane Superfund Site	Lees Lane @ Ohio River, Louisville, KY 40216	Superfund Site Community Involvement Coordinator: Angela Miller
MSD	2998 Wilkie Road, Louisville, KY 40216	700 W Liberty St, Louisville, KY 40203
MSD Central Maintenance Facility	3050 Commerce Center Place, Louisville, KY 40211	700 W Liberty St, Louisville, KY 40203
LG & E Cane Run Generating Station	5252 Cane Run Road, Louisville, KY 40258	PPL Corp. 2 N. Ninth St. Allentown, PA 18101-1179
LG & E Mill Creek Generating Station	14660 Dixie Hwy, Louisville, KY 40272	PPL Corp. 2 N. Ninth St. Allentown, PA 18101-1179

Sources: 1. Kentucky Institute for the Environment and Sustainable Development; 2. Center for Sustainable Urban Neighborhoods, formerly University of Louisville, transferred to Neighborhood Associates, Washington, D.C.

Secondly, the website has a chart showing warehouses and wholesalers of chemicals in the Louisville area, their addresses, and the address of the headquarters of the chemical company listed.

The third section lists some local properties for sale. The address, the approximate sale price, and the realtor website address are provided.

The fourth chart lists local Jefferson County Public Schools, their addresses, and a description of the level of education. The author noted that when showing the website to a class, one of the students replied to the picture used, saying that it was the middle school they had attended, and indeed that was the view they saw every day.

Table 8.2 Warehouses and Wholesalers of Chemicals

Company	Address	Headquarters
Lubrizol Advanced Materials	4200 Bells Lane, Louisville, KY 40211	29400 Lakeland Blvd. Wickliffe OH 44012
Poly One Corp	4250 Bells Lane, Louisville, KY 40211	4250 Bells Ln, Louisville, KY 40211
Zeon Chemicals	4111 Bells Lane, Louisville, KY 40211	4111 Bells Ln, Louisville, KY 40211
Zeon Chemicals	7050 Riverport Drive, Louisville, KY 40258	7050 Riverport Dr, Louisville, KY 40258
Clariant Corporation	1227 S. 12th Street, Louisville, KY 40210	1227 S 12th St, Louisville, KY 40210
Hexion Specialty Chemicals	462 S 4th St #1800, Louisville, KY 40202	Meidinger Tower 462 S 4th St #1800, Louisville, KY 40202
BASF	3400 Bank Street, Louisville, KY 40212	3400 Bank St, Louisville, KY 40212
Clariant Corporation	1600 W. Hill Street, Louisville, KY 40210	1227 S 12th St, Louisville, KY 40210

Sources: 1. Kentucky Institute for the Environment and Sustainable Development; 2. Center for Sustainable Urban Neighborhoods, formerly University of Louisville, transferred to Neighborhood Associates, Washington, D.C.

Table 8.3 Properties for Sale

Address	Approximate Sell Price	Realtor Website
3304-3306 Camp Ground Road, Louisville, KY 40216	1.22 Acres for Sale Assessed at $25,840	Loop.net
3357 Camp Ground Rd, Louisville, KY 40216	Sold $38,000	Zillow.com
3515 Camp Ground Road, Louisville, KY 40216	For Sale $104,691	Redfin.com
3619 Camp Ground Road, Louisville, KY 40216	For Sale $50,298	Trulia.com
6303 Camp Ground Road, Louisville, KY 40216	For Sale $72,900	Realtor.com
5025 Cane Run Road, 5025 Cane Run Road	For Sale $175,900	Realtor.com
3505 Lee's Lane, Louisville, KY 40216	Vacant Land for Sale $45,000	Lees Lane Smith Family Trust KCREA.com
3907 Lees Lane, Louisville, KY 40216	Sold for $93,000	Movoto.com
4208 Lees Lane, Louisville, KY 40216	For Sale $36,850	Joe Hayden Real Estate Team.com

Sources: 1. Kentucky Institute for the Environment and Sustainable Development; 2. Center for Sustainable Urban Neighborhoods, formerly University of Louisville, transferred to Neighborhood Associates, Washington, D.C.

Table 8.4 Jefferson County Public Schools in 40216

Public School	Address	Education Level
Gutermuth Elementary School	1500 Sanders Lane	Elementary
Kerrick Elementary School	2210 Upper Hunters Trace	Elementary
Waller-Williams Environmental	2415 Rockford Lane	Elem. Middle, High
Schaffner Traditional Elementary	2701 Crums Lane	Elementary
Crums Lane Elementary	3212 S Crums Lane	Elementary
Wellington Elementary	3400 Lees Lane	Elementary
Western High School	2501 Rockford Lane	High
Mill Creek Elementary	3816 Dixie Hwy	Elementary
Butler Traditional High School	2222 Crums Lane	High

Sources: 1. Kentucky Institute for the Environment and Sustainable Development; 2. Center for Sustainable Urban Neighborhoods, formerly University of Louisville, transferred to Neighborhood Associates, Washington, D.C.

RUBBERTOWN: PRODUCTS AND CHEMICALS

Page 3 of the website lists some of the products and chemicals made and/or used in Louisville.

The Rubbertown chemical complex produces a wide variety of chemicals and materials. We have made these products necessities in our society. These companies employ hundreds of citizens and bring millions of dollars into the local economy. These companies also have by-products and waste material that must be dealt with. Some of the by-products and waste materials become pollution that causes ill health. Synthetic rubber, elastomers, freon, and plexiglass are some of the products that are made in the Rubbertown chemical complex. The exchange for products we have made into "everyday necessities" is a list of chemical compounds with health concerns attached to them.

When Louisville's air was tested for levels of 250 chemicals, 9 results came back at borderline amounts. None of the nine chemicals was over their allowed maximum, but the levels of these chemicals represent an elevated risk priority.

Table 8.5 Chemicals

Chemical	Found In	Health Concern
1,3 Butadiene	Synthetic rubber	Carcinogen
Acrylonitrile	Air around chemical plants	Carcinogen
Benzene	Crude oil to make plastics	Carcinogen
Chloroform	Air/coastal waters around chlorine	Liver and kidney complications
Chloroprene	Wet suits orthopedic braces	Kidney and dermal complications
Chromium	Air/water/soil due to industrial process	Neurotoxin
Ethyl acrylate	Acrylic nails Dental materials Flavoring agents	Carcinogen
Formaldehyde	Plywood Particle board	Carcinogen
Perchloroethylene	Drycleaning process Water repellants	Carcinogen

Sources: 1. Kentucky Institute for the Environment and Sustainable Development; 2. Center for Sustainable Urban Neighborhoods, formerly University of Louisville, transferred to Neighborhood Associates, Washington, D.C.

LEGACY CHEMICALS

Legacy chemicals are those chemicals found in soil (and sometimes storage) that have been left by previous occupants. Even though no industry holds itself responsible for the contamination left behind, the chemical pollutants remain in the soil. When new industry moves in or, in some cases, when residential development moves in, the new occupants are unaware of the past chemical contamination they will now live with.

RUBBERTOWN: HEALTH

To better understand the impact of the Rubbertown chemical complex industry situated in West Louisville, the author relayed information found in the 2017 Health Equity Report for Louisville, Kentucky, prepared by the Center for Health and Equity. Since the making of the website in 2020, other equity reports have been conducted. For more updated reports, visit https://aarp -states.brightspotcdn.com/04/3d/96dc55264d7f91815f3c96d5cbb7/aarp-ky -louisville-health-equity-report-with-crops.pdf

Table 8.6 Past Contaminators and Contaminants

Company	Legacy Chemical(s)
Aetna Oil and Louisville Refinery	Fuels, gasoline, kerosene, naphtha, oil, petroleum coke
LG & E	Uncovered coal ash, mercury
National Synthetic Rubber	Styrene-butadiene
National Carbide	Calcium carbide, acetylene gas
Standard Oil Company	Oil refinery
Union Carbide	Butadiene from grain alcohol
U.S. Office of War Production	Calcium carbide, acetylene gas

Sources: 1. Kentucky Institute for the Environment and Sustainable Development; 2. Center for Sustainable Urban Neighborhoods, formerly University of Louisville, transferred to Neighborhood Associates, Washington, D.C.

Health Equity Report 2017

The Health Equity Report of 2017 is a comprehensive study written and compiled by Brandy N. Kelly Pryor, PhD; Rebecca Hollenbach, MPH, CHES; Aja Barber, MS; and T. Gonzales, MSW. The authors define health equity in Louisville to mean that all have a just and fair opportunity to attain good health and be able to reach their maximum potential as a human. The Health Equity Report suggests that inequities in practice and policy should be addressed if we want to eliminate health disparities.

Asthma

Since asthma has no known cure, those who suffer from asthma learn how to recognize triggers that set off asthmatic symptoms. Exposure to dust and air pollutants are common triggers but can be better managed by limiting those exposures. Health Equity figure 8.1 shows the count of inpatient admissions for asthma, by zip code in Louisville, KY.

The Louisville Metro Air Pollution Control District implemented the Strategic Toxic Air Reduction (STAR) program in 2005. The STAR program was intended to regulate harmful pollutants originating from large industrial emitters. UofL has conducted long-term monitoring efforts that show the effectiveness of the STAR program in lowering toxic emissions and improving air quality in parts of West Louisville.

Cancer

Cancer is the leading cause of death in Louisville. Almost 7 percent of the adult population has been diagnosed with cancer, not including skin cancer. One reason for elevated rates of death from cancer could be the lack of access

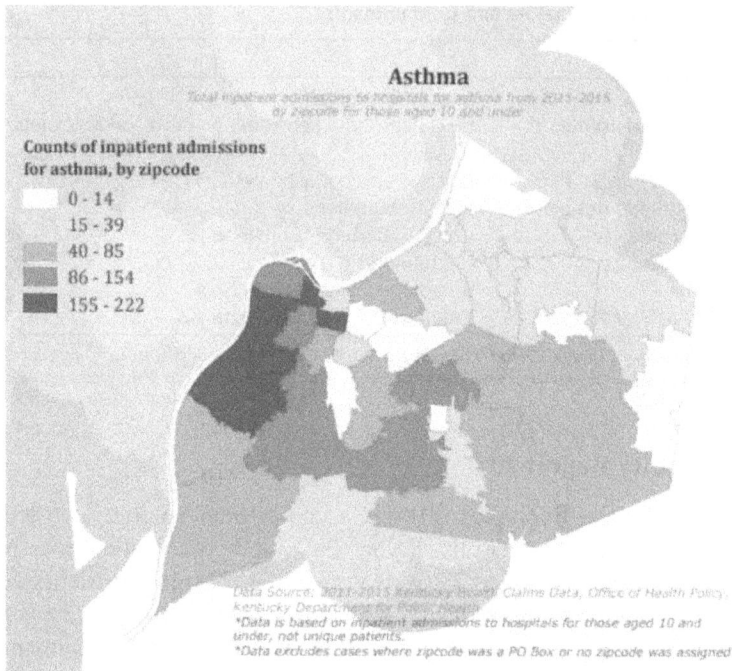

Figure 8.1   Health Equity Asthma Rates in Louisville, KY. *Source*: 2011–2015 Kentucky Health Claims Data, Office of Health Policy, Kentucky Department for Public Health.

to healthcare. If cancers are left unevaluated and untreated, they can become more serious, at which point the treatments will have more adverse effects, often resulting in premature death. Health Equity shows the age-adjusted death rate per 100,000 in Louisville, KY.

Life Expectancy by Zip Code

Our environment plays a key role in the outcomes of our health. Each of us has unique circumstances that help determine our quality and quantity of life. The Health Equity Report calls this unique circumstance our root cause of health. Go to https://www.rwjf.org/en/insights/our-research/interactives/whe reyouliveaffectshowlongyoulive.html to view the average life expectancy in your zip code. Health Equity figure 8.3 shows the life expectancy of residents of Louisville, KY.

Crime

West Louisville has a disproportionate amount of Jefferson County's homicide cases. Health Equity figure 8.4 shows the location of homicides in

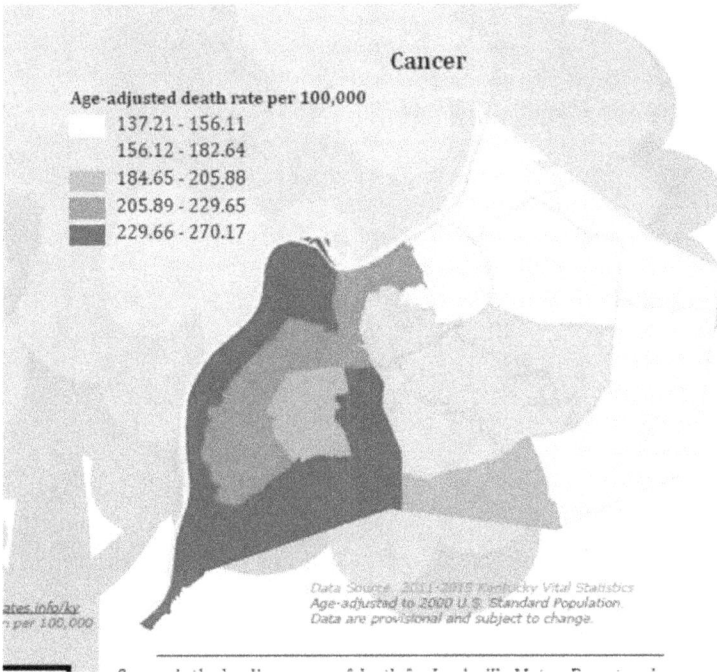

Figure 8.2   Health Equity Cancer Rates in Louisville, KY. *Source*: 2011–2015 Kentucky Vital Statistics.

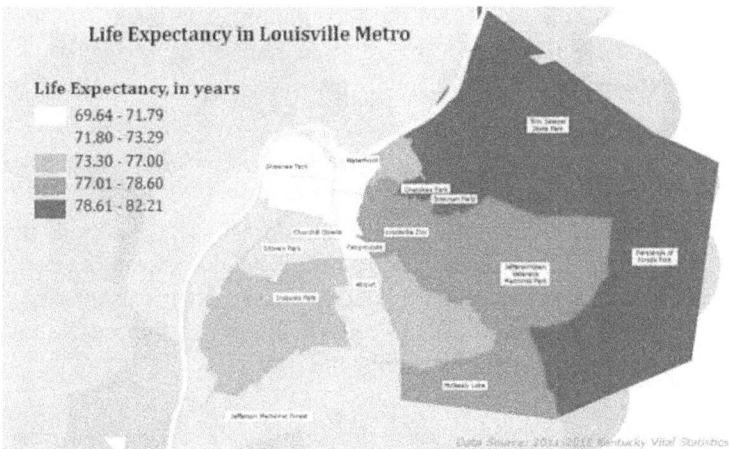

Figure 8.3   Health Equity Life Expectancy in Louisville, KY. *Source*: 2011–2015 Kentucky Vital Statistics.

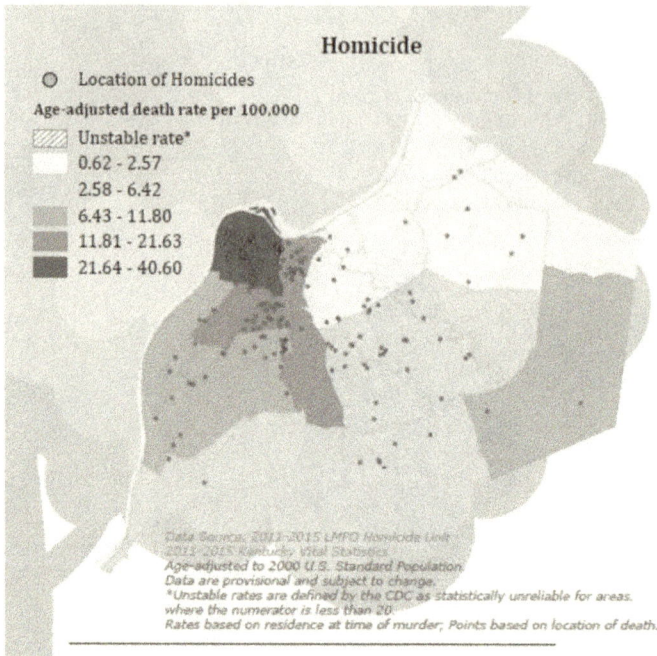

Figure 8.4   Health Equity Crime Locations in Louisville, KY. *Source*: 2011–2015 LMPD Homicide Unit, 2011–2015 Kentucky Vital Statistics.

Louisville, KY. When our communities continually lack economic development, this can lead the citizens of these communities to react with violence. When we witness gun violence, we are more likely to carry our own weapon as well as experience psychological distress.

Suicide

Suicide prevention is strongest when we can address social determinants such as access to mental health services, positive neighborhood environments, and reducing access to the means of suicide. You can visit 988lifeline.org to learn how to help others in times of need. Health Equity figure 8.5 shows the suicide death rate per 100,000 in Louisville, KY.

Lead Poisoning

In the current day, the top cause of lead poisoning in children is still dust from lead paint. There is no amount of lead in the air, water, or soil that is acceptable. There is no safe level of lead in the human system. Those who are youngest suffer the most ill effects from lead poisoning. Health Equity figure 8.6 shows the lead poisoning in Louisville, KY.

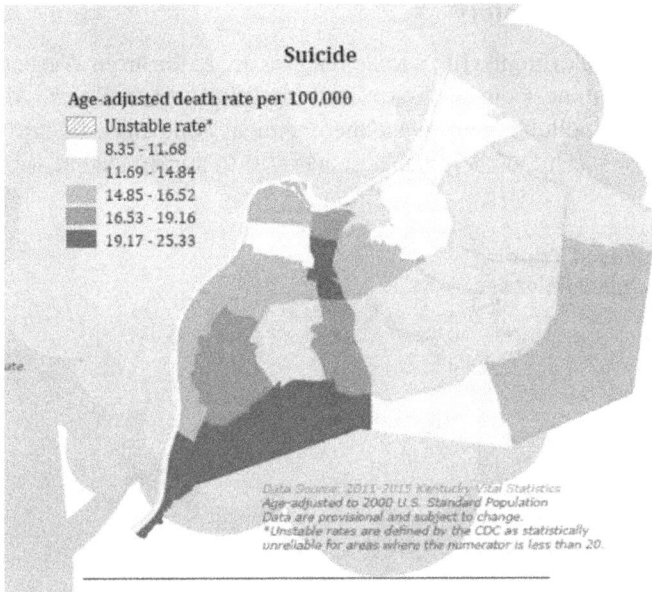

Figure 8.5   Health Equity. Suicide Rates in Louisville, KY. *Source: 2011–2015 Kentucky Vital Statistics.*

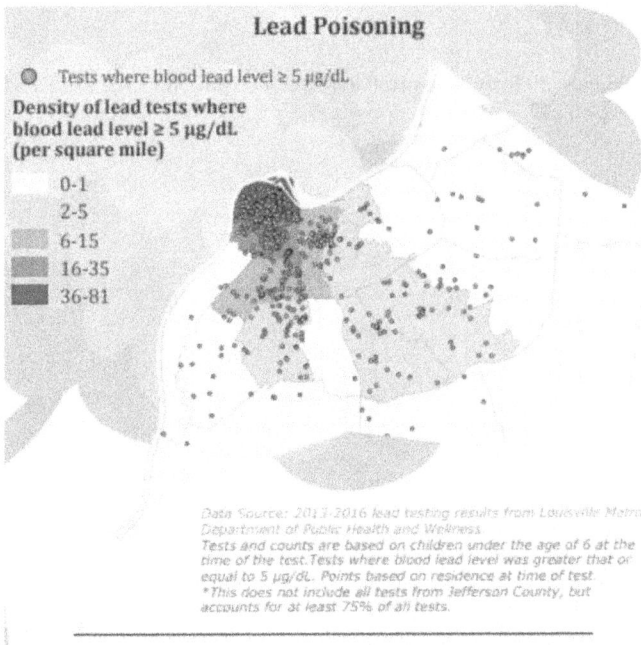

Figure 8.6   Health Equity. Lead Poisoning in Louisville, KY. *Source*: 2013–2016 lead testing results from Louisville Metro Department of Public Health and Wellness.

Toxic Release Inventory

Over 80 percent of Louisville's toxic releases are in the air in West and South Louisville, coming from 56 Toxic Release Inventory facilities. Almost 80 hazardous air pollutants are what the chemical companies release into the air. All these hazardous pollutants are harmful to human health, with at least

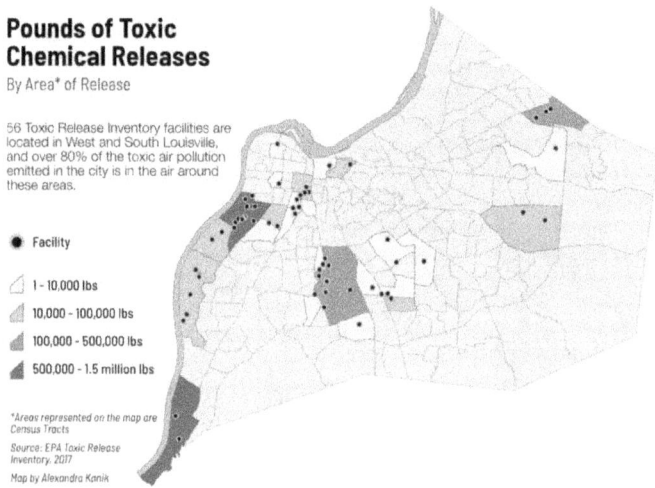

Figure 8.7 **Pounds of Toxic Chemical Releases in Louisville, KY.** *Source*: EPA Toxic Release Inventory, 2017

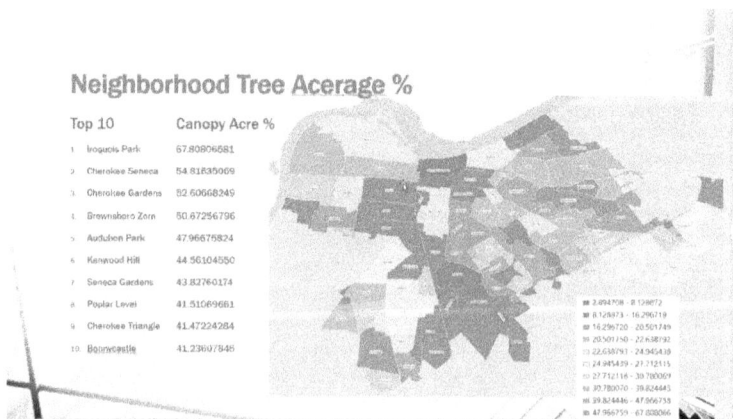

Figure 8.8 **Neighborhood Tree Acreage.** *Sources*: 1. Kentucky Institute for Environmental & Sustainable Development; 2. Center for Sustainable Urban Neighborhoods, formerly University of Louisville, transferred to Neighborhood Associates, Washington, D.C.

22 pollutants increasing the likelihood of cancer risk. Figure 8.7 shows the pounds of toxic chemical releases in Louisville, KY.

Page 2 of www.Rubbertown.sunnyways2.com ends with a trailer from Bullfrog Entertainment.

BULL FROG ENTERTAINMENT

Homo Toxicus

Homo Toxicus is an 88-minute documentary suggesting that risk management procedures do not take into account the fact that the average human body now carries almost 250 toxins. With the invention of over 100,000 chemicals since the last world war, modern-day medicine does not look at the low levels of various combinations of hundreds of chemical exposures in our systems. The trailer for the documentary can be found at https://youtu.be /Aa6V1FuF7EM

The last page of the website www.rubbertown.sunnyways2.com reports on the community advocates working toward social and environmental justice for the citizens of West Louisville.

RUBBERTOWN COMMUNITY ADVOCATES

Kentuckians for the Commonwealth
Together, we organize for a fair economy, a healthy environment, new safe energy, and an honest democracy.

West Jefferson County Community Task Force
West Jefferson County Community Task Force (WJCCTF) was established in 1988 to work on environmental justice issues in West Louisville.

Kentucky Division for Air Quality
A Kentucky government office is committed to achieving and maintaining acceptable air quality.

Air Pollution Control District
The Air Pollution Control District (APCD) has worked to improve Louisville's air quality for over 67 years. The APCD is authorized to implement the Federal Clean Air Act. The APCD collects air monitoring and emissions data locally.

Agency for Toxic Substances and Disease Registry
The Agency for Toxic Substances and Disease Registry (ATSDR) protects communities from harmful effects related to exposure to natural and man-made hazardous substances.

Louisville Department of Health and Wellness
The Louisville Department of Health and Wellness is committed to achieving health equity and improving the health and well-being of all Louisville residents and visitors.

University of Louisville Christina Lee Brown Envirome Institute
Health is a state of optimal physical, mental, and social well-being.

Louisville Metro Council
Find your Metro Council Member. Districts 1–26.

Rubbertown Emergency Action
ReAct (Rubbertown Emergency Action) group began organizing West Louisville residents to fight pollution in their neighborhood and community.

EPA Region 4
Serving Alabama, Florida, Georgia, Kentucky, Mississippi, North Carolina, South Carolina, Tennessee, and 6 Tribes

Rubbertown Community Advisory Council
Developing mutual trust between participating chemical companies and the surrounding community.

Videos and documentaries that are based on this research can be found at:

1. John Gilderbloom "Pollution in Louisville Kentucky"
   https://www.youtube.com/watch?v=vvacxkfyNSQ
2. *Rubbertown*, which is currently on Amazon and one of the authors (Gilderbloom) is featured.
3. Julian Agyeman, editor of *Local Environment*, gave a wonderful speech on Sustainability but things got a little wild between a panel of local government speakers and John Hans Gilderbloom at 1 hour and 32 minute mark when I told them pollution in Louisville's Rubbertown was a major issue. https://www.youtube.com/watch?v=467DocgXrJw
4. *Dark Waters* which is available on Amazon is an excellent film of a negligent chemical company polluting a small West Virginia town near the Ohio River. It is a true story that is written by the lawyer who defended and won a major lawsuit. The residents were given a $70

million settlement from the chemical company on the manufacturing of Teflon that caused enormous negative health impacts (death, body disfiguration, heart, and lung problems) on the citizens of a West Virginia town. This same company operates in Rubbertown.

ACKNOWLEDGMENTS

This is a revised and greatly expanded (four times bigger) version of a paper that was published in City Lab/Bloomberg originally. This article was reposted by various environmental and planning groups to over a million readers: Sierra Club, Pro-urban, Community Urban blog, ACSP, and other places as we know. We thank the editors (Jennifer Sondag and David Dudley) for giving us permission to republish parts of this article. Special thanks to Tracy Walsh, our Editor at Bloomberg, for making this even stronger, insisting on revisions and fact checking. Our thanks for additional editing from Carla Snyder and Ellen Slaten.

At the University of Louisville, we worked with graduate student Jeremy Chesler, who put together two CDC tables 8.1A and 8.2A on comparing the dirty air in West Louisville neighborhoods near Rubbertown and far-away East neighborhoods and cities with clean air. Other students at the UofL seminar, James Mains, Bunny Marie Hayes, Sait Sarr, Makayla Marr, and Sarah Catherine Haynes, gave us good feedback. Our thanks also to our colleagues who participated in our seminar discussions, Justin Mug, Russell Barnett, Robert Friedlander, and Dwan Terner.

Revised and reprinted with permission from *Local Environment*.

REFERENCES

Brenner, M. (2004). The man who knew too much. *Vanity Fair*, April 1, 2004. Retrieved online: https://www.vanityfair.com/magazine/1996/05/wigand199605

Brooks, R. (2020). African Americans struggle with disproportionate COVID death toll. *National Geographic*, April 21, 2020. Retrieved online: https:/www.nationalgeographic.com/history/2020/04/coronavirus-disproportionately-impacts-African-Americans/

Bruggers, J. (2020). "Louisville's 'Black Lives Matter' protests echo a long quest for environmental justice." *Courier Journal*, June 24, 2020 reprinted from James Bruggers 2020 "Louisville's 'Black Lives Matter' demonstrates a long quest for environmental justice." June 21, 2020. http://insideclimatenews.org

Bullard, R. D. (1990). *Dumping in Dixie: Race, Class, and Environmental Quality*. Boulder: Westview Press.

Centers for Disease Control and Prevention, National Center for Chronic Disease Prevention and Health Promotion, Division of Population Health. (2020). *500 Cities Project Data.* https://www.cdc.gov/500cities

Chow, N., Fleming-Dutra, K., Gierke, R., Hall, A., Hughes, M., Pillishvili, T., Ritchey, M., Roguski, K., Skoff, T., and Ussery, E. (2020). Preliminary estimates of the prevalence of selected underlying health conditions among patients with coronavirus disease 2019—United States, February 12–March 28, 2020. *Morbidity and Mortality Weekly Report*, 69(13), 382.

Coffman, B. (2019). UofL's Green Heart Project's large-scale tree planting underway in Louisville. *University of Louisville Press Office*, October 14, 2019.

Gilderbloom, J. I., & Appelbaum, R. (1988). *Rethinking Rental Housing.* Philadelphia: Temple University Press.

Gilderbloom, J., & Dwenger, C. (2019). *Lack of Trees is not Why People Are Dying in West Louisville.* University of Louisville, Center for Sustainable Urban Neighborhoods.

Gilderbloom, J., & Meares, W. L. (2022). The impact of walkability and pollution on rental housing prices: A study of 148 cities. *Journal of Urban Affairs*, 44(8), 1059–1075.

Gilderbloom, J. H., Meares, W. L., & Squires, G. D. (2020). Pollution, place, and premature death: Evidence from a mid-sized city. *Local Environment*, 25(6), 419–432. https://www.tandfonline.com/doi/full/10.1080/13549839.2020.1754776

Gilderbloom, J. H., & Squires, G. D. (2020). The health emergency that is coming to West Louisville. *City Lab / Bloomberg Wire Services*, April 22, 2020. https://www.citylab.com/perspective/2020/04/air-pollution-covid-19-health-environmental-justice-poverty/610366/

Gilderbloom, J., Washington, C., Quenichet, K., Manella, C., Dwenger, C., Slaten, E., Sarr, S., Altaf, S., & Frederick, C. (2020). What cities are the most dangerous to your health? Ranking the most polluted mid-size cities in the United States. *The Lancet*, January 9, 2020. Retrieved online: https://papers.ssrn.com/sol3/papers.cfm?abstract_id=3506217

Greater Louisville Project. (2013). *Building a Healthier Louisville.* Louisville: Community Foundation of Louisville.

Hauck, G., Nichols, M., Marini, M., & Pantazi, A. (2020). Coronavirus spares one neighborhood but ravages the next. Race and class spell the difference. *USA Today*, May 3, 2020. Retrieved Online: https://www.usatoday.com/in-depth/news/nation/2020/05/02/coronavirus-impact-black-minority-white-neighborhoods-chicago-detroit/3042630001/

Irfan, U. (2020). Tree planting is Trump's politically safe new climate plan. *Vox*, February 5, 2020.

Kim, K. H., Kabir, E., & Kabir, S. (2015). A review on the human health impact of airborne particulate matter. *Environment International*, 74, 136–143.

Landrigan, P. J., Fuller, R., Acosta, N. J., Adeyi, O., Arnold, R., Baldé, A. B., . . . & Chiles, T. (2018). The Lancet Commission on pollution and health. *The Lancet*, 391(10119), 462–512.

Louisville Metro Department of Public Health and Wellness. (2014). Louisville Metro health equity report. *LouisvillKY*, 2014. Retrieved online: https://www.google .com/url?sa=t&rct=j&q=&esrc=s&source=web&cd=1&ved=2ahUKEwiym97liL bpAhULd98KHZTsAK0QFjAAegQIAhAB&url=https%3A%2F%2Flouisvilleky .gov%2Fsites%2Fdefault%2Ffiles%2Fhealth_and_wellness%2Fche%2Fhealth _equity_report%2Fher2014_7_31_14.pdf&usg=AOvVaw2O3NGRMefBQyt cPOVVGL26

Luthra, S. (2020). Was the novel coronavirus really sneaky in its spread to the U.S.? Experts say no. *Kaiser Health News*, March 19, 2020. Retrieved online: https://khn .org/news/fact-check-president-trump-was-the-novel-coronavirus-really-sneaky-in -its-spread-to-the-u-s-experts-say-no/

Smith, P., Pennington, M., Crabtree, L., & Illback, R. (2011). *Louisville Metro Health Equity Report: The Social Determinants of Health in Louisville Metro Neighborhoods*. REACH of Louisville, Inc.

Ungar, L. (2014, April 10). St. Matthews tops life expectancy; Old Louisville lowest. *Louisville Courier Journal*, April 10, 2014. Retrieved online: https://www.courier-journal.com/story/news/local/2014/04/10/st-matthews-life-expectancy-highest -old-lou-lowest/7536183/

USA Today Editorial Board. (2020). Donald Trump now a tree hugger? Please. Don't be fooled by one trillion trees initiative. *USA Today*, February 12, 2020.

Wu, X., Nethery, R. C., Sabath, B. M., Braun, D., & Dominici, F. (2020). Exposure to air pollution and COVID-19 mortality in the United States. *Science Advances,* 6(45), eabd4049. Retrieved online: https://projects.iq.harvard.edu/covid-pm

Yang, J., Zheng, Y., Gou, X., Pu, K., Chen, Z., Guo, Q., Ji, R., Wang, H., Wang, Y., & Zhou, Y. (2020). Prevalence of comorbidities and its effects in patients infected with SARS-CoV-2: A systematic review and meta-analysis. *International Journal of Infectious Diseases*, 94, 91.

Yue, H., Ji, X., Zhang, Y., Li, G., & Sang, N. (2019). Gestational exposure to PM2.5 impairs vascularization of the placenta. *Science of the Total Environment,* 665, 153–161.

Chapter 9

Pollution, Place, and Premature Death

Evidence from a Mid-Sized City

John Hans Gilderbloom, Wesley L.
Meares, and Gregory Squires

Social critic Paul Goodman (1960) wrote in *Growing Up Absurd*, "A man has only one life and if during it he has no great environment, no community, he has been irreparably robbed of a human right" (p. 17). A person's neighborhood is a powerful predictor of how long he/she will live. If our residence is in the same proximity as dangerous environmental contaminants, residents might have a shorter life. While we know from studies done outside the United States that premature death is a major problem for neighborhoods close to environmental contaminants, this issue has not gotten the attention it deserves in the USA. Is Flint the exception or the rule in the United States? There have been few studies here in the USA that systematically study pollution at the neighborhood level and its impact on health by linking place to lifespan.

Worldwide, air pollution was responsible for 7 million premature deaths in 2010, according to a study published in the *Journal of The American Medical Association* and sponsored by the World Health Organization (Kuehn, 2014). By 2050, according to the Organization for Economic Co-operation and Development (OECD, 2012), bad air will be the number one cause of premature death rates—rising ahead of sanitation and dirty water, which are currently the primary causes. A recent study published in the British medical journal *The Lancet* (2013) estimates that 1.2 million premature deaths occur annually in Chinese cities due to polluted air (Wong, 2013).

RESEARCH IN THE UNITED STATES

Systematic research literature on environmental justice issues has long established that these hazardous sites—including waste and sources of air pollution—are disproportionately located in minority neighborhoods (Been, 1995; Pastor, Sadd, & Hipp, 2001; Bullard, 1990, 2000; Appelbaum, 1976; Gilderbloom et al., 2020; Pahl, 1975). Although there has been some debate as to whether hazardous sites have been located in or relocated to minority neighborhoods, or whether minority population "move-in" occurred subsequent to the location of those sites (Bullard, 2000; Pastor et al., 2001), recent studies indicate that the sites are located in minority neighborhoods, not that minorities tend to move to areas where such sites are located (Hipp & Lakon, 2010; Smith, 2009; Swanston, 2012). This can be due to the neighborhood's lack of political power or a weak real estate market (Hamilton, 1995; Banzhaf, 2012). Consistent evidence supporting the disproportionate impact of environmental contaminants on racial and ethnic minority neighborhoods should raise the level of concern about the possible link to shortened lifespans. Evidence shows that if you live in neighborhoods with high pollution, your life will be shortened by four to 13 years (Lancet, 2013; Kuehn, 2013; Wong, 2013).

Countering the Claims That Place Causes Premature Death: Louisville

City officials in Louisville have funded studies arguing that environmental degradation has little impact on lifespan. Our research into the relationship between proximity to environmental toxins and premature death rates in neighborhoods is a case study of the medium-sized city of Louisville, Kentucky. The neighborhood difference in life expectancy in Louisville is extreme—ranging from 10 to 13 years (Smith et al., 2011; Greater Louisville Project, 2013). These Louisville-based studies argue that environmental contaminants are not the cause of premature death rates but are mostly related to lifestyle choices, which are tied to socio-economic status. Very few studies have attempted to examine the relationship between environmental factors and premature neighborhood death, let alone health, in Louisville—many of which suffer from methodological or statistical problems.

There is a consensus that those who work in Rubbertown chemical factories (an area in the far western corner of Louisville located on a large strip of land with 11 plants) can face both dangerous and deadly environments. Accidents and mishaps in Rubbertown have been severe and have resulted in numerous worker deaths throughout the years. Several historic events highlight the tragedies of Rubbertown. In 1955, there was a death caused by an explosion at the Goodrich plant, and in 1961, around 1,000 residents were evacuated from a nearby neighborhood as a cloud of acrid gas formed from the former Stauffer Chemical Co. (Barnett, 2011). In 1965, the largest accident occurred when 12 workers died from explosions and fires rolling through the DuPont plant (Barnett, 2011). In 1985, a tank explosion killed three workers at Borden Chemicals and, most recently, in 2001, an explosion at Carbide Industries killed two workers (Barnett, 2011). Twenty-six former B. F. Goodrich workers have died from liver cancer (Barnett, 2011). Bill Moyers' PBS documentary *Trade Secrets* (2001) on the chemical industry brought national attention to workers in Louisville chemical plants who were dying prematurely. These workers had disintegrating bones in their hands and up to 60 non-naturally occurring chemicals in their bodies.

The history of Rubbertown is tainted with industrial accidents and unfortunate environmental devastation affecting workers in the area. However, what about residents who may not work in Rubbertown but live nearby?

Richardson (1998) argues that it is the workplace—not the place of residence—that causes premature deaths. He found that environmental conditions of "place" had no impact on premature deaths in medium-sized cities such as Louisville. He argued that the lack of correlation between place and early death is the result of several factors. First, residents are highly mobile and do not stay long enough in one place for it to impact health. Second, the workplace has a greater impact on premature deaths than the neighborhood. Richardson, in fact, argues that the neighborhood has no impact.

But Richardson committed several methodological missteps: (1) the number of cases examined was reduced to just West Louisville residents instead of the metro area, providing no comparison; (2) the regression results did not sufficiently take into account the range of other variables considered by this present study and others such as Macintyre (1989), Frank (2003), Danneberg et al. (2011), and Massey (2004); and (3) the number of census tracts studied was small, making a finding of significance difficult.

The *Louisville Metro Health Equity Report: The Social Determinants of Health in Louisville Metro Neighborhoods* (Greater Louisville Project, 2013) found that residents living in the poorest neighborhoods adjacent to the chemical industry have lower life expectancies—in certain cases over 10 years shorter than the average Louisville life expectancy (Smith et al., 2011, 5). Smith et al. (2011) analyzed cancer rates for the two zip codes in the county with similar incomes that were far away from Rubbertown. They found common cancers are significantly higher in the Rubbertown zip codes. They also found that people living in these zip codes have a 45 percent higher chance of getting lung cancer and a 31 percent higher chance of getting colorectal cancer (Ibid, 14). The problem with these studies is that zip codes are inexact and cover too much acreage as opposed to the more focused and smaller census tracts. Nevertheless, this study was an important first step in showing how the proximity of environmental contaminants to a neighborhood can have a negative impact on neighborhood premature death levels. Although not noted in the text, the maps clearly show that where pollution densities are highest, premature death rates are also higher.

The most recent study was the highly publicized report by the Greater Louisville Project (2013), which estimated years of projected life lost and found wide variation in the length of lifespan by neighborhoods. The purpose of the Greater Louisville Project is to make Louisville a more competitive city compared to 14 "peer cities," mostly located in the South. The larger goal is growth and prosperity, which hinges in part on providing residents in the city with a longer lifespan. The Greater Louisville Project, funded by the Robert Johnson Foundation along with a consortium of local foundations, businesses, and government agencies, put together an eclectic group of 18, drawing on doctors, PhD's, MPH recipients, and non-professional degree holders representing central business district (CBD) hospitals, medical insurance groups, school district administrators, and others. The quality of the study was questionable in that a non-scientist who had gotten a master's degree at a local college led it.

The Greater Louisville Project (2013, 1) found that a thirteen-year difference existed between life expectancy in mostly black West Louisville and mostly white East Louisville. They argue, "This report identifies our most challenging health factors: low educational attainment, unemployment, high rates of smoking, and obesity, moderate access to primary care and poor air quality." The study argued that years of projected life lost were due to social and economic factors (40 percent); physical environment (10 percent); health behaviors (30 percent); and access to medical treatment (20 percent). By themselves, these round numbers raise questions of how they were calculated and weighted. They read more like guesstimates and not serious science. The statistical reasoning behind these estimates is never explained, nor was the data provided concerning how these estimates were made. Absolutely no regression equations/coefficients, beta weights, or tolerance scores were shown. Do the estimates reflect local or national conditions for explaining neighborhood premature death rates? Perhaps these numbers were averaged out based on national studies, but as we shall show later, they are not locally representative of Louisville's unique circumstances.

The Greater Louisville Project (2013, 12) claims that "poor air quality" explains just one-tenth of reduced life expectancy. However, "poor air quality" is mostly defined as smoking.

Nothing was mentioned about environmental toxins existing in Rubbertown neighborhoods and 25 brownfields creating unhealthy air, water, and soil as factors leading to premature death in West Louisville. Consequently, the report makes no call for environmental remediation of environmental degradation, which, as we discussed earlier, has been systematically linked to poor health and premature death. The report, which was endorsed by the mayor of Louisville, claimed years of projected life loss were based on lifestyle choices, not circumstances, by emphasizing person personal health choices such as not smoking, eating healthy foods, exercising, and improving graduation rates. The major media spin by Greater Louisville Inc. was to claim that "low educational attainment" was a major source of shortened lives in these neighborhoods (Greater Louisville Project, 2013, 1).

However, just because education as a variable is correlated with premature deaths does not mean that the lack of education is the "cause" of premature deaths. Education does not provide a magic immunization shield that will keep a person living longer if they reside near a chemical factory or brownfield that emits toxins in the air and soil. Using education to explain the "causes" of premature deaths in neighborhoods does not make sense from a theoretical perspective.

Research Methodology: Environmental Toxins Impacts on Health in Louisville

Our research replicates, with modified methodology, previous studies that have established connections between environmental toxins and reduced life expectancy. While the authors of the majority of these studies chose to study at the country, state, region, or city level, we instead adopt an intra-city approach (Gilderbloom & Appelbaum, 1988; Appelbaum, 1978; Gilderbloom et al., 2020; Appelbaum et al., 1976). We analyze 170 neighborhoods in Louisville by constructing statistical models to explain why Louisville's neighborhoods differ by as much as 10 years in life expectancy. Our focus, demonstrated through two key independent variables, is on whether environmental toxins have a net impact on neighborhood life expectancy once the effects of other neighborhood factors are removed. This section describes the community under study, the dependent and key independent variables, and the model construction.

We chose to carry out our neighborhood-level analysis in Louisville, Kentucky, a mid-sized U.S. city that, in many ways, is representative of hundreds of other mid-sized cities (McMeekin, 1946). At about 600,000 people, the consolidated Louisville Metro (the old City of Louisville plus surrounding Jefferson County) ranks 27th in population nationally and, together with surrounding counties in Kentucky and Indiana, 44th as a Metropolitan Statistical Area (U.S. Census Bureau, 2012). It is one of approximately 160 cities in the nation with a population greater than 50,000 which is not located within 20 miles of another neighboring city of more than 50,000 (Appelbaum et al.,

1976; Appelbaum, 1978; Gilderbloom & Appelbaum, 1987; Frederick et al., 2018, Frederick & Gilderbloom, 2018, 2019; Gilderbloom et al., 2020). Louisville's relatively monocentric urban landscape, with surrounding sub-urban rings of decentralized development, is characteristic of older industrial cities grappling with the effects of deindustrialization and suburbanization. Like many U.S. cities, especially in the South, Louisville is segregated by race and socio-economic status. These minority populations, mostly African American, constitute about one-third of the inner core and about one-fifth of Louisville Metro. The medium size, geographic isolation, and urban structure of Louisville produce a better estimate of the environmental impact on health for a larger share of the U.S.'s urban and metropolitan population than would a case study of a megacity like New York or Los Angeles (Appelbaum, 1978; Appelbaum et al., 1976; Gilderbloom & Appelbaum, 1987; Frederick et al., 2018, Frederick & Gilderbloom, 2018, 2019).

Dependent Variable: Premature Death

The dependent variable measuring premature death at the neighborhood level is the years of potential life lost (YPLL) per 100,000 residents, between 2000 and 2010. The residential population used to create this measure was based on the population reported by the most recent (2010) Census. The years of projected life lost measure is based on data collected by the Louisville Metro Department of Health and Wellness, which gives the year of death, age at death, and last known address of all deceased persons in Jefferson County between the years 2000 and 2010. The process used to convert this data into the YPLL variable uses the following equation:

$YPLL = (E - A)/P$,
where E is the standardized expected age of death (=75), A is the age at death, and P is the 2010 population of each tract divided by 100,000.

Total years of potential life lost are summed by tract and divided by each tract's population, then divided by 100,000, to control for the differences in population across tracts. Higher numbers denote increases in YPLL—or in other words, decreases in life expectancy. While this present project does not consider specific causes of death, the YPLL dependent variable, as con-structed, is appropriate for this study. The age-specific population of each unit of analysis standardizes deaths. Thus, the variable highlights clusters of pre-mature death. The University of Louisville Institutional Review Board (IRB) oversaw the creation of this variable to protect the identity of the individuals in the sample.

Test Variable #1: Proximity to Brownfields

This study examines the impact of environmental conditions on years of projected life lost by using two key test variables measuring contaminants: Environmental Protection Agency-designated (EPA) brownfield sites, scattered throughout Louisville, and chemical factories (concentrated in the Rubbertown area) located in West Louisville along the Ohio River. Louisville has 25 brownfield sites (EPA, 2012) and one Superfund site on the National Priority List (NPL) (EPA, 2013). In Louisville, contaminated soil is a problem as 25 percent of downtown Louisville is classified as a brownfield (Barnett, 2011). The high numbers of industrial, chemical, and manufacturing plants in the county contribute to poor environmental quality (Smith et al., 2011, 27). The accuracy of the Louisville data is the result of a pilot program funded by the state, which uses aerial photography to identify potential brownfield areas. Once an area of potential concern is identified through aerial photography, the state launches a ground investigation to confirm if there is contamination. Congress established the Comprehensive Environmental Response, Compensation, and Liability Act (CERCLA) in 1980, in response to growing concerns over the health and environmental risks posed by hazardous waste sites (EPA, 2011). CERCLA is informally called the "Brownfield Law" to describe the program administered by the U.S. EPA in cooperation with state and tribal governments, allowing them to clean up hazardous waste sites and hold responsible parties accountable to perform cleanups (EPA, 2011). Our maps are pulled from EPA Superfund sites, which list both Superfunds and brownfields that lack the federal funds for cleanup.

For this study, we operationalized proximity to brownfields by constructing a dichotomous variable that codes neighborhoods within a 1.5-mile radius of any brownfield site as "1." All other neighborhoods are coded as "0."

This was done by geo-referencing the centroid, or geographic center, of each brownfield site and creating a 1.5-mile buffer. If any portion of a census tract fell within this buffer, the tract itself was coded as "1."

Test Variable #2: Proximity to Rubbertown

To develop the Rubbertown variable, we used geo-coding techniques to map proximity to the Rubber Town Industrial Zone, identifying all tracts within 10,000 feet north, south, or east of Rubbertown where wind currents can blow (west is the Ohio River area that borders Indiana). Rubbertown was measured by coding all census tracts with any part within 10,000 feet of any part of the Rubbertown factories as "1" and all other tracts as "0." The intent of this dummy variable is to gauge whether residence in this sub-region of the city near the Rubbertown factories has a net impact on years of projected life lost while controlling for other neighborhood factors. YPLL is significantly higher in Rubbertown neighborhoods.

Similar to Rubbertown plants, a large concentration of the most danger-ous brownfield sites, according to the former Mayor's office, is also located in western Louisville. Toxic sites are not randomly distributed throughout Louisville but are clustered in the northwestern section of the county, where a majority of minority and low-income neighborhoods are located. No matter how we created the measurement, using a buffer or, alternatively, distance to the nearest site, our findings were consistent.

Control Variables: Other Neighborhood Factors

Most of these variables are traditional predictors with a theoretical basis from the body of literature concerning neighborhood health differentials reviewed earlier (Louisville Department of Health and Wellness, Gilder-bloom et al., 2017; Smith et al., 2010; Gilderbloom et al., 2009; Gilder-bloom et al., 2020; Gilderbloom & Meares, 2020; Gilderbloom et al., 2015) (See also Agyeman, 2013). Racial composition, crime rate, income, and housing age are all variables believed to affect the lifespan of residents of a neighborhood.

Control variables are other, non-environmental factors that are thought to affect YPLL. The inclusion of these in our modeling ensures that the effect of brownfields and Superfund sites is not capturing some socio-demographic effect.

Statistical Model

We use OLS regression to predict neighborhood years of projected life lost with the two above-mentioned key test variables and other control variables. The full regression equation is as follows:

YPLL = β0 + β1*Nonwhite percent + β2* *Housing age* + β3*Income + β4*Crime rate + β5*Brownfields + β6*Rubbertown + ε,

where β1 through β6 are the coefficients to be estimated and ε is the error term.

Since the focus of this paper is on how to explain neighborhood variation in lifespan, we use OLS regression to examine the net effects of key variables that measure characteristics of each census tract.

For purposes of validity and reliability, five separate models were run with the brownfield and Rubbertown variables rotated in and out—alone (Equation 1 to 3) and then all together (see in Equation 5). All models shown were tested for multi-collinearity by calculating tolerance scores and examining zero-order correlation coefficients (Beck-Lewis, 1993–1994; Beck-Lewis, 1980; Appelbaum et al., 1991; Gilderbloom et al., 1992). All tolerance scores for variables used in the equation exceed 0.30. Using multiple equation specifications allows us to look at the impact of proximity to environmental contaminants' net impact separately and together. A reduced-form equation that runs only those variables that are significant and theoretically reasonable provides for possibly a more accurate estimate of the coefficient and could impact the statistical significance and weight of the coefficients. For the purpose of scientific integrity and standard regression protocols, we choose to show the results of all five specifications (Appelbaum et al., 1991; Gilderbloom et al., 2012; Beck-Lewis, 1993–1994; Beck-Lewis, 1980).

For good measure, we also provide regression runs of variables that were not shown to fit our original theoretical model.

Findings

Many of the control variables showed significance in the predicted direction. As noted above, the independent variables used are taken from previous studies and have been shown to be significant in measuring neighborhood effects (Smith et al., 2011). For the most part, the control variables operate as expected and raise no concerns. High-crime neighborhoods reduce life expectancy. Low-income neighborhoods reduce life expectancy, perhaps due in part to poor access to well-paying jobs, medical care, healthy food choices, and opportunities for exercise.

Equation 1, 2, and 3 found that our proxy measure for the average age of housing showed no impact at the .05 level but hovered around .11. In later equations (4 and 5), we removed the average age of housing stock because it was not significant at the .05 level, and in doing so, we obtained a more precise estimate of environmental degradation. This also suggests that older housing, which used lead paint, was not a significant cause of reduced life expectancy.

Our focus here is to control for variables that make both theoretical sense and show statistical significance so we can tease out the net effect of environmental degradation on neighborhood premature death rates. Our models all have high adjusted R-square statistics, showing that these variables collectively explain about three-fourths of the variation in YPLL.

We found that one of the major causes of premature deaths was proximity to Rubbertown's toxic contaminants, an effect that, according to beta values, ranked fourth after race, income, and crime. We felt it was important to parcel out the impact of brownfields located outside of Rubbertown. Brownfields also had an independent effect on neighborhood life expectancy, although they ranked below proximity to Rubbertown. We found these effects of our environmental measures (Rubbertown and brownfields) when run together in Equation 5; the two beta coefficients added together, measuring environmental degradation, were nearly as powerful as both race and income in shaping neighborhood lifespan.

Equation 5 was robust, with all control and test variables predicting neighborhood lifespan, explaining three-fourths of the variation among neighborhoods. Our odds that Rubbertown is not a contributing factor to premature neighborhood death rates are at least one out of a thousand (brownfields would be one out of a hundred).

Additional Regression Equations—Education and Medical Access

The argument made is that if residents had better access to quality medical care that was close by, life expectancy would be increased. However, access to medical care does not appear to prevent early deaths in neighborhoods, as Rubbertown neighborhoods are closer to the largest, highest quality medical complex in the region than other neighborhoods. Moreover, the locations of brownfields outside of Rubbertown are sometimes even closer to medical facilities. Medical access is not a true cause of how lives are cut short but a response to environmental degradation. This would be like saying large medical centers cause people to get sick instead of well—it might be a correlation but not a cause.

We have acknowledged previously that education is highly correlated with premature death rates, but it is inappropriate on theoretical grounds to put this variable into a regression equation. Simply having greater education does not make one immune to poisons in the environment—just as the Chernobyl disaster did not spare the educated class from a delayed death. However, for good measure, we ran the variable anyway, which revealed why the Greater Louisville Project places so much emphasis on education. With education in the equation, the effects of income, brownfields, and Rubbertown are eliminated as factors influencing reduced life expectancy. As expected, we found education to be highly correlated with several other control variables, including income and race. This raises multi-collinearity issues. The amount of explained variation is slightly higher, but it masks the net impacts of income and environmental degradation. Education cannot be justified as a variable linked to estimated premature neighborhood death on theoretical and statistical grounds.

Obviously, a diploma does not immunize someone from the toxic effects of dirty air, but it could be argued that education is key to enabling someone to move to a cleaner community. In a sense, this might constitute preventing problems that cannot be cured, at least for those with the education that enables them to move. However, a practical reality is that we cannot simply empty out neighborhoods with populations totaling 100,000 people. Remediation of environmental degradation gets to the root of these problems and constitutes a more practical approach to preventing premature deaths than calls for more education.

Other Statistical Tests

Since the data used in this research were nested in a spatial relationship, there is a potential for it to be biased due to spatial dependence. Considering this, we tested the model for spatial autocorrelation.[1] In doing so, the LaGrange Multiplier indicated that the model did suffer from spatial lag, and in running both a spatial lag model (SLM) and a spatial error model (SEM), we found the SLM was a better fit for the data and had higher levels of significance. However, both models' likelihood ratio tests indicated that spatial dependence is still present. The results of the model revealed no significant changes in the test variable's (Rubbertown) statistically significant. impact on neighborhood disparities in years of projected life lost. Therefore, being within the Rubbertown boundary presented here or within a one-and-a-half-mile radius of a brownfield site still has a significantly correlated relationship with years of potential life lost. We also found that the education variable washes out in terms of statistical significance on the lifespan measure, further showing it has no theoretical basis to justify its inclusion in the equation.

Additionally, in order to better understand the power of each coefficient of the control and test variables to explain years of potential life lost, we transformed scale (non-dummy) variables, including our dependent variable, into z-scores and reran the regressions.[2]

The interpretation of the standardized coefficients from these models provides a more accurate interpretation of the substantive magnitude of each variable. The weight of the coefficient is calculated by measuring for every one standard deviation change in the dependent variable, the change in the standard deviation for the independent variables. After conducting this specification, we discovered that our Rubbertown variable exceeds the impact of the crime variable, but the variables controlling for race and income remain the most powerful in the model.

We have used several statistical methods to test whether proximity to Rubbertown explains neighborhood disparities in premature death rates. We found, in each test, the same results:

1. Cluster analysis—Rubbertown was statistically significant.
2. Discriminant analysis—Rubbertown was statistically significant.
3. Regression runs—all five specifications of the model proved that Rubbertown was consistently statistically significant.
4. Z-score conversion runs of the independent variables showed statistical significance and moreover, the coefficient weight became stronger over the crime variable.
5. Spatial autocorrelation analysis showed that key test variable Rubbertown remained significant while education dropped out.

Policy Implications and Conclusions

Aside from the Greater Louisville Project/Robert Wood Johnson Foundation, we do not know of any mainstream scientific research that expresses the view that education trumps income and environmental contaminants as causes of high premature deaths in neighborhoods. This "statistical fallacy" simply goes against established science. A more problematic interpretation of the education effect is that it simply "whitewashes" environmental toxins as a cause of premature death rates in neighborhoods, allowing elected leaders to conduct business as usual. It allows business leaders to rally around education instead of focusing on industries that pollute the environment and cause thousands of premature deaths. We note that the captains of industry are often united in their calls for "more jobs" and reduced environmental protections. The apparent impact of education provides a convenient scapegoat that public and private sector leaders can readily embrace and successfully push on the public—claiming that health is a choice, not circumstance.

The Environmental Protection Agency (EPA) is, as the name suggests, the nation's primary protector of our air, water, and ground. Yet, political opponents and those in industry often write the agency off as little more than a job killer.

Particularly in the South, politicians who decry the job-killing effects of environmental protection win elections. The environmental movement needs to get wiser because it is losing ground. Proponents of environmental protection need to change the debate from jobs to healthy life expectancy. They might begin by asking how many people want to continue living lives shortened by ten years because of polluted air and environmental degradation in all its forms in southern states, in West Louisville, and no doubt in distressed neighborhoods in many other cities nationwide. Environmental issues should be debated as the life-and-death issues that they really are.

Environmental justice means that all neighborhoods, whether minority or impoverished, are given the same environmental protections as the more powerful neighborhoods. Environmental degradation is the major reason why thousands of people are prematurely dying in West Louisville and is to a certain degree shortening lives throughout Louisville—West, North, and South. Toxins in the air, water, and soil are killing Louisville residents. These findings might also help explain why Louisville has one of the shortest lifespans of any city in America with up to five years lost when compared to New York, Los Angeles, San Francisco, Santa Barbara, and Santa Rosa. Everyone who lives there pays a price, along with children bused into nearby schools and those who work there.

Lower-income communities and racial and ethnic minority populations have historically been the targets of disproportionate amounts of pollution. To provide more insight into these problems, our study builds on previously published linear regression models. This kind of modeling allows us to better understand the impacts of neighborhood environmental toxins on health. In doing so, we advance the notion that neighborhoods located on or near toxic sites shorten lives.

People live longer, up to 10 years longer, in neighborhoods that have clean air than residents of polluted places. This is an empirical reality. Inequalities are generated within and among cities that contribute to varying life chances, including the length of life itself. Instead of responding with more pills, radiation, implants, cutting, and gadgets, why not address the causes of these premature death rates? Why not remove the toxins in what we drink, breathe, and plant in our neighborhoods? We are spending billions of dollars to find a cure for diseases that kill while ignoring the causes of these problems.

We need to adjust our thinking to be more proactive instead of reactive in public health. As the Physicians for Social Responsibility (n/d) point out, we should be "preventing what we cannot cure." One such preventive measure is ensuring that our communities, including our poor inner-city neighborhoods, enjoy a clean environment. A cleaner environment also addresses the challenge of climate change/chaos. It is a win-win for everyone and improves our stewardship of the fragile earth.

One of the problems with American exceptionalism is that it might be a problem everywhere else, but in the USA, pressure is put on scientists not to attack the pollution industry. Many think that Flint, Michigan, is the exception, but there are hundreds of Flints around the nation, and Louisville is one of them. Proponents of environmental protection need to shift the spotlight to healthy life expectancy, asking why so many people needlessly die too soon near polluted neighborhoods. At least part of the cure for many of our cities is in the air, water, soil, streets, and trees.

NOTES

1. Gilderbloom, J. H., W.L. Meares, and G.D. Squires (2020) "Pollution, place and premature death: Evidence from a midsized city." *Local Environment: International Journal of Sustainability and Justice.* Published online, April 18, 2020 Volume 25 (6):419–432, https://doi.org/10.1080/13549839.2020.1754776.
2. Gilderbloom, J. H., W.L. Meares, and G.D. Squires (2020) "Pollution, place and premature death: Evidence from a midsized city." *Local Environment: International Journal of Sustainability and Justice.* Published online, April 18, 2020 Volume 25 (6):419–432, https://doi.org/10.1080/13549839.2020.1754776.

REFERENCES

Action on ambient air pollution. (2013). *The Lancet* [Editorial], 382(9897), 1000. doi: 10.1016/S0140-6736(13)61960-1

Agyeman, J. (2013). *Introducing just Sustainabilities: Policy, Planning, and Practice.* London: Zed Books.

American Lung Association. (2013). *State of the Air 2013: Most Polluted Cities.* Retrieved from http://www.stateoftheair.org/2013/city-rankings/most-polluted-cities.html

Appelbaum, R. (1978). *Size, Growth, and U.S. Cities.* New York: Praeger Publishing.

Appelbaum, R. P., Bigelow, J., Krammer, H. P., Molotch, H. L., & Relis, P. M. (1976). *The Effects of Urban Growth: A Population Impact Analysis.* New York: Praeger.

Banzhaf, S. (2012). *Superfund Taint and Market.* Stanford: Stanford University Press.

Barnett, R. (2011). *Environmental Issues: Louisville, KY.* Louisville: Kentucky Institute for the Environment and Sustainable Development.

Been, V. (1995). Analyzing evidence of environmental justice. *Journal of Environmental Land Use and Environmental Law*, 11, 1–37.

Been, V., & Gupta, F. (1997). Coming to the nuisance or going to the barrios: A longitudinal analysis of environmental justice claims. *Ecology Law Quarterly*, 24, 1–56.

Bullard, R. D. (1990). *Dumping in Dixie: Race, Class, and Environmental Quality* (Vol. 3). Boulder: Westview Press.

Bullard, R. D. (2000). *Dumping in Dixie: Race, Class, and Environmental Quality.* Rev. ed. Boulder: Westview Press.

Bullard, R. D., Mohai, P., Saha, R., & Wright, B. (2008). Toxic wastes and race at twenty: Why race still matters after all of these years. *Environmental Law*, 38, 371.

Chetty, R., Stepner, M., Abraham, S., Lin, S., Scuderi, B., Turner, N., Bergeron, A., & Cutler, D. (2016, April). The association between income and life expectancy in the United States, 2001–2014. *Journal of the American Medical Association*, 315(16), 1750–1766.

Dannenberg, A. L., Frumkin, H., & Jackson, R. J. (2011). *Marking Healthy Places: Designing and Building for Health, Well-being and Sustainability*. Washington, DC: Island Press.

Editorial. (2014). *The Lancet*, 383(9919), 785–795.

Environmental Protection Agency. (2006). Brownfield solutions series: Anatomy of a brownfields. *Brownfield Solution Series*. Retrieved from http://www.epa.gov/ Brownfields/overview/anat_bf_redev_101106.pdf

Environmental Protection Agency. (2011). *This is Superfund: A Community Guide to EPA's Superfund Program*. Retrieved from 9.13.11 Environmental Protection Agency EPA: Washington D.C., from http://www .epa .gov /envirofw /html / frs demo /geospatial data /geo data state combined .html

Environmental Protection Agency. (2012). *Facility Registry System State CSV Download File Descriptions*. Retrieved from http://www.epa.gov/envirofw/html/ frs_demo/geospatial_data/geo_data_state_combined.html

Environmental Protection Agency. (2013). *Superfund Sites Region 4*. Retrieved from http://www.epa.gov/region4/superfund/sites/sites.html#KY

Frank, L. D., Engelke, P., & Schmidt, T. L. (2003). *Health Community Design: The Impact of the Built Environment on Physical Activity*. Washington, DC: Island Press.

Frederick, C., Hammersmith, A., & Gilderbloom, J. H. (2019). Putting "place" in its place: Comparing place-based factors in interurban analyses of life expectancy in the United States. *Social Science and Medicine*, 232, 148–155.

Frederick, C., Riggs, W. W., & Gilderbloom, J. (2018). Commute mode diversity and public health: A multivariate analysis of 148 US cities. *International Journal of Sustainable Transportation*, 12(1), 1–11. doi: 10.1080/15568318.2017.1321705

Frederick, C., & Gilderbloom, J. (2018). Commute mode diversity and income inequality: an inter-urban analysis of 148 midsize US cities. *Local Environment: The International Journal of Justice and Sustainability*, 23(1), 54–76. doi: 10.1080/13549839.2017.1385001

Gilderbloom, J. (2009). Amsterdam: Planning and policy for the ideal city? *Local Environment*, 14(5), 473–494.

Gilderbloom, J. I., & Appelbaum, R. P. (1987). *Rethinking Rental Housing*. Philadelphia: Temple University Press.

Gilderbloom, J. H., & Meares, W. L. (2022). How inter-city rents are shaped by health considerations of pollution and walkability: A study of 146 mid-sized cities. *Journal of Urban Affairs*, 44(8), 1059–1075.

Gilderbloom, J. H., Meares, W. L., & Squires, G. D. (2020). Pollution, place and premature death: Evidence from a midsized city. *Local Environment: International Journal of Sustainability and Justice*, 25(6), 419–432. doi: 10.1080/13549839.2020.1754776

Gilderbloom, J. I., Squires, A. G., Hanka, M., & Kenitzer, Z. (2012). Investors: The missing piece in the foreclosure racial gap debate. *Journal of Urban Affairs, 34*(5), 559–582.

Gilderbloom, J. H., Squires, G. D., & Meares, W. (2020). Mama, I can't breathe. Louisville's dirty air has steep medical and economic costs. *Local Environment, The Conversation, 25*(8), 619–627. doi: 10.1080/13549839.2020.1789570

Gilderbloom, J., Squires, G. D., Riggs, W. W., & Capek, S. (2017). Think globally, act locally: Neighborhood pollution and the future of the earth. *Local Environment: The International Journal of Justice and Sustainability, 22*(7), 894–899. doi: 10.1080/13549839.2017.1278751

Gilderbloom, J., Riggs, W. W., & Meares, W. L. (2015). Does walkability matter? An examination of walkability's impact on housing, values, foreclosures and crime. *Cities, 4*(Part A), 13–24.

Global Burden of Disease Study. (2010). *The Lancet*, December 13, 2012.

Goodman, P. (1960). *Growing Up Absurd*. New York: Vintage Books.

Greater Louisville Project. (2013). *Building a Healthier Louisville*. Louisville: Community Foundation of Louisville.

Hamilton, J. (1995). Testing for environmental racism: Prejudice, profits, political power? *Journal of Policy Analysis and Management, 14*(1), 107–132.

Hipp, J. R., & Lakon, C. M. (2010). Social disparities in health: Disproportionate toxicity proximity in minority communities over a decade. *Health and Place, 16*(4), 674–683. doi: 10.1016/j.healthplace.2010.02.005

Kuehn, B. M. (2014). WHO: More than 7 million air pollution deaths each year. *Journal of the American Medical Association, 311*(15), 1486. doi: 10.1001/jama.2014.4031

LaCour-Little, M., & Green, R. K. (1998). Are minorities or minority neighborhoods more likely to get low appraisals? *Journal of Real Estate Finance and Economics, 16*(3), 301–315.

Lewis-Beck, M. (1980). *Applied Regression*. Newbury Park: Sage.

Lindauer, T., & Hunt, D. (2008). *2007 Real Estate Market in Review*. Louisville: Property Valuation Office.

Massey, D. S. (2004). Segregation and stratification: A biosocial perspective. *Du Bois Review: Social Science Research on Race, 1*(1), 7–25.

McCright, A. M., &. Dunlap, R. E. (2000). Challenging global warming as a social problem: An analysis of the conservative movement's counter-claims. *Social Problems, 47*(4), 499–522.

McGinnis, J. M., Williams-Russo, P., & Knickman, J. R. (2002). The case for more active policy attention to health promotion. *Health Affairs, 21*(2), 78–93. doi: 10:1377/hlthaff.21.2.78

McMeekin, I. M. (1946). *Louisville: The Gateway City*. New York: Julian Messner.

Organisation for Economic Co-operation and Development. (2012). Environmental outlook to 2050: The consequences of inaction. *OECD Publishing*, March 15, 2012. doi: 10.1787/9789264122246-en

Pahl, R. E. (1975). *Whose City?* London: Penguin.

Pastor, M., Sadd, J., & Hipp, J. (2001). Which came first? Toxic facilities, minority move-in, and environmental justice. *Journal of Urban Affairs*, 23(1), 1–21.

Peterson, E. (2013). Rubbertown and health: The whole series. [Radio series]. *Louisville Public Media*, January 27, 2013. Retrieved from http://wfpl.org/post/Rubbertown-and-healthwhole-series-0

Physicians for Social Responsibility. (n.d.). Website slogan. Retrieved May 31, 2014, from http://www.psr.org/

Richardson, T. (1998). *The Effects of Race on Public Health Planning: An Analysis of Premature Mortality in an Urban Environment.* Doctoral Dissertation University of Louisville.

Robert Wood Johnson Foundation. (2012). *County Health Rankings.* http://www.cou ntyhealthrankings.org/our-approach). Population Health Institute, University of Wisconsin, Wisconsin.

Smith, C. L. (2009). Economic deprivation and racial segregation: Comparing Superfund sites in Portland, Oregon and Detroit, Michigan. *Social Science Research*, 38(3), 681–692.

Smith, P., Pennington, M., Crabtree, L., & Illback, R. (2011) *Louisville Metro Health Equity Report: The Social Determinants of Health in Louisville Metro Neighborhoods.* REACH of Louisville, Inc.

Swanston, S. F. (2012). An environmental justice perspective on superfund reauthorization. *Journal of Civil Rights and Economic Development*, 9(2), 18.

United States Census Bureau. (2000). *American Community Survey.* Retrieved from American Community Survey. http://www.census.gov/acs/www/

United States Census Bureau. (2007). *American Community Survey.* Retrieved from American Community Survey Online: http://www.census.gov/acs/www/.Trade *Secrets: A Moyers Report (2001) www.pbs.com*

Wong, E. (2013). Early deaths linked to China's air pollution totaled 1.2 million in 2010, Data Shows. *The New York Times*, April 2, 2013.

Zimmerman, R. (1993). Social equity and environmental risk. *Risk Analysis*, 13(6), 649–666. doi: 10.1111/j.1539-6924.1993.tb01327

Chapter 10

How Brownfield Sites Kill Places and People

An Examination of Neighborhood Housing Values, Foreclosures and Lifespan

John Hans Gilderbloom, Wesley
L. Meares, and Billy Riggs

This study examines the impact of EPA Brownfield sites on neighborhood housing values, foreclosures, and premature deaths. While previous research has focused on the impact of Brownfield sites on neighborhood housing values, little research has been done on the impact of Brownfield sites on foreclosures and premature deaths (Noonan et al., 2007; Decker, Nielsen, and Sindt, 2005; Ihlanfeldt and Taylor, 2004). This study investigates 170 neighborhoods (Census Tracts) in Louisville, KY, a mid-sized city. Louisville serves as an example of the hundreds of mid-sized cities in the United States. Its dynamics are far more representative than those of large megacities such as New York or Los Angeles, and the relative isolation of Louisville produces a better estimate of costs and longevity (Appelbaum, 1978; Appelbaum et al., 1976; Ambrosius et al., 2010). Our efforts look simultaneously at economic, social, and health concerns while controlling for a number of essential variables.

The price effects of Brownfield sites on urban markets have not been extensively documented (as noted in: Decker, Neilsen, and Sindt, 2005; Greenberg and Hughes, 1992; Ihlanfeldt and Taylor, 2004; McCluskey and Rausser, 2003; Noonan, Krupka, and Baden, 2007). Many studies (Been, 1995; Pastor, Sadd, and Hipp, 2001; Bullard, 1990, 2000) identify potential limitations of a non-context-sensitive modeling approach, especially in areas of color, and we attempt to rectify that by measuring the market effects of Brownfield sites using census tracts in Louisville, KY. We employ an Ordinary Least Squares

(OLS) regression on variables previously unexamined in Brownfield site studies and illuminate a more complete depiction of the social, health, and monetary costs of inequitable—and possibly unjust—presence of Brownfield sites in largely low-income host neighborhoods, which often are predominantly African American. In particular, we examine the rate and character of proximal foreclosures, providing a window into contemporary neighborhood dynamics. Finally, this study offers policy suggestions within the context of an anecdotal case to illustrate the possible impacts of site remediation on neighborhoods with Brownfield sites.

REVIEW OF THE LITERATURE

Although public concern for environmental contamination and its subsequent impact on property values has been studied exhaustively over the last several decades, discussion regarding the impact of environmental contamination on nearby foreclosure rates has been surprisingly absent from the literature. Previous studies have most commonly employed Rosen's (1974) hedonic price analysis to estimate the cost of hazardous waste sites to host neighborhoods and have utilized regression analysis to control for demographics and other intangibles (Decker, Neilsen, and Sindt, 2005; Greenberg and Hughes, 1992; Stegman, 1969; Stegman and Sumka, 1978; Ihlanfeldt and Taylor, 2004; McCluskey and Rausser, 2003; Noonan, Krupka, and Baden, 2007; Gilderbloom and Appelbaum, 1988). Similarly, studies on foreclosure rates in close proximity to other vacant or foreclosed homes employ this methodology (Clauretie and Daneshvary, 2009; Harding, Rosenblatt, and Yao, 2009; Leonard and Murdoch, 2009; Li and Morrow-Jones, 2010). We consider this methodology to be appropriate and precise. Therefore, we do not seek to challenge this methodology; we intend to implement it to shed new light on the effects of Brownfield sites on property valuation, foreclosure, and longevity.

A limited amount of research has been conducted on housing markets in low-income African American communities, both with regard to the frequency of foreclosures (Edmiston, 2009; Li and Morrow-Jones, 2010; Gilderbloom et al., 2012) and proximity to Brownfield sites (Noonan, Krupka, and Baden, 2007; Pastor, Sadd, and Hipp, 2001; Bullard, 1990; 2000). To our knowledge, no existing study has looked at the relationship between foreclosures and proximity to environmental hazards.

Pricing the Effects of Brownfield Sites in Host Neighborhoods

Previous studies on the impact of environmental hazards on property values illuminate an array of negative price effects for host neighborhoods. Reported adjacency effects range from twelve percent to one percent of total property value, while other studies indicate price effects ranging from "insignificant" to as much as 6 percent per mile away from designated waste sites (Noonan, Krupka, and Baden, 2007, p. 667).

Longitudinal studies reflecting the price effects of Brownfield site cleanup over time attempt to build upon the original data collected. Noonan et al. (2007) find that accurate price differentiation can only be ascertained by allowing sufficient time for turnover to occur in the housing market; thereby, illustrating a more complete analysis of the affected housing stock and other neighborhood intangibles, such as neighborhood sorting, to emerge. Several longitudinal studies indicate that housing appreciation rates may be reduced at the initial listing of a Brownfield site on the National Priorities List (NPL) and will continue for many years subsequent to the site's rehabilitation (McCluskey and Rausser, 2003; Kiel and McClain, 1995; Crone and Voith, 1992).

Alternatively, indirect non-price effects on neighborhoods in close proximity to brownfield sites have also been found to be particularly relevant when estimating property values. Simple hedonic estimates may overlook certain neighborhood dynamics directly impacted by the presence of toxic sites; for example, environmentally induced migration and stigma (Noonan et al., 2007; Decker, Nielsen, and Sindt, 2005; Ihlanfeldt and Taylor, 2004).

According to Elliott-Jones (1996), adverse public opinion following the discovery of environmental contamination in the real estate market is reflected in housing prices. Stigma following awareness of environmental contamination can be permanent or temporary. This stigma may also have an impact on the speed at which housing prices revert back to pre-contamination levels following cleanup (McCluskey and Rausser, 2003). Studies show that costs associated with hazardous sites appear to be more correlative with the publicity of the hazardous site than indicative of actual danger (Kolhase, 1991; Michaels and Smith, 1990). The correlation between media exposure and housing depreciation, as well as the correlation between the timing of placement on the NPL of Brownfield sites and subsequent depreciation, support this claim (McCluskey and Rausser, 2003). Similarly, Ihlanfeldt and Taylor's (2004) study found no price differentiation prior to public announcement.

The passing of the Emergency Planning and Community Right-to-Know Act of 1986 (EPCRA) brought with it a range of negative stigmas associated with fears of cleanup costs and property devaluation (Decker et al., 2005). However, the vast majority of hazardous waste sites are not assigned to the NPL, despite severe contamination (Ihlanfeldt and Taylor, 2004). The price effects of these unlisted sites may be the subject of future study.

Brownfield Sites and Race

The literature has long established that hazardous waste sites are disproportionately located in minority neighborhoods (Been, 1995; Pastor, Sadd, and Hipp, 2001; Bullard, 1990, 2000). However, there has been some discrepancy as to whether hazardous sites have relocated to minority neighborhoods, or whether minority move-in occurred subsequent to their location (Bullard, 2000; Pastor et al., 2001). Recent studies indicate that disproportionate sites are the cause of this phenomenon (Hipp and Lakon, 2010; Smith, 2009; Swanston, 2012). This can be due to the neighborhood's lack of political power (Hamilton, 1995) or as a result of market appeal (Banzhaf, 2012). In many cases, low-income neighborhoods have lower property costs; therefore, they are substantially more attractive to industrial, profit-seeking enterprises (Been and Gupta, 1997; Pastor et al., 2001), which creates a larger social justice issue that operates along economic disparities rather than racial or ethnic biases.

However, there are some demographic effects resulting from Brownfield sites, as well as Brownfield site remediation. The reduction in property values subsequent to sites makes homes affordable to lower-income groups, encouraging minority move-in. Upon remediation, the continued devaluation in these neighborhoods attracts a larger minority population, as well as more renters (Noonan et al., 2007).

In addition to disproportionate Brownfield sites, minority neighborhoods also suffer from high foreclosure rates. Edmiston (2009) and Gilderbloom et al. (2012) confirm that foreclosure rates are highest in low-income neighborhoods, but only with regard to foreclosures originating from subprime mortgage agreements. Controlling for subprime mortgage agreements, low-income, minority neighborhoods actually show lower foreclosure rates than their higher-income counterparts (Edmiston, 2009; Gilderbloom et al., 2012). Several studies support the long-held assertion that discriminatory banking procedures award minorities with higher numbers of subprime mortgage agreements (Lauria, Baxter, and Bordelon, 2004). The higher interest rates associated with subprime mortgages further limit mobility for minorities in depressed neighborhoods as their housing stock depreciates and becomes difficult to sell (Ferreira, Gyourko, and Tracy, 2010).

High foreclosure rates have a number of detrimental neighborhood effects for an already marginalized population (Kiel and Williams, 2007). The out-migration and housing turnover in minority neighborhoods disrupt neighborhood dynamics, resident attachment to the neighborhood, political economy, and credit quality for future home loans, thereby re-segregating and disempowering minority homeowners (Li and Morrow-Jones, 2010).

Studies also indicate that proximity to existing foreclosures and vacancies further reduces neighborhood home values (Leonard and Murdoch, 2009;

Edmiston, 2009; Scheutz, Been, and Ellen, 2008; Ferriera, Gyourko, and Tracy, 2010; Clauretie and Daneshvary, 2009; Harding, Rosenblatt, and Yao, 2009). Each additional foreclosure within 250 feet of a listed sale reduces its price by approximately $1,666 (Leonard and Murdoch, 2009, 317). Other studies indicate as much as a 20 percent price reduction to an adjacent property, and discounts increase with the number of proximal foreclosures (Clauretie and Daneshvary, 2009). Property values are further reduced by the increasing supply of vacant or available properties within the entire neighborhood. Vacant properties are associated with an increased risk of property crime, which makes the housing stock less attractive to potential buyers (Edmiston, 2009; Li and Morrow-Jones, 2010).

However, many studies do not control for the indirect intangibles associated with concentrated foreclosures and vacancies. Perceived neighborhood quality, resulting from poorly maintained properties, causes as much as one-third of the negative price effects of foreclosures (Clauretie et al., 2009). Low-income residents may divert spending from routine maintenance and reinvestment as their property values decrease, thus contributing to the number of unmaintained and vacant structures (Leonard and Murdoch, 2009). This "contagion effect" occurs at the onset of physical distress, lasts through the foreclosure sale, and then stabilizes (Harding, Rosenblatt, and Yao, 2009).

Especially characteristic of minority neighborhoods is the proportion of investment or non-owner-occupied properties. Approximately 30 percent of all foreclosures are not owner-occupied and are disproportionately located in minority neighborhoods (Edmiston, 2009). The contagion effect of investment property foreclosures, combined with a lack of incentive to care for non-owner-occupied land, poses additional threats to these homeowners.

Contemporary Neighborhood Effects

Many of the previously mentioned studies sought to isolate their independent variables and apply controls to explain the variables' effect; however, these studies did not account for the important factors surrounding Brownfield site effects. While this study isolates a key independent variable—the location of the brownfield site—it further examines this phenomenon using a set of intervening variables, which might also explain the effects of brownfield sites. This aligns more closely with the current discourse surrounding the equitable allocation of urban resources (Bhan, 2009; Harvey, 2008; Mitchell, 2012), as well as underscoring the intrinsic benefits of cleaning up contaminated sites (Gamper-Rabindran, Mastromonaco, and Timmins, 2011). Even though previous research has examined many of the independent variables in this study, this unique combination of variables has not been utilized thus far.

STUDY AREA: LOUISVILLE

Louisville, Kentucky, a medium-sized U.S. city, is one of almost 162 cities in the nation with a population greater than 50,000 that is not located within 90 miles of another neighboring city of more than 50,000 (Appelbaum et al., 1976; Gilderbloom and Appelbaum, 1988; Gilderbloom, 2008; Gilderbloom et al., 2011). Louisville provides an excellent case study since it is not a megacity like Los Angeles, Chicago, or New York. Louisville's urban landscape possesses characteristics representative of many other U.S. cities, such as Indianapolis, IN, Lexington, KY, Nashville, TN, and Fresno, CA (Ambrosius et al., 2010). These representative characteristics include a minority population of approximately one-third in the dense inner core and approximately one-fifth across the Louisville Metropolitan area. Popularly held to be the "gateway from the North to the South," Louisville's location on the border of the South and Midwest sections of the U.S., and its proximity to the Ohio River, similarly result in cultural characteristics comparable to other large northern and midwestern populations. River cities such as St. Louis, Memphis, New Orleans, and Cincinnati have similar characteristics as well (Ambrosius et al., 2010; McMeekin, 1946, 256).

Prior to 2003, the city of Louisville was a separate entity from Jefferson County (Ambrosius, 2010; Savitch and Vogel, 2004). Subsequent to their merging in 2003, the consolidated "Louisville Metro" ranked 18th in population nationally and 48th as a Metropolitan Statistical Area (U.S. Census Bureau, 2007). The central business district (CBD) Census Tract provides 52,000 jobs, or roughly 13 percent of all employment in Jefferson County (U.S. Census Bureau, 2000).

Louisville may be further reduced into sub-regions that include the following: the poorer African American region in western Louisville, the wealthier white neighborhoods in eastern Louisville, and the working-class mixed-race neighborhoods in the southern region. Louisville Metro government has attempted to re-define the city's perception of its housing markets subsequent to the merge with the reclassification of three new regions: the inner beltway; the area between the beltways; and outside the outer beltway (Louisville-Jefferson County Metro, 2006). Despite these efforts, eastern Louisville has retained its reputation as a superior housing submarket, and western Louisville is still largely perceived to house undesirable land and an impoverished populace.

Data

The geography of our study is Louisville Metro, and our unit of observation is the census tract. For the purpose of our study, we use census tract and neighborhood interchangeably. There are 170 tracts located within Louisville Metro. We use several key dependent variables to measure the impact of EPA Brownfield sites on the quality of urban neighborhood life: neighborhood median assessed residential housing value for the years 2000, 2006, and 2008; the number of foreclosures in the neighborhood between 2007–2008; and the number of years of life lost to premature deaths in 2010 by neighborhood per 100,000 residents.

Foreclosure Sales

One of the unique aspects of this study is the inclusion of foreclosures. The variable for the number of foreclosures in the neighborhood was gathered from data provided by the Jefferson County Property Valuation Administration (JCPVA). This variable is a percentage of homes in the neighborhood with mortgages that are entering foreclosure in 2007–2008. The use of this data by Gilderbloom et al. (2009) and Gilderbloom et al. (2012) provides reassurance that the data is reliable or without sizeable errors and that any errors that do exist are likely to be distributed evenly.

Housing Values

The dependent variable, median assess residential housing value, is a raw dollar amount of the neighborhood median assessed value for the years 2000, 2006, and 2008. The data for this variable was collected from the JCPVA and geo-coded by the Kentucky State Data Center using the Louisville/Jefferson County Information Consortium's (LOJIC) GIS system. The prior use of this data by Lindauer and Hunt (2008) and Gilderbloom et al. (2010) assures us that it is unlikely the data contains sizeable errors and any errors that do exist are likely to be evenly distributed.

Premature Death

The dependent variable measuring premature death is the years of potential life lost per 100,000 residents in 2010. The data is contained within the individual census tracts of Louisville, as provided by the Louisville Department of Health and Wellness. The residential population used to create this measure was based on the population reported by the 2010 Decennial Census. The Institutional Review Board (IRB) oversaw the creation of this variable to protect the identity of the individuals in the sample.

EPA Brownfield Site Measurement

We used data provided by the EPA in order to locate the Brownfield sites. EPA Brownfield sites are defined as abandoned hazardous waste sites that have been designated for cleanup using funds set aside based on the Comprehensive Environmental Response, Compensation, and Liability Act of 1980 (CERCLA). It is also important for the city to identify Brownfield sites in order to protect residents and receive state and federal funding for cleanup. The accuracy of these data for the Louisville area is the result of a pilot program funded by the state which uses aerial photography to identify potential Brownfield areas in Louisville. Once an area of potential concern is identified through aerial photography, the state launches a ground investigation to confirm if there is contamination. Although this method is not perfect, it allows for more precision.

For this study, we operationalized the EPA variable as the distance in miles between the census tracts and the nearest Brownfield site. This was created by geo-referencing the central point of each Brownfield site and measuring its distance to each census tract in order to identify the closest distance. Additionally, we created a second variation of the brownfield site variable[1]. We created three dummy variables: (1) within .5 mile of a brownfield site; (2) within 1 mile of a Brownfield site; and (3) within 1.5 miles of a Brownfield site. The researchers calculated the measurement based on the center point of

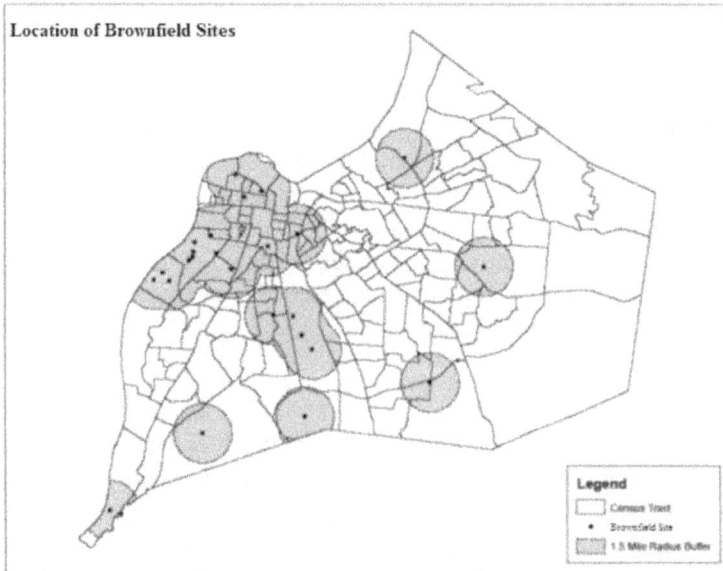

Figure 10.1 Location of Brownfield Sites. *Sources:* 1. Kentucky Institute for the Environment and Sustainable Development; 2. Center for Sustainable Urban Neighborhoods, formerly University of Louisville, transferred to Neighborhood Associates, Washington D.C.; 3. Gilderbloom, J.I., Meares, W.L., & Riggs, W. (2016). How brownfield sites kill places and people: an examination of neighborhood housing values, foreclosures, and lifespan. Journal of Urbanism: International Research on Placemaking and Urban Sustainability, 9:1, 1–18, DOI: 10.1080/17549175.2014.905488

the brownfield site. If any portion of a census tract fell within the buffer, the whole tract was included in the measurement.

Figure 10.1 shows the location of the EPA Designated Brownfield sites in Metro Louisville. Figure 10.2 shows the location of these toxic sites is not a random distribution but is clustered in the northwestern section of the county, where a majority of high-percentage minority and poor neighborhoods are located. No matter how we created the measurement, using the buffer or distance to the nearest site, our findings were consistent.

Our regression model uses control variables commonly used in regression analysis of contemporary neighborhood dynamics, using data at the census tract level to improve validity (Riggs, 2013). Regarding the independent variables, we rely on data provided by the following sources: (1a) the U.S. Census Bureau's 2000 Census (100 percent counts); (2) the Louisville Metro Police Department (LMPD) for 2004 data, the earliest year this type of data was available; (3) Louisville Department of Health and Wellness statistics for 2010; (4) and JCPVA. Table 10.1 provides a list of variables used in these models, shows how they were operationalized, and lists their descriptive statistics.

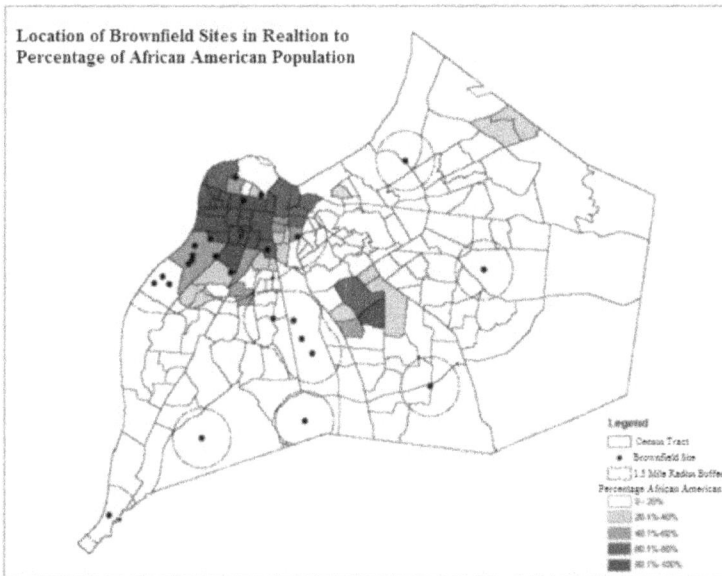

Location of Brownfield Sites in Realtion to
Percentage of African American Population

Legend

Census Tract
Brownfield Site
1.5 Mile Radius Buffer
Percentage African American
0 - 20%
20.1% - 40%
40.1% - 60%
60.1% - 80%
80.1% - 100%

Figure 10.2 Location of Brownfield Sites in Relation to Percentage of African American Population. *Sources*: 1. Kentucky Institute for the Environment and Sustainable Development; 2. Center for Sustainable Urban Neighborhoods, formerly University of Louisville, transferred to Neighborhood Associates, Washington D.C.; 3. Gilderbloom, J.I., Meares, W.L., & Riggs, W. (2016). How brownfield sites kill places and people: an examination of neighborhood housing values, foreclosures, and lifespan. Journal of Urbanism: International Research on Placemaking and Urban Sustainability, 9:1, 1–18, DOI: 10.1080/17549175.2014.905488

Table 10.1 Description of Dependent and Independent Variables

DV	Year	Source	Measure	Min	Max	Mean	Std Dev	N
Median assessed value	2000	JCPVA	$	4,105	255,510	88,594.3	49,071.5	169
Median assessed value	2006	JCPVA	$	8,820	333,765	117,455.5	61,543.9	167
Median assessed value	2008	JCPVA	$	32,030	366,070	125,925.7	66,533.4	168
Percentage of mortgaged units entering foreclosure (2007–2008)		JCPVA	#	0	12.75	5.13	2.83	170
Years of Potential Life Lost (YPLL) rate per 100k		JCHD	#	2,477.5	21,688.0	8,455.6	3,883.7	170
IV								
Distance to the central business district (CBD) tract (miles)	2000	Census	mi	0.0	18.6	7.0	4.0	170
Foreclosure sales from 07 to 08	07–08	JCPVA	#	0	197	54.8	44.6	170
Total crimes per 100,000 residents	2007	LMPD	#	193.7	51,216.6	6,500.3	5,432.8	170
Percent of nonwhite residents, 2000	2000	Census	%	1.4	99.4	25.4	29.5	170
Number of housing units, 2000	2000	Census	#	10	3,358	12,96.4	605.0	170
Median Household Income, 2000	2000	Census	$	6,086	110,472	40,524.45	19,527.82	170
Median housing age, 2000	2000	Census	#	2	60	38.7	15.1	170
Proximity to nearest Brownfield site (in miles)	2000	EPA	#	0	7.6	2.2	1.35	170

Sources: 1. Data sources included in table; 2. Kentucky Institute for the Environment and Sustainable Development; 3. Center for Sustainable Urban Neighborhoods, formerly University of Louisville, transferred to Neighborhood Associates, Washington D.C.; 4. Gilderbloom, J.I., Meares, W.L., & Riggs, W. (2016). How brownfield sites kill places and people: an examination of neighborhood housing values, foreclosures, and lifespan. Journal of Urbanism: International Research on Placemaking and Urban Sustainability, 9:1, 1–18, DOI: 10.1080/17549175.2014.905488

Notes: JCPVA= Jefferson County Property Valuation Administrator; LMPD = Louisville Metro Police Department; mi = mile.

Methods

Previous research suggests that the presence of nearby Brownfield sites puts downward pressure on housing values, but virtually no research has evaluated the impact on foreclosures and premature deaths. Our study builds on previously published linear regression models. This kind of modeling allows us to better understand the impacts of Brownfield sites on neighborhoods. In doing so, we advanced the notion that a Brownfield site's impact on urban life is an issue of proximity and propensity—the closer the Brownfield site, the greater the economic, health, and social problems. The researchers wanted to look simultaneously at these economic, social, and health concerns while controlling for a number of variables.

We constructed Ordinary Least Squares (OLS) regression models to analyze the relationship between Brownfield sites and neighborhood property valuation, foreclosures, and premature deaths. These variables illuminate a more complete depiction of the social and monetary costs of Brownfield practices within neighborhoods. Historically, housing price studies (done mostly by economists) rely on parcel-level data analysis, which allows substantial insight into the role of property characteristics and amenities (such as trees, the number of bedrooms, and the presence of a garage). The OLS approach can help understand intercity and intra-neighborhood differentials (Gilderbloom and Appelbaum, 1988; Ambrosius et al., 2010) with similar success in understanding variations in housing price and foreclosures by neighborhood census tracts.

A similar regression model can be applied in predicting premature mortality in neighborhoods by using the same socio-economic variables as used in the previous equations estimating neighborhood housing prices and foreclosures (Richardson, 1998, 131–166; Krieger and Fee, 1994; Dannenberg et al., 2011; Frank et al., 2003). While the data is not readily available for census tracts, we replicated what we had previously done by developing variables that would measure foreclosure levels (Richardson, 1998, 131–166; Dannenberg et al., 2011; Frank et al., 2003), which improved the overall explanatory power of the equation in predicting neighborhood premature deaths.

Since the focus of this paper is on neighborhood variation and effects, we feel that the approach of using OLS regression to examine differences at the neighborhood level is more prudent, helping to advance knowledge of the impacts of brownfield sites in U.S. cities. Furthermore, outside of housing values, no other dependent variable was available at the parcel/individual level; thus, in order to be consistent in our analysis, all variables were aggregated at the census tract level. A full model is represented below:

F (y)= α + Distance to the central business district (miles)+ High interest loans-foreclosure count + Number of housing units, 2000+Median housing age, 2000+Median Household Income, 2000 +Total crimes per 100000 residents 2007+ Percent of nonwhite residents, 2000 + Proximity to Nearest Brownfield site (in miles) + e (1)

Where y is the dependent variable and e is an error term

Our models proved to be robust with relatively high adjusted R squares (.64 to .86). Many of the control variables showed significance in the predicted direction. As noted above, the independent variables used are taken from previous studies and have been shown to be significant in measuring neighborhood effects (Ambrosius et al., 2010; Gilderbloom et al., 2012). For the most part, the control variables operate as expected and raise no concerns. While they are presented in the tables, space does not allow us to veer into a discussion of them—we want to keep the focus on the EPA Brownfield variable. We examined all the models presented in these tables for multicollinearity problems, and we found the zero-order correlation coefficients, variance inflation factors, and tolerance scores show no indication of violating the assumptions of OLS regression.

FINDINGS

Housing Values

EPA Brownfield sites have a negative impact on neighborhood housing values. The further away a neighborhood was from a brownfield site, the higher the median assessed value. This held true for the three years tested. Though the beta shows a modest negative impact, when we created three dummy variables identifying if a neighborhood was within .5 mile, 1 mile and 1.5 miles of a brownfield site, we were able to measure the impact. From the year 2000, there is statistical support revealing that houses within a half-mile radius of toxic sites were worth $10,342 less than houses outside of that area. When the scope is expanded to a 1-mile radius, homes inside the radius were worth $6,995 less than homes outside the radius. The researchers discovered similar results for median assessed values in the year 2006. When examining housing values for 2006, within a half-mile of a Brownfield site, we discovered that houses within this range were worth $7,615 less than houses outside of the radius. Houses located within a mile of a Brownfield site in 2006 were worth $7,556 less than other homes. Houses located within a mile and a half radius were worth $6,463 less.

The pattern continued in 2008. Homes located within a half-mile of the site were worth an average of $11,911 less, which was less than homes at a one-mile distance from the site, valued at $9,757 less. The latter homes were valued lower than homes at a one mile and a half mile distance. Homes in this radius were worth $8,261 less than homes outside of the radius.

Foreclosure

EPA Brownfield sites had a negative impact on the number of foreclosures within the neighborhood. The greater the distance from the Brownfield site, the lesser the number of foreclosures. When we conducted the analysis using the dummy variables, EPA Brownfield sites significantly increased the foreclosure rate in neighborhoods within a half-mile,, a mile and a mile and a half radius. During this time period, neighborhoods within half a mile of a Brownfield site saw 10 more foreclosures than neighborhoods outside the radius.

Premature Deaths

EPA Brownfield sites are correlated with premature deaths in neighborhoods. The analysis showed that people are more likely to die prematurely in neighborhoods with EPA Brownfield sites. This was also the case when we used a dummy variable to measure this as well. These findings contrast sharply with the work of Richardson (1998), who found that the environmental conditions of "place" had no impact on premature deaths in medium-sized cities such as Louisville. He argues that the lack of correlation between place and early death is the result of several factors. First, residents are highly mobile and do not stay long enough in one place for it to impact health. Secondly, the workplace has a bigger impact on premature deaths than residence.

The problem with these arguments is that Richardson (1998) made several methodological errors: (1) the number of cases examined was reduced to just West Louisville residents instead of the metro area; (2) the regression results did not sufficiently take into account the range of other variables considered by this study and those such as Macintyre (1989), Frank (2003), Danneberg et al. (2011) and Massey (2004); and (3) the analysis had a hypothesis that the workplace is more consequential than the quality of homes or the places that make up trips between work and home. Thus, this new study hopes to improve upon previous efforts to measure the impact of place on neighborhood health. (Table 10.2)

Table 10.2 OLS Regression Results Median Assessed Values for the Years 2002, 2006, and 2008

Specification	2000 Unst.	2000 Beta	2006 Unst.	2006 Beta	2008 Unst.	2008 Beta
(Constant)	35,250.33**		57,779.68***		57,532.24**	
Standard error	12,941.73***		17,423.59***		19,934.34**	
Distance to the central business district (miles)	−2,413.58***	−.198***	−3584.23***	−.232***	−443.626***	−.269***
High interest loans-foreclosure count	−466.88	−.069	−677.83 +	−.08 +	−1,126.44*	−.123*
Number of housing units, 2000	−8.306*	−.102*	−11.48*	−.11*	−8.67	−.078
Median housing age, 2000	−139.3	−.043	−389.07 +	−.095+	−526.5*	−.120*
Median Household Income, 2000	2.257***	.898***	2.92***	.919**	3.21***	.937***
Total crimes per 100,000 residents 2007	−.585 +	−.065 +	−.545	−.037	1.48**	.120**
Percent of nonwhite residents, 2000	−186.65***	−.112**	−143.56	−.068	−269.017**	−.118**
Proximity to Nearest Brownfield site (in miles)	3,734**	.103**	4,234.45*	.093*	4,224.73*	.086*
F	136.3		118.53		103.16	
R Square	.871		.856		.838	
Adjusted R Square	.865		.849		.830	
N	169		167		167	

Sources:: 1. Kentucky Institute for the Environment and Sustainable Development; 2. Center for Sustainable Urban Neighborhoods, formerly University of Louisville, transferred to Neighborhood Associates, Washington D.C.; 3. Gilderbloom, J. I., Meares, W. L., & Riggs, W. (2016). How brownfield sites kill places and people: an examination of neighborhood housing values, foreclosures, and lifespan. Journal of Urbanism: International Research on Placemaking and Urban Sustainability, 9:1, 1–18, DOI: 10.1080 /17549175.2014.905488*Note*: Unst.= Unstandardized coefficients. Beta = standardized. $p < 0.1+$. $*p < 0.05$. $**p < 0.01$. $***p < 0.001$.

Note: level of aggregation (tract)

Table 10.3 OLS Regression Results Percentage of Mortgage Units entering Foreclosures during 2007–2008

	Unst.	Beta
(Constant)	1.01	
Standard error	1.21	
Distance to the central business district (CBD) tract (miles)	.366***	.521***
Number of housing units, 2000	.001**	.182**
Median housing age, 2000	.062***	.327***
Median Household Income, 2000	-6.573 E-005***	−.453***
Total crimes per 100,000 residents 2007	−1.894 E-005	−.035
Percent of nonwhite residents, 2000	.058***	.603***
Proximity to Nearest Brownfield site (in miles)	−.285*	−.136*
F	44.098	
R Square	.656	
Adjusted R Square	.641	
N	169	

Sources: 1. Kentucky Institute for the Environment and Sustainable Development; 2. Center for Sustainable Urban Neighborhoods, formerly University of Louisville, transferred to Neighborhood Associates, Washington D.C.; 3. Gilderbloom, J. I., Meares, W. L., & Riggs, W. (2016). How brownfield sites kill places and people: an examination of neighborhood housing values, foreclosures, and lifespan. *Journal of Urbanism: International Research on Placemaking and Urban Sustainability*, 9:1, *1–18, DOI: 10.1080/17549175.2014.905488Notes:* Unst. = Unstandardized coefficients. Beta= standardized. p < 0.1+. *p < 0.05. **p < 0.01. ***p < 0.001. *Note:* level of aggregation (tract)

Table 10.4 OLS Regression Results Years of potential life lost (YPLL) per 100000 residents

	Unst.	Beta
(Constant)	5,610.8***	
Standard error	1,471.71**	
Distance to the central business district (CBD) tract (miles)	190.388*	.198*
High interest loans-foreclosure count	8.72	.016
Number of housing units, 2000	.112	.017
Median Household Income, 2000	−.063***	−.317***
Median housing age, 2000	62.62***	.243***
Total crimes per 100,000 residents 2007	.134***	.187***
Percent of nonwhite residents, 2000	57.67***	.438***
Proximity to Nearest Brownfield site (in miles)	−420.78**	−.147**
F	55.64	
R Square	.734	
Adjusted R Square	.721	
N	169	

Sources: 1. Kentucky Institute for the Environment and Sustainable Development; 2. Center for Sustainable Urban Neighborhoods, formerly University of Louisville, transferred to Neighborhood Associates, Washington D.C.; 3. Gilderbloom, J.I., Meares, W.L., & Riggs, W. (2016). How brownfield sites kill places and people: an examination of neighborhood housing values, foreclosures, and lifespan. *Journal of Urbanism: International Research on Placemaking and Urban Sustainability*, 9:1, *1–18, DOI: 10.1080/17549175.2014.905488Notes:* Unst. = Unstandardized coefficients. Beta = standardized. p < 0.1+. *p < 0.05. **p < 0.01. ***p < 0.001.
Note: level of aggregation (tract)

Case Study: Louisville's East Russell

From a policy standpoint, data is important but anecdotes resonate. The case study of Louisville's East Russell neighborhood can help contextualize results from our regression models and demonstrate the possible benefits of a Brownfield site remediation. Among all the former Brownfield sites that have been cleaned up, one in Louisville stands as an example of how a site cleanup can bring about change in terms of property values, foreclosures, and remaking the health of a neighborhood. The Louisville East Russell neighborhood was one of the most notoriously polluted areas in the city. It is predominantly Black and has suffered problems with crime, poverty, and abandonment for years. Contributing to these problems was an EPA Brownfield site, the site of the old Trolley Barn. Constructed in 1879, its original purpose was to house the city's mule-drawn trolleys and the animals that pulled them. Through the years, the Trolley Barn had become highly contaminated with a combination of fuel, pesticides, and chemicals. Though some former Brownfield sites look pristine, the Trolley Barn was a foreboding eyesore of high fences, barbed wire, boarded windows, crumbling bricks, and warning signs posted around the premises. The cleanup took years, since there was a risk of water-table contamination; however, once completed, the block was no longer foreboding. It was renovated into a beautiful postmodern structure that won praise for its preservation of old brick and wood beams and attention to natural light. The Trolley Barn is now the Kentucky Center for African American Heritage. It has served as an anchor institution of the community, used as an anecdote to support neighborhood revitalization. The university-community partnership resulted in quantifiable changes, including the second-largest increase in surrounding property values in Louisville, and the lowest foreclosure rate of any neighborhood—black or white.

The lesson of East Russell is exemplary—showing how a cleanup can help spark a neighborhood renaissance and reinfuse pride into a predominantly Black neighborhood. While several factors played a role in this neighborhood renaissance, the historic importance of the site was clearly a factor. The Trolley Barn was only one of several factors causing a continuous downward spiral in the quality of life among residents.

Policy Implications and Conclusion

As the authors wrote this paper, the EPA was addressing a highly contaminated Brownfield site called Black Leaf (Bruggers, 2013, B3). Currently, a multi-agency task force is examining how exposure to the site has deteriorated residents' health and affected foreclosures, property values, and the overall quality of the neighborhood. Inspired by the example of East Russell, the citizens have demanded that the government and community take action to environmentally remediate the neighborhood. Similar efforts have resulted in an award-winning "Waterfront park" surrounded by upscale new and renovated housing in the riverfront district of downtown Louisville, and the Papa John's Stadium project saw similar effects, where the cleanup occurred and neighborhood improvements followed.

But not all Brownfield site cleanups can be done successfully and result in neighborhood renewal. Unlike the Trolley Barn, where the contamination was contained on one site, the Black Leaf neighborhood has multiple properties that are owned by a number of individuals and entities. Finding these owners and getting their permission has been a daunting task that has still yet to be completed and could result in cleanup and neighborhood renewal not being fully addressed.

The data explored as a part of this study show a clear pattern of housing devaluation and increased chances of foreclosure for residents who live near Brownfield sites, which result in a reduced quality of life. This study is unique in that it shows a direct connection between neighborhood quality and the proximity of Brownfield sites, something planners and academics should note.

This relationship between environmental quality, housing value foreclosures, and premature death also provides urban policy makers with questions. How does one stop this cycle of ghettoization? How does one deal with the contagion effect and environmental handicap in communities with homes near Brownfield sites? The answers to these problems are anything but clear. It is complicated by trends in urban gentrification and chronic disinvestment in inner-city minority neighborhoods, let alone the recent housing crises (Gilderbloom, 2008; Massey, 2004; Riggs, 2011; Williams and Jackson, 2005). Yet, to paraphrase T. S. Eliot (1949), just because a problem is complicated does not mean it is not worth working to resolve—even if the answer is opaque. Put simply, environmental cleanup matters. It has value and should be prioritized, especially in areas that have experienced chronic disinvestment—where blue-collar working-class minorities are completely segregated from the urban white-collar residents.

This is the situation in many segregated urban areas, especially within the Black community, such as in East Russell. Since the 1960s, funding for urban regeneration near many Brownfield sites has dwindled steadily. Investment has come mainly from tax incentives and districts that support developer interests. Data has shown that there are far more Brownfield sites in areas of color, and fewer cleanup efforts for these areas (Zimmerman, 1993). The areas that need the most environmental cleanup receive the least amount of investment (Hird, 1993).

A confluence of issues, discussed here, frames the new face of environmental justice, which may exist in the absence of blatant racist intent—a product of economic inequality and socio-political exclusion (Mohai, Pellow, and Roberts, 2009). And, while modern society would assume that technology and globalization have helped raise awareness of many social problems, these societies have also created a distance that allows communities to overlook and disinvest in things like hazardous material cleanup and improvements to affordable housing or the streetscape (Mohai and Saha, 2006; Bullard, Mohai, and Wright, 2008). Reich suggests these clusters of inequity cause a "negative wealth effect" that can compound disinvestment (Jones, Reich, and Rogers, 2012; Reich, 2007). Places become overlooked because other locations, even within blocks, are not dependent on them—perpetuating the cycle of poverty and handicapping disadvantaged communities.

Urban planners have the power to stop this cycle of poverty and environmental NIMBYism by creating a new socioeconomic lens for Brownfield sites. Literature shows that long-term exposure to environmental problems has a negative impact on the health and welfare of minority residents (Francis, Szegda, Campbell, Martin, and Insel, 2003; Geronimus and Thompson, 2004; Gilderbloom, 2008; Goldman, Glei, Seplaki, Liu, and Weinstein, 2005; Meares, forthcoming). This extends to housing values, health, and foreclosures, suggesting that Brownfield sites are complicit in creating cycles of poverty and social injustice.

Planners need to assist poor neighborhoods by increasing the number of cleanups, which in turn will help sustain the neighborhood. This could be accomplished by using community-based and participatory planning to increase the lay understanding of, participation in, and advocacy for cleanup plans. Community-based planning allows grassroots, minority, and faith-based groups to better understand risks and potential outcomes and organize to incentivize development in places with the lowest probability of private investment. The action of empowering communities to seize their own reinvestment planning of hazardous sites might help remedy some of the apparent inequity in Brownfield policy and alleviate some of the urban market handicaps caused by environmental contamination.

Revised and reprinted with permission from *Journal of Urbanism: International Research on Placemaking and Urban Sustainability.*

ACKNOWLEDGMENTS

There are a number of people we would like to thank for their assistance in pursuing this research. Russ Barnett, Director of the Kentucky Institute for the Environment and Sustainable Development, provided information about the use of aerial photography in Louisville to identify Brownfield sites along with his leadership in the annual West Louisville environmental justice tour. Zach Kentizer (ABD at Ohio State University) gathered the original information on the location of the Brownfield sites and produced all the maps used here. Tony Lindauer, Property Value Administrator, and Jay Mickle, computer guru at the Jefferson County Property Valuation Administration, were instrumental in providing us with housing data. Also, Professor Greg Squires was helpful in obtaining data on foreclosures from the Federal Home Loan Bank. Our departments where some of this research was conducted as graduate students: U.C. Berkeley (Riggs) and the University of Louisville (Meares), whose support allowed us to conduct our research and present the findings at the 2013 Urban Affairs Association Meetings in San Francisco. Also, our new departments at University of San Luis Obispo (Riggs) and Georgia Regents University (Meares) have supported our ideas and research agendas. Lastly, we would like to thank the editors (Matt Hardy and Emily Talen) and referees whose comments only made this paper better. Each author made equal contributions, and the ordering is by alphabet.

REFERENCES

Appelbaum, R. (1978). *Size, Growth, and U.S. Cities*. New York: Praeger Publishing.

Appelbaum, R., Bigelow, J., Krammer, H., Molotch, H. and Relis, P. (1976). *The Effects of Urban Growth: A Population Impact Analysis*. New York: Praeger.

Ambrosius, J. D. (2010). *Religion and Regionalism: Congregants, Culture, and City-County Consolidation in Louisville, Kentucky*. Doctoral Dissertation. Louisville: University of Louisville.

Ambrosius, J., Gilderbloom, J. and Hanka, M. (2010). Back to black . . . and green? location and policy interventions in contemporary neighborhood housing markets. *Housing Policy Debate*, 20(3), 457–484.

Banzhaf, S. (2012). *Superfund Taint and Market*. Redwood City: Stanford University Press.

Barnett, R. (2013). Former Assistant Director of *Kentucky Department for the Environment Protection*; Current Director of the *Director of the Kentucky Institute for Environment and Sustainable Development*. Phone Interview: August 26, 2013.

Been, V. (1995). Analyzing evidence of environmental justice. *Journal of Environmental Land Use and Environmental Law*, 11, 1–37.

Been, V. and Gupta, F. (1997). Coming to the nuisance or going to the Barrios-A longitudinal analysis of environmental justice claims. *Ecology Law Quarterly*, 24, 1–56.

Bhan, G. (2009). "This is no longer the city I once knew." Evictions, the urban poor and the right to the city in millennial Delhi. *Environment and Urbanization*, 21(1), 127–142. https://doi.org/10.1177/0956247809103009

Bruggers, J. (2013). Crews remove soil at black leaf site. *Courier-Journal*, August 5, 2013.

Bullard, R. D. (1990). *Dumping in Dixie: Race, Class, and Environmental Quality* (Vol. 3). Boulder: Westview Press.

Bullard, R. D. (2000). *Dumping in Dixie: Race, Class, and Environmental Quality*. Boulder: Westview Press.

Bullard, R. D., Mohai, P., Saha, R. and Wright, B. (2008). Toxic wastes and race at twenty: Why race still matters after all of these years. *Environmental Law*, 38, 371.

Burgess, E. W. (1925). The growth of the city: An introduction to a research project. In *The City*, eds. R. E. Park, E. W. Burgess, and R. D. McKenzie, 47–62. Chicago: University of Chicago Press.

Chatman, D. and Voorhoeve, N. (2010). The transportation-credit mortgage: A post-mortem. *Housing Policy Debate*, 20(3), 355–382. https://doi.org/10.1080/10511481003788786

Clauretie, T. and Daneshvary, N. (2009). Estimating the house foreclosure discount corrected for spatial price interdependence and endogeneity of marketing time. *Real Estate Economics*, 37(1), 43–67.

Crone, T. and Voith, R. (1992). Estimating house price appreciation: A comparison of methods. *Journal of Housing Economics*, 2, 324–338.

Cullen, J. B. and Levitt, S. D. (1999). Crime, urban flight, and the consequences for cities. *Review of Economics and Statistics*, 81(2), 159–169.

Dannenberg, A. L., Frumkin, H. and Jackson, R. J. (2011). *Marking Healthy Places: Designing and Building for Health, Well-being and Sustaiability*. Washington, DC: Island Press.

Decker, C., Nielsen, D. and Sindt, R. (2005). Residential property values and community right-to-know laws: Has the toxics release inventory had an impact?. *Growth and Change*, 36(1), 113–133.

Edmiston, K. (2009). Characteristics of high-foreclosure neighborhoods in the tenth district. *Economic Review*, 94(Q2), 51–75.

Eliot, T. S. (1949). *Christianity and Culture*. New York: Harcourt.

Elliot-Jones, M. (1996). Stigma in light of recent cases. *Natural Resources and Environment*, 10(4), 56–59.

Ferreira, F., Gyourko, J. and Tracy, J. (2010). Housing busts and household mobility. *Journal of Urban Economics*, 68, 34–45.

Francis, D. D., Szegda, K., Campbell, G., Martin, W. D. and Insel, T. R. (2003). Epigenetic sources of behavioral differences in mice. *Nature Neuroscience*, 6(5), 445–446.

Frank, L. D., Engelke, P. and Schmid, T. L. (2003). *Health Community Design: The Impact of the Built Environment on Physical Activity*. Washington, DC: Island Press.

Gamper-Rabindran, S., Mastromonaco, R. and Timmins, C. (2011). Valuing the benefits of superfund site remediation: Three approaches to measuring localized externalities. *National Bureau of Economic Research*, working paper 16655. Retrieved from http://www.nber.org/papers/w16655

Geronimus, A. T. and Thompson, J. P. (2004). To denigrate, ignore, or disrupt: Racial inequality in health and the impact of a policy-induced breakdown of African American communities. *Du Bois Review: Social Science Research on Race*, 1(2), 247–279.

Gilderbloom, J. I. (2008). *Invisible City: Poverty, Housing and New Urbanism*. Austin: University of Texas Press.

Gilderbloom, J. I. and Appelbaum, R. P. (1988). *Rethinking Rental Housing*. Philadelphia: Temple University Press.

Gilderbloom, J. I., Ambrosius, J. D., Squires, G. D., Hanka, M. J. and Kenitzer, Z. E. (2012). Investors: The missing piece in the foreclosure racial gap debate. *Journal of Urban Affairs*, 34(5), 559–582.

Gilderbloom, J. I., Hanka, M. J. and Ambrosius, J. D. (2009). Historic preservation's impact on job creation, property values, and environmental sustainability. *Journal of Urbanism*, 2(2), 83–101.

Gilderbloom, J. I., Hanka, M. J. and Ambrosius, J. D. (2010). Without bias? government policy that creates fair and equitable property tax assessments. *The American Review of Public Administration*, 42(5), 591–605.

Greenberg, M. and Hughes, J. (1992). The impact of hazardous waste superfund sites on the value of houses sold in New Jersey. *Annals of Regional Science*, 26, 147–153.

Goldman, N., Glei, D. A., Seplaki, C., Liu, I.-W. and Weinstein, M. (2005). Perceived stress and physiological dysregulation in older adults. *Stress: The International Journal on the Biology of Stress*, 8(2), 95–105. https://doi.org/10.1080/10253890500141905

Goodstein, R. and Lee, Y. (2010). Do foreclosures increase crime? Rochester: SSRN.

Hamilton, J. (1995). Testing for environmental racism: Prejudice, profits, political power? *Journal of Policy Analysis and Management*, 14(1), 107–132.

Harding, J., Rosenblatt, E. and Yao, V. (2009). The contagion effect of foreclosed properties. *The Journal of Urban Economics*, 66, 164–178.

Harvey, D. (2008). The right to the city. *New Left Review*, 53(5), 23–40.

Hipp, J. R. and Lakon, C. M. (2010). Social disparities in health: Disproportionate toxicity proximity in minority communities over a decade. *Health and Place*, 16(4), 674–683. 10.1016/j.healthplace.2010.02.005

Hird, J. A. (1993). Environmental policy and equity: The case of Superfund. *Journal of Policy Analysis and Management*, 12(2), 323–343.

Ihlanfeldt, K. and Taylor, L. (2004). Externality effects of small-scale hazardous waste sites: Evidence from urban commercial property markets. *Journal of Environmental Economics and Management*, 47, 117–139.

Jones, C. N., Reich, R. and Rogers, G. S. I. T. (2012). Wealth and poverty *NYEPA*

Kiel, K. and McClain, K. (1995). The effect of an incinerator siting on housing appreciation rates. *Journal of Urban Economics*, *37*, 311–323.

Kiel, K. A. and Williams, M. (2007). The impact of Superfund sites on local property values: Are all sites the same? *Journal of Urban Economics*, 61(1), 170–192.

Kolhase, J. (1991). The impact of hazardous waste sites on housing values. *Journal of Urban Economics*, 30, 1–26.

Krieger, N. and Fee, E. (1994). Social class: The missing link in U.S. health data. *International Journal of Health Services*, 24(1), 22–44.

LaCour-Little, M. and Green, R. K. (1998). Are minorities or minority neighborhoods more likely to get low appraisals? *Journal of Real Estate Finance and Economics*, 16(3), 301–315.

Lauria, M., Baxter, V. and Bordelon, B. (2004). An investigation of the time between mortgage default and foreclosure. *Housing Studies*, 19(4), 581–600.

Leonard, T. and Murdoch, J. (2009). The neighborhood effects of foreclosure. *Journal of Geographical Systems*, 11(4), 317–332.

Li, Y. and Morrow-Jones, H. (2010). The impact of residential mortgage foreclosure on neighborhood change and succession. *Journal of Planning Education and Research*, 30(1), 22–39.

Lindauer, T. and Hunt, D. (2008). *2007 Real Estate Market in Review*. Louisville: Property Valuation Office.

Macintyre, S. (1989). The west of Scotland twenty-07 study: Health in the community. In *Readings for a New Public Health*, eds. Martin C. and D. MacQueen, 56–74. Scotland: Edinburgh University Press.

Massey, D. S. (2004). Segregation and stratification: A biosocial perspective. *Du Bois Review: Social Science Research on Race*, 1(1), 7–25.

McCluskey, J. and Rausser, G. (2003). Hazardous waste sites and housing appreciation rates. *Journal of Environmental Economics and Management*, 43, 166–176.

McMeekin, I. M. (1946). *Louisville: The Gateway City*. New York: Julian Messner.

Meares, W. L. (2014). *Uncovering Access: The Impact of Walkability on Neighborhood Composition and Housing*. Doctoral Dissertation. Louisville: University of Louisville.

Michaels, R. and Smith, V. (1990). Market segmentation and valuing amenities with hedonic models: The case of hazardous waste sites. *Journal of Urban Economics*, 28, 223–242.

Mitchell, D. (2012). *Right to the City: Social Justice and the Fight for Public Space*. New York: Guilford Press.

Mohai, P., Pellow, D. and Roberts, J. T. (2009). Environmental justice. *Annual Review of Environment and Resources*, 34, 405–430.

Mohai, P. and Saha, R. (2006). Reassessing racial and socioeconomic disparities in environmental justice research. *Demography*, 43(2), 383–399.

Noonan, D., Krupka, D. and Baden, B. (2007). Neighborhood dynamics and price effects of Superfund site cleanup. *Journal of Regional Science*, 47(4), 665–692.

Pastor, M., Sadd, J. and Hipp, J. (2001). Which came first? Toxic facilities, minority move-in, and environmental justice. *Journal of Urban Affairs*, 23(1), 1–21.

Reich, R. (2007). How capitalism is killing democracy. *Foreign Policy*, September/October.

Richardson, T. (1988). *The Effects of Race on Public Health Planning: An Analysis of Premature Mortality in an Urban Environment*. Doctoral Dissertation. Louisville: University of Louisville.

Riggs, W. (2011). Going home again. *Berkeley Planning Journal*, 23(1), 195–200.

Riggs, W. (2013). Steps toward validity in active living research: Research design that limits accusations of physical determinism. *Health & Place*, 26, 7–13. https://doi.org/10.1016/j.healthplace.2013.11.003

Rosen, S. (1974). Hedonic prices and implicit markets: Product differentiation in pure competition. *Journal of Political Economy*, 82, 34–55.

Rowan, G. T. and Fridgen, C. (2003). Brownfields and environmental justice: The threats and challenges of contamination. *Environmental Practice*, 5(1), 58–61. https://doi.org/10.1017/S1466046603030163

Sampson, R. J. and Wilson, W. J. (1995). Toward a theory of race, crime, and urban inequality. In *Race, Crime, and Justice: A Reader*, eds. J. Hagan and R. Peterson, 177–190. Redwood City: Stanford University Press.

Savitch, H. V. and Vogel, R. K. (2004). Suburbs without a city: Power and city-county consolidation. *Urban Affairs Review,* 39(6), 758–790.

Scheutz, J., Been, V. and Ellen, I. (2008). Neighborhood effects of concentrated mortgage foreclosures. *Journal of Housing Economics*, 17, 306–319.

Smith, C. L. (2009). Economic deprivation and racial segregation: Comparing Superfund sites in Portland, Oregon and Detroit, Michigan. *Social Science Research*, 38(3), 681–692.

Sooman, A. and Macintyre, S. (1995). Health and perceptions of the local environment in socially contrasting neighbourhoods in Glasgow. *Health and Place*, 1(1), 15–26.

Stegman, M. A. (1969). Accessibility models and residential location. *Journal of the American Institute of Planners*, 35(1), 22–29.

Stegman, M. A. and Sumka, H. J. (1978). Income elasticities of demand for rental housing in small cities. *Urban Studies*, 15(1), 51–61.

Swanston, S. F. (2012). An environmental justice perspective on Superfund reauthorization. *Journal of Civil Rights and Economic Development*, 9(2), 18.

Troy, A. and Grove, J. M. (2008). Property values, parks, and crime: A hedonic analysis in Baltimore, MD. *Landscape and Urban Planning*, 87(3), 233–245. https://doi.org/10.1016/j.landurbplan.2008.06.005

United States Census Bureau. (2000). *American Community Survey*. Retrieved from American Community Survey Online: http://www.census.gov/acs/www/

United States Census Bureau. (2007). *American Community Survey*. Retrieved from American Community Survey Online: http://www.census.gov/acs/www/

Williams, D. and Jackson, P. (2005). Social sources of racial disparities in health. *Health Affairs*, 24(2), 325.

Wilson, J. Q. and Kelling, G. (1982). The police and neighborhood safety: Broken windows. *Atlantic Monthly*, 127, 29–38.

Wilson, J. Q. and Kelling, G. (2003). Broken windows: The police and neighborhood safety. *Criminological Perspectives: Essential Readings*, 400.

Zimmerman, R. (1993). Social equity and environmental risk. *Risk Analysis*, 13(6), 649–666. https://doi.org/10.1111/j.1539-6924.1993.tb01327.x

Part III

NEIGHBORHOOD AND CITY EFFORTS TO COMBAT CLIMATE CHANGE

Chapter 11

Part 1: How to Make Our Schools Greener and Our Students Smarter

John Hans Gilderbloom, Stephen A. Roosa,
Isaiah Kingsberry, and Jennifer Stekardis

One of the greenest school buildings in the United States was not built in America's environmental capital of Oregon, but in the capital of coal country: Kentucky! In brief, green schools outperform other schools in the most significant metrics. We never expected to see such a powerful case for green schools. This chapter explains how, by cooperating with state and local school boards and two graduate students in my environmental policy seminar, we discovered these findings.

Conventional belief is that green schools cost more money to construct and are a greater burden on taxpayers than traditional buildings; however, we found the opposite to be true. The cost of building a green school was considerably less than that of most local schools. Many believe green energy systems are expensive and thus do not produce any real cost savings. While this might be true regarding solar panel installations on suburban houses, with effective leadership that avoided these costly obligations, the energy savings were spectacular, saving taxpayers lots of money. During the summer, the schools were selling excess electricity to the electric companies, earning payments averaging $3,500 monthly. The original intention was to go net zero—meaning any energy expended would be offset by electricity produced by the school's large solar arrays.

Academic performance in green schools was superior to brown schools. In fact, the longer students stayed in the green school, academic performance improved, especially in math. Part of the reason is that the school's lighting system uses sunlight and has a daylight harvesting system. These systems continuously vary the amount of artificial lighting by monitoring the natural light that is available from windows. These systems reduce the energy needed for lighting systems and reduce air conditioning loads. As numerous studies show, schools with day-lighting systems have higher academic performance. By utilizing a "green roof" or roof with vegetated areas and rainwater harvesting systems, the schools offer a living scientific laboratory that allows students to engage in science and math. Finally, attendance was more frequent at the green schools, likely due to the more interesting and interactive learning environments that enable students to retain information more effectively. Transforming our educational structures into green facilities is one method of enhancing the learning environments for students.

What is the impact of green schools concerning taxes, costs, pollution, educational achievement, health, and grades? Oddly, this has never been seriously studied. What are the environmental, economic, and educational outcomes of this project? For Kenton County Schools in Kentucky, the goal was to produce a cost-effective, state-of-the-art green building. Rob Haney, their facilities director, was responsible for tracking the consequences of the transition from energy-intensive schools to green schools. Haney's findings included: (1) enormous energy savings, meaning more money to invest in learning and reduced expenses for taxpayers; (2) daylighting in classrooms resulted in higher grades and test scores, less violence, and happier students; (3) the cost of building a green school is less than that of a typical school; and (4) water conservation techniques and rainwater collection systems reduce water costs. Duke Energy, the electric supplier, pays the school an average of $3,500 a month for exported electricity.

Turkey Foot Middle School was not designed with a goal of being a net zero building. However, it uses solar photovoltaic arrays to produce electricity

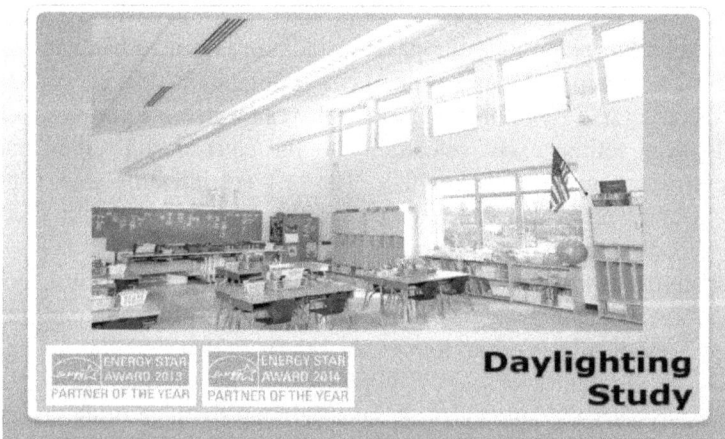

Figure 11.1 Daylighting Study. *Source*: Turkey Foot Middle School PowerPoint presentation. Our thanks to Turkey Foot Middle School for the use of this slide.

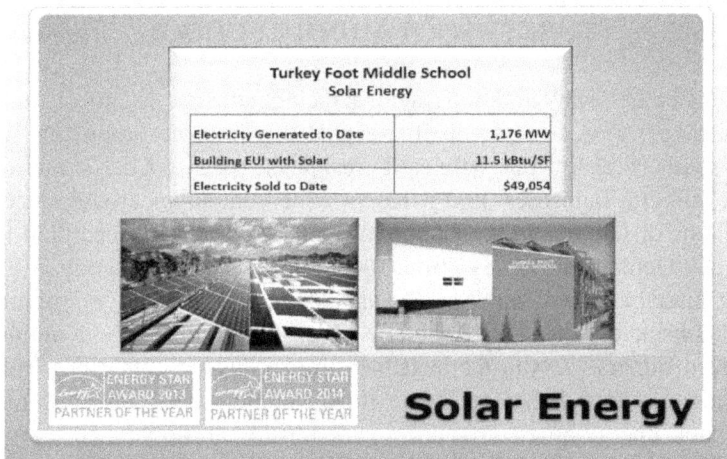

Figure 11.2 Solar Energy. *Source*: Turkey Foot Middle School PowerPoint presentation. Our thanks to Turkey Foot Middle School for the use of this slide.

Figure 11.3 Solar Collectors at Turkey Foot Middle School. *Source*: Image by Stephen Roosa.

and a ground-loop geothermal system to support heating and air conditioning requirements. Haney explains what can be learned from that moving forward. He will discuss how graduate and faculty students can collaborate to help with the performance evaluation of this landmark green school that covers energy savings, educational achievement, and health.

CASE STUDY: GREEN BUILDING TECHNOLOGIES AT
TURKEY FOOT MIDDLE SCHOOL

Kentucky boasts many traditions: a horse racing industry that attracts millions of tourists per year, bourbon distilleries that ship products around the world, and a southern charm that defines its residents' culture. Coal is one of the state's defining industries, and it has provided employment and electricity to millions of families in the Commonwealth for decades. According to the Kentucky Geological Survey, "In 2014, the coal industry employed 11,586 people directly in coal mining jobs and indirectly provided approximately 36,000 jobs across the state. Kentucky ranks second nationally in number of employed miners." Given the state's historical and economic ties to fossil fuels, it seems improbable that Turkey Foot Middle School could become one of the most energy-efficient facilities in Kentucky's history.

Turkey Foot Middle School is located in Edgewood, an area in northern Kentucky. Following Kenton County's success in constructing two high-performance school facilities with low energy use intensities (EUI), the Kentucky Department of Education asked the Kenton County School Board to consider constructing a net zero school building. Net zero buildings are designed to be either fully independent from utility power grids or connected to provide supplemental power to deficient areas. However, the challenge to build a net zero building at Turkey Foot was complicated by the fact that neither green infrastructure nor solar energy had been widely accepted in the coal-dependent region. The school board also had to consider the conservative leanings of its citizenry: the public demanded a school that was cost-effective and could demonstrate a return on investment. Even with these expectations set by the state, culture, and residents, Kenton County vigorously began designing Turkey Foot Middle School.

Turkey Foot Middle School's designers implemented a variety of architectural elements to create an environmentally conscious facility:

- Three types of roof-mounted solar PV arrays;
- Ground-loop geothermal heating and air conditioning;
- LED lighting systems with daylight harvesting in classrooms;
- Natural lighting in hallways using solar tubes with prismatic reflectors;
- Energy-efficient kitchen designs;
- Compact building architecture;
- Large windows;
- Educational programs;
- Energy education programs (sustainability courses, construction seminars, etc.);
- Use of the green room as an educational energy day annually;

- Energy awareness; and
- Recycling programs.

As a result of these efforts, Turkey Foot Middle School achieved numerous successes. The building is now one of the most energy-efficient and cost-effective facilities in Kentucky. This achievement becomes apparent when analyzing the school's construction costs, which are on par with regional construction expenses for a typical school constructed at the same time. According to the *School Planning and Management 2011 Construction Report*, the national median for the total construction cost for a school building without solar—the most common structure type at the time—was $215 per square foot. The total actual district cost with solar for the Turkey Foot project, however, was only $201 per square. It is important to note that this figure represents the cost without Federal Grant Assistance: with assistance, total expenses further decline to $186 per square foot.

Even without financial assistance, Turkey Foot exceeded national construction price standards, making it a relatively more affordable facility. It is important to consider the myriad of sustainability-focused structures and programs employed by the school, as well as the fact that Turkey Foot was among the first regional institutions of its kind to implement them. Given these considerations, it would seem that the project would have been much more expensive than the national median. However, the data demonstrate that this was not the case, and the implications of these findings are tremendous:

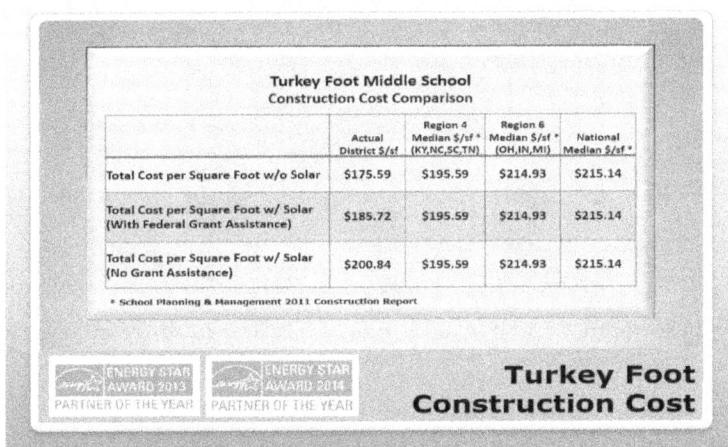

Turkey Foot Middle School Construction Cost Comparison

	Actual District $/sf	Region 4 Median $/sf * (KY,NC,SC,TN)	Region 6 Median $/sf * (OH,IN,MI)	National Median $/sf *
Total Cost per Square Foot w/o Solar	$175.59	$195.59	$214.93	$215.14
Total Cost per Square Foot w/ Solar (With Federal Grant Assistance)	$185.72	$195.59	$214.93	$215.14
Total Cost per Square Foot w/ Solar (No Grant Assistance)	$200.84	$195.59	$214.93	$215.14

* School Planning & Management 2011 Construction Report

Figure 11.4 Construction Cost Comparison. *Source*: Turkey Foot Middle School PowerPoint presentation. Our thanks to Turkey Foot Middle School for the use of this slide.

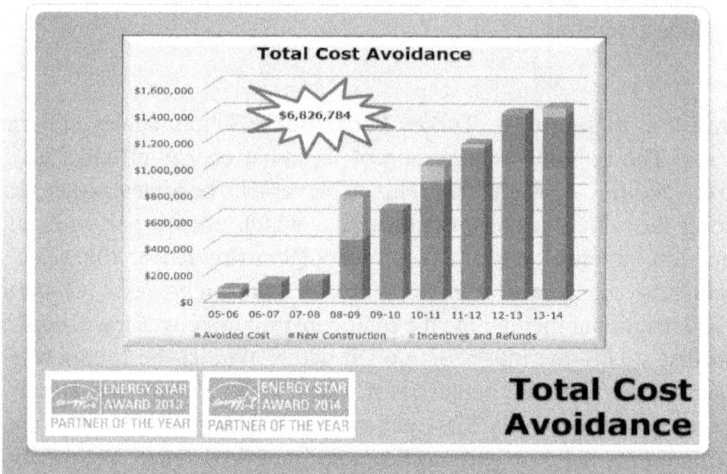

Figure 11.5 Cost Avoidance. *Source*: Turkey Foot Middle School PowerPoint presentation. Our thanks to Turkey Foot Middle School for the use of this slide.

Figure 11.6 Rooftop Solar Array at Turkey Foot Middle School. *Source*: Image by Stephen Roosa.

environmentally conscious buildings can also be economically conscious. The possibility of constructing more green structures in the future is growing. This transformation could lead not only to an environmentally aware citizenry, but to a more sustainable standard for building construction nationwide. The Turkey Foot Middle School narrative speaks to how determination and proper planning can make green architecture a reality (Haney, 2018).

Cost and Uses of Solar Energy: A Review of Data from *The New York Times*

Coal is the predominant energy source for many communities in the southern and midwestern United States, and Kentucky is often referred to as the nation's coal capital. Despite this abundance, coal can have detrimental impacts on the natural environment and human health. Coal combustion releases numerous pollutants into the atmosphere including sulfur oxides, nitrogen oxides, mercury, and particulate matter. CO_2 is a powerful greenhouse gas whose accumulation is partially responsible for trapping atmospheric heat from solar radiation. Particulate matter, due to its miniscule size, can be easily ingested by humans and other organisms leading to increased risk of respiratory disease. As the environmental and health hazards associated with coal energy become increasingly understood, renewable energy options, such as solar thermal and PV, are becoming attractive alternatives (Roosa, 2008).

Solar energy is a promising resource in some regions while still presenting challenges in others. In areas where sunlight is prevalent for significant portions of the day, solar provides an inexpensive, reliable means of powering appliances, heaters, and lights. However, controversy arises in regions where access to solar PV panels is restricted due to financial or geographic barriers. Additionally, the United States competes with foreign economies for the opportunity to create and sell solar resources: in 2018, for example, the U.S. government targeted solar PV panels imported from China by implementing tariffs of 30 percent. Here, it becomes clear that solar energy has potential for both successes and challenges.

Undoubtedly, solar energy requires an initial investment that is relatively higher than traditional electricity sources. According to data compiled by *The New York Times*, residential solar system prices for 2015 were about $5,000 whereas utility-scale systems reached nearly $3,000 (*The New York Times,* 2015).[1] These costs make solar energy appear unattainable for many consumers, especially when cheaper resources like coal are already available. Fortunately, installation costs are rapidly declining, and this downward trend is projected to continue. As technology continues to advance, coal-based electricity becomes more costly, and public awareness of the benefits of renewable energy sources increases, solar energy will become increasingly affordable. *The New York Times*' study projects that by 2022, solar prices in the United States are predicted to fall to just above $3,000 for residential units and to under $2,000 for utility-scale units. In countries where solar is more widely accepted by the general populace, prices are projected to be even lower: in India, for example, solar systems are projected to cost a mere $1,000 by 2022. Evidently, solar energy is becoming more affordable for homeowners and industries alike.

Another concern about solar PV electricity generation is its functionality on cloudy, rainy days. In regions with variable weather, consumers worry about the reliability of solar energy and whether they will receive a reasonable return on their investment. Improving storage capacity is key to resolving these intermittency concerns. Although solar energy is certainly most efficient in regions with high insolation, effective energy storage technologies exist that allow solar power to be utilized without sunlight. Many solar PV panels power Lithium-ion (LI) battery storage systems, which can then supply energy on demand. LI batteries are useful in storing solar-generated electricity, and their cost is declining. *The New York Times* reports that the battery price per kilowatt-hour has decreased from $1,000 down to approximately $400 from 2010 to 2015.[1] In addition to LI batteries, other storage technologies that have been coupled with solar PV systems include lead-acid batteries, sodium sulfur batteries, compressed air storage, and telecom flywheels. Although intermittency during periods of low insolation is a valid concern, these findings reveal that storage technology provides a promising solution.

As the costs of electrical energy storage continue to decline and hybrid renewable generation becomes increasingly available, solar power is becoming an increasingly viable alternative to traditional coal-powered electricity. Overall, solar power possesses powerful potential to become a prevalent, environmentally friendly, and cost-effective energy source for twenty-first-century consumers in the United States and globally.

REFERENCES

Gilderbloom, J. I. (April 18, 2016). The Causes of Shorter Lifespans. *The New York Times*. A-18. doi: 10.3386/w17302.

Haney, R. The Greenest Building in Kentucky: Measuring the Economic, Environmental, and Educational Rewards. Kenton County Schools. Accessed August 5, 2018.

Irwin, N., & Bui, Q. (April 11, 2016). The Rich Live Everywhere, For the Poor Geography Matters. *The New York Times*.

Jefferson County Data Portal. (2019). Main Menu, assessment.jefferson.kyschools.us /publicDatasets/publicMainMenu.aspx. University of Kentucky. Kentucky Geological Survey Coal Fact Sheet. https://www.uky.edu/KGS/education/factsheet/ coal.pdf

Part 2: The Impact of Air Pollution on Public School Achievement

John Hans Gilderbloom, Isaiah Kingsberry, Gregory D. Squires, and Charlie Zhang

Air pollution is an educational issue that leads to adverse health effects, affecting everyone in the community. The negative impact of air pollution on school children is an urgent problem that has not been effectively recognized. The early cognitive and biological development of children makes them more vulnerable than adults to air pollution. Studies have indicated that neighborhoods or schools close to hazardous waste sites are associated with lower levels of educational attainment, lower socioeconomic status, and a larger population of minority residents (Berman et al., 2018; Lu, Hackman, & Schwartz, 2021; Shier, Nicosia, Shih, & Datar, 2019). The few studies that have examined the relationship between air pollution and educational achievement have found alarming correlations of increased levels of pollutants with lower intelligence quotients (IQ) and test scores (Miller & Vela, 2013; Mohai, Kweon, Lee, & Ard, 2011; Shehab & Pope, 2019; Zhang, Chen, & Zhang, 2018).

The less time a child spends in school, the less opportunity there is to learn and receive help with academic skills. Air pollution can impact a child's health throughout the year, causing increased illness-related absenteeism (Shier et al., 2019). Fatigue and respiratory illnesses are health effects caused by air pollution that have direct impacts on illness-related absenteeism. Research in the states of Texas, Maryland, and Nevada found that higher levels of air pollution were positively associated with absenteeism (Berman et al., 2018; Chen, Jennison, Yang, & Omaye, 2000; Currie, Neidell, & Schmieder, 2009).

The relationship between air pollution and educational performance needs to be further studied, both within cities and neighborhoods, to better understand its impacts on absenteeism and educational achievement. Exposure to air pollution appears to adversely affect the brain and has been associated with decreased cognitive function in adolescents (Hedges, Erickson, Gale, Anderson, & Brown, 2020; Shier et al., 2019). Longitudinal research has measured annual air pollution levels, cumulative air pollution, and short-term air pollution. The research has found statistically significant correlations with cognitive outcomes on reading and mathematics tests for children in kindergarten through the eighth grade. Shier et al. (2019) found a significant and inverse relationship between annual ozone measures and lower mathematics test scores in third grade students. Greater exposure to air pollutants lengthens the effects they have on the body and the developing minds of children (Hedges et al., 2020). Different air pollutants adversely affect the brain in various ways. For example, lead and methylmercury have been recognized as developmental neurotoxins and are associated with lower IQs (Shier et al., 2019). A child's ability to stay focused in the classroom can lead to improved educational performance (Shehab & Pope, 2019; Zhang et al., 2018). The thalamus is the part of the brain responsible for relaying motor and sensory skills, as well as regulating sleep and alertness. Air-borne particulate matter when inhaled by children has negative effects on thalamic volume (Hedges et al., 2020).

Education can improve an individual's social mobility including the ability to access better quality neighborhoods. It is important for individuals hoping to improve their future opportunities. However, education alone does not determine physical health nor directly protect people from environmental hazards. Locally defining the problems of environmental degradation and their solutions will better address a range of social problems such as the years of projected life loss compared to average life expectancy.

While the majority of peer reviewed research suggests that air pollution has a negative impact on learning, there are other studies arguing that air pollution is not correlated with certain aspects of the educational process (Berman et al., 2018). One study found that only a marginally significant trend in absenteeism was related to pollution levels in Maryland (Shehab & Pope, 2019). However, Maryland is among those U.S. states with the highest air quality, ranking 16th in the nation; Kentucky ranks as 35th (U.S. News and World Report, 2019).

Impact of Air Pollution on Infant Health

Air pollution is potentially more damaging for infant children. While much research on the fetal origins hypothesis centers on how undernutrition, disease, and maternal health habits create lasting effects in newborns, a growing body of literature shows the links between proximity and exposure to air pollution and adverse health effects in the fetus (Almond & Currie, 2011).

Studies have confirmed the links between fetal exposure to pollution and higher infant mortality rates (Chay & Greenstone, 2003; J. H. Gilderbloom, Meares, & Squires, 2020). There are also established links between a reduction in environmental carbon monoxide and a reduction in instances of low birth weight (Currie et al., 2009; Currie & Walker, 2011). Infants born weighing less than 5.5 lbs. (2.5 kg) often suffer from early sickness or infections and may have development issues as they grow older (Centers for Disease Control and Prevention, 2023; J. Gilderbloom et al., 2019). Each of these studies supports the notion that the proximity of a pregnant mother to higher levels of air pollution can adversely impact the health of the newborn child. These effects can take place in seemingly healthy babies later in life. Past research has shown that prenatal exposure to a variety of types of air pollution negatively affects non-health endpoints later in life, such as academic test scores and economic outcomes (Prinz, Chernew, Cutler, & Frakt, 2018). Further work illustrates that adult non-health endpoints are affected as well. Both lower labor force participation and lower earnings are correlated with greater exposure to pollution at birth (Gilderbloom & Meares, 2022; Isen, Rossin-Slater, & Walker, 2017).

Local Impact of Air Pollution

According to ProPublica, there are more than 1,000 places in the U.S. in which deadly toxins are released into the atmosphere by industries; an estimated 250,000 people living in these environmental hot spots likely have higher levels of cancer risk than the EPA deems acceptable (Shaw & Younes, 2021). Louisville, Kentucky is just one example of a non-coastal city with a high rate of toxin emissions; we should be concerned about other places as well, particularly those located in southern states, where there are fewer regulations on pollution.

City of Louisville officials often mention the metropolitan area's clean air; an assertion unsupported by available evidence. One of the impacts that has not been explored is the effects of air pollution on the children attending Louisville's elementary schools, the primary focus of this study.

Rubbertown: America's Most Polluted Neighborhood

Rubbertown is a local moniker for an area in the far western section of Louisville, located in Jefferson County, Kentucky. It is located on a large strip of land adjacent to the Ohio River that has 11 large chemical plants which are responsible for over 40 percent of Jefferson County's air emissions. The history of Rubbertown is tainted with industrial accidents and unfortunate environmental devastation affecting workers in the area. There is a consensus that those who work in Rubbertown's chemical factories face both dangerous and deadly environmental hazards.

Air pollution emissions from coal-fired generation exacerbate the problems. The LGE Mill Creek electric (1,465 MW) thermal plant, operating since 1972, is located several miles south of Rubbertown and consumes about 4.8 million tons of coal annually. During warmer periods winds from the south push the air pollution from this plant directly toward west Louisville neighborhoods. It has scrubbers to remove sulfur dioxide yet emits other types of hazardous air pollutants which are unabated. Admitting to past emissions issues, LGE plans to install fabric filters or baghouses for all four coal-fired generating units to reduce particulate, mercury, sulfuric acid emissions and other hazardous air pollutants but intends to keep using coal.

Industrial accidents in Rubbertown have been severe and resulted in numerous worker deaths throughout the years. Several historic events highlight the tragedies of Rubbertown (Barnett, 2011). In 1961, approximately 1,000 residents were evacuated from a nearby neighborhood as a cloud of acrid gas formed from the former Stauffer Chemical Company. In 1965, an accident occurred resulting in the deaths of 12 workers who died from explosions and fires rolling through the DuPont synthetic rubber plant; there were at least 37 other injuries. In 1985, a tank explosion killed three workers at Borden Chemicals, and in 2001, an explosion at Carbide Industries killed two workers. Twenty-six former B.F. Goodrich workers have died from liver cancer. A Public Broadcasting Service documentary, *Trade Secrets,* in 2001 which focused on the chemical industry brought national attention to workers in Louisville's chemical plants who were dying prematurely (Moyers, Kral, Ritsher, Jones, & Camp, 2001). These workers had disintegrating bones in their hands and up to 60 non-naturally occurring chemicals in their bodies. There are similar health impacts for residents who may not work in Rubbertown but who live nearby.

A Louisville Metro health report in 2013 noted that residents living in the poorest neighborhoods adjacent to the industrial park have much lower life expectancies—in certain cases over 10 years shorter than the average area resident (Smith, Pennington, Crabtree, & Illback, 2011). This report analyzed cancer rates by local zip codes and found that rates of common forms of cancer are significantly higher in the Rubbertown zip codes. They also determined that people living in these zip codes have a 45 percent higher chance of contracting lung cancer and a 31 percent higher chance of getting colorectal cancer.

This study examines how educational achievement is impacted in neighborhoods with high levels of air pollution. Empirical evidence has long supported the idea that poor air quality has adverse impacts on public health. In fact, it was this knowledge that initially prompted the enactment of the U.S. Clean Air Act (Frank, Engelke, & Schmid, 2003). These effects can be exacerbated by living in close proximity to areas with higher levels of air pollution. While multiple variables affect the quality of life for the groups discussed, air pollution appears to be a key component. According to the

Centers for Disease Control and Prevention, students attending Louisville elementary schools located near industrial and chemical factories were two to four times more likely to have health problems such as asthma, chronic obstructive pulmonary disease, and lung or heart disease than those living in other areas of the city.

Methodology and Study Focus

The methodology of our study was multifaceted. We performed an exhaustive and thorough search of information sources to ensure that the data we used was complete. Using the data available, we performed a two-group comparison. We researched the top ten elementary schools' assignment boundaries with the worst and best Risk-Screening Environmental Indicators (RSEI) ratings (Environmental Protection Agency, 2020). We then used quantitative data to perform a simplified and multivariant regression analysis using a set of data from 82 JCPS elementary schools. Using the information from the comparative assessments and the regression analysis, we drew our key conclusions which were supported by the methodology. The study focuses on determining the impacts of air pollution on student health and education in elementary grades.

Attendance boundaries, sometimes known as school catchment areas, define the geographic extent served by a local school for the purpose of student assignments. According to the EPA's RSEI data, many are near Rubbertown's chemical factories, liquor distilleries, and other industrial plants. Using the RSEI data expands and strengthens our argument as it shows a pattern of uneven development in multiple minority and low-income neighborhoods across Louisville. Reliable and verifiable EPA, Centers for Disease Control and Protection, and Jefferson Country Public Schools (JCPS) data compared the quality of educational performance in elementary schools in heavily polluted neighborhoods to those in eastern Louisville (Jefferson County Public Schools (JCPS), 2019). Notice in figure 11.7 the heavy concentration of chemical factories, coal burning factories, and liquor distilleries that are located near elementary schools in western Louisville. Conversely, the 10 elementary schools furthest away from these plants in eastern Louisville show very little air pollution.

JCPS operates 172 public schools with approximately 96,000 students as of 2022. Table 11.1 shows the demographic comparisons between the 10 least polluted schools and the 10 most polluted schools. Geographic information system and spatial analysis methods were used to assign RSEI measures to each school based on spatial relationships assigned by the EPA. All this data is publicly available. The schools located in areas with the poorest air quality (table 11.2) have student enrollments that are 75 percent less white and 250

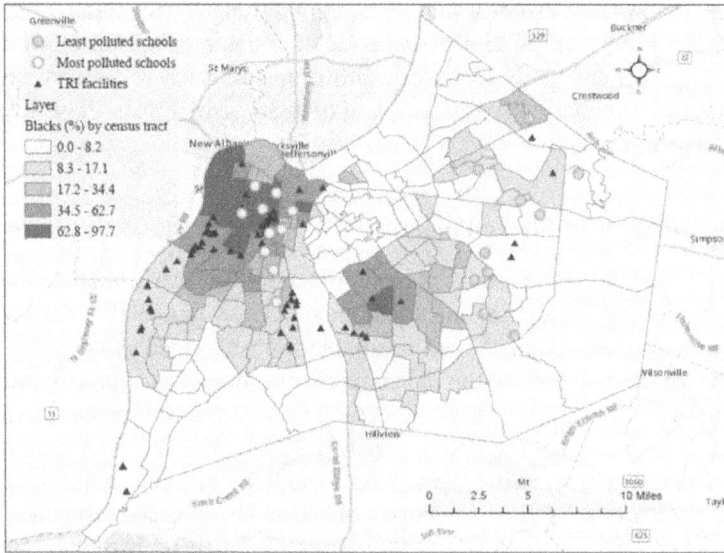

Figure 11.7 The 10 Most and Least Polluted Elementary School Attendance Zones in JCPS. *Sources*: 1. CASCADE JCPS Data Portal; 2. EPA's risk screening environmental indicator 810m x 810m micro-gridded data

Table 11.1 Demographic Comparisons between Most and Least Polluted Schools in the JCPS District

Groups	Statistics	% Black	% White	% Hispanic	% Other	Free/Reduced Lunch Rate
A. 10 least	Mean	18.4	54.8	13.0	11.	46.3
polluted schools	Std. dev.	8.7	13.7	8.7	1.2	16.4
B. 10 most	Mean	65.5	15.5	10.1	7.4	86.2
polluted schools	Std. dev.	18.4	10.7	7.2	2.8	4.5
Mean difference		47.1	−39.3	−2.9	−3.7	39.9
Student t-test	*p*-value	< 0.001	< 0.001	0.429	0.002	< 0.001

Source: CASCADE Jefferson County Public Schools (JCPS) Data Portal. (2019). Retrieved from https://www .jefferson.kyschools.us/departments/data-management-research/data-books

percent more African American than their counterparts in areas where the schools have better air quality.

There are large economic gaps between the groups as an economic proxy we used the free and reduced lunch programs. The schools in locations with poor air quality had an average of 46.3 percent of the students on reduced free lunch programs; schools in locations with better air quality averaged 86.2 percent on reduced free lunch programs. The demographic differences

Table 11.2 Risk screening Environmental Indicator Pollution Profiles for Most and Least Polluted Schools in the JCPS District

Groups	Statistics	Toxic-Weighted Pounds	RSEI Score	RSEI Score— Cancer	RSEI Score— Non-Cancer
A. 10 least polluted schools	Mean	2,184.67	283.50	273.68	16.05
	Std. dev.	454.40	188.01	181.27	11.78
B. 10 most polluted schools	Mean	76,713.80	20,928.62	20,835.75	532.81
	Std. dev.	25,889.49	7,208.18	7,183.85	331.80
Mean difference		74,529.13	20,645.12	20,562.07	516.76
Student *t*-test	*p*-value	< 0.001	< 0.001	< 0.001	< 0.001

Sources: 1. CASCADE Jefferson County Public Schools (JCPS) Data Portal. (2019). Retrieved from https://www.jefferson.kyschools.us/departments/data-management-research/data-books; 2. EPA's risk screening environmental indicator 810m x 810m micro-gridded data

between levels of air quality in school zones tell a story about the long-lasting impacts of apartheid in Louisville that left minority and low-income communities powerless to prevent the endangerment of their neighborhoods by toxic industrial air pollution.

Data Sources

Pollution data was obtained from the EPA's publicly available (RSEI) gridded model (Ash & Boyce, 2018; Berman et al., 2018). The RSEI model uses data from the Toxics Release Inventory (TRI), which measures the danger to human health. This data is reported by individual polluting facilities to estimate the risks of exposure to toxic emissions across the United States; it is provided in an 810 m by 810 m grid format (roughly a 1/4 square mile area). This publicly available data was spatially joined and averaged over the JCPS schools. The model considers the volume of chemical emissions weighted by the potential harm that the pollutants pose to humans.

Under the Emergency Planning and Community Right-to-Know Act of 1986, the EPA is required to annually collect and make publicly available information about industrial management of toxic chemicals (Christiansen & Urquhart, 1992). To complete this, industrial facilities handling toxic waste are required to report on the quantities, composition, and management of their toxic wastes. Toxicity is calculated by assessed dosage and exposure at which harmful health effects, such as cancer, reproductive and developmental toxicity, neurotoxicity, and other forms of acute illness, threaten those exposed. This is further weighted by the strength of evidence as framed by the EPA guidelines for carcinogenic risk assessments (Wiltse & Dellarco, 1996). It scales chemical toxicity as "0" for chemicals with no or inadequate evidence of toxicity to a weight of "1" for chemicals with sufficient evidence proving harmful exposure. For each school district, an average RSEI hazard score was generated by averaging the value of grids that overlapped school assignment zones.

Two Group Comparison

RSEI is a model that helps collect and explore data on toxic substances from industrial factories, specifically, the toxic chemicals released, together with factors of each chemical's relative toxicity and potential human exposure. Table 11.2 shows the air pollution profiles of the areas in Jefferson County where 10 schools (Group A) are located in areas with the lowest air quality and 10 schools (Group B) are located in areas with the best air quality. The results of student's *t*-test were used to determine the statistical significance of differences between the two groups.

When outdoor air quality is poor, it is introduced into the schools through the ventilation systems. When not in school, students continue to be exposed to the polluted air in their neighborhoods. Students attending the Group A schools located in areas with lowest air quality are 35 times more likely to be exposed to environmental toxins. They have an average of 76,714 pounds of air-borne toxins, while students attending the Group B schools, located in areas with the best air quality, have an average of 2,185 pounds of toxins. RSEI scores are designed to be comparable; a score 35 times higher than another indicates that there is a potential risk that is 35 times higher. Looking at the RSEI Scores in table 11.2, which considers the estimated population in a grid cell, students in the Group A schools located in areas with lowest air quality have a risk of exposure to air-borne toxins that is 73 times higher than the risk of exposure the students in the Group B schools located in areas with the best air quality. Of those toxins, some are known carcinogens and others are considered to be detrimental to human health. The differences are alarming as it impacts learning and overall student health. The data also show that pollution levels are significantly increasing, which is contrary to the claims of the local air pollution control district which claims the ambient air quality is improving (Air Pollution Control District, 2019).

Exposure to high levels of air pollution impacts learning ability with half the elementary students near Rubbertown factories unable to read or perform mathematics at proficiency levels compared to schools located in areas with the best air quality which are dramatically higher. Student absenteeism is three times higher, and suspensions are 14 times higher in the Group A schools located near Rubbertown as compared to those in the Group B schools located in areas with the best air quality. Children with headaches, asthma, stomach aches, trouble breathing, and difficulties sleeping have trouble maintaining focus in the classroom.

Assessing Academic Differences

Academic success was assessed using the results of state testing and absenteeism rates. The Kentucky Performance Rating for Educational Progress (KPREP) is a standardized state test administered by the Kentucky Department of Education to assess student academic achievement. Elementary students are administered the KPREP exam annually to assess mathematics and reading skills. The reading and mathematics exams for elementary students are given to students in grades 3–5. Students in grades 4 and 5 take a writing skills exam as well. In addition, students in grade four take a science exam while those in grade five take a social studies exam. Our assessment of academic differences studies elementary students from grades 3, 4 and 5 (Kentucky Department of Education, 2021).

Over half of the elementary students in the Group A schools located near Rubbertown factories read at the novice level; only one-sixth of students share the same score in the Group B schools that are located in areas with the best air quality (table 11.3). Moreover, there are three times more students who scored at the novice level in the reading section of the KPREP in the Group A schools located compared to the Group B schools. Students are much less likely to achieve distinguished scores in polluted schools. Nearly one-fourth of students in the least polluted schools scored in areas with poor air quality.

Similarly, the results of the mathematics scores mirror the reading scores (table 11.4). Over half of the students in the Group A schools received the lowest score (novice) on the mathematics section of KPREP versus a small fraction of students in Group B schools. Further, just 4 percent of students in the Group A schools received a score of distinguished while one-fourth of students in the Group B schools scored at this level. Moreover, the mathematics scores were nearly twice as high for the students in the Group B schools compared to the Group A schools. Nearly one-fifth of teachers in the Group A schools leave the following school year versus less than 5 percent leave the Group B schools.

Chronic absenteeism rates were used to assess academic performance. Chronic absenteeism is defined by JCPS as students who attended school for at least 10 days but no less than 90 percent of full-time equivalency. Chronically missing school can make it difficult for students to comprehend and complete course work. It may be impacted by medical absences caused by chronic asthma attacks and other illnesses. Students attending the Group A schools located in areas with lowest air quality had a chronically absent rate that was three times the rate of students from Group B schools located in areas with the best air quality (table 11.5). Nearly one quarter of the students from Group A schools missed 10 percent or more of the school year. The student suspension rate in the Group B schools was 1 for every 100 students, but 15 per 100 students in the Group A schools. Future research may seek to

Table 11.3 Kentucky Performance Rating for Educational Progress (KPREP) Reading Performance Comparison

Groups	Statistics	Reading— Novice	Reading— Apprentice	Reading— Proficient	Reading— Distinguished	Reading - Total Score
A. 10 least polluted schools	Mean	16.3	22.9	36.3	24.5	2.7
	Std. dev.	7.6	5.3	3.7	10.5	0.3
B. 10 most polluted schools	Mean	53.7	25.5	16.0	4.8	1.7
	Std. Dev.	12.8	3.9	7.1	4.3	0.3
Mean difference		37.4	2.6	−20.3	−19.6	−1.0
Student *t*-test	*p*-value	< 0.001	0.233	< 0.001	< 0.001	

Source: CASCADE Jefferson County Public Schools (JCPS) Data Portal. (2019). Retrieved from https://www.jefferson.kyschools.us/departments/data-management-research/data-books

Table 11.4 Kentucky Performance Rating for Educational Progress (KPREP) Mathematics Performance Comparison

Groups	Statistics	Math— Novice	Math— Apprentice	Math— Proficient	Math— Distinguished	Math —Total Score
A. 10 least polluted schools	Mean	15.2	27.9	33.7	23.2	2.6
	Std. dev.	7.3	7.3	5.5	10.4	0.3
B. 10 most polluted schools	Mean	51.8	31.2	13.1	3.9	1.7
	Std. dev.	17.3	5.2	9.3	4.6	0.4
Mean difference		36.6	3.3	−20.6	−19.3	−1.0
Student *t*-test	*p*-value	< 0.001	0.257	< 0.001	< 0.001	

Source: CASCADE Jefferson County Public Schools (JCPS) Data Portal. (2019). Retrieved from https://www.jefferson.kyschools.us/departments/data-management-research/data-books

Table 11.5 Additional Performance Indicators

Groups	Statistics	Teacher Retention (%)	Teacher Attendance (%)	Chronically Absent (%)	Suspensions (%)
A. 10 least polluted schools	Mean	94.6	95.0	8.8	1.4
	Std. dev.	5.0	1.0	4.0	1.3
B. 10 most polluted schools	Mean	81.6	94.6	24.2	15.1
	Std. dev.	10.6	0.7	6.5	13.5
Mean difference		−13.0	−0.4	15.4	13.7
Student *t*-test	*p*-value	< 0.01	0.268	< 0.001	< 0.001

Source: CASCADE Jefferson County Public Schools (JCPS) Data Portal. (2019). Retrieved from https://www.jefferson.kyschools.us/departments/data-management-research/data-books

isolate medical absences from delinquency for more reflective results. Racial demographics are provided by JCPS and expressed via the proportion of Black students.

Regression Analysis

Using a simplified linear regression analysis, we used data from all 82 elementary schools in Jefferson County Public Schools with school zone pollution levels and compared five outcomes for JCPS: mathematics and reading test performance, absenteeism, suspensions, and teacher retention (table 11.6). Although detailed socio-economic data from JCPS was unavailable, we used a proxy variable (participation in school lunch program) as a measure of income. In the regression analysis, the two specific measures of school performance, air pollution and school lunch program participation, were measured against the five dependent variables: reading, mathematics, absenteeism, suspensions, and teacher retention.

Due to RSEI hazard levels being skewed toward larger values with pollution values growing exponentially as we near TRI facilities, the logarithm of RSEI hazard scores were used as the independent variable. The logarithm linear regression showed statistical support that pollution is significantly correlated with elementary education achievement. A logarithmic linear analysis was performed and the five key dependent variables were determined to be statistically significant at the p-value < .05. Pollution's impact on both reading and mathematics scores explained about half of the variation. Chronic absenteeism, suspensions and teacher retention were all statistically

Table 11.6 Results of Simple Linear Regression: RSEI Toxic Hazard Log Score, Sample of 82 Elementary Schools in Jefferson County

Dependent Variable	Coefficients		Adjusted R Squared
	Log (RSEI)	Intercept	
Reading–Novice (%)	25.443 ($p < 0.001$)	−70.447 ($p < 0.001$)	0.581
Math–Novice (%)	25.377 ($p < 0.001$)	−71.816 ($p < 0.001$)	0.498
Chronic absence (%)	10.011 ($p < 0.001$)	−23.449 ($p < 0.001$)	0.463
Suspensions (%)	8.560 ($p < 0.001$)	−29.284 ($p < 0.001$)	0.286
Teacher retention (%)	−8.786 ($p < 0.001$)	123.481 ($p < 0.001$)	0.185

Sources: 1. CASCADE Jefferson County Public Schools (JCPS) Data Portal. (2019). Retrieved from https://www.jefferson.kyschools.us/departments/data-management-research/data-books; 2. EPA's risk screening environmental indicator 810 m x 810 m micro-gridded data

significant. These equations showed that pollution is statistically significant and potentially contributes to deficiency in learning, the number of suspensions, rates of absenteeism, and poor teacher retention.

A multivariate regression was performed using standardized pollution and school lunch rates as independent variables against the same dependent variables of novice scores, chronic absenteeism, suspensions, and teacher retention (table 11.7). For KPREP reading and mathematics, lunch rates and pollution were both found to be statistically significant in modelling test scores with lunch rates being only slightly elevated in explanatory power than pollution. For chronic absences, pollution and lunch rates were significant explanatory variables; however, absences were nearly twice as sensitive to lunch rates as to air pollution. Suspensions were not well modeled by the inclusion of reduced cost lunch participation rates, yielding a low standardized coefficient of only 0.3 and a large $p = 0.759$, while pollution correlated significantly with suspension rates. This is a potentially important finding which suggests that either behavioral issues and/or the administration of punishments are more closely correlated with the development of toxic industries than the income of their parent. Conversely, teacher retention was weakly correlated with pollution levels, producing a $p < 0.10$, while reduced lunch program participation was significantly correlated at a 0.05 alpha level. Still, both correlated negatively with teacher retention with participation in reduced cost school lunch programs being slightly more explanatory than pollution levels.

Table 11.7 Standardized Regression, $N = 82$

		Coefficients		
Dependent Variable	Intercept	RSEI - Pollution	Free/Reduced Lunch	Adjusted R Squared
Reading–Novice (%)	32.725 ($p <$ 0.001)	6.130 ($p <$ 0.001)	8.333 ($p < 0.001$)	1.720
Math–Novice (%)	31.089 ($p <$ 0.001)	6.915 ($p <$ 0.001)	7.178 ($p < 0.001$)	0.585
Chronic absence (%)	17.147 ($p <$ 0.001)	1.718 ($p =$ 0.008)	4.262 ($p < 0.001$)	0.652
Suspensions (%)	5.425 ($p <$ 0.001)	3.828 ($p <$ 0.001)	0.304 ($p = 0.759$)	0.278
Teacher retention (%)	87.855 ($p < 0.001$)	−2.125 ($p =$ 0.104)	−2.865 ($p =$ 0.029)	0.223

Sources: 1. CASCADE Jefferson County Public Schools (JCPS) Data Portal. (2019). Retrieved from https://www.jefferson.kyschools.us/departments/data-management-research/data-books; 2. EPA's risk screening environmental indicator 810m x 810m micro-gridded data

CONCLUSION

This research examines the impact of pollution on educational achievement in Louisville, Kentucky, one of most polluted cities in the United States. We proceeded to study this with proficiency score data from JCPS and information from the EPA Toxics Release Inventory which measures the dangers of air-borne toxins to human health. For our study, we performed a two-group comparison using 10 public elementary schools located in the western Louisville area known as Rubbertown (Group A) and 10 elementary schools located in eastern Louisville (Group B). It was known that the schools in western Louisville were located near industrial and chemical factories and experienced high levels of toxic air pollution; those schools in eastern Louisville were located the furthest away from the industrial and chemical factories and experience much better air quality. The test score data we used represented hundreds of students in these 20 schools.

Our conclusions challenge the silence and misinformation from the Louisville Air Pollution Control District and newly established University of Louisville Center for Healthy Air, Water, and Soil, entities that claim air quality in Louisville is improving. Our research found that: (1) air pollution getting "better" does not mean that the air is healthy; (2) the air quality is not improving or, at a minimum, is staying about the same, according to the EPA TRI; and (3) the quantities of air-borne toxins for schools near chemical factories, which are attended by mostly minority and low-income children, is 35 times higher than the amount of toxins around elementary schools attended by mostly white children elsewhere in the city where the air is clean.

Consequently, student proficiencies in reading and math and attendance are lower, and suspensions more pervasive in schools near the factories causing the air pollution. Novice scores in reading are 54 percent and in math are 52 percent near chemical factories compared to 16 percent in reading and 15 percent in math for schools with better air quality. There is a wide difference of 20 percent in reading and math scores between schools with high levels of air pollution compared to those with clean air: 36 percent of the students are proficient in reading in locations with less-polluted air schools versus 16 percent for those with higher levels of air pollution; 34 percent of the students are proficient in math in locations with less-polluted air versus 13 percent for those with higher levels of air pollution. Suspensions are fourteen times higher in schools with higher levels of air pollution versus only 1 percent in schools in locations with less-polluted air. Sixty-six percent of children attending polluted schools are Black and only 16 percent are white; 55 percent percent of children attending schools in locations with better air quality are white and only 18 percent are black. Moreover, teacher retention levels in schools located in areas with higher levels of air pollution are problematic, with one out of five leaving the toxic schools every year as opposed to one out of twenty leaving the clean air schools. Our statistical regression analysis determined that air pollution explains from 59 percent to 76 percent of the variation—a very powerful indicator. Air pollution offers less clarification in terms of suspensions (28 percent) and teacher retention (22 percent).

Linking School Policies to Air Pollution

Horace Mann, a Massachusetts educator, famously wrote in 1848, "Education, then beyond all other devices of human origin, is the great equalizer of the condition of men—the balance wheel of social machinery" (Mann, 1848). While many have used education for upward mobility, providing equal educational opportunity remains a major challenge. At the K-12 level, reliance on property taxes to fund public schools, tracking systems, and the resources wealthy parents can provide (SAT preparation tutorials, summer travel, music lessons, etc.) all bring social and societal inequities into the schools. The same dynamics are operative in institutions of higher education. They are manifest in legacy and development admissions (i.e., children of prospective donors have preference), early admissions programs requiring students confirm admission decisions prior to knowing the status of financial aid, and unpaid internships.

School policies often fail to fully consider the context in which schools operate. As education historian Jean Anyon once asserted: "to solve the systemic problems of urban education, we need not only school reform but reform of these public policies as well" (Anyon, 2005). While this has long been known, what has not been recognized is the contribution of air pollution to inequities in education. This study focuses on Louisville, Kentucky but Louisville is not unique. Air pollution and its many costs are widely understood to be major challenges to the nation's metropolitan areas though local officials often publicly minimize the extent or costs. These data demonstrate that toxic air pollution in Louisville has negative impacts on children's learning and behavioral abilities. It reduces rates of teacher retention. Yet the Air Pollution Control District final report claims that the "air is cleaner than ever" and that the air quality is not harmful to health (Air Pollution Control District, 2019).

This reveals deep injustice in regard to pollution and socio-economic inequality. Those who can afford to move to neighborhoods with schools in areas with less polluted air tend to do so; those who cannot afford to move or desire to live in predominantly minority neighborhoods are faced with the consequences of living with toxic air pollution.

The larger point is that the neighborhoods in western Louisville (including just south of west Louisville), which include more than 62,000 residents of which 75 percent are Black, would benefit from tougher enforcement of pollution regulations. This would not only improve children's learning abilities but also allow home values to increase, reducing the costly abandonment of homes. It would stimulate greater investment in local businesses, reducing medical costs and crime. Why can't minority communities have equal justice and fairness when it comes to air quality? Our data provides strong evidence that toxic air pollution negatively impacts educational achievement.

Reducing air pollution helps change the lives of the economically disadvantaged, providing them with better educational opportunities. One solution to solving the inner-city educational crisis: expanded and tougher regulation of toxic air pollution, which in the case of Louisville is worse than most other cities in the nation. Another possible but costly solution is to relocate polluting factories far away from residential areas and public schools. Air pollution is costly to human lives and some forms create high levels of greenhouse gases. Pollution is not the only factor creating these challenges, but it is one that city leaders have often ignored.

In the words of Larissa Lockwood, Director of Clean Air at Global Action Plan,

> Children have the right to breathe clean air (J. H. Gilderbloom, Squires, & Kingsberry, 2021). As our Clean Air Day research showed, air pollution doesn't just affect a child's health, but can also affect their working memory and hence their ability to learn. To safeguard the future of our children's health and educational potential, we need to work together to take urgent actions [. . .] to eliminate harmful pollutants.

The first step to resolving any problem is to describe it. When a nine-year old girl in London is declared dead from air pollution, it makes us wonder how many more children are also being harmed (Lockwood, 2020). As London's Mayor Sadiq Khan stated,

> Toxic air pollution is a public health crisis, especially for our children. Ministers and the previous mayor have acted too slowly in the past, but they must now learn the lessons from the coroner's ruling and do much more to tackle the deadly scourge of air pollution in London and across the country.

It makes us wonder how many other children are also being harmed. This research indicates that exposure to high levels of air pollution also impacts the educational performance of young children.

Opportunities for Future Research

This is exploratory research using publicly available data which provides preliminary but fairly significant (statistically and otherwise) evidence on the impacts of air pollution on educational achievement. Using this research study, we ask JCPS to provide additional information to enable us to assess student achievement scores. This data would consider achievement scores by gender, race, gender, socioeconomic status, housing tenure (renter or home-owner), one or two parents in the household, educational levels, and students' eligibility for free or reduced cost lunch. Further research should seek data in which the interschool socio-economic variability may be assessed in conjunction with academic achievement to further isolate the impact of wealth on academic achievement from pollution.

In the future we hope to explore the impact of low proficiency scores. A 2012 study found a strong correlation between low reading proficiency scores and decreased high school graduation rates (Hernandez, 2011). More alarming are the strong connections between low literacy skills and incarceration rates which require further study. For example, 85 percent of all juveniles who interface with the juvenile court system and more than 60 percent of all prison inmates are functionally illiterate (United Way of Hays & Caldwell Counties, 2023).

ACKNOWLEDGMENTS

We are thankful to a Jefferson County Public School teacher and Metro Louisville planner who participated in this research but chose to remain anonymous. We thank Stephen Roosa, PhD, our primary editor, for his exhaustive editorial work and detailed contributions. Our thanks to Ellen Slaten, PhD, and Carla J. Snyder, MSL, for editing and content contributions. Our thanks to the late Patty Gilderbloom, who taught in two of these polluted schools and alerted us to health and cognitive issues for teachers and students. (See tables 11.1, 11.2, 11.3, 11.4, 11.5, 11.6, and 11.7.)

The Impact of Air Pollution on Public School Achievement was originally published in the AEE *International Journal of Strategic Energy and Environmental Planning* (*IJSEEP*). The Association of Energy Engineers (AEE), founded in 1977, is a non-profit professional association of more than 17,000 members and over 32,000 certified professionals in more than 100 countries worldwide dedicated to serving members and other industry professionals to save energy, reduce GHGs, make buildings perform better, and help reach global goals for Net-Zero. For more about AEE, visit www.aeecenter.org. Revised and reprinted with permission from *IJSEEP* editor, Dr. Stephen Roosa.

REFERENCES

Air Pollution Control District. (2019). *Air pollution control district final report.* Retrieved from apcd-2019-annual-report (louisvilleky.gov)

Almond, D., & Currie, J. (2011). Killing me softly: The fetal origins hypothesis. *Journal of Economic Perspectives, 25*(3), 153–172.

Anyon, J. (2005). *Radical possibilities: Public policy, urban education, and a new social movement.* New York: Routledge.

Ash, M., & Boyce, J. K. (2018). Racial disparities in pollution exposure and employment at US industrial facilities. *Proceedings of the National Academy of Sciences of the United States of America, 115*(42), 10636–10641. doi:10.1073/pnas.1721640115

Barnett, R. (2011). *Louisville, Kentucky: Environmental issues.* Kentucky Institute for the Environment and Sustainable Development.

Berman, J. D., McCormack, M. C., Koehler, K. A., Connolly, F., Clemons-Erby, D., Davis, M. F., . . . Curriero, F. C. (2018). School environmental conditions and links to academic performance and absenteeism in urban, mid-Atlantic public schools. *International Journal of Hygiene and Environmental Health, 221*(5), 800–808.

Centers for Disease Control and Prevention. (2023). *Particle pollution.* Retrieved from https://www.cdc.gov/air/particulate_matter.html

Chay, K. Y., & Greenstone, M. (2003). The impact of air pollution on infant mortality: Evidence from geographic variation in pollution shocks induced by a recession. *The Quarterly Journal of Economics, 118*(3), 1121–1167.

Chen, L., Jennison, B. L., Yang, W., & Omaye, S. T. (2000). Elementary school absenteeism and air pollution. *Inhalation Toxicology, 12*(11), 997–1016. doi:10.1080/08958370050164626

Christiansen, S. J., & Urquhart, S. H. (1992). The emergency planning and community right to know act of 1986: Analysis and update. *Brigham Young University Journal of Public Law, 6*, 235.

Currie, J., Neidell, M., & Schmieder, J. F. (2009). Air pollution and infant health: Lessons from New Jersey. *Journal of Health Economics, 28*(3), 688–703.

Currie, J., & Walker, R. (2011). Traffic congestion and infant health: Evidence from E-ZPass. *American Economic Journal: Applied Economics, 3*(1), 65–90.

Environmental Protection Agency. (2020). *EPA's Risk-Screening Environmental Indicators (RSEI) methodology RSEI version 2.3.9.* doi:https://www.epa.gov/sites/default/files/2020-12/documents/rsei_methodology_v2.3.9.pdf

Frank, L., Engelke, P., & Schmid, T. (2003). *Health and community design: The impact of the built environment on physical activity.* Island Press.

Gilderbloom, J., Washington, C. B., Quenichet, K., Manella, C., Dwenger, C., Slaten, E., . . . Frederick, C. (2019). What cities are the most dangerous to your health? Ranking the most polluted mid-size cities in the United States: Ranking the most polluted mid-size cities in the United States (12/13/2019). *SSRN Preprint article.* Retrieved from https://ssrn.com/abstract=3506217 or http://dx.doi.org/10.2139/ssrn.3506217

Gilderbloom, J. H., & Meares, W. L. (2022). How inter-city rents are shaped by health considerations of pollution and walkability: A study of 146 mid-sized cities. *Journal of Urban Affairs, 44*(8), 1059–1075. doi:10.1080/07352166.2020.1803751

Gilderbloom, J. H., Meares, W. L., & Squires, G. D. (2020). Pollution, place, and premature death: Evidence from a mid-sized city. *Local Environment, 25*(6), 419–432. doi:10.1080/13549839.2020.1754776

Gilderbloom, J. H., Squires, G. D., & Kingsberry, I. (2021). How many more children must be hurt by pollution? http://info.primarycare.hms.harvard.edu/blog/children-hurt-pollution. *Harvard Medical School Primary Care Review, March 3*. Retrieved from http://info.primarycare.hms.harvard.edu/blog/children-hurt-pollution

Hedges, D. W., Erickson, L. D., Gale, S. D., Anderson, J. E., & Brown, B. L. (2020). Association between exposure to air pollution and thalamus volume in adults: A cross-sectional study. *PLoS ONE, 15*(3), e0230829. doi:10.1371/journal.pone.0230829

Hernandez, D. (2011). *How third-grade reading skills and poverty influence high school graduation.* Annie E. Casey Foundation. https://eric.ed.gov/?id=ED518818, accessed January 28, 2023.

Isen, A., Rossin-Slater, M., & Walker, W. R. (2017). Every breath you take—Every dollar you'll make: The long-term consequences of the clean air act of 1970. *Journal of Political Economy, 125*(3), 848–902. doi:10.1086/691465

Jefferson County Public Schools (JCPS). (2019). *Jefferson county public schools data portal.* Retrieved from https://www.jefferson.kyschools.us/departments/data-management-research/data-books

Kentucky Department of Education. (2021). *Kentucky Performance Rating for Educational Progress (K-PREP).* Retrieved from education.ky.gov/AA/Assessments/Pages/K-PREP.aspx

Lockwood, L. (2020). *Mayor unveils plans for further reduction in pollution at schools London assembly.* Retrieved from https://www.london.gov.uk/press-releases/mayoral/mayor-unveils-plans-to-reduce-toxic-air-at-schools

Lu, W., Hackman, D. A., & Schwartz, J. (2021). Ambient air pollution associated with lower academic achievement among US children: A nationwide panel study of school districts. *Environmental Epidemiology, 5*(6), e174. doi:10.1097/ee9.0000000000000174

Mann, H. (1848). *The 12th annual report of Horace Mann as secretary of Massachusetts state board of education.* Retrieved from https://www.republic-online.com/opinion/horace-mann-knew-the-value-of-education/article_aac951f7-eeb0-5b22-95f7-248bfc2cf6f5.html

Miller, S., & Vela, M. (2013). *The effects of air pollution on educational outcomes: Evidence from Chile.* https://publications.iadb.org/en/publication/11349/effects-air-pollution-educational-outcomes-evidence-chile, accessed January 28, 2023.

Mohai, P., Kweon, B.-S., Lee, S., & Ard, K. (2011). Air pollution around schools is linked to poorer student health and academic performance. *Health Affairs, 30*(5), 852–862. doi:10.1377/hlthaff.2011.0077

Moyers, B. D., Kral, K., Ritsher, D., Jones, S., & Camp, J. (Writers). (2001). *Trade secrets: A Moyers report.* In P. A. T. F. Films for the Humanities (Firm), Films Media Group. (Producer). USA.

Prinz, D., Chernew, M., Cutler, D., & Frakt, A. (2018). *Health and economic activity over the lifecycle: Literature review.* National Bureau of Economic Research.

Shaw, A., & Younes, L. (2021). The most detailed map of cancer-causing industrial air pollution in the US. *ProPublica,* November 2, 2021.

Shehab, M., & Pope, F. (2019). Effects of short-term exposure to particulate matter air pollution on cognitive performance. *Scientific Reports, 9*(1), 8237.

Shier, V., Nicosia, N., Shih, R., & Datar, A. (2019). Ambient air pollution and children's cognitive outcomes. *Population & Environment, 40*(3), 347–367.

Smith, P., Pennington, M., Crabtree, L., & Illback, R. (2011). *Louisville Metro health equity report: The social determinants of health in Louisville Metro neighborhoods.* https://louisville.edu/cepm/westlou/louisville-wide/lmph-health-equity -report-2011, accessed January 29, 2023.

United Way of Hays & Caldwell Counties. (2023). *Literacy statistics.* Retrieved from https://www.unitedwayhaysco.org/literacy-statistics#

U.S. News and World Report. (2019). *Pollution rankings measuring exposure to pollution and related health risks.* https://www.usnews.com/news/best-states/rankings /natural-environment/pollution, accessed February 12, 2023.

Wiltse, J., & Dellarco, V. L. (1996). US Environmental Protection Agency guidelines for carcinogen risk assessment: past and future. *Mutation Research/Reviews in Genetic Toxicology, 365*(1–3), 3–15.

Zhang, X., Chen, X., & Zhang, X. (2018). The impact of exposure to air pollution on cognitive performance. *Proceedings of the National Academy of Sciences, 115*(37), 9193–9197. doi:10.1073/pnas.1809474115

Chapter 12

Will Planting Eight Billion More Trees Solve Climate Chaos? No!

Elliott Grantz, John Hans Gilderbloom, Justin Mog, Charlie Zhang, and Avery Kahl

The Inflation Reduction Act provides a billion dollars to increase trees in urban neighborhoods. Are trees a miracle cure? Do we need not worry about pollution as we have documented in this book?

As you walk down an urban street, what do you hear? Smell? See? You hear cars driving by and people chattering on their stoops and talking on their cell phones. You smell car exhaust along with the other smells of city life. You might be walking past tall buildings or people's homes. Now imagine the same street but lined with trees. Your walk is now completely different; you hear birds chirping above your head, the air is fresh, and the trees are beautiful, tall, and green. As you walk under their canopies, you feel the temperature shift back and forth when you are protected from the sun's heat. Trees play a major role in urban settings. From cleaning of air to the sequestration of carbon to the essential symbolism of life, trees are a constant reminder that life is beautiful and forever evolving. Trees! What a wonderful thing!

AIR FILTRATION

Trees have the ability to filter air, thus improving overall air quality. In 2012, nearly 6.5 million deaths were associated with poor air quality worldwide (WHO, 2018). As cities become denser and more pollution is emitted, air quality becomes harder to control. In a survey conducted throughout metropolitan areas of the United States, it was found that, out of seven topics, people believed trees' abilities to filter the air to be among the top three reasons why trees are important to urban areas (Lohr & Pearson-Mims, 2005).

Being able to breathe clean air is essential for all life on land, and community members obviously understand the importance of air filtering trees.

An interesting aspect of air pollution is that it diffuses, which means that larger spaces are impacted, not just the area surrounding the pollution. Thus, it is important that when planting trees in urban areas, they are distributed widely and not just concentrated in one area. This could help ensure that all areas of a city receive similar amounts of air filtration and that there is little to no air quality disparity. Trees can decrease the air pollution rates in urban areas by filtering the air. Research conducted in Nottingham, England showed that a woodland is "estimated to reduce concentrations of sulfur dioxide and nitrogen oxides in the air by 4–5%" (Freer-Smith & Broadmeadow, 1996). While 4–5 percent may not seem substantial, when it comes to the amount of toxins, it is very significant in relation to the atmosphere. When planting trees, it is important to plant them close to areas of high pollution, as they work most effectively together, not spread apart (Tyrvainen et al., 2005). Because trees work better together, they should be dispersed in groups throughout an urban area to ensure that the entire community benefits from the trees.

Another pollutant that is of concern is dust. Dust can be an allergen for people and can lead to dust mites. Dust mites can become an issue and can be harmful to human health (Sundell, 1995). Trees are great at capturing dust, which is especially important in urban areas. Dust can be harmful to humans and other species and can decrease property values as it affects the aesthetics of homes and buildings. Planting trees can filter out dust, decrease pollution-related annoyances, and increase air quality (Tyrvainen et al., 2005). Trees are also a cost-effective way of filtering particles and pollutants out of our air, making it unnecessary to have in-home filters. They also help save energy if in-home filters are not needed because of the trees. Having more trees to capture dust and other pollutants can be beneficial for urban areas and human health.

With the increase in air pollution, health problems like asthma have increased. Since 1980, there has been an increase of nearly 50 percent of people and children diagnosed with asthma in the USA (Lovasi et al., 2008). By including more trees in cities, they can filter the air and have the potential to decrease the number of cases of asthma and other air and pollution-related illnesses. According to the National Institutes of Health, the increase in asthma primarily occurs in poor urban communities; this may be due to the lack of trees and plants in these areas (Lovasi et al., 2008). The study also found that as tree density increased, the frequency of the cases of asthma decreased. To ensure the health and wellness of future generations, it would behoove planners to make trees an essential part of planning urban areas, as they help decrease the likelihood of asthma, filter out pollution, and provide fresh air for people.

Water Reduction and Decreased Soil Erosion

Urban trees also play a vital role for our cities underground through runoff and erosion management. Branches and limbs help reduce runoff by breaking the fall of rain. Trees' long-reaching roots absorb the rainwater, which helps prevent flooding from occurring in our streets, streams, and lakes. A single mature tree can absorb up to 1,400 gallons of rainwater per year (USDA Forest Service n.d.). When trees absorb water, they are also filtering pollutants, preventing them from going into our lakes, streams, and ground water (USDA Forest Service, n.d.). With effective planning, trees can help with storm water management and treatment costs (Dwyer, 1992). In Minneapolis, over $9 million is saved in annual stormwater treatment because of street trees (USDA Forest Service n.d.). Planting more trees is a cost-effective way of dealing with potential flooding issues.

When trees block the wind and rain from direct impact on the ground, they are also assisting in the prevention of soil erosion. Topsoil that is continually being washed away pollutes waterways and destroy nutrient rich soil. Erosion can be avoided by trees' long root systems grabbing onto the earth, helping hold it into place (Coder, 1996). Roots protect landscapes around cities, especially where there is a slope, by reinforcing the strength of the ground. Planting more urban trees can help mitigate future issues of water pollution, flooding, and urban soil erosion.

Recreation, Public Health, and Safety

Urban trees provide a better quality of life within cities in many ways. They enhance the attractiveness of urban settings, they add esthetic value to residential streets, and they provide a tranquil, natural setting for city parks. Other social and health benefits of green spaces and urban forests are offering the opportunity to unplug from the fast-paced lifestyle of the city and connect with the natural world (Dwyer et al., 1991). Connecting with nature can help lower stress and is beneficial to mental health. People of all ages enjoy watching nature grow and change through the seasons.

Trees create an inviting atmosphere for socializing and recreation. Green spaces encourage people to be more physically active. Whether walking, running, or riding a bike, having access to these areas can contribute to healthier behaviors and overall wellbeing (Douglas et al., 2017). Children who live near green space can have an 11–19 percent lowered overweight/obesity rate (Douglas et al., 2017). In a European study, residents in green residential areas were three times as likely to be physically active and 40 percent less likely to be overweight or obese than those in less green areas (Ellaway et al.,

2005). Enjoying time in urban green spaces supports a healthier lifestyle for people in the city.

As we spend more time outside, our streets become safer. When there are more trees along streets, especially between roads and sidewalks, people feel more secure and are more inclined to walk in their city neighborhoods. This sense of security encourages socialization amongst community members. Street trees in Portland, Oregon significantly reduced violent crimes in neighborhoods with lower median household incomes, especially while there was a positive impact from the Grey to Green initiative across the city (Burley, 2018). Added tree canopies provide more walkable streets, which are vital to safety and public health for everyone.

Benefits of Natural Air Conditioning for People and Real Estate

When people are in a city park on a hot summer day and in need of a break from the heat and the sun, trees offer a shady place to rest. Understanding trees in urban areas is vital to implement current development trends, with nearly 80 percent of United States citizens living in the urban areas (US Census Bureau, 2000). Trees provide people with natural shade and air conditioning. When asked what is the most beneficial part of having trees in cities, "shade and cooling" was ranked number one in a study conducted in the larger metropolitan areas in the United States (Lohr et al., 2005). As a result of cities becoming more dense, urban heat islands become more likely, thus making the need for trees even greater. If trees are properly incorporated into urban design, they would decrease heat islands in cities.

As cities continue to grow, the amount of heat that people give off through pollution increases as well. Research has shown that in areas with dense populations of trees, like urban parks/forests, the air in these areas tends to be 2–3 degrees Celsius less than areas without trees (Tyrvainen et al., 2005). According to Tyrvainen, the temperature reducing abilities of trees come from two factors: direct shading and evapotranspiration cooling (2005). While parks and urban forests cannot cool an entire city, they do have the ability to cool up to 200–400 meters within their space (Tyrvainen et al., 2005). This is important to note, as it shows that trees need to be planted all over to ensure that all people equally benefit from their cooling effects. As cities continue to grow, the effects of trees and their cooling properties should be considered during the planning process.

Furthermore, trees can decrease the amount of money that households have to spend on energy. Having a single south facing tree has been shown to decrease energy costs significantly. Dr. John Hans Gilderbloom states that having a south facing tree by his home helped decrease his cooling costs by nearly 50 percent (http://www.sunlouisville.org). Overall, having trees and urban forests "can reduce demands for natural resources producing food and conserving energy, water and carbon dioxide" (McPherson, 1992). Not only can trees reduce the demand for natural resources, they can also make a home more valuable.

Having more trees around one's home has the potential to increase the value of the home. Yard and street trees can raise property values 2–15 percent with an average of 7 percent (Wolf, 2008). Wolf also states in her article that an inner-city home can have a 10 percent increase in property value if it is within a quarter mile from a park (2008). The Virginia Cooperative Extension suggests a 20 percent increase of property value with the presence of vegetation and trees (2009). Data interpreted by Calvin Dwenger on PVA valuations in Louisville, Kentucky show neighborhoods with more canopy coverage have higher property values than those with less canopy cover. Having thicker green views may lure buyers and benefit sellers. Planting even a few trees around a house or living in green areas can benefit homeowners in urban areas in many ways.

Not only can trees increase the value of homes, they can also save homes money. The cooling effects of shade trees result in homes using less energy, which saves people money and decreases the amount of energy needed to cool a home (Donovan & Butry, 2009). A study conducted in Sacramento, California showed that there can be an average of 31 percent reduction in carbon emissions during the summertime over 100 years if trees are planted on the west side of a house (Donovan & Butry, 2009). These emissions are primarily sequestered from electricity generation. This indicates that having trees on the southward and westward side of the home can result in saving energy and money for homeowners.

It has been shown that trees' cooling abilities are not only important to people for individual comfort reasons, but also if trees are naturally cooling the home, then that can reduce energy use. Furthermore, planting trees in urban areas would have two impacts: reduction of energy use, which would decrease the cost to cool a home. It is important to note that more trees do reduce air flow in a given area, so when planting trees, planning should be thought out with air flow in mind. A study conducted in Sacramento, California with over 450 single family homes showed that there was a significant decrease in the amount of energy used in homes with trees planted on the west and south side of the homes (Donovan & Butry, 2009). To go further, the authors also found that if the trees were planted on the north side of the house, electricity use increased by 1.5 percent (Donovan & Butry, 2009). The authors were able to measure a 26 percent–47 percent reduction in summertime cooling costs in homes that had trees from 2.4 m–6m tall (Donovan & Butry, 2009). Being able to save between one-quarter to one-half of seasonal energy costs is just another reason to include trees in the planning of urban areas. In Louisville, this author found that putting the sky lights on the south side of a Victorian next to 100-year-old homes reduced energy use to an average of $35 per month from 2010 to 2020 for a one bedroom 700 square foot home. Normally for homes in Louisville the electricity would be around $150 per month without trees. So, trees are renewable and they can reduce the cost of cooling a home. From cleaning our air to beautifying our world, trees are important in so many ways. It is amazing how much power a single tree has. One single tree can absorb up to 48 pounds of CO_2 a year (AEA, 2008). One single tree can provide daily oxygen for up to four people and one single tree can absorb 14,000 pounds of rainwater a year (USDA Forest Service, n.d.). Trees have the ability not only to filter our air but also add beauty to our cities. As a result of the many benefits associated with trees and their ability to improve urban areas, it is imperative that city officials and planners start incorporating trees into cities and urban areas on a much larger scale. Similarly, city inhabitants need to raise awareness and support the planting of more trees.

How Chemical Companies Use Trees to Deflect Attention from the Real Problem of Pollution

Trees are great but they should not be used as a means to reduce pollution. It is a losing strategy. At the University of Louisville Center for Healthy Air, Water, and Soil, which is funded by people connected to major polluters, *The New York Times* has identified tobacco companies funding research. But the major funder is a cheerful heir to a large fortune who promised $50 million based on performance criteria. The Center took the place of several scientific environmental research organizations like the Kentucky Center for the Environment and Sustainable Development. The problem is there is no proof that planting more trees or being surrounded by more trees will improve your health. Aruni Bhatnagar, University of Louisville Professor of Medicine, states that this multi-million-dollar project supported by past President Trump and Democrats has yet to produce a "peer reviewed article" providing evidence that tree canopy can improve people's health (Sand, 2024). Each tree cost $1,000 to plant and many have died.

We looked at tree density and found no correlations among neighborhoods or cities in quality-of-life indicators. More specifically, the size of a tree canopy in a neighborhood does not reduce pollution levels. In terms of health indicators, tree canopy size is not correlated with cancer rates or an increase in lifespan. We did this by measuring the amount of tree canopy in 194 Louisville neighborhoods and found no impact on lifespan. We also found no correlation between tree density and a reduction in pollution. In

Table 12.1 Tree Canopy Size Has No Impact on Lifespan

Cities with High Tree Equity Score				Cities with Low Tree Equity Score			
City	State	Tree Equity Score	Average Life Expectancy	City	State	Tree Equity Score	Average Life Expectancy
Seattle	Washington	91	81.3	Amarillo	Texas	70	76.3
Portland	Oregon	88	78.8	Fort Smith	Arkansas	72	74.9
Tucson	Arizona	87	76.7	El Paso	Texas	66	78.9
Atlanta	Georgia	93	76.2	Sacramento	California	78	78.5
Austin	Texas	93	79.7	Fresno	California	71	77.6
Denver	Colorado	90	78.5	Bakersfield	California	68	77.4
Charlotte	North Carolina	93	78.5	Lubbock	Texas	76	76.8
Washington	DC	88	77	Brownsville	Texas	74	79.7
New York	New York	89	80.6	Vancouver	Washington	78	78.7
Boston	Massachusetts	89	80	Riverside	California	76	78.7
San Francisco	California	87	82	Miami	Florida	69	79.5
Minneapolis	Minnesota	89	78.6	Hialeah	Florida	58	81.1
Chicago	Illinois	87	77.2	San Bernadino	Florida	69	76
Pittsburg	Pennsylvania	89	77.2	Laredo	Texas	74	79
Philadelphia	Pennsylvania	89	76.1	Las Cruces	New Mexico	69	78.3
Baltimore	Massachusetts	83	75.3	Las Vegas	Nevada	71	76.5
Oakland	California	84	73.3	Corpus Christi	Texas	73	77.3
Chandler	Arkansas	88	79	Moreno Valley	California	63	78.4
Albuquerque	New Mexico	87	79.9	San Diego	California	78	81.1
Louisville	Kentucky	86	75.9	Chula Vista	Texas	45	81.2
Mean		88.5	78.09			69.9	78.295

Source: Center for Sustainable Urban Neighborhoods, formerly University of Louisville, transferred to Neighborhood Associates, Washington D.C.

other words, tree growth in West Louisville neighborhoods was high in the most polluted neighborhoods with lowest lifespan (http://www.sunlouis-ville.org).

A Google search engine was used by students to measure tree density in 20 cities. As you can see from the averages in table 12.1, there was no difference in average lifespan in cities with high tree density or low tree density. In a separate research project, we looked at 144 cities that we've discussed in prior chapters and a multiple regression analysis found no difference between tree canopy impacting cancer rates or lifespan. These are tentative findings yet to be published in a major. We presented this study at an academic conference and are preparing it for publication.

Russ Barnett, the former Director of the Kentucky Institute for the Environment and Sustainable Development, said, "it would take 100 years to grow a forest and we just don't have the time to do that." In another study we conducted with a national sample of 144 semi-isolated medium-sized cities, no correlation was found between county tree canopy and increase in lifespan or reduction in cancer (http://www.sunlouisville.org). To grow eight billion trees would require a space three times the size of India. But where are the trees being planted? In Louisville, the city charges $25 per tree, and they seem to be planted in middle- and upper-class neighborhoods because the homeowner must water and fertilize. But a good, healthy fully mature tree costs $1,000 (Sand, 2024).

Scientists are skeptical. John Sherman, a professor at the MIT Sloan School of Management, and Andrew P. Jones, executive director of the non-profit Climate Interactive, state that planting eight billion trees would only reduce the temperature by one-fourth of 1 degree Fahrenheit. They also found that a trillion trees planted would only reduce carbon dioxide by 6 percent (Joselow, 2023: 8A). Planting more trees is not a solution, but reducing pollution is a much more practical, smart, and powerful way to reduce carbon emissions.

Trees are not a miracle cure for reducing toxic pollution or expanding our lifespan. Trees will not by themselves reduce greenhouse gas emissions while chemical companies continue to spew out greenhouse gases. According to the head of the Louisville Air Pollution Control District, who gave a guest lecture in Dr. Gilderbloom's class in Spring 2024, Louisville is number one in per capita greenhouse gases. It is a fool's gold approach that is unproven by scientists.

ACKNOWLEDGMENTS

Our thanks to UofL Sustainability students Avery Kahl and Abby Bills for their assistance with the tree canopy table.

REFERENCES

Alliance for Community Trees. (2011). *Benefits of trees and urban forests: A research list [PDF file].* Retrieved April 11, 2019, from http://www.actrees.org/files/Research/benefits_of_trees.pdf

Arbor Environmental Alliance. (2008). *Neighborhood revitalization initiative.* Retrieved April 11, 2019, from http://www.arborenvironmentalalliance.com/carbon-tree-facts.asp

Burley, B. (2018). Green infrastructure and violence: Do new street trees mitigate violent crime? *Health & Place, 54,* 43–49. doi:10.1016/j.healthplace.2018.08.015

Coder, R. D. (1996). *Identified benefits of community trees and forests [PDF file].* Retrieved April 11, 2019, from https://nfs.unl.edu/documents/communityforestry/coderbenefitsofcommtrees.pdf

Donovan, G. H., & Butry, D. T. (2009). The value of shade: Estimating the effect of urban trees on summertime electricity use. *Energy and Buildings, 41*(6), 662–668. doi:10.1016/j.enbuild.2009.01.002

Douglas, O., Lennon, M., & Scott, M. (2017). Green space benefits for health and well-being: A life-course approach for urban planning, design and management. *Cities, 66,* 53–62. doi:10.1016/j.cities.2017.03.011

Dwyer, J. F., McPherson, E. G., Schroeder, H. W., & Rowntree, R. A. (1992). Assessing the benefits and costs of the urban forest. *Journal of Arboriculture, 18,* 227–234.

Dwyer, J. F., Schroeder, H. W., & Gobster, P. H. (1991). The significance of urban trees and forests: Toward a deeper understanding of values. *Journal of Arboriculture, 17,* 276–284.

Ellaway, A., Macintyre, S., & Bonnefoy, X. (2005). Graffiti, greenery, and obesity in adults: Secondary analysis of European cross sectional survey. *BMJ: British Medical Journal (International Edition), 331*(7517), 611–612. doi:10.1136/bmj.38575.664549.F7

Evans, E. (n.d.). *Benefits of trees.* Retrieved April 11, 2019, from https://projects.ncsu.edu/project/treesofstrength/benefits.htm

Freer-Smith, P. H., & Broadmeadow, M. S. (1996). The improvement of urban air quality by trees. *Department for Environment, Food and Rural Affairs.* Retrieved from http://agris.fao.org/agris-search/searchIndex.do?query=center:(GB)centerkey:0&request_locale=en&sortField=3&sortOrder=2

Joselow, M. (2023). GOP wants to plant a trillion trees. Scientists are skeptical. *Sacramento Bee,* August 7, 2023.

Kane, B., & Kirwan, J. L. (2009). Value, benefit, and costs of urban trees. *Virginia Tech.* Retrieved from http://hdl.handle.net/10919/48050

Lohr, V. I., & Pearson-Mims, C. H. (2005). Urban residents rate the benefits of trees in cities highly and are unconcerned with problems associated with trees. *HortScience, 40*(4), 28–35. doi:10.21273/hortsci.40.4.1136b

Lovasi, G. S., O'Neil-Dunne, J. P., Lu, J. W., Sheehan, D., Perzanowski, M. S., Macfaden, S. W., . . . Rundle, A. (2013). Urban tree canopy and asthma, wheeze, rhinitis, and allergic sensitization to tree pollen in a New York City birth cohort. *Environmental Health Perspectives, 121*(4), 494–500. doi:10.1289/ehp.1205513

McPherson, E. G. (1992). Accounting for benefits and costs of urban greenspace. *Landscape and Urban Planning, 22*(1), 41–51. doi:10.1016/0169-2046(92)90006-l

McPherson, E., Simpson, J., Peper, P., Maco, S., Gardner, S., Vargas, K., . . . Xiao, Q. (1970). City of Minneapolis, Minnesota municipal tree resource analysis. Retrieved April 11, 2019, from https://www.fs.usda.gov/treesearch/pubs/4598

Neighborhood Revitalization Initiative. (2008). Retrieved April 11, 2019, from http://www.arborenvironmentalalliance.com/carbon-tree-facts.asp

Sand, B. (2024). Everyone says tree are good us. This scientist wants to prove it. *Washington Post,* January 29, 2024.

Sundell, J., Pershagen, G., Wickman, M., & Nordvall, L. (1995). Ventilation in homes infested by house dust mites. *Allergy, 50,* 106–112.

Tyrväinen, L., Pauleit, S., Seeland, K., & de Vries, S. (2005). Benefits and uses of urban forests and trees. In Konijnendijk, C., Nilsson, K., Randrup, T., & Schipperijn, J. (eds.), *Urban forests and trees.* Springer, Berlin, Heidelberg, 81–114. doi:10.1007/3-540-27684-X_5

U.S. Census Bureau. (2000). *Census 2000 summary file 1: GCT-PH1. Population, housing units, area, and density.* Retrieved April 1, 2003, from factfinder.census.gov/bf/_lang=en_vt_name=DEC_2000_SF1_U_GCTPH1_US10_geo_id=01000US.html

US Forest Service. (n.d.). *Benefits of trees.* Retrieved April 11, 2019, from https://www.fs.fed.us/learn/trees

Wolf, K. (2008). City trees and property values. *Arborist News, 16*(4), 34–36. Retrieved from https://www.naturewithin.info/Policy/FMJ_CityTrees_PropertyValues.pdf

World Health Organization (Ed.). (2018). *How air pollution is destroying our health.* Retrieved from https://www.who.int/air-pollution/news-and-events/how-air-pollution-is-destroying-our-health

Chapter 13

Does Walkability Matter? Exploring the Relationship between Walkability and Housing, Foreclosure, Health, and Reducing Greenhouse Gases

John Hans Gilderbloom, William W. Riggs, and Wesley L. Meares

The world's most famous urban planner, Jane Jacobs, in *The Death and Life of Great American Cities*, argues that the ideal neighborhood is one that is walkable. We want to know whether walkability matters in terms of impacting crime, foreclosures, and housing values in neighborhoods. For 50 years, we never had a reliable measure of the social and economic impact of walkable neighborhoods. This changed dramatically when walkability measurement tools were introduced. Walkability tools measure how accessible daily living activities are by foot and the likelihood that one will be car-dependent. In general, inner-city neighborhoods built before the mass production of cars are more walkable than a sprawling "residential only" suburban neighborhood that is isolated from basic necessities of everyday life.

This concept of walkability is an important emerging topic in the context of the growing dialogue on neighborhood sustainability and resilience. Although many U.S. neighborhoods had walkable origins (with local stores and shops and connections via streetcar between housing and jobs), many developments completed in the twentieth century do not have such attributes. They were designed for automobiles, with little connectivity; thus, the ease of moving via walking or cycling to destinations such as schools, stores, and workplaces is limited (Sallis, Frank, Saelens, & Kraft, 2004). This creates a dependency on private autos that has been connected with reduced physical activity and increased chances of obesity (Cervero & Kockelman, 1997; Ewing, Brownson, & Berrigan, 2006; Ewing, Schmid, Killingsworth, Zlot, & Raudenbush, 2003; Frank, Andresen, & Schmid, 2004).

The "Great Recession" of 2008–2010 offers a comparison of the economic resilience of these different types of neighborhoods. Many neighborhoods have struggled with a glut of foreclosures, where the only solution is to find creative strategies for demolition (Saito, 2011). By contrast, anecdotal evidence suggests that walkable neighborhoods, closer to the urban core, have weathered the economic crisis more easily (Lloyd, 2008). This paper aims to explore this relationship. It assesses the association of walkability, housing values, foreclosures, and crime within Louisville, KY, in the context of the idea that walkability is an important economic and social resource that, in the spirit of Rawls (1975), should be shared and allocated equitably.

DEFINITION OF WALKABILITY

Since many individuals define walkability using different terms (e.g., proximity, accessibility, suitability), it is important to establish an operational definition for the purposes of the study. Many associate walkability with suitability factors such as street width, number of lanes, safe speeds, bulb outs, crossing improvements, presence of trees, etc., and other pedestrian level-of-service and suitability factors (NCHRP Report 616, Dowling et al., 2008). Others cite perception issues with safety, such as fear of crime or heavy traffic, that some pedestrian audit tools attempt to uncover. While these are important factors for walking behavior, this study uses the concept of accessibility based on the importance of destination, land use, and population characteristics (Ewing & Cervero, 2010). It is not dismissive of suitability factors that improve the quality of the environment nor of Southworth's (2005) emphasis on safety, quality of path, and path context, but focuses on measurable factors including various destinations and street connectivity.

Walkability and Health

A large body of research has correlated neighborhood walking with higher density, street intersections, higher land use mix, and closer access to resources (Frank, Andresen, & Schmid, 2004; Frank, Schmid, Sallis, Chapman, & Saelens, 2005; Moudon et al., 2006). Studies have found that neighborhoods classified as more walkable (using walkability benchmarking tools) have higher levels of incidental walking and a lower likelihood of obesity (Frank, Kerr, Chapman, & Sallis, 2007)—a clear health issue on the rise in the United States and abroad.

A 2006 study by researchers at the Harvard School of Public Health reports that obesity is responsible for 2.6 million annual deaths worldwide (Ezzati, Martin, Skjold, Hoorn, & Murray, 2006). Other recent studies also indicate that 65 percent of American adults are overweight and 30.5 percent are considered obese (American Obesity Association, 2007), with rates expected to double within the next 10 years (U.S. Center for Disease Control). Being overweight and obese increases the risks of high blood pressure, high cholesterol, heart disease, stroke, cancer, gall bladder and respiratory disease, joint and bone disease, and many other afflictions, including diabetes (Pi-Sunyer, 1993).

This is, however, mitigable through the increased activity associated with a more walkable environment. Studies have shown that even light-to-moderate activity is associated with a substantially reduced risk of developing disease (Hu, Li, Colditz, Willett, & Manson, 2003; Thompson, Edelsberg, Colditz, Bird, & Oster, 1999), and that the built environment influences physical activity levels (Ewing, 2005; Handy, Cao, & Mokhtarian, 2005; Handy, Cao, & Mokhtarian, 2006). Furthermore, environments that are more walkable have been associated with *higher* levels of walking (Doyle, Kelly-Schwartz, Schlossberg, & Stockard, 2006; Saelens, Sallis, & Frank, 2003), and correlated with decreased risk of obesity and related illnesses (Frank, Andresen, & Schmid, 2004; Frank, Schmid, Sallis, Chapman, & Saelens, 2005).

Walkability, Economic Value, and Demand

In addition to these health benefits, there is a growing body of work that illustrates that walkable neighborhoods have intrinsic economic value by encouraging economic transactions and social exchanges (Litman, 2003; 2011) and bolstering real estate property values (Cortright, 2009; Diao & Ferreira, 2010). Research by Matthews and Turnbull (2007) shows that a more grid-like street pattern increases value in more pedestrian-oriented neighborhoods and decreases value in auto-oriented neighborhoods. Other work finds that each incremental increase in walkability can be associated with an increase in property value of up to 9 percent (Pivo & Fisher, 2011).

Improved walkability can entice consumers to purchase more local goods and facilitate greater economic efficiency (and therefore economic resilience) (Litman, 2006). The attributes associated with walkability may also have the capacity to improve safety and decrease crime (Foster & Giles-Corti, 2008; Leslie et al., 2005; Troy & Grove, 2008), having an indirect effect on real estate values.

This emergent idea that real estate values are bolstered by walkability is propped up by a recent study that indicates increased consumer demand for walkable attributes (Handy, Sallis, Weber, Maibach, & Hollander, 2008), a finding that contradicts studies suggesting that many (especially Caucasians) prefer large, single-family homes (Bajari & Kahn, 2005) and racial homogeneity (Farley, Schuman, Bianchi, Colasanto, & Hatchett, 1978; Farley,

Fielding, & Krysan, 1997; Krysan & Farley, 2001; Meen & Meen, 2003; Quigley, 1985).

There is limited evidence showing how much (or if) walkability factors enter into these housing decisions. However, research surveying a broad spectrum of real estate professionals calls out walkability as "a major amenity" (Riggs, 2011). Subject area experts such as Dr. Larry Frank of the University of British Columbia have been quoted as saying, "There is no question of a large unmet demand for walkable neighborhoods . . . (which drives) price increases in central cities."

Walkability and Equity

While all of these studies may control for racial dynamics, it is important to emphasize how they are bound by the limitations associated with self-selection and the disproportionate resource choices available to the poor and minorities. Research indicates that self-selection of housing is often related to income (Nakosteen & Zimmer, 1980) and that active behaviors are shaped by neighborhood self-selection (Handy, Cao, & Mokhtarian, 2006; Ioannides & Zabel, 2008).

Many minorities remained unable to find adequate housing in cities and cannot afford to purchase nicer housing in the suburbs for a variety of reasons, including predatory lending and insurance practices (Cutler, Glaeser, & Vigdor, 1999). Some studies suggest a housing markup of approximately 7 percent for Blacks as compared to Whites (Kain & Quigley, 1972) and continued mortgage discrimination resulting in lower ownership and higher segregation (Ellen, 2008). This is important in that many health problems that might be mitigated through more walkable neighborhoods are focused in areas with higher proportion of minorities as opposed to the general population (Galea, Freudenberg, & Vlahov, 2005; Geronimus & Thompson, 2004; Williams & Jackson, 2005).

Walkability and Forclosures

Given the possible economic and social-equity stabilization effects that walkable neighborhoods can provide, it is not surprising that recent studies are showing the resilience of these areas in the face of economic crises. Multiple studies show high patterns of foreclosure both inside the urban core and on the suburban fringe (Gilderbloom & Meares, 2020). Studies have correlated some of these foreclosures with higher transportation costs and longer trips to work (NRDC, 2010), but also with having a disproportionate impact on central city neighborhoods (Immergluck, 2009), which are presumably more walkable but have a high minority population.

The suggestion of the disproportionate impact of foreclosures is closely aligned with recent studies that assess reduced foreclosure probability with higher walkability. One such study concluded that the probability of foreclosure was different based on income (Rauterkus, Thrall, & Hangen, 2010). In high-income areas with higher walkability, the probability of foreclosure is lower, but it increases in low-income areas with higher walkability.

Given the limited publications on the topic and the variability of results, more research is needed to fully uncover the effects of increased walkability on housing appreciation and the risk of foreclosure; the results do suggest that the concept of location-efficiency and its related policies are not dead (Chatman & Voorhoeve, 2010). The idea could potentially be put to use to advocate for better mortgage rates and underwriting flexibility in lower-risk walkable areas.

METHODOLOGY

Sample: Case Study of 170 Neighborhoods in a Mid-Size City

For a number of reasons, we chose Louisville as fairly representative of a typical mid-size city. With a population of roughly 741,000 spread over 385 square miles along the Ohio River, Louisville, KY, provides a model city for study. Louisville is one of 375 metropolitan areas identified by the U.S. Census and ranks as the 47th largest metropolitan area. Unlike the extremes of larger mega-regions on the east and west coasts of the United States, the Louisville housing market is relatively stable and finds itself to be more representative of the neighborhood dynamics of the South and Midwest than New York, Washington D.C., San Francisco, and Los Angeles. Louisville, along with 144 other cities (Cincinnati, Indianapolis, Lexington, Nashville), was ranked as mid-size and a good place to study growth dynamics because it has over 50,000 persons and is not located within another 20 miles of a large city (Appelbaum et al., 1976; Appelbaum, 1978; Molotch, 1976). Because of the stability of its market and the distance it has from other large urban areas, Louisville has been used in a number of studies investigating mid-size cities (Appelbaum, Bigelow, Kramer, Molotch, & Relis, 1976; Gilderbloom & Appelbaum, 1987; Ambrosius, Gilderbloom, & Hanka, 2010; Frederick, 2018a, 2018b, 2019). The city provides a range of neighborhood types with a large variation in density and walkability across the region—including examples of both more and less walkable neighborhoods. Louisville has a simple, relatively mono-centric format (as opposed to the poly-centric models of many larger municipal areas) that is ringed by two freeways. It has one central business district (CBD) with approximately 52,000 jobs (13 percent of the total) forming an inner beltway with high-density housing, an in-between area with smaller homes, and then an outside beltway where there has been increased building of larger, more suburban homes (Ambrosius, Gilderbloom, & Hanka, 2010). This is an important distinction to make, not only because of the differences in physical form at the neighborhood level that might influence walking but because of the underlying behavioral/driving habits for those who live outside of the CBD. Research has shown that although areas of higher density may have more walking for transportation purposes, lower-density areas offer more opportunities for leisure walking, making behavior very important (Forsyth, Oakes, Schmitz, & Hearst, 2007). Those who want to live in a location where they can walk for transportation may choose locations that support this, but those whose preference is to walk for leisure may prioritize housing with more access to leisure opportunities—Louisville offers both such environments.

As demonstrated below, another reason for choosing Louisville is the treasure trove of data that city leaders eagerly provided. This data is much harder to access in other cities, which might explain why research is stalled in many other areas. Another barrier to access is that if the data exists, it might not be available for political reasons (Gilderbloom & Appelbaum, 1987; Frederick, 2018a, 2028b, 2019; Ambrosius et al., 2009).

Some agencies charge thousands of dollars for the data and/or do not provide a computer file but paper-only printouts—making it costly and labor-intensive to format back to computer. Certain government agencies would not allow the data to be released, but the data would be sent to us from an anonymous source, or a website would be provided that had the data. Sometimes it is just a matter of luck and/or having a good public reputation for having integrity and fairness.

Walkability scores, census data, and historic preservation data are easy to procure, but due to the political nature of other data, it was much harder to get census tract information covering foreclosures, low-interest loans, crime, Section 8, EPA Superfund toxic waste sites (covered in another paper), and premature death rates (which is covered in another research paper). Once data is procured, sensitive data must be approved for analysis by the Institutional Review Board, guaranteeing that no identities are revealed in the data set (which is not an issue when we create averages for each census tract).

Data

The geography of our study is Louisville Metro, and our unit of observation is the census tract. The population of all Census tracts in our geography contains 170 tracts. We have several key dependent variables we use to measure the impact of walkability on the quality of urban neighborhood life: neighborhood house prices and foreclosures, and crime.

Our dependent variable is the Median Assessed Residential Housing Value. We operationalized this variable in three different ways: (1) Median Assessed Residential Housing Value in 2000, 2006, 2008, and 2010; and (2) Median Assessed Residential Housing raw dollar change (2000 to 2006, 2000 to 2008, 2000 to 2010). For these variables, data on all residentially-zoned properties in Louisville Metro, the geography of our study, for the years 2000 and 2010 was collected from the Jefferson County Property Valuation Administrator (PVA) and geo-coded by the Kentucky State Data Center using the Louisville/Jefferson County Information Consortium's (LOJIC) GIS system. This data has been established as accurate by the state and in academic articles (Lindauer & Hunt, 2008; Gilderbloom, Meares, & Squires, 2020; Gilderbloom, 2017; Frederick, 2018a, 2028b, 2019). While the data complements the findings for the 2000 census, it also is available to measure changes in neighborhoods in 2006, 2008, and 2010.

Kentucky Revised Statute 134.385 requires that local property assessments be at least 80 percent of the fair market value and provides for state oversight in case of a discrepancy between the assessment and the market value. In previous analyses, we found that assessments are on average 86 percent of sales prices with no large variation among neighborhoods based on neighborhood indicators (although there is variation based on employment density) (Gilderbloom, Riggs, & Meares, 2015; see also Pollakowski, 1995). Nevertheless, there is the public perception of assessors undervaluing properties.

In Louisville, assessments are undertaken at a minimum every three years. Our analysis spans 10 years and a high proportion of properties were assessed once, twice, or three times during our analysis period. The assessor's office does not only depend on its assessments (which can be political) but also on sales prices—whenever a home is sold, the assessor's office takes the sales price as an update for its database. We assume that assessed values, the basis of our dependent variable, are a good proxy for market values (see also Clapp & Giacotto, 1992; Gilderbloom, Squires, & Meares, 2020; Gilderbloom, Squires, Riggs, & Capek, 2017). Our analysis period ranges from 2000 to 2010, which includes the years when the United States experienced a house price bubble (Shiller, 2007; Gilderbloom, Riggs, & Meares, 2015). However, during that time, the Louisville metropolitan area experienced only a modest increase in house prices (Lindauer & Hunt, 2008).

Walkability Measurement

While there have been many methods established to objectively measure the walkable environment, in order to quantify neighborhood walkability, our study takes advantage of the tool established by Walk Score™ in its analysis as the key test variable. Developed by Frontlane in partnership with academics including Larry Frank and Reid Ewing, it uses a method similar method to Frank's SMARTRAQ model, combining land use mix, density, and street grid density based on geo-location. Google is used to index adjacent amenities as a proxy for land use mix, density comes from U.S. Census figures, and intersections are counted using an algorithm on a street network. The weighted scores are summed and normalized to 100, yielding a score from 0–100. These methods place the highest priority on land use mix as the leading predictor of walking behavior.

Using the Walk Score™ tool has clear advantages. While many tools employ surveys, self-reporting, audits, and observational data measures, the Walk Score™ tool provides a direct and replicable manner of assessing geospatial, population, and land use characteristics to benchmark walkability (Brownson, Hoehner, Day, Forsyth, & Sallis, 2009; Heath et al., 2006). Studies have concluded that the Walk Score™ suffices as a reliable tool for measuring the walkability of an area and may be more accurate than other tools as Google tends to be reviewed and updated on a more frequent basis than other static databases (Carr, Dunsiger, & Marcus, 2010, 2011).

While this method provides a simple research solution to determine a location's walkability, it also has limitations. First, the tool is limited based on its reliance on Google Maps as an underlying database, with potential flaws in the exact geo-location and classification of use categories (which are many times user-contributed). Second, until recently, the tool did not incorporate network characteristics but used straight-line distance calculations that did not take into account street connectivity. Third, the tool does not account for street quality (such as the presence of trees, sidewalk width, etc.), safety (from traffic or crime), and terrain characteristics (slope). Lastly, similar to other metrics, the tool may suffer from aggregation errors based on the unit of spatial analysis.

In addition to the raw Walk Score™ (Range 0–100) we used a dummy variable as an alternative to the ratio variable. This variable isolates the top 33 percentile known, at the very minimum, to have some amenities within walking distance. Walk Score™ makes the case that, at this level, most errands can be accomplished on foot, calling a score of 92 "a walker's paradise." This score indicates that most of the areas in the bottom 66 percentile are car-dependent and that individuals would be disinclined to run most errands on foot without other means of transportation. We operationalize our dummy variable as 1 equals walkable neighborhoods and 0 equals car-dependent. In Louisville we have 32 walkable neighborhoods.

We developed a hedonic price equation that controls for recognized independent variables that predict our dependent variables and added in our test variable, walkability. Our regression model used control variables that are regularly used in regression analysis of contemporary neighborhood dynamics to look at neighborhood housing, foreclosures, crime, and health

(Ambrosius, Gilderbloom, & Hanka, 2010; Appelbaum, 1978; Gilderbloom, Anaker, Squires, Hanka, & Ambrosius, 2011). Regarding the independent variables, we rely on data provided by 1a) the U.S. Bureau of the Census's 2000 and 2010 Census; 1b) the U.S. Bureau of the Census's 2000 Transportation Planning Package[1] and by 2) the Louisville Metro Police Department (LMPD) for 2004 data, the earliest year this type of data was available, City of Louisville Health Statistics, PVA. We relied on the following variables to control while measuring the net impact of walkability. This required the approval of the Institutional Review Board to ensure anonymity.

What makes a walkable neighborhood? The amount of adjusted explained variation at 55 percent is satisfactory. This test revealed that four independent variables were successful at predicting walkable neighborhoods. Areas that are closer to the Central Business District tend to have higher walkability. This can be explained by the agglomeration effect that occurs in the urban core, making it easier to access necessities and entertainment activities on foot. The second variable that predicted higher walkability was neighborhoods that had a higher percentage of Caucasian residents. This is because people with higher incomes tend to live in suburban areas. Another variable that predicts walkable neighborhoods is the median household income. The variable reveals that households with higher incomes tend to have lower scores. Furthermore, the test revealed that historic neighborhoods have higher walk scores. Most of these neighborhoods were designed and built before the mass production of automobiles. Consequently, people had to walk or bike for short trips and take trains and trolleys for longer distances. This meant that shops, grocery stores, and entertainment activities were all located within a 10 minute walk of these neighborhoods, giving them better access (Duany, Plater-Zyberk, & Speck, 2000). The founders of new urbanism argue that the model for new urbanism is based on these traditional neighborhoods. New urbanism makes walkability the foundation for the sustainability of neighborhoods that incorporates social, health, and economic concerns.

Regarding walkability and its impact on neighborhood house values, four out of five dependent variables measuring neighborhood house values showed a significant relationship with walkability. The amount of explained variation from the adjusted R square was excellent and above 83 percent in the equations. The standardized beta coefficient appeared to increase in a monotonic relationship between 2000, 2006, and 2008 in size. However, in 2010, this standardized coefficient was not statistically significant.

We then took out the ratio variable on walkability and used a dummy variable that compared the most walkable neighborhoods (the top 33 percentile) to more car-dependent neighborhoods. In this model, the dummy variable walkable became significant for 2010 and revealed that walkable neighborhoods were worth $33,443; in 2008 it was $25,871; in 2006 where it was

$22,749, and 2000 where it was $11,605. These differing results could be caused by several factors. First, the operationalization of walkability was different. Second, real estate neighborhood prices nearly doubled and tripled in Louisville neighborhoods between 2000 and 2008 and then stagnated in the metropolitan area as a whole. However, prices continued to increase in the walkable neighborhoods as noted. An additional run looking at the percentage change in neighborhood housing values found that walkability predicted an increase in property values for the dependent variables from 2000 to 20006 and 2000 to 2008. As expected, we did not find statistical significance between 2000 and 2010 because of the housing collapse.

When examining the effect of walkability and foreclosures, we found that the more walkable a neighborhood, the less foreclosures it had on average. Again, the amount of explained variation was excellent. We found this to be true for equation one, which measures the raw number of foreclosures from 2004 to 2008 and for 2007 and 2008. More precisely, a highly walkable neighborhood had 11 fewer total foreclosures from 2004 to 2008; similarly, foreclosures for 2007–2008 are five fewer.

When we used our standard regression equation to examine the impact of walkability on four measures of crime in a neighborhood, there was no significance found. We looked at four measures property crime, murder, violence, and total number of crimes. The amount of explained variation was satisfactory, between .40 and .61. End of the story? Not so fast. We then did a split regression analysis by removing neighborhoods that were either 50 percent or more African American or 25 percent or more African American. When we removed all census tracts that had a majority minority population within them, we found that walkability deterred property crime, murders, and violent crime. When we refined the selection variable even more pointedly to examine only neighborhoods that are 75 percent or more white, we found that the results were similar.

POLICY IMPLICATIONS AND CONCLUSIONS

The automobile changed how and where Americans lived in the second half of the twentieth century. Americans left behind the walkable mixed use of old downtown and first-generation suburbs for homogenized residential suburban development. Older historic downtown neighborhoods were built for active users (bikers and walkers) along with trolleys, cable cars, and a mix of purposes that incorporated employment, grocery stores, places of worship, restaurants, schools, medical services, and recreational space. These areas displayed economically diverse residents where the poor and working class lived together in the same neighborhoods with a sprinkling of the rich. These particular neighborhoods worked best and were in the highest demand. Remnants of their DNA can still be seen in many historic neighborhoods around the United States: the West Village/Greenwich Village in New York; Beacon Hill in Boston; Shadyside in Pittsburgh; Old Louisville and the Highlands in Louisville; West Cleveland/Tremont in Cleveland; Rogers Park in Chicago; the Mission District and Noe Valley in San Francisco; downtown Ballard in Seattle. Increased use of the automobile allowed the suburbs to happen and caused traditional neighborhoods to fall into decline.

The quality and essence of several healthy neighborhoods still exist because a generation of activist-citizens, the first advocates of the modern sustainable city, stood up against adverse changes to the traditional fabric of American cities. In places like Greenwich Village, a mother and freelance writer named Jane Jacobs became involved in stopping a freeway from ripping through her beloved neighborhood. In Cleveland and San Francisco, individuals Norman Krumholz and Catherine Bauer Wurster worked as urban advocates for neighborhood preservation and social equity in the face of pushes to modernize their cities. Wurster was a member of the prominent Telesis group, a research collective that believed in the importance of community in urban planning and in the importance of every urban environment to support living, working, and playing there. Jacobs ascribed to many of the same values and became notably prominent in her organizing efforts—which inspired her to write the "Bible" on urban planning, *The Death and Life of Great American Cities*. The book argued that the ideal neighborhood was one that was walkable with lots of mixed uses and a diverse community.

In this study, we tested Jacob's premise of the value of the walkable, car-free environment empirically by using high-power statistical analysis of 170 neighborhoods with a simple question: Does walkability matter? Our results seem to indicate that it does. When it comes to neighborhood stability, economic resilience, and livability, the aspects that make up a walkable community have value. This may not come as a surprise to many in the planning community; however, for 50 years, those in the field have not had a reliable measure to quantify the social and economic impact of walkable neighborhoods—only the instincts of planners such as Jacobs to trust. The computational ability to benchmark communities using a walkability measure meets that need. Walkability measures now assist planners and policymakers in gauging how accessible daily living activities are by foot, and the likelihood of being car-dependent where one lives. Many inner-city neighborhoods built before the mass production of cars are more walkable than sprawling "residential-only" suburban neighborhoods that are isolated from the basic necessities of everyday life.

We show for the first time with an array of neighborhood indicators how walkability has a net significant impact on neighborhood resilience and

consistently positive impacts on walkability. The independent variable walkability is as strong as race when it comes to predicting neighborhood housing values and foreclosures. Based on this, we show that neighborhood hedonic equations must take into consideration walkability.

To test this, we developed a hedonic price equation that controls for recognized independent variables that predict our dependent variables and added in our test variable, walkability. We find that walkability is statistically significant in predicting an increase in neighborhood housing values and has a significant negative correlation with foreclosures in the neighborhoods of Louisville. Finally, although our initial results show no association between walkability and crime, when we exclude neighborhoods that are mostly black and rerun the regression, we find that walkability is also associated with reduced crime in several measures.

These kinds of outcomes have not been extensively documented in hedonic regression analysis of mid-sized cities, and the significant association of walkability in such models indicates specific policy implications for sustainable neighborhood design. The associations suggest that walkability is a value proposition—that those areas that are more walkable have higher housing values, are less prone to foreclosure, have improved longevity, and experience lower crime rates. These facts are especially timely when put into the context of the housing collapse of 2008–2009. This implies that neighborhoods that have been designed in the spirit of Jane Jacobs, with safe and connected streets, high levels of density, and varied land uses, not only benefit residents through the intrinsic health and environmental benefits of more walking but also offer owners more resilient and possibly more stable economic value.

We recognize that the reasons people choose neighborhoods to live, work, and play in are different today than in the past. Soaring gas prices have many Americans seeking housing that is closer to employment, central city locations, cultural amenities (such as museums and sports complexes), and historic areas that have been preserved. Gentrification is just one expression of this change in preferences, starting in New York, London, and Paris in the late 1950s and later in San Francisco and Chicago in the 1970s (Ambrosius, Gilderbloom, & Hanka, 2010; Gilderbloom & Appelbaum, 1987; Savitch, 1981; Zukin, 1982). Gasoline prices may continue to rise as supplies fall and demand increases from other emerging industrialized countries,[2] and this may continue to influence how to get to school, work, stores, and worship. However, as our models indicate more and more, when individuals weigh neighborhood choices, the ability to walk will factor into that choice alongside schools, crime rate, proximity to jobs, downtown etc. The American Automobile Association states that, on average, the cost of owning a car comes to $8,000 a year when you factor in depreciation, financing, maintenance, gas, and insurance (Gilderbloom, Riggs, & Meares, 2015; Frederick, 2018a, 2028b, 2019). And since walkability matters and is of increasing importance in housing choice, a clear question arises for policymakers—how can we best promote and create walkable places?

Three policy areas dealing with (1) land use, (2) street design, and (3) affordability offer tools that might be used to make neighborhoods more walkable. First, regarding land use, neighborhood walkability is heavily tied to the number and variation of amenities or destinations available within a short walking distance. Unfortunately, many times planning and zoning codes are restrictive of land use variation. Policies that relax zoning and allow for ground-floor commercial or retail conversions may encourage transactional uses that increase neighborhood interactions and walking destinations (even in suburban locations). Examples of this are *poparillas* in Latin America, which allow small entrepreneurs to establish pop-up or informal businesses that provide neighborhood services selling everything from "a couple of light bulbs, some matches, some cokes, and cheese and bread . . . like a 7–11, just out of their closet" (Riggs, 2011).

Secondly, investment in streetscape improvements is of parallel importance. Many cities across the United States have adopted "complete streets" policies to encourage safer and more bike-accessible and pedestrian-friendly streets. The recognition of both the economic and health benefits of walkability and policies that support this environment are important. These policies might include providing sidewalk bulb-outs and safety refuges, reducing crossing distances/street widths; installing crosswalks and road markings for pedestrians; planting trees; offering benches; building shelters; and displaying art. These improvements could be funded through neighborhood benefits districts or other forms of self-imposed financing, counteracting the trends of public disinvestment that many areas have seen in the wake of the recent economic crises.

Lastly, there is a need to reinforce the affordability and diversity of these walkable areas—especially as they gain in attractiveness and importance during the course of housing decisions. Since simply investing in an area is likely to exacerbate inequities that may already exist, policies are needed to increase affordable housing for those who may face increasing prices due to on-street public investment. The authors recognize that the high economic value of walkable areas may present difficulties for renters and minorities who may not be able to capture the economic value of owned housing. Opportunities to counteract this trend might germinate from policies that would increase or establish required below market rate (BMR) unit thresholds for new construction, or policies expanding density thresholds in order to increase the number of units available by right—driving down prices by increasing supply. Simple techniques might involve the conversion of warehouses into housing, or the conversion of basements, garages, and attics infilling historic neighborhoods, mixed-use homes, and a denser population of individuals living closer to downtown business centers.

Each of these areas, (1) increasing the number of land uses, (2) increasing investment in the streetscape, and (3) increasing affordability options and standards, are goals that neighborhood organizations may advocate and implement to increase neighborhood walkability. Done in parallel, they could provide neighborhood investment and increase opportunities for a diverse socio-cultural cross-section of individuals to live and play in areas that are walkable.

The future of urban planning is creating neighborhoods that are sustainable. Does walkability matter? Yes! People will demand more walkable neighborhoods that are closer to the necessary conveniences of life. Gas prices will continue to rise, forcing folks to rethink the neighborhoods they choose. When they choose a neighborhood, they consider characteristics such as crime rate, walkability, gentrification, green living, mixed use, and proximity to downtown. This was not always the case in the past, and the walk score was not a significant factor in choosing neighborhoods as it is now, with gas prices soaring.

To offset high gasoline prices, people will migrate closer to downtown historic neighborhoods to save on gas and start using alternative transportation. Relocating to a sustainable neighborhood means a better return on their investment, reducing air pollution, reducing car dependency, and living in denser and more diverse neighborhoods. Active transportation will have a positive impact on personal health by walking or biking to work or school and will allow the residents to live green.

Revised and reprinted with permission from *Cities: The International Journal of Urban Policy and Planning.*

NOTES

1. The Census Transportation Planning Package (CTTP) data, while useful for transportation analysis, has limitations, including its reliance on decennial census data, which may not capture changes in commuting patterns and transportation needs in non-census years. Additionally, the aggregation of data at larger geographic levels, such as census tracts or block groups, may mask variations in smaller, localized areas. Temporal discrepancies between the CTTP and other datasets, such as those from the LMPD or City of Louisville Health Statistics, may also affect the alignment of trends across the different data sources.

2. A recent survey of 2000 students and employees found that one-fourth will move to a closer location to school or work if gas prices rise to $5 or more per gallon (see chapter 14).

REFERENCES

Ambrosius, J. D., Gilderbloom, J. I., & Hanka, M. J. (2010). Back to black . . . and green? Location and policy interventions in contemporary neighborhood housing markets. *Housing Policy Debate*, 20(3), 457–484.

American Obesity Association. (2007). *American obesity fast facts. American obesity association obesity statistics.* Retrieved from http://obesity1.tempdomainname.com/subs/fastfacts/obesity_US.shtml

Appelbaum, R. (1978). *Size, growth, and U.S. cities.* New York: Praeger Publishing.

Appelbaum, R., Bigelow, J., Kramer, H., Molotch, H., & Relis, P. (1976). *The effects of urban growth: A population impact analysis.* New York: Praeger.

Bajari, P., & Kahn, M. E. (2005). Estimating housing demand with an application to explaining racial segregation in cities. *Journal of Business and Economic Statistics*, 23(1), 20–33.

Brownson, R. C., Hoehner, C. M., Day, K., Forsyth, A., & Sallis, J. F. (2009). Measuring the built environment for physical activity: State of the science. *American Journal of Preventive Medicine*, 36(4), S99–S123.

Carr, L. J., Dunsiger, S. I., & Marcus, B. H. (2010). Walk ScoreTM as a global estimate of neighborhood walkability. *American Journal of Preventive Medicine*, 39(5), 460–463.

Carr, L. J., Dunsiger, S. I., & Marcus, B. H. (2011). Validation of Walk Score for estimating access to walkable amenities. *British Journal of Sports Medicine*, 45(14), 1144–1148.

Cervero, R., & Kockelman, K. (1997). Travel demand and the 3Ds: Density, diversity, and design. *Transportation Research Part D: Transport and Environment*, 2(3), 199–219.

Chatman, D., & Voorhoeve, N. (2010). The transportation-credit mortgage: A postmortem. *Housing Policy Debate*, 20(3), 355–382.

Clapp, J. M., & Giaccotto, C. (1992). Estimating price indices for residential property: A comparison of repeat sales and assessed value methods. *Journal of the American Statistical Association*, 87(418), 300–306.

Cortright, J. (2009). Walking the walk: How walkability raises home values in U.S. cities. *CEOs for Cities.* Retrieved from http://www.ceosforcities.org/pagefiles/WalkingTheWalk_CEOsforCities.pdf

Cutler, D. M., Glaeser, E. L., & Vigdor, J. L. (1999). The rise and decline of the American ghetto. *The Journal of Political Economy*, 107(3), 455–506.

Diao, M., & Ferreira, J. (2010). Residential property values and the built environment. *Transportation Research Record: Journal of the Transportation Research Board*, 2174(1), 138–147.

Dowling, R. G., Reinke, D. B., Flannery, A., Ryus, P., Vandehey, M., Petritsch, T. A., Landis, B. W., et al. (2008). *NCHRP report 616: Multimodal level of service analysis for urban streets.* Washington, DC: Transportation Research Board of the National Academies.

Doyle, S., Kelly-Schwartz, A., Schlossberg, M., & Stockard, J. (2006). Active community environments and health: The relationship of walkable and safe communities to individual health. *Journal of the American Planning Association*, 72(1), 19–31.

Duany, A., Plater-Zyberk, E., & Speck, J. (2000). *Suburban nation. The rise of sprawl and decline of the American dream.* New York: North Point Press.

Ellen, I. G. (2000). Race-based neighborhood projection: A proposed framework for understanding new data on racial integration. *Urban Studies,* 37(9), 1513–1533.

Ellen, I. G. (2008). Continuing isolation: Segregation in America today. In *Segregation: The rising costs for America*, James, H.C. and Nandinee, K. K. (eds), 261–277. New York: Routledge.

Ewing, R. (2005). Can the physical environment determine physical activity levels? *Exercise and Sport Sciences Reviews*, 33(2), 69–75.

Ewing, R., Brownson, R. C., & Berrigan, D. (2006). Relationship between urban sprawl and weight of United States youth. *American Journal of Preventive Medicine*, 31(6), 464–474.

Ewing, R., & Cervero, R. (2010). Travel and the built environment—A meta-analysis. *Journal of the American Planning Association,* 76, 265–294.

Ewing, R., Schmid, T., Killingsworth, R., Zlot, A., & Raudenbush, S. (2003). Relationship between urban sprawl and physical activity, obesity, and morbidity. *Urban Ecology*, 567–582.

Ezzati, M., Martin, H., Skjold, S., Hoorn, S. V., & Murray, C. J. (2006). Trends in national and state-level obesity in the USA after correction for self-report bias: Analysis of health surveys. *Journal of the Royal Society of Medicine*, 99(5), 250.

Farley, R., Fielding, E. L., & Krysan, M. (1997). The residential preferences of blacks and whites: A four-metropolis analysis. *Housing Policy Debate*, 8(4), 763.

Farley, R., Schuman, H., Bianchi, S., Colasanto, D., & Hatchett, S. (1978). Chocolate city, vanilla suburbs: Will the trend toward racially separate communities continue? *Social Science Research,* 7(4), 319–344.

Forsyth, A., Oakes, J. M., Schmitz, K. H., & Hearst, M. (2007). Does residential density increase walking and other physical activity? *Urban Studies*, 44(4), 679.

Foster, S., & Giles-Corti, B. (2008). The built environment, neighborhood crime and constrained physical activity: An exploration of inconsistent findings. *Preventive Medicine*, 47(3), 241–251.

Frank, L., Kerr, J., Chapman, J., & Sallis, J. (2007). Urban form relationships with walk trip frequency and distance among youth. *American Journal of Health Promotion*, 21(4 Suppl), 1–8.

Frank, L. D., Andresen, M. A., & Schmid, T. L. (2004). Obesity relationships with community design, physical activity, and time spent in cars. *American Journal of Preventive Medicine*, 27(2), 87–96.

Frank, L. D., Schmid, T. L., Sallis, J. F., Chapman, J., & Saelens, B. E. (2005). Linking objectively measured physical activity with objectively measured urban form: Findings from SMARTRAQ. *American Journal of Preventive Medicine*, 28(2), 117–125.

Frederick, C., & Gilderbloom, J. (2018b). Commute mode diversity and income inequality: An inter-urban analysis of 148 midsize US cities. *Local Environment:*

The International Journal of Justice and Sustainability, 23(1), 54–76. https://doi .org/10.1080/13549839.2017.1385001

Frederick, C., Hammersmith, A., & Gilderbloom, J. H. (2019) "Putting "place" in its place: Comparing place-based factors in interurban analyses of life expectancy in the United States. *Social Science and Medicine,* (232), 148–155.

Frederick, C., Riggs, W. W., & Gilderbloom, J. (2018a). Commute mode diversity and public health: A multivariate analysis of 148 US cities. *International Journal of Sustainable Transportation,* 12(1), 1–11. https://doi.org/10.1080/15568318 .2017.1321705

Galea, S., Freudenberg, N., & Vlahov, D. (2005). Cities and population health. *Social Science & Medicine,* 60(5), 1017–1033.

Geronimus, A. T., & Thompson, J. P. (2004). To denigrate, ignore, or disrupt: racial inequality in health and the impact of a policy-induced breakdown of African American communities. *Du Bois Review: Social Science Research on Race,* 1(2), 247–279.

Gilderbloom, J., Anaker, K., Squires, G., Hanka, M., & Ambrosius, J. (2011). *Why foreclosure rates in African American neighborhoods are so high: Looking at the real reasons.* ERSA conference papers.

Gilderbloom, J., Riggs, W. W., & Meares, W. L. (2015). Does walkability matter? An examination of walkability's impact on housing, values, foreclosures and crime. *Cities,* 42(Part A), 13–24.

Gilderbloom, J., Squires, G. D., Riggs, W. W., & Capek, S. (2017). Think globally, act locally: Neighborhood pollution and the future of the earth. *Local Environment: The International Journal of Justice and Sustainability,* 22(7), 894–899. https://doi .org/10.1080/13549839.2017.1278751

Gilderbloom, J. H., & Meares, W. L. (2022). How inter-city rents are shaped by health considerations of pollution and walkability: A study of 146 mid-sized cities. *Journal of Urban Affairs,* 44(8), 1059–1075.

Gilderbloom, J. H., Meares, W. L., & Squires, G. D. (2020). Pollution, place and premature death: Evidence from a midsized city. *Local Environment: International Journal of Sustainability and Justice.* Published online April 28, 2020, 25(6), 419–432, https://doi.org/10.1080/13549839.2020.1754776

Gilderbloom, J. H., Squires, G. D., & Meares, W. (2020). "Mama, I can't breathe." Louisville's dirty air has steep medical and economic costs. *Local Environment, The Conversation,* 25(8), 619–627. https://www.tandfonline.com/doi/full/10.1080 /13549839.2020.1789570

Gilderbloom, J. I., & Appelbaum, R. P. (1987). *Rethinking rental housing.* Philadelphia: Temple University Press.

Handy, S., Cao, X., & Mokhtarian, P. (2005). Correlation or causality between the built environment and travel behavior? Evidence from Northern California. *Transportation Research Part D: Transport and Environment,* 10(6), 427–444.

Handy, S., Cao, X., & Mokhtarian, P. L. (2006). Self-selection in the relationship between the built environment and walking: Empirical evidence from Northern California. *Journal of the American Planning Association,* 72(1), 55–74.

294 *John Hans Gilderbloom et al.*

Handy, S., Sallis, J. F., Weber, D., Maibach, E., & Hollander, M. (2008). Is support for traditionally designed communities growing? Evidence from two national surveys. *Journal of the American Planning Association*, 74(2), 209–221.

Heath, G. W., Brownson, R. C., Kruger, J., Miles, R., Powell, K. E., & Ramsey, L. T. (2006). The effectiveness of urban design and land use and transport policies and practices to increase physical activity: A systematic review. *Journal of Physical Activity and Health*, 3(1), S55–S76.

Hu, F. B., Li, T. Y., Colditz, G. A., Willett, W. C., & Manson, J. A. (2003). Television watching and other sedentary behaviors in relation to risk of obesity and type 2 diabetes mellitus in women. *The Journal of the American Medical Association*, 289(14), 1785.

Hudnut, W. H. (1998). *Cities on the rebound*. Washington, DC: Urban Land Institute.

Immergluck, D. (2009). The foreclosure crisis, foreclosed properties, and federal policy: Some implications for housing and community development planning. *Journal of the American Planning Association*, 75(4), 406–423.

Ioannides, Y. M., & Zabel, J. E. (2008). Interactions, neighborhood selection and housing demand. *Journal of Urban Economics*, 63(1), 229–252.

Kain, J., & Quigley, J. (1972). Housing market discrimination, home-ownership, and savings behavior. *The American Economic Review*, 62(3), 263–277.

Krysan, M., & Farley, R. (2001). Residential preferences of Blacks: Do they explain persistent segregation, *The Social Forces*, 80, 937.

LaCour-Little, M., & Green, R. K. (1998). Are minorities or minority neighborhoods more likely to get low appraisals? *Journal of Real Estate Finance and Economics*, 16(3), 301–315.

Leinberger, C. B. (2008). *The next slum?* Atlantic Monthly.

Leinberger, C. B. (2009). *The option of urbanism: Investing in a new American dream*. Washington, DC: Island Press.

Leslie, E., Saelens, B., Frank, L., Owen, N., Bauman, A., Coffee, N., & Hugo, G. (2005). Residents' perceptions of walkability attributes in objectively different neighbourhoods: A pilot study. *Health & Place*, 11(3), 227–236.

Lindauer, T., & Hunt, D. (2008). *2007 Real estate market in review*. Louisville: Property Valuation Office.

Litman, T. A. (2003). Economic value of walkability. *Transportation Research Record: Journal of the Transportation Research Board*, 1828(1), 3–11.

Litman, T. A. (2006). *Socially optimal transport prices and markets*. Victoria Transport Policy Institute (www. vtpi. org).

Litman, T. A. (2011). Economic value of walkability. *World Transport Policy & Practice*, 10(1), 5–14.

Lloyd, C. (2008). *Is suburbia turning into slumburbia*. San Francisco Chronicle, 14.

Matthews, J. W., & Turnbull, G. K. (2007). Neighborhood street layout and property value: The interaction of accessibility and land use mix. *The Journal of Real Estate Finance and Economics*, 35(2), 111–141.

Meen, D., & Meen, G. (2003). Social behaviour as a basis for modeling the urban housing market: A review. *Urban Studies*, 40(5–6), 917–935.

Molotch, H. (1976). The city as a growth machine: Towards a political economy of place. *American Journal of Sociology,* 82(September), 309–322.

Moudon, A. V., Lee, C., Cheadle, A. D., Garvin, C., Johnson, D., Schmid, T. L., Weathers, R. D., et al. (2006). Operational definitions of walkable neighborhood: Theoretical and empirical insights. *Journal of Physical Activity & Health,* 3, 99.

Nakosteen, R. A., & Zimmer, M. (1980). Migration and income: The question of self-selection. *Southern Economic Journal,* 46(3), 840–851.

Newman, P., Bealey, T., & Boyer, H. (2009). *Resilient cities: Responding to peak oil and climate change.* Washington, DC: Island Press.

NRDC. (2010). Reducing foreclosures and environmental impacts through location-efficient neighborhood design.

Owen, D. (2009). *Green metropolis: Why living smaller, living closer and driving less are the keys to sustainability.* New York: Riverhead Books.

Pi-Sunyer, F. X. (1993). Medical hazards of obesity. *Annals of Internal Medicine,* 119(7 Part 2), 655.

Pivo, G., & Fisher, J. D. (2011). The Walkability premium in commercial real estate investments. *Real Estate Economics,* 39(2), 185–219.

Pollakowski, H. O. (1995). Data sources for measuring house price changes. *Journal of Housing Research,* 6(3), 377–387.

Quigley, J. M. (1985). Consumer choice of dwelling, neighborhood and public services. *Regional Science and Urban Economics,* 15(1), 41–63.

Rauterkus, S., Thrall, G., & Hangen, E. (2010). Location efficiency and mortgage default. *The Journal of Sustainable Real Estate,* 2(1), 117–141.

Rawls, J. (1975). Fairness to goodness. *The Philosophical Review,* 84(4), 536–554.

Riggs, W. W. (December 16, 2011). *Walkability and housing: A comparative study of income, neighborhood change and socio-cultural dynamics in the San Francisco Bay Area* (Doctoral dissertation). University of California, Berkeley.

Riggs, W. W., & Gilderbloom, J. (2017). How multi-lane, one-way street design shapes neighborhoods life: Collisions, crime and community. *Local Environment: The International Journal of Justice and Sustainability,* 22(8), 917933. https://doi .org/ 10.1080/13549839.2017.1303666

Saelens, B., Sallis, J., & Frank, L. (2003). Environmental correlates of walking and cycling: Findings from the transportation, urban design, and planning literatures. *Annals of Behavioral Medicine,* 25(2), 80–91.

Saito, M. (August 29, 2011). *"Land bank" knocks out some foreclosure problems : NPR.* Retrieved from http://www.npr.org/2011/08/29/139971310/land-bank -knocks-out-some-foreclosure-problems

Sallis, J. F., Frank, L. D., Saelens, B. E., & Kraft, M. K. (2004). Active transportation and physical activity: opportunities for collaboration on transportation and public health research. *Transportation Research-Part A Policy and Practice,* 38(4), 249–268.

Savitch, H. V. (1981). *Post-industrial cities: Politics and planning in New York, Paris, and London.* Princeton: Princeton University Press.

Shiller, R. J. (2007). *The subprime solution: How today's global financial crisis happened, and what to do about it.* Princeton: Princeton University Press.

Southworth, M. (2005). Designing the walkable city. *Journal of Urban Planning and Development*, 131(4), 246–257.

Thompson, D., Edelsberg, J., Colditz, G. A., Bird, A. P., & Oster, G. (1999). Lifetime health and economic consequences of obesity. *Archives of Internal Medicine*, 159(18), 2177.

Troy, A., & Grove, J. M. (2008). Property values, parks, and crime: A hedonic analysis in Baltimore, MD. *Landscape and Urban Planning*, 87(3), 233–245.

Williams, D., & Jackson, P. (2005). Social sources of racial disparities in health. *Health Affairs*, 24(2), 325.

Zukin, S. (1982). *Loft living: Culture and capital in urban change.* Baltimore: John Hopkins University.

Chapter 14

Biking Is the Best Choice for Health, Safety, and Zero Emissions

John Hans Gilderbloom, Justin Mog,
and Zachary E. Kenitzer

Biking is one of the most viable transportation choices for healing a broken earth. Compared to cars, it releases none of the dangerous greenhouse gases that wreak havoc on the environment. Choosing a bike over a car means better health for bikers, longer life spans, decreased medical expenditures, and an average of $8,000 per year in savings. A bike-friendly neighborhood also increases community and housing values.

Historically speaking, in the late 1800s and early 1900s, the bicycle was one of the most popular forms of transportation in the United States. Moreover, the sociological importance of the bicycle has been noted by renowned figures including Annie Londonderry, Albert Einstein, and Friedrich Nietzsche. In 1896, for example, feminist icon Susan B. Anthony noted that bicycles were one of the most important tools for the liberation of women, giving women a "feeling of freedom and self-reliance" (Flottorp, 2023).

Further, in 1880, the League of American Wheelmen was formed to promote bicyclists' interests. Over the next decade, local clubs of cyclists— or "wheelmen"—formed throughout the country. By 1898, according to the League of American Bicyclists, its membership numbered more than 102,000, including the Wright Brothers, Diamond Jim Brady, and John D. Rockefeller. Many clubs had meeting halls and admitted women as well as men. Since that time, the bicycle has largely become limited to recreation instead of transportation, inhibiting the creation of the cycling infrastructure we desperately need today. During the coronavirus pandemic, biking lanes in Paris were increased by 420 miles.

The mass production of the automobile starting in the early twentieth century made private cars more affordable to the working class, which, in turn, led to a steady decline in bicycle use. Bikes, along with light rail and pedestrians, were pushed off roadways in favor of cars. Urban development patterns changed to meet the needs of ubiquitous car use, resulting in central business districts full of parking lots and extensive highway systems, ultimately contributing to suburban sprawl. Fast forward to today: the cost of car ownership and operation has skyrocketed, and biking has, in turn, experienced the largest mode share increase of any method of transportation from 2000 to 2010, rising by 40 percent, with the lowest socio-economic groups reporting the largest growth rates (McKenzie, 2014).

Lastly, less attention has been paid to the potential local economic benefits to cities and urban neighborhoods from their becoming more bike-friendly. Increased bicycling as transportation—or at least the infrastructure allowing for it—is now a green economic development strategy in cities as diverse as Chicago, New York City, Portland, and San Francisco. Growing demand for bike-friendly urban biking has its roots in responding to young people who want to live car-free near downtown, reduce their carbon footprints, improve the safety and quality of downtown neighborhoods, and enhance personal health and finances. This chapter seeks to answer the question: What kind of green economic return might a community expect to see by compelling car-dependent commuters to alter their commuting behavior and increase bike commuting?

PREVIOUS RESEARCH ON THE IMPACT OF BICYCLING

Operating, owning, and accommodating vehicles is expensive, representing a sizable portion of household income as well as expenditures for businesses, universities, and cities. Much has been written about these costs and the health and environmental benefits derived from reducing car dependence. Little, however, has been published on the potential impact of the investment on the returns generated from those cost savings because of altered individual commuting behavior. Of the published articles discussing the economic returns from encouraging biking through the promotion of bike infrastructure, some note that the number of cyclists in the United States is too small to have any significant or measurable impact (Gilderbloom, 2012; Pucher, Buehler & Seinen, 2011).

Others note that the recent surge in biking is *too* recent and is therefore a phenomenon that has yet to be studied by scholars, thus limiting the availability of peer-reviewed literature on the topic (Weigand, 2008). Weigand's review of studies on the economic impact of the bicycling industry argues that many of the published studies must be utilized with caution. Weigand explains that much of the published work was conducted by cities and states with already robust bicycling-related activities and industries, rather than new pro-biking initiatives. The studies reviewed by Weigand do not discuss an individual's motivation for using the bicycle as a mode of transportation, nor do they report on the economic benefits specifically witnessed by urban neighborhoods. Much of the remaining literature does not address the economic impacts of cost savings on urban neighborhoods; however, the literature does provide calculations of the savings generated from substituting a bike for automobile transportation. First, however, a detailed look at the cost of using private cars is necessary.

Cost of Driving for Citizens, Cities, and Universities

The Surface Transportation Policy Project, a nationwide coalition for safer communities and smarter transportation choices, is an expense second only in magnitude to housing. The average American household devotes 18 cents out of every dollar to transportation. In some metro areas, households are spending more on transportation than on housing. Ninety-eight percent of that spending is for the purchase, operation, and maintenance of automobiles. Some American families even spend more on driving than on healthcare, education, or food. The poorest families spend the greatest proportion, sometimes in excess of one-third of their income. Estimates on how much the average person spends vary. The Surface Transportation Policy Project found that households in automobile-dependent communities devote 50 percent more to transportation (more than $8,500 annually) than households in communities with multi-modal transportation systems (less than $5,500 annually) (The Surface Transportation Policy Project, 2000).

The Bureau of Labor Statistics estimates the average cost of owning and operating a car has risen in cities to $8,689 (BLS, 2009). According to the Automobile Association of America (AAA), excluding loan payments, Americans driving a medium-sized sedan 15,000 miles per year can expect to spend $9,151 per year, or approximately $0.61 per mile driven; those driving large sedans can expect to spend about $0.75 per mile, or a total of $11,248 per year (AAA, 2013). These figures include average expenses on fuel, routine vehicle maintenance, wear and tear on tires, insurance coverage, license and registration fees, loan finance charges, and vehicle depreciation costs. Total car ownership and operation costs for both medium and large sedans are detailed in table 14.1 (AAA, 2013).

AAA also reports that these costs are expected to rise, as they did from 2009 to 2010 when the cost of driving increased by nearly three and a half percent (about $290 per vehicle) despite a national average decrease in maintenance and insurance rates. Biking and walking, meanwhile, are considered the least expensive modes of transportation. Biking is only a fraction of the cost, with one estimate at just $120 per year (Portland Plan Partners, 2012, 125).

Table 14.1 Total Costs (per year and per mile) of Car Ownership and Operation for Medium and Large Sedans (AAA, 2013)

Total Cost Per Mile (15,000 miles/year)	Medium Sedan	Large Sedan
Cost per mile × 15,000 miles	$3,164	$3,522
Cost per day × 365 days	$5,987	$7,726
TOTAL Cost per year	$9,151	$11,248
TOTAL Cost per mile	$0.61	$0.75

Source: American Automobile Association (AAA). (2013). 2013: Your Driving Costs.

The reliance on the car for transportation creates a heavy financial burden on cities and universities as well. For universities and businesses, parking structures are an extremely costly venture. Steve Lawrence, Associate Vice President of Facilities Management at Central Michigan University, estimated a cost range of $11,500 to $13,000 per parking space in a garage, while a typical surface lot costs $1,800 to $2,000 per parking space (Finneren, 2005). Others estimate the cost of constructing one car parking space in a paved lot to be as high as $22,000 and the cost of constructing one car parking space in a garage to be $20,000 to $30,000. Conversely, the cost to purchase and install one bike rack is approximately $1,500 (Cascade, 2010).

Lastly, spending on highway construction and expansion that raises peak-period speeds discourages the use of transportation alternatives, including public transit, alternate routes, and off-peak travel times (Downs, 1962). Downs explains that this is a result of induced demand. Such commuting behavior results in maximizing road capacity during peak travel times, which ultimately leads to congestion (Downs, 1962; Schrank, Lomax & Eisele, 2011). Other costs associated with congestion that could be lowered with changes in commuting travel mode are motorists' opportunity costs and the increased operating costs of trucking and shipping companies (Schrank, Lomax & Eisele, 2011; Local Government Commission, 2000).[1]

The 2007–2008 increases in gas prices and the housing market collapse are also accelerating the growth in bicycle ridership (Mapes, 2009). When deciding where to live, individuals in the United States will often accept a longer commute in exchange for affordable housing (Lovins & Cohen, 2011; Sermons & Seredich, 2000).[2] An economic tradeoff between home location and commuting time, however, may be substituted for a tradeoff between transportation modes (Boarnet & Sarmiento, 1998; Duany, Plater-Zyberk & Speck, 2001). McAnn (2000) estimates that in substituting bike, walking, or public transit travel for a car, commuters could save enough money to locate closer to the central city. McAnn further reports that households in car-dependent communities spend on average $1,200 to $6,000 more annually on surface transportation than households in walkable areas with access to transit.[3]

Housing Value and Transportation Cost

Microeconomic theory predicts that housing value is related to transportation costs. So, as the distance to destinations like work, shopping, or entertainment declines, home value should increase (Matthews, 2006). Moreover, a 1999 study by the Urban Land Institute of four new pedestrian-friendly communities determined that homebuyers were willing to pay a $20,000 premium for those homes compared to similar houses in surrounding non-pedestrian-friendly areas (Local Government Commission, 2000). In 2001, Charles Tu and Mark Eppli found that homebuyers are willing to pay more to live in new urbanist traditional neighborhood developments. They found the price premium to be 14.9 percent in Kentlands, MD (outside of Washington D.C.), 4.1 percent in Laguna West (near Sacramento, CA), and 10.3 percent in Southern Village (in Chapel Hill, NC) (Tu & Eppli, 2001).

Retail locations, in addition to private residences, can experience a significant increase in value when located in pedestrian- and bike-friendly business districts (Litman, 2003). Five years after a major downtown redevelopment project, which focused on pedestrian and bike accessibility, commercial rents in West Palm Beach, Florida, rose from $6 per square foot to $30 per square foot (Lockwood & Stillings, 1998). Property values also increased from a range of $10–$40 per square foot to $50 - $100 per square foot (Lockwood & Stillings, 1998). Participating towns in California's "Main Street Program" saw a 167 percent increase in commercial property values, a 65 percent increase in retail rental rates, and a 71 percent increase in office rental rates (Eichenfeld and Associates and the Local Government Commission, 2002). Transportation mode also affects the amount of money spent by consumers visiting business districts. Sztabinski (2009) notes that when analyzing the spending categorized by mode of travel, walkers and cyclists spend more than users of public transit and motorists. Participants in California's Main Street Program found that their retail sales increased by 105 percent.

Economic Benefits of Bikeable Communities

In literature, the recent upsurge in bicycling and its associated benefits is addressed at two levels: household and community. Mapes (2009) attributes the recent increase in bike ridership to a renewed interest in urban neighborhoods and walkable cities. The age groups most likely to commute by bike are 16–24 years (1 percent commute by bike) and 25–34 years (0.8 percent commute by bike); other age groups have a bike commute rate of less than 0.7 percent. It is this demographic (16–34 years), more than any other, that is moving into city neighborhoods.

Geller (2013) calculated costs for Multnomah County (where Portland is located) and found that each dollar-year increase in gasoline (one-dollar increases in cost sustained over the course of a year) results in $240 million not spent on other goods and services, many of which would eliminate the environmental detriment created by car commutes in Multnomah County (Geller, 2013). Those dollars spent on commuting result in fewer new local jobs and no wage increases. Nationally, according to the Texas Transportation Institute, a 30 percent increase in non-motorized travel in the United States would reap estimated monthly savings of $168.3 billion, or roughly $2 trillion per year (Schrank, Lomax & Eisele, 2011).

Kenworthy and Laube (1999) and Hartgen and Fields (2009) report that most automobile-dependent cities are less wealthy and less economically competitive than cities oriented to a greater diversity of transportation. To resolve this issue, Duoma and Cleveland (2008), Rietveld and Daniel (2004), and Topp (2008) argue that cities can reduce their car dependency by increasing access to or improving the condition of existing bicycle facilities.

The socio-economic group most likely to commute via bike is those making $0–$24,999 (up to 1.5 percent commute by bike) (McKenzie, 2014). It is this same socio-economic group that tends to live in the central city. This is partially explained by the fact that car ownership and operation function under a regressive tax on lower-income groups, as the proportion of household income spent on cars is higher for the poor than for upper-middle- and high-income households. Table 14.2 below shows the bike commuting rates for 2008–2012 by the Census Bureau for bike commuting by household income group.

Table 14.2 Bike Commuting by Household Income Group (2008–2012, ACSM)

Household Income	% of Workers	Margin of Error
$0–$9,999	1.5	0.1
$10,000–$14,999	1.1	0.1
$15,000–$24,999	1.0	Z
$25,000–$34,999	0.7	Z
$35,000–$49,999	0.6	Z
$50,000–$74,999	0.5	Z
$75,000–$99,999	0.4	Z
$100,000–$149,999	0.4	Z
$150,000–$199,999	0.5	Z
$200,000 +	0.5	Z

Sources; 1. ACSM American Fitness Index™: Health and Community Fitness Status of the 40 Largest Metropolitan Areas. (2010 Edition). American College of Sports Medicine. http://www.americanfitnessindex .org/docs/reports/2010AFIReport-Final.pdf 2. Census Bureau *Modes Less Traveled: Bicycling and Walking to Work in the United States 2008–2012.* Z = rounds to 0 McKenzie, B. (2014, May).

University populations have shown to be especially responsive to alternative modes of transportation. Data from the 2008 American Community Survey, an annual one-percent sample of the U.S. population conducted by the U.S. Census Bureau, show that the top nine metropolitan statistical areas where individuals bicycle to work have major state universities as major staples of their local economies (see Table 3: Means of Transportation to Work by Metropolitan Statistical Area). These areas are geographically, climatically, and economically diverse in nature, but all house a major state university with at least 17,000 students.

Regarding university campuses, since 2008, more research into various aspects of increasing non-motorized transportation has been conducted. This research has concluded that transportation demand management programs can both reduce the costs of increasing parking capacity and strengthen the use of non-motorized and non-auto-based transportation (Riggs, 2014; Tudela-Rivadeneyra et al., 2015), and that systematic documentation and benchmarking of commuter behavior should be conducted (Riggs, 2014). Scholars know less, however, about the economic impacts of increasing alternative transportation options. Pucher and Buehler's (2012) *City Cycling*, one of the best original scholarly reads on the topic, has 15 chapters written by leading experts, but none address the impact of bike riding on consumer spending. Nor is there an entry in the index concerning community development or the green economic benefits of bicycling. Similar omissions on the economic impact of urban biking can be observed in other scholarly works (Wray, 2008; Heying, 2010; Mapes, 2009) and applied planning publications (Birk, 2010; Byrne, 2010). While the literature does provide estimations of the monetary savings generated from substituting bicycles for automobile transportation, it does not address the potential economic benefits of cost savings on urban neighborhoods specifically. This chapter aims to fill this gap in knowledge by utilizing survey results and calculated estimates to demonstrate the potential local "green dividend" that can result from increased urban biking.

According to the Texas Transportation Institute, a 30 percent increase in non-motorized travel in the United States would reap an estimated monthly savings of $168.3 billion, or roughly $2 trillion per year (Schrank, Lomax & Eisele, 2011). These savings could be realized as discretionary purchases. Decreasing motorized travel not only generates cost savings but also encourages public and private investment in local commercial and retail locations (Lockwood & Stillings, 1998). Reducing reliance on automotive travel leads to a willingness in consumers to spend more in walkable and bikeable communities because of a change in economic tradeoffs (Boarnet & Sarmiento, 1996; Lawton, 2001, 203). A consumer's market radius decreases when the consumer substitutes automobile travel for walking or biking (Krizek, 2001,

9), meaning a greater proportion of a consumer's spending is done closer to his or her residence (Krizek, 2001, 17).

Health Benefits of Cycling

Public health is influenced by the interactions between people and their surrounding built environments. Environmental factors may reduce or increase the risk of injury or death, as well as influence the frequency and type of physical activity one engages in. The Tompkins County Planners of New York define the built environment as "the part of the environment formed and shaped by humans, including buildings, structures, landscaping, roads, signs, trails, and utilities" (Tompkins County Planning, 2010). Design elements of the built environment can provide opportunities to improve public health through increased physical activity. Improving the quality of bicyclists' travel experience through an improved sense of safety, comfort, and accessibility will encourage more physical activity and therefore improve the overall health of Metro residents.

Physical activity decreases mortality and the risk of cardiovascular disease, certain cancers, diabetes, obesity, and asthma (Wendell, Tom & Rohm, 1998). In 2005, life expectancy in Kentucky was 75.5 years compared to 78.4 years for the United States, ranking 43rd in the nation (Health Status—Kentucky, 2010). Jefferson County fares no better, with a life expectancy of 75.3 years (Community Health Status Indicators, 2010). During the same period, life expectancy was 78.8 years in the Netherlands (Netherlands Life Expectancy at Birth, 2010), a nation notorious for being bike and pedestrian friendly (Gilderbloom, Hanka & Lasley, 2009). Non-motorized travel accounts for 40 percent of all trips in the Netherlands, nearly six times greater than the U.S. rate of seven percent (Pucher & Dijkstra, 2003).

In 2006, Kentucky was ranked sixth highest in the nation for heart disease mortality with a rate of approximately 1 in 425 people, compared to the national rate of approximately 1 in 500 people (Health Status—Kentucky, 2010). Heart disease mortality for Jefferson County is estimated to be approximately 1 out of 483 people (Kentucky Health Facts, 2010). Cancer incidence in Kentucky was the fourth highest in the nation in 2004 (Health Status—Kentucky, 2010). The prevalence of diabetes in Kentucky was 9.8 percent in 2008, the ninth highest in the United States (Health Status—Kentucky, 2010). The prevalence of diabetes is even higher for Jefferson County at 10 percent (Kentucky Health Facts, 2010). The Netherlands, however, a more bike and pedestrian-friendly country, had less than half the diabetes prevalence of Kentucky, with only 3.9 percent (StatLine, 2008).

In 2007, Kentucky ranked sixth highest in the United States for adult obesity, as measured by a body mass index of 30 or greater (Health Status – Kentucky, 2010). Obesity prevalence in Kentucky is just above the national rate, with values of 66.6 percent and 63 percent, respectively (Health Status – Kentucky, 2010). For childhood obesity, Kentucky ranked third highest in the United States, with a statewide prevalence of 37 percent, compared to a national prevalence of 32 percent (Health Status – Kentucky, 2010). In comparison with 2005 obesity rates in the Netherlands, U.S. obesity was slightly over 20 percent higher (Gilderbloom, Hanka & Lasley, 2009). Also, in 2007, Kentucky had the 14th highest asthma prevalence at nine percent, compared to 8.2 percent for the nation (Health Status—Kentucky, 2010). In Louisville, the asthma prevalence is even higher at 11 percent (Kentucky Health Facts, 2010). Clearly, this is a nationwide epidemic.

In 2010, Louisville was ranked nearly last (46th out of the 50 largest U.S. cities) for the amount of physical activity in the American College of Sports Medicine's annual fitness index ranking. While the city was credited for its per capita park acreage, farmers' markets, golf courses, and tennis courts, challenges included minimal state physical education requirements and the small percentage of the population bicycling or walking to work (ACSM, 2010).

NutriStrategy (n.d.), a nutrition and fitness organization, estimates that light cycling (10–11.9 miles per hour) could burn between 300–500 calories per hour, 500–700 calories per hour for moderate cycling (12–13.9 miles per hour), and 600–800 calories for vigorous cycling (14–15.9 miles per hour). Marshall reports that "reducing the average energy imbalance (caloric intake minus metabolic activity) among persons in the United States by approximately 100–165 kcal/day would prevent average weight gain" of about 2.2 lbs/year (Marshall, Brauer & Frank, 2009).

In addition to reducing the risk of chronic diseases and weight gain, regular physical activity also decreases stress levels (Fox, 1999). It is estimated that 47 percent of adults suffer from adverse effects related to stress (APA Stress Survey: Children More Stressed than Parents Realize, 2009), and 75 to 90 percent of primary care physician visits are related to stress (America's No.1 Health Problem, 2010). Sixty percent of work absences are attributed to stress, which is estimated to cost companies over 57 billion dollars yearly (Clark, 2010). Stress contributes to mortality from heart disease, cancer, lung disease, accidents, depression, cirrhosis, and suicide. The many physical and mental health symptoms associated with stress provide a stronger argument for increasing access to physical activity because of its stress-reducing qualities.

Benefits to the Environment

More biking and fewer cars mean a cleaner environment and healthier people. Increased rates of greenhouse gas (GHG) emissions are warming the earth, and carbon dioxide (CO_2) is the primary culprit. In the United States, CO_2 accounts for over 80 percent of total GHG emissions (NHTS, 2009). Activities such as burning fossil fuels have increased atmospheric CO_2 levels to 35 percent higher than those of the Industrial Revolution. While some in colder climates might lightheartedly welcome "balmier" weather, the reality is that the effects of climate change could pose serious threats to public health, economic stability, and even national security.

In 2004, transportation accounted for nearly one-quarter of the global energy-related CO_2 emissions; of those emissions, motor vehicles accounted for approximately three-quarters (Kahn et al., 2007). In the United States, approximately one-third of CO_2 emissions are transportation-related, with automobile travel accounting for 90 percent of all trips (NHTS, 2009; Kahn et al., 2007). The principal source of these emissions is gasoline. Private cars and trucks burn 40 percent of the total oil consumed in the United States, equivalent to 10 percent of world demand (Gotschi & Mills, 2008). The United States has been binge drinking oil, and the massive hangover is just setting in: the combustion of each gallon of gasoline emits approximately 20 pounds of CO_2 or 23 pounds if refinement and distribution are included (Glaeser, 2008). Annually, in the United States, personal transportation accounts for approximately 136 billion gallons of gasoline, or 1.2 billion tons of CO_2 (Gotschi & Mills, 2008), which amounts to approximately one-fifth of global CO_2 emissions (Ewing et al., 2008).

The thirst for oil and consumption of fossil fuels is not expected to decrease. Rather, global transportation-related carbon emissions are projected to increase by 80 percent by 2030 (Kahn et al., 2007). Although great strides have been made to improve the fuel efficiency of motor vehicles, individuals are now taking more trips and traveling farther to reach their destinations. In fact, in recent decades, the number of vehicle miles traveled (VMT) in the United States has increased three times faster than population growth (Gotschi & Mills, 2008). These negative external costs resulting from increased fuel efficiencies (i.e., Jevons Paradox: increased fuel efficiency often yields increased vehicle miles traveled), as well as the many indirect benefits of travel reductions (i.e., less congestion, reduced emissions, and health benefits), are often excluded from analyses in programs aimed at reducing transportation-related emissions (Litman, 2010).

To reduce GHG emissions to a level that will help mitigate climate change, a multifaceted approach will be required. The paradigm of the past several decades has been to increase fuel efficiency in the hopes that it will offset demand for oil and lessen environmental destruction. However, it is clear from our current predicament that increased fuel efficiency alone will not suffice. Although technology possesses the capacity to spur the industrial growth that contributes to global climate change, this power alone will not be the solution. It is irrational to think of a solution to climate change that does not involve significant changes to our transportation system and our commuting choices. If we are to seriously pursue sustainability, it is time for a paradigm shift. We must understand that we are all contributors to global climate change and must do our part to reduce our dependence on burning fossil fuels. One way of making a significant impact is by biking within our communities (to work, school, and for other short trips). Think globally, bike locally.

Carbon dioxide emissions from the transportation sector can be thought of as a three-legged stool: a function of vehicle fuel efficiency, fuel carbon content, and VMT (Ewing et al., 2008). Increasing fuel efficiency and finding alternative sources of fuel will be critical in developing an effective transportation program, but one of the simplest things that can be done is to drive less. About half of all car trips are less than five miles (Maibach, Steg & Anable, 2009), which could instead be easily completed with a 20-minute bike ride.

Local infrastructure, density, and the spatial structure of the built environment influence the amount of potential GHG mitigation possible from reduced VMTs. For example, smart growth development patterns (i.e., increased density, walkability, etc.) can create up to 35 percent less VMT than sprawling suburban-type growth (Ewing et al., 2008). Efforts to reduce CO_2 emissions by driving less can reap significant benefits. For example, a 30 percent reduction in vehicle miles traveled could result in a 28 percent reduction in CO_2 emissions (Ewing et al., 2008).

Reducing the number of VMTs, driving fuel-efficient vehicles, carpooling, using public transportation, chain trips, walking, and bicycling are all important components of reducing the overall carbon footprint of daily travel. However, it can be argued that none are more fun, exciting, and rewarding than riding a bike. A bicycle commuter who rides five miles to work, four days a week, avoids 2,000 miles of driving per year, which is equivalent to 100 gallons of fuel and 2,000 pounds of CO_2 emissions avoided. Such savings would have reduced the average American's carbon footprint by approximately 4 percent (Gotschi & Mills, 2008). Total savings from shifting short trips from driving to methods like bicycling could amount to 2.4–5 billion gallons of fuel per year, or 21–45 million tons of CO_2 (Gotschi & Mills, 2008). While bicycling may not solve the problem of climate change on its own, it must be part of the solution.

CASE STUDY: ECONOMIC BENEFITS OF INCREASED
BICYCLE USAGE IN A MID-SIZED CITY

This study examines the potential economic benefits of increased bicycle usage by auto-commuters to the University of Louisville. Increased bicycle usage can be attributed to several factors: shorter distances between work and home, increasing costs of automobile transportation, rising home prices, and the presence of bike-friendly infrastructure. The authors implemented a survey tool to better understand these factors, which has also been done in other studies of alternative transportation choices (Goetzke & Rave, 2010; Pucher & Buehler, 2012; Clifton, 2012). Currently, Louisville ranks especially low among cities in which citizens choose bicycling, walking, and other non-automotive travel. Louisville ranks 153rd of 384 metropolitan areas in individuals who typically bike to work, 182nd in individuals who walk to work, and 188th in the percentage of individuals who do not use an automobile to commute. Moreover, the League of American Cyclists recently ranked the Commonwealth of Kentucky (where Louisville is the largest city) as second to last in being bike-friendly (Broken Sidewalk, 2015).

One of the largest employers in the downtown area is the University of Louisville. Our study found that, within the university community, the proportion of the survey population using alternative means of transportation to get to the campus is minimal compared to other university populations. This is demonstrated by the high percentages of university students and employees in other "college towns" that walk and/or bike to work, and the higher rates of walking and/or biking by the best educated (McKenzie, 2014; Shoup, 2005; Riggs, 2013). Davis, CA, has a biking-commuting ratio of 18 percent of workers. Isla Vista, a student community next to the University of California at Santa Barbara, has the highest known bike ratio of 32 percent for any community.

Universities such as the University of California at Santa Barbara and Davis, the University of Colorado (Boulder), Indiana University, and Portland State University have been proactive in creating campuses that are conduits of alternative transportation modes to help lower the cost of education and increase retention rates (Pucher & Buehler, 2012). Our study concentrated on the exogenous and endogenous factors that affect transportation mode selection in hopes of identifying barriers that could be overcome through investments and incentive programs.

BICYCLE USE AND SAFETY

Bike Lane Collisions

Road design positively influences cyclist safety when it accommodates all users of the road, including cyclists and pedestrians. The presence of bike lanes in road design designates space for cyclists on roadways and reserves sidewalks for pedestrians, thereby accomplishing this objective. Bike lanes use signs, color designations, and pavement striping to provide a visual indication to share the road, cautioning drivers that cyclists may be in the vicinity. However, there are critics of whether bike lanes truly reduce bicycle accidents, arguing they do more harm than good. To quell this contention, Lott and Lott (1976) compared roads with and without bike lanes and determined roads with bike lanes had 53 percent fewer bike accidents. Moritz (1998) investigated the danger indices (number of crashes divided by commute distance) and found they were over twice as high for roads without bike facilities (i.e., provisions to accommodate cycling, like bike lanes) (Transportation Toolkit, 2010). A 2006 New York City safety report revealed that, over the previous ten years, only one of 225 bicycle deaths had occurred in a bike lane (Mapes, 2009).

Despite their myriad of benefits, it is important to acknowledge that bike lanes are not a panacea and that there is much room for improvement in bike lane infrastructure (Smith & Walsh, 1988) compared collision counts pre- and post-bike lanes and found that there was an increase in bicycle collisions in the initial year following the bike lane addition. Bike lanes are often responsible for an increase in cyclists because of an increased safety perception (Byrne, 2009), which may in turn contribute to a rise in bike accidents. Depending on the design, bike lanes can also present new dangers. Oftentimes, bike lanes are placed between the driving lane and the parallel street parking, which can create dangers for cyclists when car doors are opened. This area of a bicycle lane subject to encroachment by an opened car door is referred to as the door zone. Bicyclists report that they frequently feel the need to swerve to avoid a suddenly opened car door, causing them to enter the driving lane. Ultimately, this phenomenon can lead to cyclist-motorist collisions. For some cyclists, the door zone in bike lanes causes them to ride as close as possible to the driving lane to feel safest, which can make the bike lane less effective.

While discrepancies exist over whether bike lanes genuinely contribute to reductions in bike collisions, there is a consensus that they, at the very least, increase driver awareness of cyclists' presence. The perception of safety goes a long way toward increasing ridership. Take it from a rock music legend: In *Bicycle Diaries*, David Byrne reports he feels safer in bike lanes because he is no longer as paranoid that a driver will swerve into his lane (Byrne, 2009).

Bike Lane Alternatives

Former Louisville Mayor Jerry Abramson instituted a policy to include bike lanes as part of every road that is built or reconfigured in the city, with about thirty miles being striped so far (Mapes, 2009). 2010 Louisville Mayoral Candidate Jackie Green, an avid cyclist and local business owner, is among those who do not think the city should stop there: he supports alternate bike facilities that would separate cyclists from danger zones completely. Alternatives to bike lanes include cycle tracks, bicycle boulevards, and multi-use paths. These are described below:

- Cycle tracks are like marked bike lanes but separated from the road by a physical barrier such as a curb (Reynolds et al., 2009), which reduces the likelihood of motorists entering bike paths. However, some argue against cycle tracks because they protect cyclists in the middle of the block, but not at the intersections where most bike collisions occur.
- Bike boulevards are "low-traffic streets that discourage all but neighborhood auto travel while providing good through routes for cyclists" (Mapes, 2009). They are streets that accommodate bike travel over motorist travel via traffic-calming techniques (Dill, 2009). For example, stop signs placed on the intersecting streets can be used as a traffic-calming measure to allow bike traffic to flow with limited motorist interruption. Additionally, bicycle boulevards sometimes have speed bumps with bicycle-sized cutouts to accommodate cyclists only.
- Multi-use paths are paved or unpaved off-road bike facilities shared among non-motorized users (Reynolds et al., 2009). Because they are separate from both motorists and the door zone, they too provide a safe alternative to bike lanes. While some research indicates an increase in bicycle accidents on multi-use paths, because of the slow speeds, the resulting injuries are less severe (Mapes, 2009; Rivara et al., 1997).

Cyclists opposing the implementation of bike lanes or other bicycle facilities believe cyclists operate best when they act and are treated as drivers of vehicles on roads (Mapes, 2009). Some believe bike facilities, such as bike lanes, cycle tracks, and multi-use paths, are intended to simply keep cyclists out of motorists' way (Mapes, 2009). As fervent bicycle-rights advocate John Forester states, "typical Americans believe that cyclists are inferior to motorists in legal status and in competence, that cyclists should defer to motor traffic, and that failure to defer to motor traffic is dangerous" (Forester, 2009).

Whether you advocate cycling infrastructure or believe cyclists should act as motorists while on city streets, integrating cyclists and automobiles into the same transportation network will require adjustments for users operating at very different speeds.

Collision Speed

The speed of motor travel on streets greatly influences bicycle safety and is a clear indicator that roads are designed not for the most vulnerable users, but for motorists. As Dan Burden, a bike advocate, explains, "The human body is not designed to move faster than fifteen miles per hour. Our sight, our ability to interpret things, to process things, is bicycling speed" (Mapes, 2009). Klop and Khattak (1999) demonstrated that increased automobile speed limits were associated with increased severity of cyclist injuries in accidents. Similarly, a study by Kim et al. (2005) found that the likelihood of severe injury increases as vehicular speed increases, and the fatality risk for cyclists more than doubles when motorist speed is above 30 miles per hour: if a cyclist is hit by a car traveling 20 miles per hour, there is a five percent chance that the crash will result in cyclist fatality. This figure jumps to 45 percent when the automobile is traveling at 30 miles per hour, and to 80 percent fatality at 40 miles per hour (Gowin, Designing Streets for Bicyclists, 2010).

Reducing speed limits is one way to reduce motorist speed, although it would likely spark fierce opposition from motorists. Approaches to reduce speeds that would draw less resistance include reducing the number of traffic lanes and narrowing the street widths (Pedestrian and Bicyclist Intersection Safety Indices Final Report).

Collisions on One-Way Roads, Riding Against Traffic, and Sidewalk Riding

One factor influencing traveling speed and bicycle safety is street design. Allen-Munley et al. (2004) found more severe cyclist injuries were reported in collisions on one-way streets than on typical two-way streets (Reynolds et al., 2009). Wachtel and Lewiston (1994) report that cyclists traveling in the wrong direction are 3.6 times more likely to have collisions than those following the direction of traffic (*How Not to Get Hit by Cars*, 2009). Cyclists tend to ride against traffic on one-way roads rather than accessing safer, yet less convenient, distant street routes or busier arterial roads (Alrutz et al., 2002). Traveling against the flow of traffic is dangerous because motorists making right turns from side streets could lead directly into the cyclists' paths, thereby increasing the chance of head-on collisions (Wrong Way Cycling, 2010; *How Not to Get Hit by Cars,* 2009). In head-on collisions, the traveling velocity of the bike and motorist are combined, leading to a more forceful impact and increasing the likelihood of injury compared to if the cyclist were flowing with the direction of the motorist (*How Not to Get Hit by Cars*, 2009).

Reducing the number of one-way streets and providing cycling network connections to desirable locations (to prevent cyclists from circumventing one-way street routes) will reduce the likelihood of cyclists engaging in convenient but unsafe travel. Watchel and Lewiston (1994) found that sidewalk riding is twice as dangerous for cyclists. Within the geographical boundary limits of Louisville Metro, no person 11 years of age or older is permitted to operate a bicycle on any sidewalk, and no one is permitted to ride on the sidewalk downtown. This is unsafe for cyclists and pedestrians. If a car makes a right turn, the car may cross directly into the cyclist's path. Motorists do not expect to encounter cyclists in crosswalks. Pedestrians also do not expect cyclists on sidewalks and underestimate the speed at which bicycles travel, leading to disastrous consequences when they interact.

Why Aren't More People Biking?

Despite the potential savings and the increased costs associated with driving, the number of Americans owning and operating automobiles continues to grow (Schrank, Lomax & Eisele, 2011; Federal Highway Administration, 2009 and 2003; United States Bureau of Transportation, 2009).[4] So, why is it that less than one percent of the overall U.S. population uses a bike as its primary mode of transportation? Pucher and Dijkstra (2003) cite that the greatest deterrents to cycling are (1) the lack of appropriate facilities for cycling and walking and (2) the real or perceived danger of cycling and walking in American cities. The world's best cycling cities use both penalties and rewards to encourage increased bicycle use. Carrots represent such amenities as better bike facilities and infrastructure; sticks represent such tactics as higher gas prices, adjusting congestion "fees," and tolls. In the United States, we focus too much on carrots and not enough on sticks.

Providing adequate bike facilities can do much to eliminate the dangers associated with cycling (Burden, 2001). Pucher and Buehler (2008 and 2007) argue that providing separate cycling facilities on high-traffic roads at intersections, along with implementing traffic-calming measures in residential areas, is the key to achieving high cycling rates. Pucher and Buehler (2007 and 2008) also note that providing adequate bike parking, integrating the cycling facility network into public transportation, and conducting comprehensive traffic education and training sessions for cyclists and motorists have increased cycling rates in European cities. The United States, Pucher and Buehler (2008) suggest, could adopt similar strategies to increase ridership.

Goetzke and Rave (2010) argue that the addition of bike lanes in German cities did not increase the number of individuals biking. They did, however, find that social networking does increase the number of people cycling; when people see or hear about others biking, they are more likely to bike themselves.

Contrary to Goetzke and Rave's study of German cities, we discovered that additional bike lanes would significantly increase the number of bike commuters in the United States (Gilderbloom, 2002). According to McMahon (2012), the cost of adding bike lanes is minimal compared to building freeways. For the $50 million spent building a four-lane freeway that is one mile in length, 150 miles of bike lanes could be constructed (McMahon, 2012). It is estimated that bike lane construction costs even less for urban streets: according to Walkinginfo.org (2011), installation—the striping or addition of signage—of a bicycle lane on a roadway costs approximately $5,000 to $50,000 per mile. The cost range accounts for variations in pavement conditions, as well as the need for removal and repainting of lines, and the adjustment of existing signage.

In physically implementing these bicycle lanes, the site advises that it is most cost-efficient to create bicycle lanes during street reconstruction, street resurfacing, or at the time of original construction. Roadways can also be narrowed to accommodate the addition of bicycle lanes. Narrowing a roadway to 10 or 11 feet (the standard road lanes are 12 feet) and striping excess asphalt can cost from $1,000 per mile to $100,000 per mile (Walkinginfo.org, 2011). This cost range represents the variation in projects, from simply adding striping to a shoulder to constructing a raised median or widening an existing sidewalk (Walkinginfo.org, 2011). Regardless of how the expenditures are spent, a one-million-dollar investment in bike lanes would bring back a return of 25 to 100 times in community spending.

Reducing the number of people commuting via automobile saves taxpayers and entities land and money otherwise used to construct parking garages and surface lots. Even ignoring the considerable ongoing costs of maintaining, lighting, cleaning, and enforcing regulations for parking areas, a substantial amount of money is saved by not constructing the additional parking structures and surface lots that would be needed to meet expected future demand. If roughly ten percent of the campus community were to start commuting by bicycle instead of by car, the University of Louisville and other colleges could save an estimated $59 million. These savings would come from deferring the construction of the 3,000 additional parking garage spaces required to meet projected university needs, as stated in the latest revision of the campus master plan (University Planning, Design, & Construction, 2009).

Case Study of University of Louisville

Revealed from the University of Louisville Survey, most people drive a car alone from home to campus; roughly one-third of the students surveyed walk, bike, or take the bus. Many students report wanting to ride a bike or walk to improve their health but are reluctant to do so because of the risks perceived from the lack of bike lanes. In addition, the survey found that even more people would bike if they were given a free bicycle in exchange for not purchasing a parking pass. Some argue that people do not bike because the four-season climate is inhospitable to commuting by bike, but, based on a comparison of Louisville to three cities known for biking, weather does not seem to be a primary factor.

The importance of the built environment is further illustrated by the fact that cycling in the United States is more dangerous for cyclists than in the Netherlands, a widely accepted bike-friendly city. The U.S. fatality rate per 100 million bike trips is 12.5 times higher than in the Netherlands. Additionally, the fatality rate per 100 million km traveled is also higher in the United States than in the Netherlands, with values of 7.2 and 2.0, respectively. A comparison of the injury rate per 500,000 km traveled reveals a considerable difference between the United States and the Netherlands: 25 and 0.4, respectively (Pucher & Dijkstra, 2003). By comparing figures in the United States to those of the Netherlands, it becomes apparent that America has much more work to do to become a bike-friendly city.

In our analysis, we calculated a conservative cost-saving projection based on the number of individuals who would cycle to campus following changes in bike infrastructure and services provided by the University of Louisville. If roughly ten percent of the campus community, or 3,041 individuals, gave up their cars and biked to the University of Louisville, there would be an estimated annual savings of $24 million. If that percentage increased to 30 percent, or 7,603 individuals, the savings would be close to $61 million. If bike lanes were installed within a 10-minute bike commute of the campus to nearby homes, the savings within that commute radius would be $26 million. If the bike lanes were extended out to a twenty-minute commute radius, the savings would increase to $55 million. A significant portion of that savings would feed back into the community.

Altogether, if bike lanes were installed throughout Metro Louisville, the economic return for just faculty, staff, and students who chose a bike as their primary mode of transportation would be around $78 million. This does not include others who would use the bike lanes for home to work commuting outside of the university community, in which case the savings would be even greater.

When asked how they would spend the estimated $3,000–$8,000 savings generated from giving up automobile transportation, one in three respondents of the University of Louisville survey said they would improve their housing. Roughly one in four would buy better quality groceries, and one in five would buy clothes and accessories. One-tenth of respondents noted that they would plant a garden, while another one-tenth said they would join a fitness club. One-twelfth of the population said it would splurge by eating out more (see figure 14.3). The positive impact of this deferred spending could be significant for the local economy and for overall quality of life.

Methods

In the spring of 2010, the university's Sustainability Council requested a comprehensive transportation survey be conducted of the university community as part of the greenhouse gas emissions' reporting required through the American College & University Presidents' Climate Commitment. A professor of Urban and Public Affairs collaborated with the assistant to the Provost for sustainability initiatives (both of whom are co-authors of this paper) to administer the survey. The lead author of this paper, with permission from the Provost, used the survey development and administration process as a teaching tool for a graduate-level seminar course at the university. The university granted the professor and the students a small sum of money to cover the costs of survey administration.

The purpose of the survey was to better understand the commuting behaviors of the university community and its willingness to consider transportation alternatives. The survey included questions about survey recipients' commutes to and from campus, their willingness to pay more for parking permits and gasoline, their views on bicycling as a means of transportation, and their opinions on cycling and transportation safety. The questions were composed with input from students, faculty, staff, the community, and the Sustainability Council, and benchmarked against previous surveys cited by national bicycling organizations. The survey was reviewed and approved with several small changes in the survey wording by the Institutional Review Board (IRB). Controls for the survey included housing location, gender, race or ethnicity, occupation at the university, and educational attainment.

The survey was distributed via email to a random sample of 9,653 students, faculty, and staff at the university. Of those surveyed, 2,032 responded, with faculty and staff responding at a rate of 31 percent (+/− 1 percent margin of error) and students responding at a rate of 11 percent (+/− 5 percent margin of error). The higher rate of response from faculty and staff is interesting, suggesting that they may be more responsive to requests from the provost. In an age where the validity of survey results is called into question because random dialing surveys do not include cell phone numbers, having the weight of the provost's request for participation helped deliver a strong response. The response rate does, however, raise the question of potential bias in respondents; it is possible that only those interested in transportation issues responded to the survey request. However, for a survey response, this is a large sample, which provides greater representation and precision in terms of estimates. The questionnaire was constructed following the design guidelines of the University of Chicago National Opinion Research Center and follows conventional social science guidelines for survey instruments (Babbie, 1990, 2012; Schutt, 2010).

Finally, this study was conducted in a university environment and thus may not be directly applicable to the larger urban environment. Furthermore, Louisville and the Old Louisville neighborhood containing the University of Louisville may be unique. This limitation suggests further research in the area is warranted.

Data Analysis

University of Louisville is a highly car-dependent campus community. Nearly four out of five university employee respondents and nearly two-thirds of student respondents report that they primarily drive to campus. When accounting for carpooling, we found that nearly nine in ten employees are dependent upon automobiles for their regular commute. Only 4 percent of students and two percent of employees reported that they bike to campus on a regular basis. Students are more likely to walk to campus than faculty, though only one out of five do so regularly. This includes the roughly one-quarter of the student body which lives in university-affiliated housing on or near campus. More than half of students, faculty, and staff commute more than 30 minutes to campus. This immense car dependence is typical of U.S. cities, as explained in the literature review (Pucher, Buehler & Seinen, 2011).

Our survey results suggest that many of the car-dependent faculty, staff, and students would be more willing to consider commuting by bike if additional bike infrastructure were provided. This number varies according to the estimated cost of gas, nearness to the university, and type of bike infrastructure in question. In general, however, 16 percent of students and faculty responded approvingly to the question, "If a dedicated bike lane was provided from my neighborhood to campus, I would be more likely to bike to campus." We also asked respondents whether they were more likely to bike if they could ride with another person and found that 9 percent of the students and 7 percent of the employees indicated they would do so. Our findings show that both physical infrastructure and social factors are key to achieving an increase in urban biking.

Consumer preference surveys can also be tested by the reality of where existing bike infrastructure has been installed. The 2010–11 Campus Travel Survey for the University of California, Davis found that 47 percent of undergraduate students and 55 percent of graduate students commute to class by bike, while 46 percent of faculty and 40 percent of staff commute to the campus by bike (Handy, 2011). Furthermore, the University of California, Santa Barbara reports that 57 percent of undergraduate students, 35 percent of graduate students, 22 percent of faculty, and 7 percent of staff members use bikes to commute to campus (Santa Barbara Bicycle Coalition, 2010). For UC Santa Barbara, these numbers have grown significantly over the past forty years. In 1970, 38 percent of the student population commuted to campus by bike; now, 52 percent choose bikes. Additionally, UC Santa Barbara finds that for every three students bicycling, there is approximately one walking to campus.

When asked what would most encourage people in the University of Louisville community to bike, the top answer given by respondents was the installation of bike lanes, bike routes, and signage to make the commute safer. Figure 14.1 shows the exact results obtained on this question. A sizable percentage of respondents said they would be more willing to bike to campus if secure, out-of-the-elements bike parking, maps, information about safe bike routes to campus, and bike racks were provided. Specifically, respondents want more separate/protected bike lanes (facilities and safety), maps, way-finding signs (facilities and safety), and secure places to store bikes (facilities). It should be noted that a less popular response was "training in safe and confident cycling in traffic," suggesting that bikers recognize the dangers of riding in traffic,

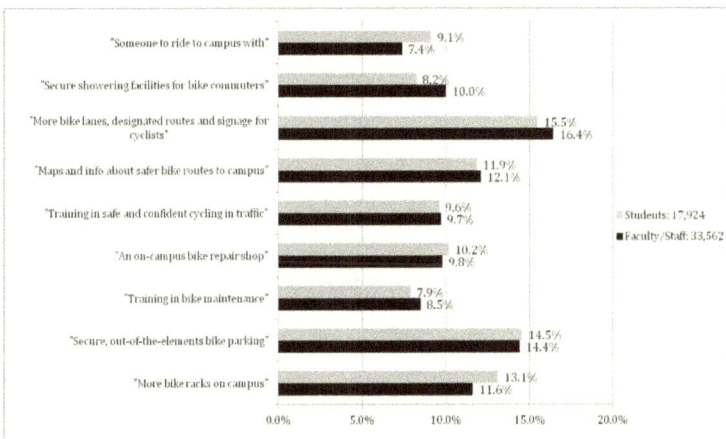

Figure 14.1   Survey Question: What Would Incentivize You to Bike to Campus More Often? *Source*: Center for Sustainable Urban Neighborhoods, formerly University of Louisville, transferred to Neighborhood Associates, Washington D.C.

fast multi-lane one-way streets, and non-segregated bike lanes. Bikers prefer to ride on streets that are calm and slow, even without a bike lane. In fact, Louisville's most dangerous biking area is considered a multi-lane one-way street with an unprotected bike lane (Riggs & Gilderbloom, 2015).

Another potential impetus for altering commuting habits is an increase in the price of gas. Our survey revealed that, if the price of gas per gallon increased, more people would consider moving closer to campus, thus making bike commuting more likely. If the price were to exceed $5 per gallon, more than one-third of the survey respondents reported they would consider such a move. Also of note are the endpoints of the response trend lines. Many respondents already live near campus, and many will live wherever they desire regardless of the price of gasoline. This suggests that neighborhood qualities other than the cost of transportation are most important in deciding one's place of residence. In other words, this suggests that qualities such as neighborhood vibrancy, safety, walkability, and transportation mode diversity—as found by Deaken et al. (2004) and Gilderbloom et al. (2014) affect choice of residence location just as the cost of driving does. Figure 14.2 shows the results obtained on this question.

The impact of this re-distributed spending could be significant for the local economy, especially considering that 10 of the 14 answer choices occur at the local level and would be more likely satisfied by local sources.

How does this compare to "average" American spending habits? A 2009 study by the Bureau of Labor Statistics (BLS) estimated that the average American household spends 21 cents of every dollar on shelter.[5]

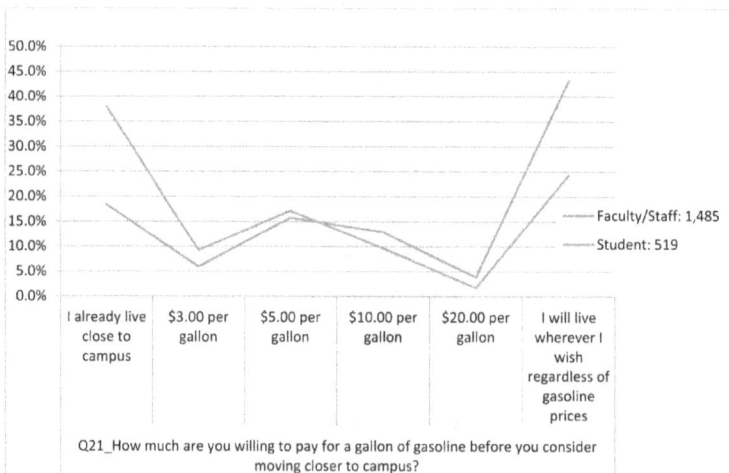

Figure 14.2   Results of Survey Question on Residential Choice Based on Gas Prices.
Source: Center for Sustainable Urban Neighborhoods, formerly University of Louisville, transferred to Neighborhood Associates, Washington D.C.

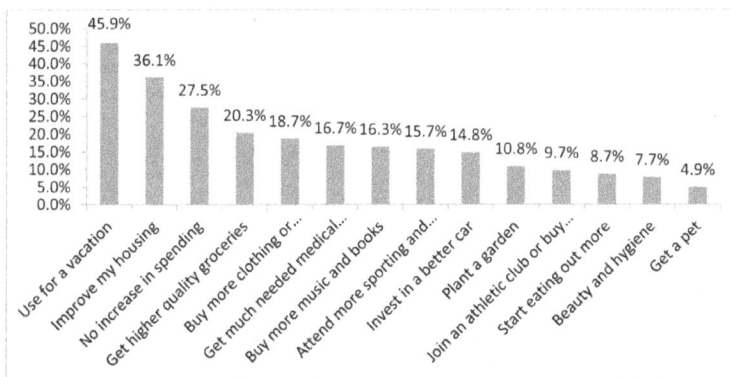

Figure 14.3   Results of Survey Question on How Respondents Would Spend Savings from Reduced or Discontinued Auto-based Commuting "Economists have estimated that non-car commuters from home to campus can save thousands of dollars--roughly $3,000 to $8,000 a year. If such savings were to be realized, what would you do with this savings?" *Source*: Center for Sustainable Urban Neighborhoods, formerly University of Louisville, transferred to Neighborhood Associates, Washington D.C.

Transportation is the second largest spending category, roughly 16 cents of every dollar spent. Food accounts for 13 cents. Insurance and pension (excluding home insurance) amount to seven cents, equal to the seven cents spent on utilities. Other household needs account for another seven cents, healthcare is roughly seven cents, and entertainment adds up to 6 cents. Apparel, services, and cash contributions each account for four cents, while education and reading consume just two cents out of every dollar spent.

In our analysis, we calculated a conservative cost-saving projection based on the number of individuals who indicated that they would ride a bicycle to campus because of changes in bike infrastructure and services provided by the university. This calculation is performed by multiplying the number of potential cyclists by the amount of money saved by an individual not owning a car. According to respondents, a significant portion of savings would be spent in the community; thus, monies previously sent outside Metro Louisville, to industries such industries as national finance, car manufacturing, and oil companies, none of which are in most cities, would instead be available to support local economies. If bike lanes were extended out from the university to within an 11-20-minute commute radius, the savings would increase to $55 million. Altogether, if bike lanes were installed throughout Louisville, the economic return for just faculty, staff, and students who would choose a bike as their primary mode of transportation would be around $78 million—table 14.3 provides details of these calculations. This does not include others outside of the university community who would also use bike lanes for

Table 14.3 Saving Estimates Given Variation in the Number of People and Amount of Reduction in Driving

Commute Time	Would Bike	Would Not Bike	Individual Savings	Cumulative Total Savings
Within 10 minutes	3,226	2,483	$ 25,804,680	$ 25,804,680
11–20 minutes	3,670	5,505	$ 29,360,000	$ 55,164,680
21–30 minutes	1,906	6,347	$ 15,251,544	$ 70,416,224
31–40 minutes	581	2,918	$ 4,645,344	$ 75,061,568
41–60 minutes	318	1,400	$ 2,541,160	$ 77,602,728
Over 1 hour	95	223	$ 763,200	$ 78,365,928

Source: Center for Sustainable Urban Neighborhoods, formerly University of Louisville, transferred to Neighborhood Associates, Washington D.C.
Cost Calculation Formula: "Would Bike" X $8,000

commuting. Consumer preference surveys cannot accurately predict an exact amount of savings and reinvestment but provide us with an approximate measure.

Another, more nuanced way of calculating the potential economic impact of bicycling is to create alternative scenarios based on variable estimates of the annual costs of owning and operating a car. The conventional AAA estimate of the annual cost of operating a car is now approximately $8,000. One might ask how much could be saved if someone did not completely give up their car but instead limited their number of automobile trips. Figure 14.4 takes this scenario into consideration and estimates the total potential savings ($2,000, $4,000, $6,000, and $8,000) based upon variations in how much individuals reduce driving and the percentage of employees and students at the University of Louisville that bike. We estimate that even if individuals reduced rather than eliminated car dependency, saving $2,000–$4,000 a year, the potential overall impact would be more significant.

Not surprisingly, our survey also suggests that people who live close to work or school are more willing to consider bicycling as a practical means of transportation. Considering that Louisville is a monocentric city (Gilderbloom & Appelbaum, 1988) where the highest concentration of jobs and educational institutions is still in the central business district (Gilderbloom & Appelbaum, 1988), those who live close to downtown are more likely to favor alternative modes of transportation if proper infrastructure exists. Therefore, to maximize bicycle usage, the most logical first step is to build infrastructure in the central business district and then extend out to nearby neighborhoods. There is a powerful green dividend to urban biking: economy, environment, and ecology.

Our research finds support for the argument that increased bike commuting is dependent on social and infrastructural factors. We found no support for Goetzke and Rave's (2010) assertion that bike infrastructure is not correlated

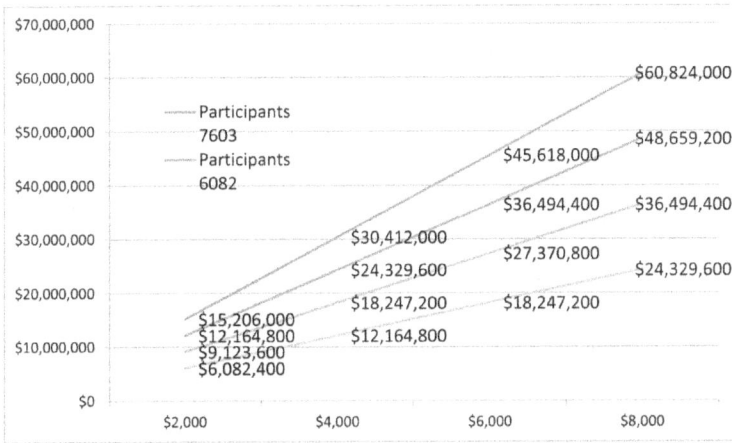

Figure 14.4   Projected/Estimated Savings Based on Varying Levels of Savings Resulting from Reduced or Discontinued Auto-Commuting. *Source*: Center for Sustainable Urban Neighborhoods, formerly University of Louisville, transferred to Neighborhood Associates, Washington D.C.

with an increase in biking. We also test claims by bike advocacy groups that an increase in urban biking results in greater neighborhood consumer spending on various essential and non-essential consumer products. Bureau of Labor Statistics (2009) data on consumer spending help bolster our findings that savings from urban biking will result in greater local spending through reallocation of monies not spent supporting car-based transportation. Our research suggests that increased urban biking means neighborhoods will become more prosperous and sustainable. More scholarly research is needed to further explore how urban biking creates and expands neighborhood sustainability.

A few related points are worth making. First, it appears the potential positive economic impact of providing more bike infrastructure could be quite large. As it stands now, the average nationwide rate of bike commuting is roughly 1 percent of the population, or 3,000,000 people. If another 1 percent could be added to that total, and each saved just $2,000 per year from doing so, it could divert as much as $6,000,000,000 from a few national industries and instead flow to businesses with more beneficial impacts on local economies.[6]

Finally, our estimates of savings do not reflect positive externalities obtained through the wide-scale adoption of bike lanes: improved air quality, reduced health care costs, added employment, and pollution control costs (i.e., higher prices for gasoline), among others. More research focused on determining comprehensive cost savings and quality of life benefits resulting from bike infrastructure investments should be conducted.

CONCLUSION

The local economic benefits are significant in shifting car-centered transportation toward a more bike-friendly culture in urban neighborhoods. Fewer cars and more bikes and bike-oriented infrastructure could translate into higher property values, more jobs, decreased traffic congestion, and, ultimately, more money in the consumer's pocket. Approximately one-fourth of those sampled reported they would start biking to work or school if more bike infrastructure—bike lanes, signage, secure parking facilities—were provided. Survey respondents also noted that they would be more likely to bike to campus if bike safety courses, bike shop vouchers, secure bike parking, and bike repair shops were made available. Another one-fourth of respondents said that they would switch their transportation mode of choice from a car to a bike, and another one-fourth of respondents said that they would move closer to the university if the cost of gas were to exceed $5 per gallon. Consequently, we estimate that the Old Louisville neighborhood could see an infusion of up to $26 million in new community spending simply through a reduction in car dependency. This spending on housing repairs, installation of community gardens, and locally operated grocery stores and restaurants— expected results of the cost savings created by the reduction or discontinuance of car-based commuting—represents the potential green dividend of investments in urban bicycling. Another important finding is that people who ride bikes will be healthier due to increased physical activity and improved air quality. But, most importantly, the net zero impact of carbon emissions from bicycles means improved stewardship of Mother Earth with a reduction of greenhouse gases and, hopefully, a way to heal the earth back to normalcy.

NOTES

1. Arnott and Small (1994) estimate that for the 39 metropolitan areas in the United States having a population over one million, nearly 33 percent of all auto travel is in congested conditions, during which the average speed is one-half of its free-flow value. Arnott and Small (1994) also report that about half of the congested driving takes place on expressways, causing delays of six-tenths of a minute per kilometer. The congestion on arterials causes about 1.2 minutes per kilometer in delays (Arnott & Small, 1994).

2. For the typical American household, transportation expenses are second only to housing expenses, but in some cases, transportation expenses can be higher than housing expenses. In a study of 28 metropolitan areas, it was reported that residents pay one percent more in transportation expenses than they do in housing expenses (Bernstein, Makarewicz & McCarty, 2005). According to a study completed by the Surface Transportation Policy Project (Bernstein & Mooney Bullock, 2000), a nationwide coalition for safer communities and smarter transportation choices, the average American household devotes 18 cents of every dollar earned to transportation. Furthermore, the report finds that 98 percent of that spending is designated for the purchase, operation, and maintenance of automobiles.

3. (Lovins & Cohen, 2011) claim that if a family relocates from a suburban neighborhood that requires maintaining only one vehicle while having access to other transportation alternatives, the family's annual housing costs could increase by $5,000. The family's transportation costs will be cut by an estimated $12,000 per year (Lovins & Cohen, 2011). These redistributions of expenses represent a net decrease in the cost of living by 18.3 percent (Lovins & Cohen, 2011). Should a family requiring two cars and 25,000 miles driven per vehicle of annual travel reduce to one car, that family could reduce living costs by 9.4 percent (Lovins & Cohen, 2011). These figures shed light on the potential cost savings of reducing motor vehicle travel and dependence, but do not accurately reflect the cost savings of substituting automobile travel for bike travel alone, as the calculations include allowances for heavy public transit use.

4. According to a study by the U.S. Department of Energy (2010), from 1969 to 2009, the U.S. experienced a 16 percent decrease in the number of persons per household and a 66 percent increase in the number of vehicles per household. According to the National Household Travel Survey conducted by the U.S. Department of Transportation Bureau of Statistics (2002), 71 percent of all Americans commute more than six miles to work one way each day, with 33 percent of Americans commuting more than 15 miles to work one way each day. Pucher and Dijkstra (2003) cite that 49 percent of all trips made by Americans in metropolitan areas are shorter than three miles, 40 percent are shorter than two miles, and 28 percent are shorter than one mile. More than 14 percent of commuters travel each day from the suburbs to the city (U.S. Department of Transportation, 2002).

5. BLS notes that shelter includes mortgages, taxes, insurance, rent, and maintenance.

6. Another indication of the positive community economic development impact of biking is to count the number of businesses added when a "friendly bike infrastructure" of bike lanes, signage, education, bike parking, and conversion of fast multi-lane one-way streets to calmer two-way streets is put into a neighborhood. This was done in the Old Louisville neighborhood, which is adjacent to the University of Louisville. First, we found that a local "green developer" responded by building a 255-unit mixed-use development of new and renovated housing adjacent to the university community. This development houses 13 businesses, 10 of which are locally owned. Additionally, several blocks away on a recently converted two-way street with a well-designed bike lane, several other businesses (including a coffee house, 24-hour diner,

and a grocery store) were added or expanded. All this new business equates to 150 jobs created by the existence and operation of the 34,650 square feet of commercial space in the new mixed-use development and 27 jobs in the nearby businesses (sunlouisville.org; see also Clinch, 2011).

REFERENCES

AAA. (2011). *2011 Your Driving Costs*. Retrieved from http://www.aaaexchange .com/main/Default.asp?CategoryID=16&SubCategoryID=76&ContentID=353.

ACSM American Fitness Index™: Health and Community Fitness Status of the 40 Largest Metropolitan Areas. (2010 Edition). *American College of Sports Medicine*. Retrieved from http://www.americanfitnessindex.org/docs/reports/2010AFIReport -Final.pdf.

Alrutz, D., Angenendt, W., Draeger, W., and Gündel, D. (2002). *Traffic Safety on One-Way Streets with Contra Flow Bicycle Traffic*. Retrieved February 15, 2010, from http://www.bikexprt.com/: http://www.bikexprt.com/research/contraflow/ gegengerichtet.htm

American Automobile Association (AAA). (2013). *2013 Your Driving Costs*. Retrieved from http://www.aaaexchange.com/main/Default.asp?CategoryID=16 &SubCategoryID=76&ContentID=353.

America's No.1 Health Problem. (2010). Retrieved from http://www.stress.org/americas.htm

APA Stress Survey: Children More Stressed than Parents Realize. (November 23, 2009). Retrieved March 24, 2010, from *American Psychological Association Practice Organization*. http://yalestress.org/upload/docs/APA%20Stress%20Survey %20brief%202009%20includes%20teen%20stress.pdf

Arnott, R., and Small, K. (1994). "The Economics of Traffic Congestion." *American Scientist, 82*, 445–455.

Babbie, E. (2012). *The Basics of Social Research*. Belmont: Wadsworth Publishing Co.

Babbie, E. (1990). *Survey Research Methods*. Belmont: Wadsworth Publishing Company.

Bernstein, S., Makarewicz, C., and McCarty, K. (June 2005). *Driven to Spend: Pumping Dollars Out of Our Households and Communities*. Washington, DC: The Surface Transportation Policy Partnership. Retrieved from http://www.transact.org /library/reports_pdfs/driven_to_spend/Driven_to_Spend_Report.pdf.

Bernstein, S., and Mooney Bullock, R. (2000). *Driven to Debt*. Washington, DC: The Surface Transportation Policy Partnership.

Birk, M. (2010). *Joyride: Pedaling Toward a Healthier Planet*. Portland: Cadence Press.

Boarnet, M., and Sarmiento, S. (1998). Can Land Use Policy Really Affect Travel Behavior? A Study of the Link between Non-Work Travel and Land Use Characteristics. *Urban Studies, 35*, 1155–1169.

Broken Sidewalk. (2015). *League of American Cyclists Rank Kentucky 49th in New Bike-Friendly State Rankings.* Retrieved May 12, 2015, from http://www.borken-sidewalk.com.

Burden, D. (January 2001). *Building Communities with Transportation.* Washington, DC: The Transportation Research Board. Retrieved from http://www.walkable.org /assets/downloads/trbpaper.pdf.

Burden, D. (March 5, 2010). Session 9. (U. 6. Class, Interviewer).

Byrne, D. (2009). *Bicycle Diaries.* New York: Viking.

Byrne, D. (2010). *Bicycle Diaries.* New York: Penguin.

Calories Burned Bike Riding or Cycling. (n.d.). Retrieved from http://www.nutristrat-egy.com/fitness/cycling.htm.

Cascade Bicycle Club Education Foundation. (2010). *Commuting Statistics.* Retrieved from http://www.cbcef.org/bike-commuting-statistics.html.

Clark, M. C. (2010). The Cost of Job Stress. Retrieved March 24, 2010, from *Winning Workplaces*: http://www.winningworkplaces.org/library/features/the_cost_of _job_stress.php.

Clifton, K. J., Morrissey, S., and Ritter, C. (2012a). Business Cycles: Catering to the Bicycling Market. *TR Transportation Research News, 280*(May –June 2012), 26–32.

Clifton, K. J., Morrissey, S., and Ritter, C. (2012b). *Mode Choice and Consumer Spending: An Examination of Grocery Store Trips.* Presented at 91st Annual Meeting of the Transportation research Board, Washington, D.C.

Clifton, K. J., Muhs, C., Morrissey, S., Morrissey, T., Currans, K., and Ritter, C. (2012c). *Consumer Behavior and Travel Mode Choices.* Portland: Oregon Transportation Research and Education Consortium.

Clinch, R. (2011). *The Economic and Workforce Development Impacts of the Cardinal Towne Project.* Prepared for 3rd Street Development, LLC.

Community Health Status Indicators. (2010). Retrieved March 31, 2010, from http://www.communityhealth.hhs.gov/SummaryMeasuresOfHealth.aspx?GeogCD =21111&PeerStrat=3&state=Kentucky&county=Jefferson.

Cook, A., Skimming, J., and Stanford, J. (2009). "Environmental Statement on the Benefits of Cycling in London with a Focus on the Gainsborough/Windermere Pathway Connection." City of London, Ontario. Summer 2009.

DeCorla-Souza, P., and Jensen-Fisher, R. (1997). Comparing Multimodal Alternatives in Major Travel Corridors. Transportation Research Record. No. 1429, 15–23.

Dill, J. (2009). Bicycling for Transportation and Health: The Role of Infrastructure. *Journal of Public Health Policy, 30,* 95–110.

Downs, A. (1962). The Law of Peak-Hour Expressway Congestion. *Traffic Quarterly, 16,* 393–409.

Duany, A., Plater-Zyberk, E., and Speck, J. (2001). *Suburban Nation: The Rise of Sprawl and the Decline of the American Dream.* New York: North Point Press.

Eichenfeld and Associates and the Local Government Commission. (July 2002). Strategies for Revitalizing Our Downtowns and Neighborhoods: Evaluating California Main Street Programs.

Ewing, R., Bartholomew, K., Winkleman, S., Walters, J., and Anderson, G. (2008). Urban Development and Climate Change. *Journal of Urbanism, 1*(3), 201–216.

Federal Highway Administration. (2005) or (2009). *Crash Cost Estimates by Maximum Police-Reported Injury Severity within Selected Crash Geometries.* Retrieved from http://www.fhwa.dot.gov/publications/research/safety/05051/.

Federal Highway Administration and the United States Department of Transportation. (2003). *Licensed Drivers and Vehicle Registrations.* Retrieved from http://www.fhwa.dot.gov/.

Finneren, K. (2005). Construction Chaos. Central Michigan Life. 8/28/2005. Page 3a.

Flottorp, M., and Anthony, S. B. (2023). The Fierce Crusader for the Bicycle and Women's Rights. *We Love Cycling.* Retrieved March 7, 2024 from https://www.welovecycling.com/wide/2023/02/15/susan-b-anthony-the-fierce-crusader-for-the-bicycle-and-womens-rights/.

Forester, J. (2009). "*National Bicycle Transportation Policy Contradicts Itself*" June 17, 2009. http://www.johnforester.com/Articles/Government/Policy%20Issue.pdf.

Fox, K. R. (1999). The Influence of Physical Activity on Mental Well-Being. *Public Health Nutrition, 2*, 411–418.

Geller, R. (2013). *What Does the Oregon Household Activity Survey Tell Us About the Path Ahead for Active Transportation in the City of Portland?* Salem: Department of Transportation: White paper.

Gilderbloom, J. I. (2008). *Invisible City: Poverty, Housing, and New Urbanism.* Austin: University of Texas Press.

Gilderbloom, J. I., and Appelbaum, R. P. (1988). *Rethinking Rental Housing.* Philadelphia: Temple University Press.

Gilderbloom, J., Hanka, M. J., and Lasley, C. B. (2009). Amsterdam: Planning and Policy for the Ideal City? *Local Environment, 14*(6), 473–493.

Gilderbloom, J., Meares, W. L., and Riggs, W. (December, 2014). How Toxic Waste Sites in Neighborhoods Kill Places and People. *Journal of Urbanism, 22*(8), 917–933.

Glaeser, E., and Kahn, M. (2008). *The Greenness of Cities: Carbon Dioxide Emissions and Urban Development.* Harvard Institute of Economic Research Discussion Paper No. 2161.

Goetzke, F. R. (2010). Bicycle Use in Germany: Explaining Differences between Municipalities with Social Network Effects. *Urban Studies, 48*(2), 427–437. Retrieved from http://usj.sagepub.com/content/early2010/06/03/0042098009360681.

Gotschi, T., and Mills, K. (2008). Active Transportation for America: The Case for Increased Federal Investment in Bicycling and Walking. *Rails to Trails Conservancy.* https://www.railstotrails.org/resource-library/resources/active-transportation-for-america-report/.

Gowin, D. (February 9, 2010). Countywide Countermeasures 02.

Gowin, D. (January 29, 2010). Designing Streets for Bicyclists. (U. 6. Class, Interviewer).

Green, J. (February 12, 2010). (U. 6. Class, Interviewer).

Handy, S. L., and Emond, C. R. (2012). Factors Associated with Bicycling to High School: Insights from Davis, CA. *Journal of Transport Geography, 20*(1), 71–79.

Hartgen, D., and Fields, G. (2009). Gridlock and Growth: The Effect of Traffic Congestion on Regional Economic Performance. *Reason Foundation Study No. 371 Summary.* Retrieved from www.reason.org.

Health Status – Kentucky. (2010). Retrieved February 28, 2010, from *Kaiser State Health Facts.* http://www.statehealthfacts.org/profilecat.jsp?rgn=19&cat=2.

How Not to get Hit by Cars? (2009). Retrieved February http://bicyclesafe.com/.

Kahn Ribeiro, S., Kobayashi, S., Beuthe, M., Gasca, J., Greene, D., Lee, D. S., Muromachi, Y., Newton, P. J., Plotkin, S., Sperling, D., Wit, R., and Zhou, P. J. (2007). Transport and Its Infrastructure. In *Climate Change 2007: Mitigation. Contribution of Working Group III to the Fourth Assessment Report of the Inter-governmental Panel on Climate Change* [B. Metz, O.R. Davidson, P.R. Bosch, R. Dave, L.A. Meyer (eds)]. Cambridge and New York: Cambridge University Press.

Kentucky Health Facts. (2010). Retrieved March 31, 2010, from http://www.kentuckyhealthfacts.org/data/location/show.aspx?county=Jefferson.

Kenworthy, J., and Laube, F. (1999). Patterns of Automobile Dependence in Cities: An International Overview of Key Physical and Economic Dimensions with some Implications for Urban Policy. *Transportation Research Part A: Policy and Practice, 33*(7–8), 691–723.

Krizek, K. (December 2001). *Neighborhood Services, Trip Purpose, and Tour-Based Travel.* Retrieved from www.hhh.umn.edup/people/kkrizek/TrippercentTours,per cent20Transportation.pdf.

KSP. (2008). Retrieved February 13, 2010, from http://www.kentuckystatepolice.org /pdf/KY_Traffic_Collision_Facts_2008.pdf.

Lawton, K. (2001). *The Urban Structure and Personal Travel: An Analysis of Portland, Oregon Data and Some National and International Data. Supplementary Materials:* Papers and Analyses for the E-Vision Conference. Retrieved from www .rand.org/scitech/stpi/Evision/Supplement/.

Litman, T. (2003). Economic Value of Walkability. *Transportation Research Record: Journal of the Transportation Research Board, 1828,* 3–11.

Litman, T. (2010). *"Smart Transportation Emission Reductions. Identifying Truly Optimal Energy Conservation and Emission Reduction Strategies."* Victoria Transport Policy Institute. February 3, 2010.

Local Government Commission. (2000). *The Economic Benefits of Walkable Communities.* Retrieved from http://www.lgc.org/freepub/docs/community_design/ focus/walk_to_money.pdf.

Lockwood, I., and Stillings, T. (1998). *Traffic Calming for Crime Reduction and Neighborhood Revitalization.* West Palm Beach: West Palm Beach Transportation Planning. Retrieved from http://www.ite.org/traffic/documents/AHA98A19 .pdf.

Lott, D. F., and Lott, D. Y. (1976). Effect of Bike Lanes on Ten Classes of Bicycle-Automobile Accidents in Davis, California. *Journal of Safety Research, 8*(4), 171–179.

Lovins, H. L., and Cohen, B. (2011). *Climate Capitalism: Capitalism in the Age of Climate Change.* New York: Hill and Wang.

Maibach, E., Steg, L., and Anable J. (2009). Promoting Physical Activity and Reducing Climate Change: Opportunities to Replace Short Car Trips with Active Transportation. *Preventative Medicine, 49,* 326–327.

Mapes, J. (2009). *Pedaling Revolution: How Cyclists are Changing American Cities.* Corvallis: Oregon State University Press.

Marshall, J. D., Brauer, M., and Frank, L. D. (2009). Healthy Neighborhoods: Walkability and Air Pollution. *Environmental Health Perspectives, 117*(11), 1752–1759.

Matthews, J. W. (2006). *"The Effect of Proximity to Commercial Uses on Residential Prices."* Atlanta Georgia Tech: Georgia State University and Georgia Institute of Technology.

McAnn, B. (2000). *Driven to Spend: The Impact of Sprawl on Household Transportation Expenses.* Washington, DC: The Surface Transportation Policy Partnership. Retrieved from http://www.transact.org/progress/jan01/driven.asp.

McKenzie, B. (2014). *Modes Less Traveled: Bicycling and Walking to Work in the United States 2008–2012. American Community Survey Reports.* Washington, DC: U.S. Census Bureau. Retrieved from http://www.census.gov/prod/2014pubs/acs-25.pdf.

McMahon, E. T. (2012). *Bicycles Belong: A Growing American Way.* Retrieved from http://citiwire.net/columns/bicycles-belong-a-growing-american-way/.

"Money is Parking Garage Obstacle." *Central Michigan Life.* October 10, 2005. http://www.cm-life.com/2005/10/10/moneyisparkinggarageobstacle/.

Moritz, W. E. (1998). Adult Bicyclists in the United States: Characteristics and Riding Experience in 1996. *Transportation Research Record, 1636*(1), 1–7.

Netherlands Life Expectancy at Birth. (n.d.). Retrieved February 20, 2010, from *Index Mundi*: http://www.indexmundi.com/netherlands/life_expectancy_at_birth.html.

NHTS. (2009). *"The 'Carbon Footprint' of Daily Travel" National Household Travel Survey.* U.S. Department of Transportation. Federal Highway Administration. March 2009.

Partnership for a Green City. (2009). *Climate Action Report.* Retrieved from http://www.jefferson.k12.ky.us/Departments/EnvironmentalEd/GreenCity/ClimateActionRpt.pdf.

Pedestrian and Bicyclist Intersection Safety Indices Final Report. (n.d.). Retrieved February 15, 2010, from *US Department of Transportation Federal Highway Administration* http://www.tfhrc.gov/safety/pedbike/pubs/06125/chapt7.htm.

Portland Plan Partners. (2012). *The Portland Plan: Prosperous, Educated, Health and Equitable.* Portland: City of Portland.

Pucher, J., and Buehler, R. (2012). *City Cycling.* Cambridge: MIT Press.

Pucher, J., and Buehler, R. (2007). Cycling for Everyone: Lessons from Europe. *Transport Research Record, 2074,* 58–65.

Pucher, J., and Buehler, R. (2008). Making Cycling Irresistible: Lessons from the Netherlands, Denmark and Germany. *Transport Reviews, 28*(4), 495–528. Retrieved from http://www.vtpi.org/irresistible.pdf.

Pucher, J., Buehler, R., and Seinen, M. (2011). Bicycling Renaissance in North America? An Update and Re-appraisal of Cycling Trends and Policies. *Transportation Research A, 45*(6), 451–475.

Pucher, J., and Dijkstra, L. (2003). Promoting Safe Walking and Cycling to Improve Public Health: Lessons From The Netherlands and Germany. *American Journal of Public Health, 93,* 1509–1516.

Reynolds, C. C., Harris, M. A., Teschke, K., Kripton, P. A., and Winters, M. (2009). The Impact of Transportation Infrastructure on Bicycling Injuries and Crashes: A Review of the Literature. *Environmental Health, 8*(1), 47. http://www.ehjournal.net/content/pdf/1476-069X-8-47.pdf.

Rietveld, P., and Daniel, V. (2004). Nonmotorized Modes in Transport Systems: A Multimodal Chain Perspective for the Netherlands. *Transportation Research A, 38*(7), 531–550.

Riggs, W. (2013). Reflections on Campus Planning: Lessons for Professional Practice. Focus. *Journal of the City and Regional Planning Department, 10*(1), 32.

Riggs, W. (2014). Dealing with Parking Issues on an Urban Campus: The Case of UC Berkeley. *Case Studies on Transport Policy, 2*(3), 168–176.

Riggs, W., and Gilderbloom, J. (2015). How Multi-Lane One Way Street Design Shapes Life Chances: Collisions, Crime and Community. *Journal of Education Planning and Research, 36*(1), 105–11.

Schrank, D., Lomax, T., and Eisele, B. (2011). "Texas Transportation Institute's 2011 Urban Mobility Report." *The Texas Transportation Institute, the Texas A&M University System.* Retrieved from http://mobility.tamu.edu/ums/report/.

Schutt, R. (2010). *Investigating the Social World.* 6th Edition. Thousand Oaks: Pine Forge Press.

Sermons, M., and Seredich, N. (November 2001). Assessing Traveler Responsiveness to Land and Location Based Accessibility and Mobility Solutions. *Transportation Research Part D: Transport and Environment, 6*(6), 417–428.

Shoup, Donald C. (2005). *Parking on a Smart Campus: Lessons for Universities and Cities.* University of California Transportation Center. UC Berkeley: University of California Transportation Center.

StatLine. (2008). Retrieved February 28, 2010, from *Central Bureau of Statistics.* http://statline.cbs.nl/StatWeb/publication/?DM=SLEN&PA=03799eng&D1=91-92,139&D2=0&D3=0&D4=a&LA=EN&HDR=T&STB=G1,G2,G3&VW=T.

Stillings, T., and Lockwood, I. (2005). *West Palm Beach Traffic Calming – The Second Generation.* TRB Circular E-C019: Urban Street Symposium.

The Surface Transportation Policy Project. (2000). *"Driven to Spend: A Transportation and Quality of Life Publication."* http://www.transact.org/wp-content/uploads/2014/04/Driven_to_Spend-2000.pdf.

Sztabinski, F. (February 2009). *Bike Lanes, On-Street Parking and Business: A Study of Bloor Street in Toronto's Annex Neighborhood.* Toronto: The Clean Air Partnership. Retrieved from www.clearnairpartnership.org.

The World Factbook. (n.d.). Retrieved February 20, 2010, from *Central Intelligence Agency.* https://www.cia.gov/library/publications/the-world-factbook/rankorder/2102rank.html.

Tompkins County Planning. (2010). Retrieved from www.co.tompkins.ny.us/planning/vct/glossary.html.

Topp, H. (2008). Can MeetBike Replace the Car? *World Transportation Policy & Practice, 14*(3), 24–31. Retrieved from www.eco-logica.co.uk.pdf/wtpp14.3.pdf.

Transportation for America. (2010). Retrieved from http://t4america.org/resources/dangerousbydesign/.

Transportation Toolkit. (2010). Retrieved from http://www.cflhd.gov/ttoolkit/flt/ Definitions%20and%20Terms/KBDefinitionsAndTerms.htm.

Tu, C. C., and Eppli, M. J. (2001). An Empirical Examination of Traditional Neighborhood Development. *Real Estate Economics, 29*(3), 485–501.

Tudela-Rivadeneyra, A., Shirgaokar, M., Deakin, E., and Riggs, W. (2015). *The Cost versus Price for Parking Spaces at Major Employment Centers: Findings from UC Berkeley.* In Transportation Research Board 94th Annual Meeting (No. 15-3640).

United States Bureau of Transportation. (2009). *Table 1–11: Number of U.S. Aircraft, Vehicles, Vessels, and Other Conveyances.* Retrieved from http://www.bts.gov/ publications/national_transportation_statistics/html/table_01_11.html.

University Planning, Design & Construction. (2009). *Campus Master Plan, University of Louisville.* Retrieved from https://louisville.edu/updc/masterplan.

Walkinginfo.org. (2011). *Bicycle Lanes.* http://www.walkinginfo.org/engineering/ roadway-bicycle.cfm.

Watchel, A., and Lewiston, D. (1994). Risk Factors for Bicycle-Motor Vehicle Collisions at Intersections. *Institute of Transportation Engineers Journal, 64*(9), 30–35.

Weigand, L. (2008). *A Review of Literature: The Economic Benefits of Bicycling. Portland State University* (Publication No. CUS-CTS-08-03). Portland: Center for Transportation Studies, Portland State University.

Weigand, L. (2008). *A Review of Literature: The Economic Benefits of Bicycling.* Center for Urban Studies. Portland: Portland State University.

Wendell, T., Tom, B., and Rohm, Y. (1998). Physical Activity Interventions in Low-Income, Ethnic Minority, and Populations with Disability. *American Journal of Preventive Medicine, 15*(4), 334–343.

Wray, J. H. (2008). *Pedal Power: The Quiet Rise of the Bicycle in American Public Life.* Boulder/London: Paradigm Publishers.

Wrong Way Cycling. (2010). Retrieved February 15, 2010, from http://www.kenkifer .com/bikepages/traffic/wrong.htm.

Chapter 15

Designing and Building Affordable and Attractive Housing for Working People

John Hans Gilderbloom, Wesley L. Meares, and Ra'Desha Williams

There is little doubt that Covington HOPE VI is one of the best green housing developments for low- and moderate-income people in the country. In an era of climate change, this deserves high praise. River's Edge will become LEED Certified and is further pursuing certification at the Silver Level, making it one of the only affordable green housing developments in Kentucky. Because people with higher incomes generally utilize LEED, River's Edge is a major accomplishment. It provides a compass for affordable and green

Figure 15.1 Public Housing in Covington Built in 1935. *Source*: Image by John Hans Gilderbloom.

Figure 15.2 HOPE IV Housing that Replaced the Old Public Housing in Covington.
Source: Image by John Hans Gilderbloom.

housing advocates for how to get this done. Put simply, the promise of LEED is to cut energy costs by one-half of what the poor normally pay. As evaluators, we have a great deal of praise for the LEED model, but we also suggest ways that LEED certification guidelines can be improved to decrease energy consumption.

Our praise for River's Edge is also notable because of what was there before, which was an outdated, barracks-like housing development filled with crime, environmental toxins, and a sense of hopelessness. The old Jacob-Price Housing encouraged this hopelessness; River's Edge reverses it. The new development consists of 11 beautiful brick buildings consistent with the historic fabric of the German-style neighborhood.

By twenty-first-century standards, the old Jacob Price development was unlivable and needed to be replaced with modern, energy-efficient structures for mixed-income housing, and with a new-urbanist design. This was evident in the HOPE VI Jacob Price Application (2009).

Virtually every major mechanical, electrical, and plumbing system at Jacob Price Homes was either original to the 1939 construction or had exceeded its useful life expectancy. Like the units already demolished or taken out of service, many of the currently occupied units suffer from deteriorated systems, design deficiencies, inaccessibility, and density far exceeding any modern standard for urban living. The extent and nature of infrastructure and project site deficiencies, and fundamental design flaws preclude these facilities from rehabilitation. The structures contain inefficient mechanical systems that do not provide air conditioning or adequate envelope insulation; and unit interiors that are inadequate to meet the needs of today's families, with undersized rooms, grossly inadequate closet space, and no washer/dryer hook-ups; and

Figure 15.3 HOPE VI Housing in Covington for Seniors and People with Disabilities.
Source: Image by John Hans Gilderbloom.

the site's outdoor spaces are dehumanizing, lacking any semi-private or private spaces.

The former Jacob-Price apartment units were not constructed under a framework of sustainability. They were poorly insulated and installed with inefficient appliances, lighting, and heating systems, and were in poor physical condition. The revitalization project was to tackle these design flaws and enhance the lives of HOPE VI residents and the Eastside.

The redevelopment of Jacob Price Homes through the HOPE VI Program provides Jacob Price residents, the City of Covington, and the Housing Authority with a unique opportunity to reclaim and revitalize the Jacob Price site, create a new mixed-income community, catalyze the redevelopment of the surrounding neighborhood, and provide employment and educational opportunities to Jacob Price residents (HOPE VI Jacob Price Application, 2009).

Demolition of the former Jacob Price Homes was necessary for the Eastside to climb out of a physically distressed environment. HOPE VI revitalization accomplished this by replacing standard public housing with a new urbanist-style mixed-income community.

The Eastside Neighborhood, as demonstrated by the following data, has suffered a long and continuous downward spiral of social, economic, and physical distress. Jacob Price Homes, built 70 years ago, has been a contributor to this decline and distress, as made evident by the significant number of vacant properties immediately adjacent to the site and on surrounding blocks. The demolition of Jacob Price Homes and its replacement with a new mixed-income community, combined with new construction and renovation off-site in the neighborhood, will remove this negative influence and create a positive catalyst for neighborhood renewal and regeneration (HOPE VI Application, 2009).

Before demolition, the crime rate in the Jacob Price community had been high for years. Crime has decreased since then, largely due to the reductions in the number of tenants. One explanation for the high rate of crime is the high rate of unemployment among residents. A vast number of residents were, and still are, unemployed. A likely cause of the high rate of unemployment is that half of the residents interviewed do not have high school diplomas, and many are without vocational training. When under-education is mixed with the absence of a computer or email account, employment opportunities become scarcer—many minimum wage job applications require a computer to apply. The Community Building will have a computer center, and units will have infrastructure for Internet connections.

Nearly everyone who had relocated to other public housing communities indicated a desire to move back to the new HOPE VI development. The residents who lived in the old Jacob Price development support the new urbanist design. This support was evident not only from interviews of former residents but in the application for HOPE VI, in which Resident Council President Brenda Ramsey endorsed the development. For roughly three years, residents, neighbors, consultants, city officials, social workers, preservationists, designers, planners, and HUD and business leaders met with 20–40 people to haggle, debate, discuss, vote, and come to a consensus on how to create a dynamic neighborhood that would deliver pride, ownership, hope, and a sustainable future.

For this program to be approved and implemented, it needed the endorsement of the residents. Residents who were relocated to other Covington public housing communities expressed a strong desire to move back into an attractive housing development closer to downtown and public transit, and with more living space. Some residents also perceived that the new development would offer a safer and healthier environment without mold, toxins, or poor insulation. Residents appreciated the fact that the new development would include more attractive landscaping and more green options such as nearby community gardens and opportunities to walk and bike. They welcomed the opportunity to live in a mixed-income neighborhood that would take away the stigma of living in a housing "project."

One of our more significant findings is that most residents do not own cars. Many use bikes for transportation, which supports recent U.S. Census data indicating the poor are most likely to commute via bike. River's Edge is an ideal community for walking and riding bikes.

One important concern raised by the survey was that roughly three-fourths of the residents interviewed showed signs of smoking. While a majority of respondents interviewed opposed a smoke-free policy in the buildings, we will argue that such a policy is appropriate for a development that promotes itself as "green." If the building is called "green" with better insulation, removal of mold, and toxins in the building materials and appliances, Integral Property Management (IPM) has an opportunity to remove the carcinogens and other toxins found in cigarette smoke through thoughtful tenant policies that will protect the health of residents as well as property values. To encourage a healthy lifestyle, the Community Center (building #2) will also have a fitness center.

From blighted to beautiful, community rebirth is also the promise of HOPE VI. The city should build or improve sidewalks within and around the perimeter of the development to ensure easy pedestrian access to surrounding educational, cultural, and recreational amenities. Additionally, the city should add bicycle lanes to make it easier for those who own bikes to use them for work or school. Since only 20 percent of public housing residents own an automobile, the number of parking spaces allocated for the new development may be reduced to incorporate more green spaces and bicycle parking. The city may choose to allocate nearby street parking for residents only. This would increase the amount of green space while limiting the number of outsiders coming into the community.

HOPE VI ENVIRONMENTAL PROMISE

The HOPE VI program has four important objectives: (1) change the stigmatization of barracks-style housing with new urbanist design, (2) reduce the concentration of poverty, (3) provide support services, jobs, education, and health services to residents, and (4) develop partnerships between public and private agencies and organizations (HUD, 1999; HUD, 2000; Schwartz, 2006; Popkin et al., 2000, 2004; Popkin, 2002; Gilderbloom and Hanka, 2006; Brazley and Gilderbloom, 2007; Gilderbloom, 2008). Affordable housing is seen as the key to sustainable neighborhoods.

How do we define housing affordability that includes transportation and energy costs? Most urban experts believe the answer is in our downtowns, where energy and transportation costs can be lowered. In the past, it was simply measured as someone who pays 25 percent or less of their monthly income into housing costs (President Reagan later raised this to 30 percent of their income). Using this percentage of income-to-rent, study after study showed that half the poor and a large percentage of moderate-income house-holds had unaffordable housing. Affordable housing is more prevalent farther from downtown, yet the higher costs of transportation and energy consumption are not factored into the affordability equation.

Rather, the definition of affordable housing must be expanded to include both transportation and environmental costs. That is, for an affordable housing development to be sustainable, it is dependent upon transportation costs and energy costs. This chapter shows how River's Edge is today's ideal low- to moderate-income housing development. The Housing Authority of Covington (HAC) is productively working toward healthier, more affordable, and more equitable neighborhoods. This is illustrated by summarizing HAC's efforts in sustainable environmental design.

The HOPE VI program involved the demolition of the old Jacob Price public housing community and replacing it with a new urbanist-style, mixed-income community. HAC realized the need to address these issues, and in 2010 entered a partnership with Integral Development, LLC to replace the Jacob-Price development with River's Edge at Eastside Pointe. Several "sustainability" frameworks are being implemented, including Leadership in Energy and Environmental Design (LEED), Enterprise Green Communities (Enterprise), and Energy Star-approved appliances. For a public housing development, this is a productive step in the direction of sustainable neighborhoods. Still, there is room for critique that supplements efforts for increased sustainability in physical design. The core of the critique falls on the LEED model; we explore how LEED supports or deters healthy, afford-able, and equitable communities.

River's Edge will be LEED Certified and is also pursuing certification at the Silver Level, making it one of the only affordable housing developments in Kentucky that has health and sustainability in its plans. Trends show that wealthier people and groups seek LEED certification, as opposed to lower-income groups (Beidelman, 2011). River's Edge represents a small step in reversing this trend: a major accomplishment in the sustainability paradigm. The accomplishment does not go unnoticed, yet more could be done to support sustainability in Covington. We recommend a more holistic approach to thinking about sustainable neighborhoods by decreasing LEED's monopoly on the sustainable design movement.

River's Edge at Eastside Point

The River's Edge development model illustrates a case of how to reduce energy costs both inside and outside of the home. For example, locating the development in a walkable neighborhood ensures a person has the ability to access their needs, despite any transportation-related deficiencies.

Walkability reduces car dependency and provides active transportation infra-structure options for walking, biking, and bus usage. It is a measure of vegeta-tive design (e.g., tree-lined streets) in that aesthetic sidewalks and bike lanes encourage active transportation. River's Edge is far ahead of other HOPE VI developments in many aspects, namely, those that are located far away from institutions of higher education, gainful employment, and bus and bike lanes. Most importantly, car dependency is less prevalent for lower-income citizens; thus, sustainable design must incorporate alternative transportation methods in an increasingly car-dependent society.

Urban Green Design: Enterprise, LEED, and Energy Star

The new urbanist green design attacks the stigma of conventional public housing and incorporates a pleasant ambiance into respective neighborhoods (e.g. bike racks, community gardens, better insulation, trees, modern archi-tecture, attractive accommodations, and many other features). Covington's HOPE VI offers a new start for public housing tenants who desire a lift out of poverty with opportunities to improve employment, housing, education, and health. HOPE VI takes blighted developments and turns them into new, energy-efficient, beautifully landscaped housing neighborhoods.

The LEED Evaluation Team has claimed that River's Edge has the poten-tial to reduce average energy costs to $65/month. This is important since HAC public housing tenants do not currently pay utility charges to the utility company. We agree that having tenants pay for energy will make them more frugal in their energy consumption. Proper insulation in walls, ceilings, and floors, as well as Energy Star mechanicals, will cause significant reductions in energy usage. For example, ceiling fans reduce the need for heating and cooling and can create a reduction of up to five degrees in air temperature. The final LEED evaluation is planned for April 2015, and HAC's goal is to receive LEED Silver certification.

Many cost-saving and ecological methods of building design can have indirect effects on health and can lead to market shifts with their programs. Although the recommendations are voluntary and based on developer pref-erence, they are becoming highly visible in the construction world, as they recognize the impact of physical design and the green environment on human health. Site selection for new structures must be made in a manner sensitive to the ecosystem. Buildings should be designed using footprints that minimize sprawling land use while improving the well-being and fitness of building occupants. LEED certification offers "credits" for both density and mix of uses, and improved indoor air quality, temperature control, and ambient light-ing. The program also encourages the simple inclusion of showers, changing rooms, and bike storage.

A well-designed community is a healthy community because regular physical activity such as walking, running, or cycling becomes desirable. Similar in purpose to LEED, the mission of Enterprise Green Communities (Enterprise) is, "To create opportunity for low- and moderate-income people through affordable housing in diverse, thriving communities." River's Edge is required to conform to Enterprise, which specifies the need for conjunctions of green design and affordable housing. To operationalize the requirement, Enterprise is structured around a certification and "credit," or "point" based system. Specifically, an Enterprise community must meet all mandatory affordable housing criteria as well as achieve an additional 35 points. The structure of Enterprise is similar to LEED's structure: meet several of the criteria in the design and construction stages and the development becomes "sustainable."

LEED and Enterprise serve distinct functions to provide a framework for sustainable, green-building design. The two organizations are distinct in their practice, yet LEED is generally cited as the higher-level certification program. Thus, most of the research focuses on LEED. Before explaining the basic structure of LEED certification, as well as HAC and River's Edge at Eastside Pointe's role in certification, it is important to understand the process by which credits are approved and codified. All people can "propose a credit" through the USGBC online interface. The USGBC considers these proposals–and creates their own–at which point the proposal is reviewed at three levels: the technical advisory group, the board of directors, and the steering committee. The last step is for the USGBC members to vote to accept or reject the proposal.

LEED certification is categorized into four levels based on the number of points a project achieves: LEED Certified (45 points), LEED Silver (60 points), LEED Gold (75 points), and LEED Platinum (90+ points). Because River's Edge is not complete, we are unable to report on the development's level of certification; the goal of Silver is plausible, but the construction team should double-check LEED's minimum requirements for tobacco control to ensure their certification projection is legitimate (see chapter on sustainable physical design below).

Energy Star specializes mostly in energy-efficient technology design for residential appliances (e.g., refrigerators, ovens, washers/dryers, etc.). The appliances installed at River's Edge include washing machines, dishwashers, refrigerators, ceiling fans, light fixtures, HVAC systems, and windows. The list excludes incidental fixtures in service areas, such as attics and mechanical rooms. The use of Energy Star-qualified products reduces the electricity bills for residents while offering higher-quality products. On the construction side of things, LEED evaluators predict that average energy costs per unit will reduce to $65/month. Monthly savings of this nature would exceed expectations. Our research is unable to substantiate these claims until, at a minimum, the development is complete.

Improving LEED Guidelines: Emphasizing Nature over Mechanicals

That said, LEED is not the be-all and end-all for environmental design. A LEED-certified development or neighborhood suggests the essence of sustainability; still, there are items that LEED misses that reduce energy consumption.

While LEED provides extensive benefits in sustainability, health, equitability, and design, we can identify missing elements of sustainable design. A few examples include expanding the interplay between city policy and solar-oriented development, simple mechanics that reduce energy consumption as opposed to Energy Star products, comprehensive effects of well-planned landscaping, and scale.

Neither does LEED catalog their methodological approach to weighting various criteria. It is a fairly subjective process—not based on science, but on the needs of the industry to consume products that are energy efficient.

Most notably, because LEED is bound by its focus on engineering solutions that consume electricity, rather than harnessing nature (e.g., wind, sun, and trees) to reduce energy costs, LEED gives minimal credit for recycling an old building, which is also a notable sustainable practice. For example, the new development does not receive equal credit for reusing the brick waste of the old Jacob-Price Homes. When the old units were demolished, the dump truck drivers were asked where the bricks were being taken; we were told the bricks were being placed along the riverbed. Several palettes of the old bricks have been saved, and there are plans to use them in constructing the gazebo and memorial garden for Jacob Price Homes. After all, the former units provided Covington residents with public housing for 75 years.

Given the past problems of Jacob-Price housing that were polluted with toxic carcinogens, it is ironic that LEED gives minimal credit for banning smoking. As of the latest version, "environmental tobacco smoke control" is required and is, therefore, not even counted as a credit. This is important to our critique because, formerly, a development could be LEED maximum certified, replete with chain smokers in all units; whereas now, a building with smoking allowed in the units cannot become certified, regardless of how many other credits it receives. Furthermore, there are ambiguities and deviations from the smoking requirement that introduce confusion into its purpose. All to say, LEED is an excellent and productive model to follow for environmental design, but it does not capture the complete spectrum of environmentally healthy and sustainable design over time and space.

Nature versus High-tech Engineering

We invariably acknowledge—while quantitatively unknown—the benefits that have come from Energy Star products, as well as LEED's partnership and advocacy of Energy Star products. Yet, one element of energy savings and sustainability that is exceedingly difficult to measure is behavior. The vast majority of people will flip the light switch on regardless of whether the light bulb is Energy Star approved or not. Still, the question remains: why incentivize energy waste via LEED credits and Energy Star? In fact, the appliances do not require usage at all; despite the use of energy-efficient products, energy is still used and contributes to energy waste. Simply put, why give credits for clothes dryers that waste energy, for example, instead of giving credits for clotheslines that harness nature?

Many private communities have banned the use of clotheslines, citing concerns over aesthetics and property values. Energy Star's own research indicates that electric clothes dryers account for 6 percent of *all* household electricity use. Utilizing clotheslines reduces the carbon footprint by 4.4 pounds for all laundry loads done in one day (Ball, 2009). The point being that Energy Star products are great to a certain extent, but they are not the final say in energy sustainability, and use of their products still results in burning fossil fuels, although in smaller quantities. It would be in IPM's favor to incentivize and supply clothesline use by providing the necessary infrastructure.

Harnessing the Solar Resource

The solar resource is underutilized and misunderstood by most in the development industry. Harnessing the sun's energy falls into two primary categories: passive solar and active solar. Several implementation methods and technologies exist for each category, but they all relate to either (1) designing a structure to utilize irradiation *as is*, or (2) capturing and transferring the irradiation into a generated form of energy (e.g. electricity and heated liquid). Aside from the harnessing technique used, solar energy is captured at various scales—from a single residential unit to solar farms that reside on thousands of acres.

Unfortunately, a great deal of passive solar benefit is already forgone, as the construction at River's Edge nears completion. Designing for the utilization of the sun in a passive manner has tremendous benefits and paybacks over the life cycle of the housing unit. All that is required is a devoted planning team during the design phase. The City of Covington could have played a preventative role in this issue, and it is something that HAC and city decision makers should work toward for future development. Cities, through their policing power in zoning, have the legal capacity to dictate that all new developments should be solar-oriented, and the city gets to define "solar-oriented development." The power and effect of such a policy enable residents to save money on energy bills, but most importantly, it mobilizes a person in their ability to harness the solar resource for the future. That is, a development that is not solar-oriented cannot be retrofitted to become solar-oriented without demolition. Retrofits for expanded window surface area can be accomplished, but the baseline of the structure cannot be shifted to increase its sun-angle. If the City required solar-oriented development, it would eliminate the need and burden of "policing" sustainability design elements from HAC's shoulders. It is consistent with HAC's broader interests to advocate for a solar orientation policy with the City Council, and we encourage HAC to engage in such advocacy.

Active solar is a different ball game. It is fair that HAC has valued the benefits of energy savings in favor of Energy Star appliances and against, for example, solar photovoltaic (PV) installation. Solar PVs are not cheap, but HAC has managed to install PVs on the main Community Building. Prices for PV installation have fallen significantly over the past several years and are now a viable option for long-term, financial, and energy savings. HAC expects significant energy savings because of Energy Star products; still, savings could be extended in the long term with the complete installation of solar PVs—especially because Kentucky permits net metering (getting paid for excess energy sent to the grid).

Landscaping: Creating a Healthy and Vibrant Community

The original Jacob Price Homes development had no landscaping other than turf and several trees on the property. Landscaping is another critical component of creating a vibrant community. Landscaping serves to attract native wildlife, beautify neighborhoods, raise property values, and encourage residents to interact with their environment by drawing them outdoors.

Research indicates that landscaping improves the health and well-being of people when properly executed. A report titled *Public Health and Landscape: Creating Healthy Places* explores the dimensions in which landscaping has a positive impact on health and community life. In five principles, they state that *Healthy Places*:

1. Improve air, water, and soil quality by incorporating measures that help us adapt to, and where possible mitigate, climate change.
2. Help overcome health inequalities and promote healthy lifestyles.
3. Make people feel comfortable and at ease, increasing social interaction and reducing antisocial behavior, isolation, and stress.

4. Optimize opportunities for working, learning, and development.
5. Are restorative, uplifting, and healing for both physical and mental health conditions.

HAC and the developers have included landscaping in the comprehensive development design. Before construction began, the site had to be clear-cut due to site-grade issues. The presence of trees is crucial to sustainable development for several reasons; to mitigate the loss of the existing trees, new trees are being planted along sidewalks, roads, and around the new buildings (see figure 15.4). The trees will improve soil and water quality, decrease air pollution, provide added drainage, cool the surrounding area (thereby reducing energy bills spent on air conditioning), provide outside shade, minimize the heat island effect, encourage walking and biking during the summer months, and increase property values in surrounding neighborhoods.

The River's Edge landscape plan calls for the planting of non-invasive, drought-tolerant plants and the creation of green spaces for resident use and enjoyment. Drought-tolerant plants are a good choice for landscaping because they require less water, are more resistant to pests and disease (cutting down on the use of herbicides and pesticides), and in general require less maintenance than non-drought-tolerant plants. Drip irrigation will be used on less water-intensive plantings to conserve water.

Figure 15.4 HOPE VI Housing in Covington Facing the Street Using New Urbanist Principles. *Source:* Image by John Hans Gilderbloom.

At River's Edge, the landscape design will be park-like, with benches, a gazebo, a playground, and outdoor grills. These social spaces will encourage physical activity and social interaction between residents, and provide a safe place for children to play. Green spaces will encourage residents to explore their surroundings and interact with each other.

Landscaping is often overlooked as a component in creating vibrant communities; however, it is one of the most cost-effective methods of enhancing the value of homes and improving the quality of life. In this regard, the River's Edge development demonstrates a strong commitment to creating a beautiful neighborhood that will provide economic, environmental, and social benefits.

Sustainable Drainage Systems

Many parking lots and roadways are designed with useful gutters and storm drains to manage rainwater runoff, but this method of stormwater control also has many downsides. Rainwater mixes with various harmful chemicals and pollutants, carrying them into sewers. When sewers overflow into streams and rivers, the pollutants contaminate waterways. A more sustainable option is directing drainage toward quasi-natural features (e.g., bioswales, rain gardens, and rainwater cisterns). Rainwater runoff collected in bioswales and rain gardens infiltrates the soil, recharging groundwater and removing pollutants, and rainwater captured in cisterns can be used for irrigation of the landscape. Bioswales were recommended in the baseline report along with rain barrels.

A garden located at River's Edge would provide physical activity suitable for all ages and allow residents to socialize. There is evidence that suggests community gardens yield more than just fruits and vegetables. Research conducted by the Colorado School of Public Health and Denver Urban Gardens indicates that those who garden tend to be healthier and are involved in more social activities. Community gardens also stimulate citizen engagement. Workshops would allow residents to not only learn how to grow their own food but also prepare delicious and healthy dishes using it.

Additional Recommendations & Insights

There are few examples of affordable, energy-efficient housing developments that are not car-dependent in the United States. Rivers Edge might be one example of housing that is sustainable not only because it is energy efficient and affordable, but also because of where it is located—near downtowns and transit nodes. As a model for affordable and sustainable housing, Rivers Edge has many positive qualities. Still, there is room for improvement that would help to reverse the stigma of public housing.

- *Environmentally preferred products.* For all future construction/renovations, we urge the use of environmentally preferred products that reduce the developer's carbon footprint. These include locally manufactured, low-emission, and environmentally friendly products/materials. Granite, for example, is often used in housing development projects. It is shipped from Africa or India using "slave" labor or child labor. Countertops are developed here in the USA and can be more attractive than foreign materials like granite. We commend HAC and IPM for not using foreign materials in the development. A great organic product for kitchen and bathroom countertops is using a product from Louisville Tile called pebble countertops, which are extracted in the United States, and easy to install, with 12" by 12" tile that costs only $17 a square foot. Thus, the normal cost of a granite countertop that is granite is $2,000 as opposed to the cost of pebble countertops at $450 for the same job.
- *Rooftop gardens* could have been incorporated into the design as well. Not only are landscaped rooftops attractive, but they are also sustainable in terms of filtering storm water and temperature control. Traditional roofing materials, whether light or dark, absorb and radiates energy, contributing to the urban heat island and increasing energy costs via air conditioners. Constantly running air conditioners also contribute to the heat island. Rooftop gardens are also appealing to all income groups, which is important when competing against other middle-income developments (see figures 15.5 and 15.6 for sustainable IPA development in Indianapolis).
- *Dog parks* contribute to sustainable neighborhoods via aesthetics, safety, and increasing property values. They also present a welcoming environment for dogs, which could be important for health as it has been shown that dogs, as companions, correlate with decreasing blood pressure. The presence of dogs can also decrease crime and robbery. The presence of dog parks appeals to residents who want to stay for longer periods of time—a key element of sustainability (see figure 15.6 for IPA development).

Figure 15.5 Dog Park that Follows New Urbanist Principles. *Source*: Image by John Hans Gilderbloom.

Figure 15.6 Community Gardens Should Be in Every Affordable Housing Development.
Source: Image by John Hans Gilderbloom.

Areas for improvement not addressed by LEED or Developers:

• *Reflective surfaces.* Dark-colored materials absorb more heat than light-colored materials and release it throughout the night, contributing to the urban heat island effect and higher energy costs. By converting dark-colored tiles to light-colored shingles on roofs and using light-colored materials for sidewalks, parking lots, and roadways, more sunlight is reflected back, reducing temperatures in the surrounding area and lowering AC costs during the summer. Several experts (including the co-author of this report) have found that it can lead to a 20 percent reduction in energy costs. Yet, LEED only gives one point toward certification for light-colored roofing, and River's Edge has black roof tiles instead of the recommended white tiles. Black tiles encourage energy waste. (See figure 15.7.)
• *Permeable surfaces:* Instead of asphalt parking lots, we urge the developers to use porous concrete for all paving. Not only is it permeable (allowing stormwater to seep through it), but it also has a high albedo (solar reflectance level), reducing the urban heat island effect.
• *Beautiful buildings* are rarely demolished. River's Edge is by all measures beautiful and fits into the fabric of the community. These buildings are preserved and reused and transformed for different purposes. They are rarely demolished because their beauty holds value in the community. However, if buildings are ugly, dirty, and look like prison cells, they will likely have

Figure 15.7 An Example of a Home with Energy Efficient White Roof Tiles that Reflect Back the Sun Rather than Absorbing It. *Source*: Image by John Hans Gilderbloom.

a short life. Building demolition is a waste of resources and not an example of a sustainable community.

- *Protected and artistic bike racks.* Weather-protected bike racks should be installed, with an optional surveillance camera to deter thieves. Promoting biking for recreation and transportation will make River's Edge a more vibrant, healthy community. (See figure 15.8.)
- *Protected bike lanes:* Bike lanes should be created with a formal barrier between the rider and traffic. This will make biking safer in the Eastside Neighborhood and promote the community as being not only walkable but bikeable, too. Bike parking will make the unit more attractive to renters. LEED again gives nearly zero credits for this.
- *Roof rainwater management:* Early on we had recommended that roof guttering be redirected from sewers to the vegetation and community garden. We also recommended rain barrels for the development. This was not done.
- *Reducing heat costs with fans.* Kentucky housing laws require all new housing developments to have fans installed in the bedrooms. Fans are known to reduce the temperatures in a room by five degrees and increase temperatures by two degrees during winter. In the South, before air conditioning was used, fans and high ceilings were installed to cool homes. River's Edge included these items in because they were part of the housing building code law and part of Southern culture. LEED does not give credits for this, and, in fact, a LEED representative said she was against fans and high ceilings.

Figure 15.8 Bike Rack Provides Parking for Eight Bicyclists in Place of One Car. *Source*: Image by John Hans Gilderbloom.

- *Tall ceilings.* In conjunction with fans reducing energy use, tall ceilings (nine-foot minimum) help reduce room temperatures. Warm air rises and keeps lower areas cooler, thus also reinforcing less energy use. LEED is neutral on ceiling height, but tall ceilings are a proven way of lowering energy costs. High ceilings also give a look of middle class and are consistent with a culture of Southern charm.
- *Three-story face of the building lining the sidewalk/street.* We disagree with some officials who wanted the front of the building to be pushed back to have a large parking lot there instead of the building. Research shows that there will likely be less loitering, crime, and vice because of the absence of a parking lot. LEED-ND provides points for building frontage that is adjacent to and facing the street. This is good new-urbanism practice. Balconies are absent from this green development. In Amsterdam, the policy is that every new unit built must provide a balcony because of its many benefits. Balconies would have made River's Edge more attractive to moderate- and middle-income renters. There are several important reasons for this: (1) it helps reduce crime by having people on a porch observing what is going on in the setting; (2) it provides a space for crafts, art, hobbies, painting, gardening, bike storage, a small barbeque, and growing plants (Mother-in-law's tongue) that can be used to improve the air quality of an apartment; (3) it creates additional space that does not require additional energy usage; and (4) it creates greater socialization among residents and a feeling

of we-ness. Many high-end apartment units have found a high consumer demand for balcony space.

• *Traffic calming.* Traffic calming devices such as two-way streets, bike lanes, and tree planting have been shown to reduce crime, increase business, increase housing values, and decrease housing foreclosure as we later document (see Louisville case study; Gilderbloom et al., 2014). Slower traffic increases walkability and pedestrian traffic, which facilitates a prosperous business environment. Both the mayor and the police chief of Covington support traffic calming because it reduces crime. They do not want to revisit the past where one-way streets facilitated prostitution, violence, and drive-by shootings (see figure 15.9). A key to green residential community development is turning chaotic multi-lane one-way streets into calm two-way streets with bike lanes and narrower traffic lanes. Community participation

Figure 15.9 One-Way Streets Should be Converted to Two-Way Streets to Reduce Disinvestment. *Source:* Image by John Hans Gilderbloom.

in this effort is also key. One unanticipated finding was that contrary to the conventional claim made by traffic engineers, traffic counts increased on the streets that were calmed. This is because of the perception that it was a safer drive, and some prefer a pleasant drive to racing. There are more businesses and thus more customers during morning, afternoon, and evening commutes. But is this old historic neighborhood the exception or the rule? This new study expands our research by looking not just at one neighborhood but 190 of them in the same mid-size city of the original study, and we more fully measure how traffic calming approaches such as lower speed limits, speed bumps, walkability, bike lanes, and narrower streets might positively impact urban living.

We use mixed methods in our analysis including qualitative interviews with neighborhood leaders that pushed and opposed traffic calming. Standard regression analysis was also used to produce models estimating which traffic-calming measures appear to be effective in these environments. This paper contributes to an understanding of how to make neighborhoods more livable and sustainable. LEED does not give much credit to buildings placed in areas that don't require cars because the neighborhood is walkable, bikeable, or near public transit. That is a leading reason why New York City's carbon footprint is one-third the size of car-dependent Atlanta.

Sustainable Neighborhoods: Complete Streets and Neighborhoods for All

The city should build or improve sidewalks within and around the perimeter of the development to ensure easy pedestrian access to surrounding employment, educational, cultural, and recreational amenities. Additionally, the city should add bicycle lanes to make it easier for those who own bikes to use them for commuting to work or school. Finally, since only 20 percent of Eastside's public housing residents own cars, the number of parking spaces allocated for the new development may be reduced to incorporate more green spaces and bicycle parking. The city may choose to allocate nearby street parking for residents only. This would increase the amount of green space while limiting the number of outsiders coming into the community.

A growing body of work shows that walkable neighborhoods have intrinsic economic value created by encouraging economic transactions and social exchanges (Litman, 2003, 2011; Leinberger and Alfonso, 2012) and bolstering real estate property values (Cortright, 2009; Diao and Ferreira, 2010) in addition to promoting health benefits (Meares, 2014). A new and useful tool for determining walkability is Walk Score: "Walk Score measures walkability based on distances to nearby restaurants, grocery stores and other amenities, plus other analyses of pedestrian friendliness" (Cortright, 2009). Although street quality variables are important to consider, the Walk Score methodology does not take these variables into account. River's Edge at East Pointe (1044 Greenup St.) has a Walk Score of 77, meaning it is Very Walkable, and that most errands can be accomplished on foot. Additionally, River's Edge has a Transit Score of 48, which means it has some transit, and that there are a few nearby public transportation options.

A walkable area has a greater potential for spillover effects, especially as it relates to economic transactions and housing values. Recent surveys by the National Realtor Association have discovered the desire for neighborhoods that are conducive to walking has trumped traditional housing preferences, creating a high demand for a product that has low availability (Leinberger and Alfonso, 2012). Furthermore, neighborhoods that are considered walk-able encourage more economic transactions. Individuals who use alternative modes of transit are more likely to spend more money on groceries, eating at restaurants, and shopping per month than those who use automobiles (Swanstrom, 2012), though users of alternative transportation modes tend to spend less per trip. Additionally, Leinberger and Alfonso (2012) concluded from their study of Washington D.C. that walkable neighborhoods perform better economically.

River's Edge will incorporate the following pedestrian-friendly features, which we expect will bolster its Walk Score. These design features will encourage residents to walk, creating a pedestrian-oriented community.

• Continuous sidewalks linking apartments with amenity areas
• Buffer zones between sidewalks and the street (i.e., grass and trees)
• Trees and other landscaping to provide shade and visual interest
• Sitting areas to encourage older residents to walk

Sidewalks will have curb cuts and safer crosswalk designs, which are American Disabilities Act (ADA) compliant. Another option that allows residents to live car-free is public transportation. Transit availability at River's Edge will allow residents to choose between walking, biking, and riding the local bus. The Transit Authority of Northern Kentucky (TANK) indicates that all their buses are wheelchair accessible, which ensures all residents of River's Edge have the option of using public transit. Inclement weather affects the use of modes of transportation; bike riders and walkers will tend to use public transit more, while drivers will tend to use it less during inclement weather. Covered bus stops can help retain riders that might drive otherwise, and capture bike riders and walkers as well. Currently, it is not clear whether a covered bus stop will be provided for residents of River's Edge.

With the new Licking Greenway & Trails running through the Eastside Neighborhood just a short distance away, the lack of cycling infrastructure within River's Edge is a missed opportunity. Not only is biking a healthy form of recreation, but it is also a viable means of transportation. Residents of bike-friendly neighborhoods could potentially save up to $8,000 per year by ditching their cars and commuting by bike (The Sustainable City and Biking: Implications for Health, Environment, and Economy). The money residents save on car-based transportation would likely be spent in the Eastside, boosting the local economy.

Some argue that bike riders should use the sidewalks in the absence of dedicated bike lanes and other bike infrastructure; however, Watchel and Lewiston (1994) suggest that riding on the sidewalk is twice as dangerous due to poor line of sight between the road and sidewalk. Yet, the motorist poses the most threat to bike riders; the risk of fatality increases as motorist speed increases, rising to 80 percent at speeds of 40 mph (Pucher and Dijkstra, 2003). Simply put, bike riders need the safety of bike lanes. By providing appropriate bicycle facilities, River's Edge can market itself as a bike-friendly community, encourage more residents to ride their bikes, and promote the health of its residents.

Designing Homes for the Elderly and Disabled

Livable communities are those that consider and address the basic needs of all people in the city. Planning for the needs of the future is at the crux of the concept of sustainability. Yet, LEED is weak in addressing the needs of all people, whether elderly or disabled. This is generally applied to the needs of future generations, but it is highly relevant to the elderly and disabled. Thus, sustainability in physical design must convey a welcoming environment for the gamut of social groups. Both the elderly and disabled require basic infrastructures (e.g., ramps, railings, access to cabinets/closets, an additional bathroom, kitchen, etc.) that, when not available, can deter participation in employment, health care, shopping, and social and recreational activities (Gilderbloom, 2008). In fact, the majority of the elderly and disabled require grab bars and ramps in their homes to assist with mobility—it is predicted that by the time a person reaches the age of 85, 75 percent will require these basic structural requirements (Ibid., 2008).

It is important to note that elderly and disabled-infrastructural requirements extend beyond HAC's jurisdiction. The City of Covington should ensure a safe environment by improving transportation-related infrastructure (e.g., sidewalks, curb-cuts, and bus shelters) in line with the Access Board's Right of Way guidelines for the disabled as well as planned zoning policies that locate parks, recreational areas, shopping, and employment options in proximity to elderly and disabled residences. Such policies, in turn, result in safer environments concerning violent crime—while the elderly and disabled may not be at greater risk for violent crime, they self-perceive the inability to react effectively, which hinders their contribution to society and the economy (Ibid., 2008).

Architectural and Location Needs

"Aging in place" is a senior movement to create and design housing for seniors so they can live out their lives in a familiar setting filled with familiar family, neighbors, and medical maintenance personnel. The phrase refers to living where you have lived for years, not typically in a health care environment or nursing home, using products, services, and conveniences which allow you to remain in your home as your circumstances change. In other words, "You continue to live in the home of your choice safely and independently as you age" (Aging in Place, 2014). People being forced to move out of their homes because of a developing condition of one kind or another can create physical and mental traumas. Experts have long agreed that a person who is forced to move from his or her neighborhood can undergo considerable psychological stress from the loss of community or reference groups (Harvey, 1973, 82–86; Fried, 1963; Wechsler, 1961). Depression is correlated with sudden shifts or changes in social support networks, and this may be particularly true for lower-income groups. According to Harvey (1973, 85), "low income groups . . . often identify closely with their housing environment and the psychological costs of moving are to them far greater than they are to the mobile upper middle class."

Place matters. But place is not just about location in a city or a neighborhood (see Dreier, Mollenkopf, and Swanstrom, 2004). Place is also about the kind of housing in which we live and how it shapes us as people. At River's Edge, six of 120 units will be built to the Uniform Federal Accessibility Standards (UFAS) and three will be built to suit the visually impaired. Additionally, all ground-floor units (roughly 50 units) will meet the requirements of the Fair Housing Act. The UFAS-compliant homes will have grab bars installed, elevated toilets, lowered kitchen and bathroom cabinets, and areas for wheelchairs to move under countertops. All other ground-floor units have blocking installed for any prospective resident that may require elderly and disabled infrastructure installation in future years. At this stage in the development, however, it is unclear whether residents will be responsible for the installation costs. Overall, portions of the development are adequately suited for disabled and elderly residents. Ready-to-install infrastructure is a critical component for all ground-floor units. Until the development is complete, we are unable to fully evaluate the environment for elderly and disabled residents.

Mayor Carran of Covington told us the low number of accessible units at River's Edge disappointed her, and we agree. We recommend that 10 percent (not 5 percent) of the units be made accessible, and senior units should all be designed with grab bars, elevated toilets, lower cabinets, and ramps so residents can live in place. Six units are designed for disabled residents; this includes extra space and railing for wheelchairs. All seniors will eventually need bars and ramps to move around in their housing. Moreover, physical disabilities increase with age, and more so among low-income persons. There are too few independent living homes at River's Edge. All the senior units and first-floor units should be accessible to the disabled.

A substantial number of elderly and disabled persons require architectural modifications in their homes. Roughly one-third of Covington's seniors and more than one half of the disabled needed grab bars in their homes. At least twenty percent (20 percent) of the elderly and disabled wanted ramps placed in their homes. Not surprisingly, the desire for ramps and rails increases with age and the severity of disability. The desire for ramps and rails will double as elders move between the 60–65 age group and the 75 years and older category. Our research found that for seniors over the age of 75, almost one out of three desired ramps and over one-half wanted rails. For persons with severe disabling conditions, one-third needed rails and two-fifths required ramps.

Inaccessible cabinets, closets, and bathtubs further amplify the problem of architectural barriers within the household. Even with the provision of stools, one out of three disabled people and one out of seven elderly people were unable to use cabinets and closets within their homes. Bathtub grab bars are installed in the six handicap-accessible units, but the tub itself is difficult to get in and out of with the standard height of the tub (see figure 15.10).

One out of every 20 residents had a disabling condition. One out of every 10 Covington residents requires special architectural modifications in their homes to enable them to have complete access to all parts of their home. Close to one-third of the disabled, and 15 percent (15 percent) of the elderly, could not use cabinets and closets in their own homes. Outside of the home, substantial transportation and environmental barriers often prevented the disabled and elderly from participating in the economic and social life of the community.

When away from their residences, the elderly and disabled desire certain amenities that foster greater mobility. Over three-fourths of the elderly and disabled have indicated that the presence of sidewalks and first-floor locations is important. Persons with a severe disability strongly desire a first-floor location. Location is a major issue for elderly and disabled persons when considering a residential move. In general, our research indicated that a significant number of disabled and elderly persons want the amenities found inside the city. Close to one-half wanted to be located near Metro bus stops, and over

Figure 15.10 Grab Bars Prevent Dangerous Falls and Should be Mandatory in Affordable Housing. *Source*: Image by John Hans Gilderbloom.

three-fourths wanted to be near medical services. The desire to be close to medical facilities increases with age and the severity of disability. Eighty percent (80 percent) wanted to be near shopping areas. Two-fifths of the disabled and one-third of the elderly indicated that being close to work and near a public park were important in their decision to choose a residence. A final important factor for more than eight out of every ten persons interviewed was proximity to family.

Indoor Air Quality: The Issue of Tobacco Smoke

Jacob Price was evaluated by Johns Hopkins University; they found that the housing had dangerous and deadly carcinogens from first, second, and third-hand smoke. Smoking was a major reason why Jacob Price had to be torn down because of all the carcinogens found in the units—even when empty. The high rates of smoking in Kentucky are part of the reason why Kentucky has one of the shortest healthy lifespans in the nation at 62 years of age. We get conflicting reports on whether or not smoking will be allowed or tolerated at River's Edge. Initially, there was a push for a no-smoking policy. Unfortunately, after a lot of uncertainty, we have learned that the developers will not ban smoking. HAC does not own River's Edge; rather, the developers (IPM) have the final decision, and a no-smoking policy is not advantageous to their business model. While HAC is striving for sustainability in design—and working toward obtaining LEED Silver certification—this single variable that the developers control may prohibit River's Edge from receiving any LEED certification at all, as there are requirements for non-smoking environments. HAC has instituted a smoke-free policy in their new leases for residents (see Appendix III), but this will not hold back the decreasing values of the development from an overall smoking-allowed policy.

Because tobacco smoke kills and maims, it affects more people than just the smoker; it also negatively affects future tenants and current neighbors, especially young children and the elderly. Studies in 2004 by the Urban Institute estimated that 25 percent of HOPE VI households, double the national estimate, had children with asthma. Again, with more than 9 percent of adults having asthma, Kentucky is among the top-ranking states with a high adult asthma population (Behavioral Risk Factor Surveillance System, 2009).

Our prior survey found that Jacob Price residents were against restrictions on indoor smoking; however, we believe it would be a prudent decision for Integral to eliminate smoking indoors and outside in public spaces. The dangers of smoking have been well established for decades, but a large body of research has also concluded that second-hand smoke exposure diminishes the health of non-smoking adults and children. It has been proven that smoke and its harmful effects can travel through floors, ceilings, and walls, causing additional health hazards. In addition, emerging research on the residue of tobacco smoke that lingers in a room long after smoking has occurred indicates an additional health hazard known as third-hand smoke. Walls, ceilings, and carpets absorb the harmful particulate matter from cigarette smoke, and chemical compounds are created that continue to emit dangerous toxins even when smoking has stopped. Further, there is evidence that when smoking occurs indoors, it decreases the value of the home. Eliminating smoking in the development would also decrease the risk of fires and fire-related deaths and injuries since cigarettes are the number one cause of house fires.

A recent Harvard study, which examined 49 low-income multi-unit residences, detected nicotine in 89 percent of non-smoking homes studied (Schoenmarklin, 2010). The article also concludes that eliminating smoking in public housing units positively impacts the bottom line. Smoking is the leading cause of fires in multi-unit housing, and these fires kill more people than any other type of fire. Additionally, prohibiting smoking indoors saves on cleaning and maintenance costs and helps Integral Property Management (IPM) save money for other projects that can improve living conditions for their tenants.

Another study conducted by UCLA estimates that California Apartment Association members (n = 343) could save up to $18 million a year if they were to ban smoking on their properties (Rivero, 2011). In closing, Schoenmarklin (2010) states,

> We owe our elderly, our children, and those with chronic illness a safe shelter that does not include secondhand tobacco smoke. When we adopt a smoke-free policy, we have a rare opportunity to do what is right while saving money and preventing lawsuits.

It is not only legal to prohibit smoking in public and HUD-assisted housing, but it also protects against lawsuits from tenants exposed to secondhand smoke or harmed or killed by fires caused by smokers.

In short, tobacco smoke is a known health hazard and is detrimental to the well-being of both smokers and non-smokers. By enacting a smoke-free policy, a positive difference can be made for residents' health and the developer's bottom line at River's Edge. We recommend that the Mayor and City Council consider an ordinance that bans smoking in housing developments of at least 100 units or more.

CONCLUSION

This is an example of what is possible for a livable urban environment. New urbanism provides a vision for creating sustainable, prosperous, livable, healthy, and safe communities. At the last Congress for New Urbanism (May 15–18, 2024) in Cincinnati, Ohio, Andres Duany, the principal advocate of designing Hope VI as a new urbanist community that would last forever with walkability, bike-ability, community gardens, a variety of incomes, and good density, said Hope VI created over 186,000 affordable homes. The late Marilyn Melkonian of Telesis Corporation seized on this opportunity of a partnership between government and private developers and built 16,000 Hope VI-like homes that were colorful and beautiful for an investment of $2.5 million.

There is much to learn from this model of green development in terms of how to reduce energy costs in the home. Moreover, there is another important lesson to learn as well: locate the development in a walkable neighborhood that has everyday needs within a 10-minute walking distance. Walkability reduces car dependency for seniors and provides active transportation infrastructure options for walking, biking, and bus usage. Tree-lined streets with sidewalks and bike lanes/paths also encourage active transportation. River's Edge is far ahead of most other HOPE VI developments, which are car-dependent and located far away from higher education, jobs, bus networks, and bike lanes.

In an era of climate change, building, siting, and designing sustainable, affordable homes work to the advantage of residents, neighborhoods, cities, and the world. The United States has only a handful of multi-family housing communities that are affordable, environmentally friendly, and energy efficient. Covington, Kentucky's HOPE VI development provides one of the best examples to date on how to design energy-efficient, affordable developments—both inside the home and in locating the development in a place that reduces car dependency by encouraging walking, biking, and public transit. This chapter illustrates both the highlights and shortcomings of Covington's HOPE VI green development. The demolition of the original public housing was wise because it was dysfunctional and had physical design flaws, cancerous toxins, and an unsafe environment. We have illustrated measures to improve the quality of life with green infrastructure by exploring the three environmental/energy programs used in the development: LEED, Enterprise, and Energy Star. While we acknowledge that LEED is a valuable force in teaching sustainable design, we also note its drawbacks by giving marginal credit for "old school" sustainable design. Engineering products that consume electricity are favored over solutions found in nature that don't consume electricity. Harnessing the sun's energy can do what many of the recommended mechanicals can do: drying our clothes, cooking our food, heating our homes, and keeping the lights on. In other words, why give credits for dishwashers or dryers that waste energy, and not for clotheslines and roofs that don't use energy by harnessing nature? Or why reward a home through LEED credits in the countryside that encourages the use of cars and eliminates any small fuel savings, as opposed to rewarding developments built close to work and school? We show what works well and what could be improved. Our efforts here applaud the transition of an unhealthy and dangerous public housing development into a safe and sustainable place. HOPE VI provides a compass for future affordable housing and community efforts.

ACKNOWLEDGMENTS

We would like to start by thanking Mayor Carran of Covington for allowing us the opportunity to examine data, ask questions, and continue our work in Covington. We would also like to thank Michael "Spike" Jones, Police Chief of Covington, and Brian Carter, Assistant Police Chief of Covington, for their help in gathering crime data and sharing their views on community infrastructure improvements that can help reduce crime. Furthermore, we want to thank Aaron Wolfe-Bertling, former Executive Director of the Housing Authority of Covington, along with the newly installed Director, Jeff Rieck, for supporting our serious and objective study of HOPE VI. We would like to give special thanks to Chris Bradburn, HAC Director of Resident Services, and Linda DiGirolamo, HAC Construction Manager, for reviewing the drafts to provide input and excellent editing. They were both very helpful. We also presented an earlier draft to HAC staff, which included Chris Bradburn, Linda DiGirolamo, Jeff Rieck, Diana M. Strauss, Sheryl L. Schneider, and Jennifer Bennett; their invaluable suggestions and clarifications have informed our revisions. There was no effort on their part to delete any of our suggested midcourse corrections. That does not mean they endorse our suggestions, but hopefully, they will use them to create a discussion on the design and planning of Rivers Edge at Eastside Pointe and the Eastside Neighborhood. The heart of this report is going beyond the numbers for goals and objectives by showing a human face to the positive impact of HOPE VI. We appreciate HAC's direction, support, tough criticisms, and willingness to let us write an independent report. Sheryl Schneider made sure we were invited to observe all the key meetings and have access to all key documents. The Departments of Community Development, Business Development, and Property Valuation in Covington were vital in collecting the baseline count of businesses in the Eastside neighborhood. Nick York and Emmanuel Frimpong-Boamah assisted us in gathering data; their help in preparing the report was essential to this project. Keith Wresinski produced all the maps and tables in this year's report. Carrie Alles did the graphic design of the report, as she did the previous year. We write this Year Four progress report with the hope that it will help guide officials in their policymaking. Unlike others in official city roles, we have the ability to say the flat honest truth. We hope that this report helps point to a positive direction in housing and community development policy.

REFERENCES

Abt Associates. (2003). *Exploring the impact of the HOPE VI program on surrounding neighborhoods*. Prepared for Office of Public Housing Investments, U.S. Department of Housing and Urban Development. Cambridge, MA. Retrieved from http://www.abtassociates.com/reports/ES_NeighCaseStudies_1-14-03_FINAL-A.pdf

Alward, G. S., & Palmer, C. J. (1983). IMPLAN: An input-output analysis system for forest service planning. In R. Seppala, C. Row, & A. Morgan (Eds.), *Forest sector models: proceedings of the first North American conference* (pp. 131–140). Oxford: AB Academic Publishers.

Bergstrom, J. C., Cordell, H. K., Ashley, G. A., & Watson, A. E. (1990). Economic impacts of recreational spending on rural areas: A case study. *Economic Development Quarterly, 4*(1), 29–39.

Brazley, M. E. (2002). *An evaluation of residential satisfaction of HOPE VI: A study of the Park DuValle revitalization project*. Louisville: University of Louisville.

Brazley, M., & Gilderbloom, J. I. (2007). HOPE VI housing program: Was it effective?. *American Journal of Economics and Sociology, 66*(2), 433–442.

Burby, R. J., & Rohe, W. M. (1989). Deconcentration of public housing effects on residents' satisfaction with their living environments and their fear of crime. *Urban Affairs Review, 25*(1), 117–141.

Byrne, J. P. (2003). Two cheers for gentrification. *Howard Law Journal, 46*(3), 405–432.

Carroll, M. C., & Smith, B. W. (2006). Estimating the economic impact of universities: The case of Bowling Green University. *The Industrial Geographer, 3*(2), 1–12.

Center for Urban and Public Affairs (CUPA). (2013). *Raider country creative industries economic impact analysis*. Retrieved from http://corescholar.libraries.wright.edu/cgi/viewcontent.cgi?article=1004&context=cupa_econdev

Cisneros, H., & Engdahl, L. (Eds.). (2009). *From despair to hope: HOPE VI and the new promise of public housing in America's cities*. Washington, DC: Brookings Institution Press.

Clinch, R. (2011). *The economic and workforce development impacts of the Cardinal Towne Project*. Prepared for 3rd Street Development, LLC. Louisville, KY.

Cortright, J. (2009). Walking the walk: How walkability raises home values in U.S. cities. *CEOs for Cities*. Retrieved from http://www.ceosforcities.org/pagefiles/WalkingTheWalk_CEOsforCities.pdf

Diao, M., & Ferreira, J. (2010). Residential property values and the built environment. *Transportation Research Record: Journal of the Transportation Research Board, 2174*(1), 138–147.

Dodd, E. P., Bryant, F. C., Brennan, L. A., Gilliland, C., Dudensing, R., & McCorkle, D. (2013). An economic impact analysis of south Texas landowner hunting operation expenses. *Journal of Fish and Wildlife Management, 4*(2), 342–350.

Doeksen, G. A., Johnson, T., Biard-Holmes, D., & Schott, V. (1998). A healthy health sector is crucial for community economic development. *The Journal of Rural Health, 14*(1), 66–72.

Douglas, A. J., & Harpman, D. A. (1995). Estimating recreation employment effects with IMPLAN for the Glen Canyon Dam region. *Journal of Environmental Management, 44*, 233–247.

Dreier, P., Mollenkopf, J., & Swanstrom, T. (2004). *Place matters: Metropolitics for the twenty-first century. Studies in government and public policy.* Lawrence: University Press of Kansas.

Enterprise Green Communities. (n.d.). *About us.* Retrieved online from: www.enterprisecommunity.com

Fried, M. (1963). Grieving for a lost home. In L. J. Duhl (Ed.), *The urban condition*, pp. 151–171. New York: Simon and Schuster.

Galster, G. C., & Killen, S. P. (1995). The geography of metropolitan opportunity: A reconnaissance and conceptual framework. *Housing Policy Debate, 6*(1), 7–43.

Galster, G. C., & Mikelsons, M. (1995). The geography of metropolitan opportunity: A case study of neighborhood conditions confronting youth in Washington, DC. *Housing Policy Debate, 6*(1), 73–104.

Garrett-Peltier, H. (2011). *Pedestrian and bicycle infrastructure: A national study of employment impacts.* Amherst: Political Economy Research Institute, University of Massachusetts, Amherst.

Gilderbloom, J. I. (2008). *Invisible city: Poverty, housing, and new urbanism.* Austin: University of Texas Press.

Gilderbloom, J. I., & Hanka, M. J. (2006). *Newport's HOPE VI project evaluation: Volume IX.* Louisville: University of Louisville, Center for Sustainable Urban Neighborhoods.

Gilderbloom, J. I., Hanka, M. J., & Ambrosius, J. D. (2009). Historic preservation's impact on job creation, property values, and environmental sustainability. *Journal of Urbanism, 2*(3), 83–103.

Gilderbloom, J. I., Hanka, M. J., & Lasley, C. B. (2008). *Newport's HOPE VI project evaluation: Final volume.* Louisville: University of Louisville, Center for Sustainable Urban Neighborhoods.

Gilderbloom, J. I., & Meares, W. L. (2012). *Covington HOPE VI baseline report.* Louisville: University of Louisville, Center for Sustainable Urban Neighborhoods (SUN).

Gilderbloom, J. I., & Meares, W. L. (2013). *Measuring the impact of Muncie housing proposal on jobs and environment.* Louisville: University of Louisville, Center for Sustainable Urban Neighborhoods (SUN).

Gilderbloom, J. I., Meares, W. L., & Riggs, W. (2016). How brownfield sites kill places and people: An examination of neighborhood housing values, foreclosures, and lifespan. *Journal of Urbanism: International Research on Placemaking and Urban Sustainability, 9*(1), 1–18.

Gilderbloom, J. I., & Mullins, R. L., Jr. (2005). *Promise and betrayal: Universities and the battle for sustainable urban neighborhoods.* Albany: State University of New York Press.

Gilderbloom, J. I., Wresinski, K. W., Grooms, C. W., & Meares, W. L. (2014). *From blighted to beautiful: Covington HOPE VI evaluation: Year four.* Covington: Housing Authority of Covington.

Goetz, E. (2013). *New Deal ruins: Race, economic justice, and public housing policy.* Ithaca: Cornell University Press.

Gotham, K. F. (2003). Toward an understanding of the spatiality of urban poverty: The urban poor as spatial actors. *International Journal of Urban and Regional Research, 24*(3), 723–737.

Hanka, M. J. (2009). *From vice to nice: A case study of Newport, Kentucky's HOPE VI program.* (Doctoral dissertation). Louisville: University of Louisville.

Hanka, M. J., Kumaran, M., & Gilderbloom, J. I. (October 2007). *Estimating the economic effects, consequences, and impacts of president Clinton's community economic empowerment programs: An analysis of empowerment zones in enterprise communities.* Paper presented at the 28th Annual Southern Industrial Relations and Human Resources Conference, Louisville, KY.

Harvey, D. (1973). *Social justice and the city.* Johns Hopkins University Press, Baltimore. Behavioral Risk Factor Surveillance System, 2009.

Holmes, G. M., Slifkin, R. T., Randolph, R. K., & Poley, S. (2006). The effect of rural hospital closures on community economic health. *Health Services Research, 41*(2), 467–85.

Hotvedt, J. E., Busby, R. L., & Jacob, R. E. (1988). *Use of IMPLAN for regional input-output studies.* Buena Vista: Southern Forest Economic Association.

Housing Authority of Covington. (2009). *Jacob Price Homes HOPE VI Application.* Prepared for the Department of Housing and Urban Development.

Huppertz, C. E., Bloomquist, K. M., & Barbehenn, J. M. (1994). *EFIS 5.0 Economic impact forecast system: User's reference manual.* U.S. Army Corps of Engineers Construction Engineering Research Laboratories. USACERL Technical Report TA-94/03.

Imbroscio, D. (2010). *Urban America reconsidered: Alternatives for governance and policy.* Ithaca: Cornell University Press.

Johnson, R. L., & Moore, E. (1993). Tourism impact estimation. *Annals of Tourist Research, 20*(2), 279–288.

Lees, L., Slater, T., & Wyly, E. (2008). *Gentrification.* New York: Routledge.

Leinberger, C. B., & Alfonso, M. (2012). Walk this way: The economic promise of walkable places in metropolitan Washington, D.C. Washington, DC: The Brookings Institution.

Leistritz, F. L. (1994). Economic and fiscal impact assessment. *Impact Assessment, 12*(3), 305–317.

Lindall, S. A., & Olson, D. C. (1996). The IMPLAN input-output system. Stillwater: MIG Inc.

Litman, T. A. (2003). Economic value of walkability. *Transportation Research Record: Journal of the Transportation Research Board, 1828/2003,* 3–11.

Litman, T. A. (2011). Economic value of walkability. *World Transport Policy & Practice, 10*(1), 5–14.

Lynch, T. (2000). *Analyzing the economic impact of transportation projects using RIMS II, IMPLAN and REMI.* Report for Office of Research and Special Programs, U.S. Department of Transportation.

Mandich, A. M., & Dorfman, J. H. (2014). *The impact of hospitals on local labor markets: Going beyond IMPLAN.* Paper presented at Agricultural and Applied Economics Association 2014 Annual Meeting, July 27–29, 2014, Minneapolis, Minnesota.

Meares, W. L. (2014). The walkable dividend: The impacts of walkability on housing and socio-economic composition in Louisville, KY. Dissertation. *University of Louisville Libraries.*

National Center for Preservation Technology and Training. (2001). *Preservation economic impact model user guide.* Natchitoches: National Park Service.

O'Sullivan, A. M. (1993). *Urban economics* (2nd ed.). Homewood: Irwin Publishing.

Popkin, S. J. (2002). *The HOPE VI Program – What about the residents?* Washington, DC: The Urban Institute. Retrieved from http://www.urbaninstitute.org/UploadedPDF/310593_HopeVI.pdf

Popkin, S. J., Buron, L., Levy, D. K., & Cunningham, M. K. (2000). The Gautreaux: What might mixed-income and dispersal strategies mean for the poorest public housing tenants. *Housing Policy Debate, 11*(4), 911–42.

Popkin, S. J., Katz, B., Cunningham, M. K., Brown, K. D., Gustafson, J., & Turner, M. (2004). *A decade of HOPE VI; Research findings and policy challenges.* Washington, DC: The Urban Institute and The Brookings Institution. http://urban.org/uploadedPDF/411002_HOPEVI.pdf

Pucher, J., & Dijkstra, L. (2003). Promoting safe walking and cycling to improve public health: Lessons from the Netherlands and Germany. *American Journal of Public Health, 93,* 1509–1516.

Reid, C. K. (2007). Locating the American dream – Assessing the neighborhood benefits of homeownership. In W. M. Rohe & H. L. Watson (Eds.), *Chasing the American dream: New perspectives on affordable homeownership* (pp. 233–277). Ithaca: Cornell University Press.

Richardson, H. W. (1985). Input-output and economic base multipliers: Looking backward and forward. *Journal of Regional Science, 25*(4), 607–661.

Rickman, D. S., & Schwer, R. K. (1995). A comparison of the multipliers of IMPLAN, REMI, and RIMS II: Benchmarking ready-made models for comparison. *The Annals of Regional Science, 29*(4), 363–374.

Rivero, E. (2011). Smoke-free policies could save landlords up to $18 million a year in cleaning cost. *UCLA   Health & Medicine News – One Article.* Accessed October 30, 2014.

Rose, A. (1995). Input-output economics and computable general equilibrium models. *Structural Change and Economic Dynamics, 6*(3), 295–304.

Rosenbaum, J. E. (1995). Changing the geography of opportunity by expanding residential choice: Lessons from the Gautreaux program. *Housing Policy Debate, 6*(1), 231–269.

Rypkema, D. D. (1997). *The economics of historic preservation: A community leaders' guide.* Washington, DC: Trust for Historic Preservation.

Salama, J. J. (1999). The redevelopment of distressed public housing: Early results from HOPE VI projects in Atlanta, Chicago, and San Antonio. *Housing Policy Debate, 10*(1), 95–142.

Schoenmarklin, S. (2010). Secondhand smoke seepage into multi-unit affordable housing. *Tobacco Control Legal Consortium.* Retrieved from http://publichealthlawcenter.org/sites/default/files/resources/tclc-syn-secondhand-2010_0. pdf

Schwartz, A. F. (2006). *Housing policy in the United States: An introduction.* New York: Routledge Press.

Schwartz, A. F. (2010). *Housing policy in the United States: An introduction* (2nd ed.). New York: Routledge Press.

Shierholz, H., & Mishel, L. (2009). Highest unemployment rate since 1983. *Economic Policy Institute.* Retrieved from http://www.epi.org/publication/jobspict_2009_july_preview/

Siegel, P. B., & Leuthold, F. O. (1993). Economic and fiscal impacts of a retirement/recreation community: A study of Tellico Village, *Tennessee. Journal of Agriculture and Applied Economics, 25*(2), 134–147.

Turner, M. A., Woolley, M., Kingsley, G. T., Popkin, S. J., Levy, D., & Cove, E. (2007). *Estimating the public costs and benefits of HOPE VI investments: A methodological report.* Washington, DC: The Urban Institute.

U.S. Bureau of the Census. (2000). *Census of population and housing.* Data compiled by Eric Schneider, Kentucky State Data Center. Louisville: University of Louisville.

U.S. Department of Housing and Urban Development (HUD). (1992). *Final report of the national commission on severely distressed public housing: A report to congress and the secretary of housing and urban development.* Washington, DC: HUD.

U.S. Department of Housing and Urban Development (HUD). (1999). *HOPE VI: Building communities, transforming lives.* Washington, DC: HUD.

U.S. Department of Housing and Urban Development (HUD). (2000). *Community building makes a difference.* Washington, DC: HUD.

U.S. Government Accountability Office (GAO). (2002). *Public housing: HOPE VI leveraging has increased, but HUD has not met annual reporting requirement.*

U.S. Government Accountability Office (GAO). (2007). *Public housing: Information on the financing, oversight, and effects of the HOPE VI Program.* Statement of David G. Wood, Director Financial Markets and Community Investment.

Voith, R. (2011). *Public housing stimulus funding: A report of the economic impact of Recovery Act Capital Improvements.* Philadelphia: Eco-consult Corporation.

Wachtel, A., & Lewiston, D. (1994). Risk factors for bicycle-motor vehicle collisions at intersections. *ITE Journal(Institute of Transportation Engineers), 64*(9), 30–35.

Waters, E. C., Holland, D. W., & Weber, B. A. (1997). Economic impacts of a property tax limitation: A computable general equilibrium analysis of Oregon's measure 5. *Land Economics, 73*(1), 72–89.

Wechsler, H. (1961). Community growth, depressive disorders, and suicide. *American Journal of Sociology, 67*(1), 9–16.

Weiss, S. J., & Gooding, E. C. (1968). Estimation of differential employment multipliers in a small regional economy. *Land Economics, 44*(2), 235–244.

Wyly, E. K., & Hammel, D. J. (1999). Islands of decay in seas of renewal: Housing policy and the resurgence of gentrification. *Housing Policy Debate, 10*(4), 711–771.

Zielenbach, S., & Voith, R. (2010). HOPE VI and neighborhood economic development: The importance of local market dynamics. *Cityscape, 12*(1), 99–131.

Chapter 16

Portland, the Best Livable City in America

Equity, Health, and Safety

John Hans Gilderbloom, Garlynn Woodsong, and Porter Stevens

This chapter is about how you and I, our neighbors, our cities, and our country can act against climate change. It might be happening soon in your city. In 2017, Hurricane Harvey in Texas and Hurricane Irma in Florida devastated thousands of lives, caused numerous deaths, destroyed billions of dollars' worth of property, and caused long-term damage to two of America's most economically important states. This was followed by horrific fires in Northern California in 2017 that burned 8,500 structures and caused 42 fatalities and more than $1 billion in damage. In 2012, one of the most destructive storms in history hit the New York City region and caused death and destruction. Hurricane Sandy came on the heels of record heat, droughts, destructive fires, and the 2008 financial meltdown. We already had a wake-up call in New Orleans from Hurricane Katrina in 2005. All these events are tied to climate change, which is no longer an abstract concept discussed by scientists; it is having a direct impact on the health, welfare, and safety of every American. But as urbanists and planners, we are not helpless to address these issues.

Today, Portland is considered one of the most livable and sustainable mid-sized cities in the United States. If one looks at some basic statistical data, Portland appears to be an average American city. Founded on February 8, 1851, the city currently has a population of 654,741 people, with 2,289,800 people residing in the Portland Metropolitan Statistical Area. The average income in the city is $71,005, and the unemployment rate is about 6.9 percent—a few points above the U.S. average. Source: U.S. Census Bureau Estimate (https://www.census.gov/programs-surveys/popest/data/tables.2019 .html)

Despite these ordinary stats, Portland is different from most other U.S. cities in one crucial way: it has reduced its per capita carbon emissions. Portland has not only become a city with a steadily shrinking environmental footprint, but one that is experiencing an economic boom based on the "green" economy, a massive influx of young and college-educated residents, and steadily rising rates of walking, bicycling, and transit use. While visiting Portland, one can experience a vibrant craft economy, an active downtown, and revitalized (i.e., attracting new investment, new construction, and new residents) inner-city neighborhoods.

How has the city accomplished all of this? By embracing urbanism. Over the last 40 years, the city of Portland has instituted policies that incentivized the construction of dense, mixed-use development. Combined with generous investments in streetcars, light rail, and bicycling infrastructure, Portland has created a dense, walkable urban environment that has been attracting new residents and new investments ever since. In this chapter, we will put the city of Portland under the microscope and learn what makes it tick.

HISTORY

Though Portland's modern status as the Mecca of Green Urbanism is well established, at the beginning of the 1970s it was no different from any other city in the United States. It was also facing many of the problems common to urban centers in that era. Its downtown was dying, its older neighborhoods were suffering from the effects of white flight and disinvestment and automobile use was rapidly growing.

It was at this point in history that Portland, in a sense, got lucky. Its people (and the people of the state of Oregon) had the good sense to elect a series of progressive political leaders who were able to create several visionary pieces of legislation that forever changed the path of Portland's future. The most prominent of these was Tom McCall, who was the governor of Oregon from 1966 to 1974. Governor McCall is widely known as the primary author of SB 100, the bill that created Oregon's progressive statewide land use planning system, including Portland's now-famous Urban Growth Boundary.

It is important to note that Gov. McCall was a Republican; he was a member of the curious Nixonian brand of Republicans that existed back in the 1970s, the brand of Republicanism that was progressive and would most likely be branded as traitorous proponents of socialism by today's conservatives. So why did a Republican create such an un-Republican piece of legislation? Not because he had some change of heart and became an environmentalist; it was because he wanted to protect farm and forest land from urban development, and through that protection, farming and logging jobs that made up most of his voting constituents.

While (depending on your perspective) he did not create SB 100 for the right reasons, it had the right effects on the city of Portland. The Urban Growth Boundary served to slow and stop low-density suburban development and pushed developers to start looking at vacant or underutilized properties in the city center.

It is in that city center that we find the next group of leaders responsible for Portland's resurgence. In the early 1970s, Portland's downtown (again, like many of its urban counterparts) was declining rapidly. Residents were flooding out of the old city to live in shiny new suburbs; downtown shops and department stores were being run out of business by competition from suburban shopping centers; downtown Portland was clearly dying. But unlike many other city centers that were written off and bulldozed for commuter parking lots, leaders in Portland's private business community deemed the old city worth saving. Working with city leaders, they started to create a series of Urban Renewal Projects to generate new interest in downtown Portland.

It is important to note that such efforts did not conform to the traditional American ideal of "Urban Renewal." They did not involve the widespread demolition of older buildings to make way for large-scale convention centers and hotels. Instead, Portlanders chose to make relatively small but much more effective investments. They rehabbed historic buildings, made parts of downtown more pedestrian-friendly, and invested in park space and public transportation. Such projects served to make downtown Portland a destination, with points of interest to draw suburban shoppers back to its stores and restaurants.

Portland DID engage in widespread demolition of older buildings in the name of urban renewal. It bulldozed the old Jewish-Italian neighborhood of south Downtown and South Portland, to create an urban renewal district that many modernists still celebrate to this day (without mentioning the cost in terms of human lives). It bulldozed the Albina neighborhood, because it was the black neighborhood, and literally did build a hospital, a freeway, two stadiums, and... a convention center where a totally functional urban neighborhood used to exist.

This early push to revitalize downtown Portland produced several projects that are still impacting the city today. In 1977, a downtown "transit mall" and adjoining shopping mall opened; several blocks in the middle of downtown were closed to automobiles and turned into a haven for pedestrians and mass transportation. In 1978, the city voted to remove its waterfront expressway and replace it with a waterfront park (ironically, they named it after Governor Tom McCall). Then, in 1986, Portland celebrated the completion of its first light rail line, connecting downtown with the suburban enclave of Gresham, the first of many light rail lines to follow.

When the efforts of Governor McCall and the Portland Business Community are viewed together, their cumulative effect on Portland becomes readily apparent. By redirecting development back toward the urban core and jumpstarting a downtown renaissance, Portland gave itself a head start; it was already establishing its reputation as a place for progressive/green urbanism as other cities were just starting to explore its ideals.

The Green Economy

In the intervening 40 years, Portland has reaped a progressively larger economic dividend from its commitment to green infrastructure. Since the McCall Administration, the city has committed itself to constructing alternative transportation infrastructure to serve its increasingly dense downtown and urban neighborhoods. For example, in 1996, Portland started construction of a 250-mile network of separated bicycling facilities, based on the model of Amsterdam (which is spatially similar).

Projects such as this created a now-famous "green reputation" that has attracted new businesses and residents to Portland ever since; today, the city benefits greatly from its well-known adherence to progressive urban ideals. There are 10,000–15,000 jobs in the "clean tech" industries, jobs that have weathered the recession relatively well. Many of the companies in this sector got their start in Portland, developing technical or policy solutions to common environmental problems. They now bolster the local economy by exporting those products and/or expertise to the outside world; some of those companies now do up to 90 percent of their business in other cities, regions, or even other countries.

Most crucially, Portland's progressive reputation has served as a powerful magnet for young, educated people. The population in Portland aged 25–34 years old with a college degree increased by 50 percent from 1990 to 2000. This is one of the most important statistics in this entire chapter; whenever planners or city leaders talk about revitalizing a downtown or urban core, one of the questions that is asked is "How do we attract and/or retain more young people?" In Portland, they have found the magic formula for doing just that.

Gentrification, Portland Style

When one visits Portland and walks its streets, it becomes obvious that the powerful economic and social transformation process known as "gentrification" has had a profound impact on the central city and its surrounding neighborhoods. Portland features an active 24/7 downtown, in which one would be hard-pressed to find a vacant lot or underutilized building; the neighborhoods directly adjacent to downtown exhibit similar features.

As was touched on earlier in the chapter, the enviable state of Portland's urban core is the result of years of hard work by private and public-sector actors. Since Portland is not a very wealthy city, the "secret sauce" to its urban success can be attributed to collaboration between public and private entities, sharing ideas, staff, and resources to accomplish a common goal. This process also includes a lot of public involvement, especially at the neighborhood level; city officials work extensively with neighborhood stakeholders to ensure that they make infrastructure investments with maximum positive impact. Additionally, the Portland Metro (that runs Portland's MSA) is the only elected regional authority in the United States.

Prosper Portland (formerly the Portland Development Commission), a city authority that manages "Urban Renewal Districts" throughout the city, runs the bulk of Portland's urban renewal efforts. The modus operandi of Prosper is Tax Increment Financing (TIF): making small, strategic investments in a predefined area, and feeding any resulting rise in property taxes back into that same area, creating a positive feedback loop. This method has raised over $415 million for the City of Portland, which is used for several different purposes. These include rehabbing historic buildings, remediating brownfields, and ensuring housing affordability. Small business development is also important; programs like Façade Renovation Loans and Minority Business Development are key to making the TIF districts work.

It is also important to discuss the impact of the sales tax (or lack thereof) on Portland's commercial/retail scene. In most states, the majority of city government revenue comes from the sales tax; as a result, city planning commissions are incentivized to approve as much commercial zoning as possible. This also causes them to chase after big-box stores and chain restaurants to maximize revenue. What makes Portland interesting is that it exists in a state with no sales tax; as a result, the city has little incentive to approve that glut of commercial zoning or to chase after big-box stores. This creates a constrained supply and a more ideal environment (i.e., less competition from chain stores) for small and/or local businesses to thrive.

The Portland Plan and the Complete Neighborhood

On April 25, 2013, the Portland City Council passed resolution #36918, which officially adopted "The Portland Plan" as the official planning guide for the city. The Portland Plan is a long-range planning document that includes a 25-year vision for the city, implemented through 5-year action plans. It is the result of over two years of public meetings (the organizers collected more than 20,000 public comments and suggestions), workshops, and research.

Portland subsequently adopted the Portland Comprehensive Plan, which the Portland Plan was supposed to be but wound up not being. The Portland Comprehensive Plan included a Missing Middle Housing policy, which is a critical element of the storyline.

The plan attempts to address many of Portland's (relatively) pressing problems, utilizing innovative and distinctly Portland-style solutions. One of the most promising is the concept of the "complete" or "20-minute" neighborhood. A complete neighborhood is one where residents can safely walk to all amenities they need during their daily lives; this means anything from restaurants and boutiques to hardware stores and groceries. The key component of the idea is that all amenities can be safely reached by walking and are located no more than a 20-minute walk from any point in the neighborhood.

The complete neighborhood concept is a powerful one. The requirements for a complete neighborhood (safe, walkable streets, proximity of amenities to housing) can serve as a guide for making infrastructure improvements, installing new sidewalks or crosswalks, and working to calm vehicle traffic. It can also be used as a guide for private development, encouraging the construction of mixed-use developments to bring housing and amenities closer together. Currently, the Portland Plan estimates that less than half of all Portlanders live in complete neighborhoods; it has set the ambitious goal to increase that share to 80 percent by 2035.

Streetcars and Light Rail

While the cityscape of Portland is impressive, it is in the arena of alternative transportation that it has truly staked its international reputation. A typical resident has a multitude of choices: more conventional options like streetcar, light rail, and bus, as well as more unique modes, like an Aerial Tram (which connects a riverfront residential district to a hilltop hospital complex) and Car2Go. Car sharing has been in Portland for over 20 years, and indeed it was the U.S. birthplace of car-sharing in the late 1990s.

Portland's transit statistics are particularly impressive. The city boasts a 45 percent daily rideshare; 320,000 riders per day take transit to and from work. This is similar to many larger U.S. cities like San Francisco and New York; Portland clearly punches above its weight class. Portland has the seventh highest per capita ridership in the United States. Only 60 percent of Portlanders drive by themselves to work, and the Metropolitan Area Express (MAX) light rail carries 26 percent of the commuter traffic alone. The city has invested $3 billion in the light rail system since 1986, resulting in a 108 percent increase in ridership. See https://www.google.com/url?sa =t&rct=j&q=&esrc=s&source=web&cd=&cad=rja&uact=8&ved=2ahUKE-wj0jLzUhs_uAhWTuZ4KHZP8AvUQFjAAegQIARAC&url=https%3A

%2F%2Fbikeportland.org%2F2019%2F09%2F26%2Fus-census-portland
-bike-commuting-hits-lowest-rate-in-12-years-305326&usg=AOvVaw3ErL2
qyTEp6kfT5rFo0py. Unfortunately, during and after the pandemic, the
ride share went from 45 percent to 13 percent, according to planners. The
causes of this dramatic drop were fear of contracting COVID-19 by being in
crowded enclosed places, rampant crime in downtown Portland, the George
Floyd riots, homeless encampments, and rampant drug use.

The popularity of transit in Portland has had several significant impacts on
the city. The most important statistic is that Portland is one of the few U.S.
cities (like New York, Berkeley, or San Francisco) to reduce its per capita
carbon emissions. Portland's emissions have declined by 6 percent below the
1990 baseline, while U.S. per capita emissions have increased by an average of 12 percent. Additionally, Portland is one of the few U.S. cities where
Vehicle Miles Traveled per capita has declined. Research has shown that a
Portland resident living near transit will travel 9 miles per day, compared with
22 miles in transit-poor areas.

Transit investments have also had a huge impact on Portland's real estate
market; over $10 billion has been invested in new construction (of all types)
within walking distance of MAX lines since 1980. Such developments also
retained their value during the recession. The growth of Transit-Oriented
Development (TOD) in Portland has been astonishing; one-quarter of all
housing built over the next two decades will be located within a half-mile of
a rail transit station. New development within three blocks of a streetcar or
light rail line currently makes up 75 percent of all new development in the
city. The proliferation of TODs has a significant impact on the travel habits
of their residents: 23–33 percent use mass transit as their primary mode of
transportation, and car use among TOD residents is 50 percent lower.

Unfortunately, despite this success, Portland's transit systems have fallen
on hard times. The Great Recession that occurred between 2007 and 2009 had
a significant impact on the quality of service and on financing as, like many
other municipal transit systems, it faced significant cuts in federal and state
funding. In 2009, TriMet (Portland's regional transit authority) posted a $20
million budget shortfall. This forced it to eliminate the downtown "Fare Free
Zone" that had existed since the 1970s and cut back on service frequency;
both the MAX light rail and several high-frequency bus lines had to cut back
on service frequency, and four bus routes had to be eliminated entirely.

Thankfully, the future is looking bright for mass transit in Portland. The
post-recession economy has rebounded to the point that TriMet feels confident enough to start investing in system improvements. In 2013, it earmarked
$2.1 million to restore services cut during the downturn and is currently
exploring the use of Bus Rapid Transit for future system expansions. The
MAX Orange Line opened up a 7.3-mile light rail connection to the suburban

enclave of Milwaukie. Additionally, TriMet is proposing to connect the commuter haven of Vancouver, Washington, with a commuter rail link. Projects like these clearly show that, despite the difficulties of the Great Recession, Portland still considers investments in mass transportation to be important and economically beneficial.

Unfortunately, as TriMet and other authorities continue to explore new modes and plan new routes, they are starting to encounter a social phenomenon that has dogged planners and developers alike for decades: Not-In-My-Back-Yard (NIMBY), better known as Nimbyism, that age-old belief that the community needs higher density development, clean energy facilities, and more mass transit, as long as none of them are built anywhere near my property line. As the MAX light rail system has expanded into Portland's suburbs, it has run into increasing public opposition; primarily from residents who do not want their lives (and property values) disrupted by the noise of construction and transit operation, and have their idyllic communities invaded by malicious and grimy transit riders. This attitude holds despite the reality that many of their historic downtowns were originally built around streetcar lines; they already live in transit-oriented communities.

But in places like Gresham and Milwaukee, the NIMBY mindset has dictated that the new light rail lines be built away from the historic (and denser) centers. This is a disadvantageous result because it minimizes the positive impact that the new transit projects will have on these communities; your property will have a much higher value if it is directly adjacent to a transit station as opposed to being located several blocks away. The commuter enclave of Vancouver, Washington (full of residents priced out of the Portland housing market) took NIMBY a step further; they voted down a proposal to connect their town with a new commuter rail link to Portland.

Despite these unfortunate setbacks, mass transit has secured its rightful place in Portland's urban environment. If the city continues to make smart investments in its light rail, streetcar, and bus systems, it will continue to reap substantial dividends in the form of increased property taxes, new urban development, and improved air quality.

The Craft Economy

Though Portland's reputation as a green, progressive city with twenty-first-century infrastructure has proven to be extremely attractive to new residents, what makes the city such a pleasure to visit and live in is its thriving craft culture. The city boasts over 60 craft breweries and has more restaurants per capita than any other similarly sized metro in the United States (some of which the author can personally attest are quite good). What is particularly impressive about Portland's homegrown food scene is its incredible variety of food carts.

Though the food cart craze is rapidly becoming a national phenomenon, Portland is by far the frontrunner; the city has on file 600–700 food cart licenses. Many of these carts are grouped into outdoor food courts known as "pods." Pods are usually set up in vacant lots or parking lots in downtown Portland and are home to about 20–30 food carts each. Such an arrangement is very beneficial to the cart owners, as the wider selection attracts a much higher number of customers than any food cart would on its own.

In addition to enhancing the quality and selection of street food, food carts are having a significant impact on Portland's restaurant scene; several food cart enterprises have become so successful that they traded their carts in for brick-and-mortar storefronts. Though some hold the belief that pressure from food carts hurts the restaurant industry, Portland exhibits the opposite effect. Food carts and cart-to-restaurant conversions have fostered an atmosphere of relatively friendly competition for traditional restaurants, encouraging them to improve the quality of their food and service. Ultimately, the hungry Portlander is the victor.

But what is the cause of all this? What factors or policies encouraged the creation of such a rich craft culture? As always seems to be the case with Portland, the answer may surprise you. It turns out that Portland's tough job market is partly responsible. The city's reputation as a progressive, creative, healthy, and livable city has become so powerful that members of the creative class are moving there without first finding a job or having any other form of financial/economic security; they are not moving to Portland because of a job, they're moving there because they want to enjoy the city's high quality of life.

It is often said that people move to Portland for the lifestyle, not the money. "They come to retire at age 30" is the often-told joke. Another novel explanation put forth by Heying (2010) is that people are rejecting the alienating corporate model of choosing employment that pays the most instead of being the most satisfying. According to both Heying (2010) and Abbott (2011), many people who come to Portland want to be involved with the craft industry of beer making, restaurants, clothes, music, and computer programming. According to Abbott (2011:174), "a large number of entrepreneurs and artisans are building upscale bicycles, designing clothing, brewing beer, and roasting coffee."

Why are people moving to a city with no job and poor chances that they will find one? According to economist Joe Cortright, this phenomenon can be attributed to the changing dynamics of American society. In the twenty-first century, people are moving to where they feel at home and where they can interact with citizens who have similar beliefs and ideas. Doing so gives that person the opportunity to create a network of friends and associates, a network that will help them when they are in need and support them in social or business ventures. In years past, the factory or the company filled this role; but as the United States has painfully de-industrialized, it has left a void that has begun to be filled by friends and family.

It is this supportive social network of creative individuals that draws people to Portland, and it is this network that keeps them there even if they cannot immediately find employment. If they cannot find a traditional full-time office job, many of these people, to stay in beautiful, livable Portland, get creative; they start a food truck, a brewery, or some other craft business for income.

Urban Delights

In addition to its thriving craft economy, walking the streets of Portland is simply an enjoyable experience. This pleasure comes from several visually interesting features that have been installed on Portland's streets and side-walks over the years and serve to exhibit the city's quirky character. Public art is everywhere in the form of murals on building walls and sculptures placed along the streets, in parks, and public spaces. There are also interac-tive installations, the most prominent of which is the placement of pianos at various locales; anyone and everyone can play these pianos and become a living public art installation. Small installations like these lend Portland's streets a lively and unique quality that increases the appeal of non-motorized transportation and makes the city that much more enjoyable to visit.

The Power of the Bicycle

In the American sustainability community, Portland is often touted as the bicycling capital of the United States. Though several other cities could effec-tively dispute that claim, it is obvious that, like so many other progressive transportation modes, riding a bicycle is a common and publicly accepted part of daily life. Embracing the bicycle as a practical mode of daily trans-portation is something that the city of Portland is proud of, and it is a move that has paid handsomely in social and economic dividends.

Bicycling is big business in Portland; it is estimated that bicycle-related businesses (everything from clothing and tourism to manufacturing) gener-ate $90 million in annual revenue. In the state of Oregon, bicycle tourism is a $400-million-a-year industry. Bike-related industries in the city are also major employer, creating 850 to 1,150 good-quality jobs. The Portland busi-ness community is very aware of such figures and agrees that creating and promoting bicycle infrastructure has been good for business; in a 2006 survey,

82 percent of businesses said that Portland's bicycle-friendly reputation has been beneficial for the bottom line, while only 3 percent said otherwise.

It is easy to see why the local business community is so receptive to having bicyclists as customers; a prime example of this positive effect is the bike corral, a parking infrastructure improvement that is becoming the latest bicycle-related craze in Portland. A bike corral is created when a single on-street parking space is closed to automobile use and converted into parking space for bicycles. Not only do such installations ensure that sidewalks remain clutter-free, but they also increase the number of customers that can use the same amount of space for parking; a typical bike corral can accommodate up to 12 bicycles (therefore 12 riders) in the same space used by a single car. Additionally, bicyclists typically shop more often, purchase more items overall, and are typically hungrier and thirstier than your average customer.

The positive economic impact of the bicycle industry on Portland's economy is a real-world demonstration of another Joe Cortright theory: The Green Dividend. The idea speaks to the culture of localism and how investing in bicycling (and other "sustainable") infrastructure can generate a "ripple effect" that greatly benefits the local economy. He writes: "It's time to replace the cliché of green policy as sacrifice and instead recognize that for progressive regions and their residents, being green pays handsome economic dividends."

His analysis centers on the problem posed by Americans' extensive use of gasoline and the (usually) foreign oil from which it is derived. Since this product is controlled by large companies, the revenue from gasoline sales leaves a locale and is never seen again; Cortright estimates that 73 percent of the retail price of gas immediately leaves the economy. Therefore, by switching from cars to bicycles, citizens are ensuring that more of the money they spend on transportation is staying in and benefiting the local (rather than the global) economy. Cortright estimates that if Portlanders reduce their overall driving by about 4 percent, they will keep about $800 million from leaving the city. By investing in bicycle and pedestrian infrastructure, the city of Portland could increase local spending by $54,002,341 over 15 years.

Building the infrastructure needed to safely accommodate new and existing cyclists is also a great job creator. It is estimated that for every $1 million spent on new bike lanes and other infrastructure, 11–14 jobs are created. This contrasts with road construction projects that, due to the use of small crews using numerous large machines, only create seven jobs for every $1 million spent. In Portland, the Vera Katz Eastbank Esplanade created 1,050 jobs alone; several adjacent buildings were also renovated as a result, and now house companies that employ an additional 185 people.

In addition to being a great job creator, bicycling infrastructure projects have an attractive cost/benefit ratio; multiple sources confirm that the typical margin is between 1.3 and 3.8 times the cost. Painting bicycle lanes and installing priority signaling is, relative to other types of transportation investment, relatively inexpensive; when compared to the cost per mile for highway construction, bicycling infrastructure is cheaper than municipal tap water. In 2008, Portland Mayor Sam Adams summed it up perfectly: "In 1993 we weren't the bicycling capital of America. Seventeen years later, for the equivalent cost of a single mile of freeway ($57 million), we have bike infrastructure (274 miles)."

Did transportation habits change when Portland added bike infrastructure? Indeed, research shows that Portland now has the largest share of bike commuters of any medium-sized or larger U.S. city. Researchers see signs that this mode share will continue to grow. Currently, 5 percent of Portland's working population rides a bike to work, compared to the national average of 1 percent. Moreover, Portland planners estimate that with infrastructure friendly to biking, walking, and public transit, only 60 percent drive alone to work. Six percent work at home; 5 percent walk; 12 percent use transit; 9 percent carpool; and 1 percent take a cab or motorcycle. A number of these changes occurred when Portland invested over $100 million in new bike infrastructure (Portland Plan Partners, 2012; Mapes, 2009).

The Department of Labor's Consumer Expenditure Survey shows that Portland has the second lowest rate of spending on transportation out of the 28 largest U.S. metro areas (Cortwright, 2007: 1). Furthermore, it finds that Portlanders on average drive 20 percent fewer miles than drivers in other cities and spend about 4 percent less on transportation than those in other cities.

Cortright (2007) estimates that for every 1-point increase in Walk Score, a Portland home increases $700 to $3,000 in value. Research also supports the notion of a "green dividend" in other cities, including the city of Louisville. In addition to the positive impacts on the local economy, researchers studying Portland have found major savings in health care costs. This is mainly attributed to the health benefits of increased exercise, principally biking and walking. Gotschi (2012) estimates that a $138–$605 million investment in bike infrastructure will yield health savings of between $388 and $594 million and fuel savings between $143 and $218 million. Lastly, investing in bike infrastructure creates more jobs. Building a freeway will generate roughly seven jobs per million dollars spent compared to 11 to 14 jobs per million when building bike lanes. Research by Clifton et al. (2012) has shown that bike commuters spend more per month on grocery shopping compared to those who travel by car, and they also generate more repeat business. Surveys of Portland businesses found that 82 percent consider bike commuters as good for business (Flusche, 2012; Swanson, 2012: 180), and that high-density parking for bikes, known as bike corrals, is seen as a "pro-business amenity"

(Clifton, 2012: 30; Flusche, 2012: 15; Swanson, 2012: 180). Alta Planning and Design found the direct impact of bicycle-related economic activity was $90 million dollars in 2008, which is a 38 percent increase from 2006 (Swanson, 2012; Alta Planning and Design, 2008).

Investments in bicycling infrastructure can be extremely beneficial to neighboring property owners, especially those who own homes. New residents who are shopping for their first home in Portland often consider access to major bicycling infrastructure to be a plus, and this has the resulting effect of increasing selling prices. Homes in the vicinity of neighborhood greenways (streets that are designed to be bicycling corridors) have been shown to increase in value; in some areas, there is a 10 percent price difference. Before the Great Recession, homes located in the vicinity of bicycling infrastructure were sold for an average of $2,000 more.

Additionally, realtors are picking up on the growing popularity of bicycling and are starting to use a property's vicinity to bicycle infrastructure as a selling point. They are also starting to use the popular website Walk Score as a tool to help them price their properties. Though the accuracy of this site is subject to debate, the effect that a high score can have on a home's price is not: each additional point on Walk Score adds an average of $700 to $3,000 to the value of a home.

Portland is also integrating the bicycling culture into its educational institutions, which has resulted in a dramatic increase in ridership rates. Portland State University was able to increase its bike ridership among students from 6 percent in 2005 to 12 percent in 2012. Portland State University (2013) expects that this number will eventually rise to 25 percent. Since land at its urban campus is scarce, the university has found it to be much more beneficial to invest in alternative forms of transportation than building expensive parking facilities for cars. The local non-profit organization Street Trust sponsors a program that gives every grade school class a 10-hour bike safety class. This effort, combined with substantial investments in bicycle racks and road safety near schools, has had spectacular results: elementary student non-auto share has increased from 8 percent to 42 percent in the last 12 years.

As with many other progressive trends in Portland, bicycling has a bright future. The city's Bicycle Master Plan has set the illustrious goal of increasing the bicycle rideshare from the current 5 percent to 25 percent of all daily trips; this would put the city on par with many cities in Europe. Additionally, the people of Portland will continue to reap the health benefits of their forward-thinking investments. A 2011 study concluded that Portland would see healthcare cost savings of $388 to $594 million attributable to new bicycling infrastructure by 2040; this works out to a $1–$3.40 investment/benefit ratio.

Equity and Gentrification

Though the phenomenon popularly known as "gentrification" has breathed new life and prosperity into Portland's downtown and urban neighborhoods, it has not had a positive impact on everyone; it has proven to be a double-edged sword. Critics have complained that gentrification has had the adverse effect of driving out the communities that existed before the new investments. Low-income minority communities often occupied such neighborhoods; these communities have been slowly pushed farther away from the city center and now live in more distant suburbs. This trend has had a significant impact on Portland's small but tight-knit African American community; the number of blacks living in the inner-city neighborhoods has been cut in half over the last 40 years.

Many sources have claimed that Portland's Urban Growth Boundary has also had a negative effect on low-income communities. A study produced by the Reason Foundation (a libertarian think tank) claims that "In less than a decade, Portland has transformed itself from one of the most affordable to one of the least affordable housing markets on the West Coast." Other sources have also pointed to the large homeless population (which can often be found camping out on downtown Portland streets), as evidence that renters and low-income homeowners are being literally thrown out onto the street due to increasingly unaffordable housing.

Additional (and unbiased) research needs to be conducted to more precisely understand the impact that an urban growth boundary has on home prices. However, a simple comparison of median home values shows that the Reason Foundation's claim is a *post hoc, ergo propter hoc* fallacy. When compared to other major cities in the United States, Portland is relatively inexpensive, as the table below shows. Taken from Standard and Poor's Case-Shiller Home Price Index, it summarizes the changes in median home prices for 20 major metropolitan areas from June 2006 to June 2012; it also tracks the average year-by-year change in median home values. The data clearly show that home prices in Portland are completely un-exemplary.

Though it would be foolish to claim that the Urban Growth Boundary has had no effect on median home prices, the claim that it has transformed Portland into a place where only the wealthy can purchase a home is simply not supported by current data. Portland is much less expensive than several other cities, both with and without UGBs. There are homeless people in downtown Portland because they are attracted to the city by its stable climate and its social services, not because they were forced onto the street by unaffordable housing.

Interviews with residents and officials reveal that other factors contributed to this demographic shift. As public investments in infrastructure caused property values to rise, many low-income residents (especially renters) were simply priced out of the neighborhood. Also, construction companies working on new developments in inner-city neighborhoods would often refuse to hire workers from within the community and hired outside help instead; such practices exacerbated unfavorable economic conditions in these low-income neighborhoods and gave existing residents further incentives to leave.

Additionally, stories have been told of realtors who would go door to door and offer to purchase homes from low-income owners for a large cash sum. Such homeowners would often have little financial training and would be unaware of the true value of their property, and the substantial equity that their home represents. As a result, the owners, needing the money, would take the offer and sell their home for a fraction of its actual value.

This practice is part of a larger trend of private real estate companies making substantial and targeted investments in certain neighborhoods to accelerate the gentrification process. According to an article in the Wall Street Journal (2013), a breed of investor has arisen in the post-recession real estate market; instead of just "flipping" foreclosed homes, real estate entities "have assembled billions of dollars to acquire homes and upgrade them. Their aim is to gentrify communities and profit later when rents and property values rise."

This trend is most prevalent in places like Oakland, Washington, and Philadelphia, where working-class neighborhoods are located close to major employment centers. The private sector has even taken over the role of the city in some places by repaving roads, planting street trees, and building sidewalks, all aimed at driving up the values of their property portfolios. Such major private investment has served to transform many neighborhoods from blighted and crime-ridden into new and exciting places to invest and live, but it has had the unfortunate side effect of driving out the working-class residents that called those neighborhoods home for many generations. Portland has been no exception.

Gentrification threats have justly earned a negative image in Portland's minority communities. As a result, proposed bicycling and transportation projects in the few remaining "ungentrified" inner-city neighborhoods have been met with fierce public opposition, mostly out of fear that they might be the beginning of the end of the existing community.

Is gentrification wreaking havoc throughout the United States? Not according to Douglas Massey, the demographer/sociologist who estimates only a tiny number of neighborhoods throughout America receive investment dollars, while many downtown neighborhoods are deteriorating. Massey calls it hypocritical on the part of academics to condemn white people for fleeing the city for the suburbs and then to turn around and denounce white people for moving back into historic neighborhoods. Richard Florida cites numerous studies that indicate "relatively few people" are displaced by gentrification. Minorities who suffer from segregation attend the worst schools, experience higher levels of crime, endure poor housing and health conditions, and suffer from high rates of housing decline.[6] Moreover, in cities receiving new investment to renew neighborhoods, gentrification has not been catastrophic, but has stabilized these neighborhoods, as in New York City, which is often pointed out as the epicenter of gentrification. Homeowners gain equity by higher-income classes moving nearby, and renters' higher housing costs are offset by improved infrastructure that reduces car dependency and provides easy access to needed goods.

To Portland's credit, the city has been trying to get ahead of the gentrification process and ensure that new development benefits the existing community residents instead of displacing them. After all, what's worse than gentrification? No gentrification. New investment is beneficial to homeowners, as their home value increases, and having more housing, shops, bicyclists, and pedestrians on the streets creates better and safer neighborhoods. The key is not to "stop" gentrification (if that is even possible), but to harness it to benefit the existing community. The goal is gentrification without displacement of existing residents or businesses.

In this spirit, the city has enacted several programs in minority neighborhoods. The state of Oregon has placed a cap on property taxes to reduce the impact of rising property taxes on renters and homeowners. As mentioned earlier, Prosper Portland/The Portland Development Commission has several programs working with minority-owned businesses to help them market their enterprises to the influx of new residents. Finally, communities are collaborating with the city and Verde, a non-profit, to ensure that neighborhood residents are being hired to work on new construction projects. In the Cully neighborhood, this has enabled residents to find well-paying jobs.

Why Do Marxists Hate Green Cities Like Portland and Amsterdam?

Matt Hern, in his book *What a City Is For: Remaking the Politics of Displacement*, takes the Marxist position that green urbanism is a plot to push black people out of Portland, Oregon to make it all "White" (pp. 7–19). Marxists have never been on board with green thinking and economic development. They still embrace fairytale Marxism of 175 years ago and think that places like Cuba and Venezuela are ideal.

Portland is not becoming an all-white city because people are riding bikes to improve the health of their bodies and planet earth and save money. According to Portland State Demographer Lisa Bates, going green has not pushed black people from Portland; the same small percentage continues to live in the city, which is a healthy, prosperous, and environmentally friendly place where black people have one of the longest lifespans in the nation. The 2016 U.S. Census finds that the percentage of non-white people is roughly 13 percent or in hard numbers 76,000 people out of 574,000 are non-white.

Hern quotes a neighborhood leader who says, "I knew Black people were f---d as soon as I saw the bike lanes. That's when we knew Black people weren't welcome here anymore." In reality, more black people than white people bike. Bike lanes provide safety and protect against injury. Biking is an economic benefit that saves $8,000 in out-of-pocket expenses annually. Hern also assumes that black people do not want to share neighborhoods with white people, despite surveys that suggest black people like the idea of integrated neighborhoods.

Hern suggests that cities should reserve pockets of housing exclusively for black and white people. Hern argues, "Academic researchers have been relentless, voluminous, and often brilliant at documenting successive waves of gentrified displacement, statistically, empirically, and ethnographically. Gentrification is accepted, assumed, and constantly included in discussions about any city" (p. 44). When people are anti-gentrification, they are unconsciously pro-segregation of races and classes. What is ironic and odd is Hern brags that he wrote and lived in several Native American communities, yet he preaches that white people should not move into minority communities.

Hern is overwrought on the issue of displacement. It is only an issue in a handful of global cities and would be welcomed in most mid-sized cities throughout the Midwest and South. Hern uses the radioactive term of gentrification, which is now more ideological than scientific and conjures up simplistic race wars between "haves and have nots." Sociologists, who invented the term in the 1960s to describe London, are quick to condemn any kind of gentrification, even if it means greater income and racial diversity or regeneration of a historic neighborhood, or more important, saving the city itself. A much better term is *inversion,* which neatly counters a more serious issue of suburban sprawl, also known as *dispersion.*

It is not just the rich but also the young who are wise enough to recognize that downtown living is an economic and cultural gain over suburban living. Many of the newest urban residents are weighed down with college loan debt, which puts a strain on their finances; factor in the rising cost of living, and the relatively low cost of urban living is preferable to suburban living. Downtown represents real estate opportunities with higher returns, especially in the renovation of historic buildings. Additionally, downtown provides the infrastructure to go car-free and save a resident around one-fifth of their paycheck on transportation costs.

Hern never addressed how inversion means less sprawl, fiscal solvency, greater vibrancy, and less pollution. It also means financial windfalls for working-class and minority homeowners who stay in the neighborhood. Inversion means more jobs: the Center for Sustainable Urban Neighborhoods (http://sunlouisville.org), estimates that for every one million dollars spent on development, 17 jobs are created (renovation of older buildings creates more jobs than new construction). Four out of five of these jobs go to plumbers, electricians, carpenters, masons, and clerks. Inversion creates lots of jobs for groups with the highest levels of unemployment, and a large percentage of those jobs are obtained by Hispanics and Blacks.

Black neighborhoods should have the same kind of infrastructure and benefits as white neighborhoods: clean air, parks, trees, calm streets, good housing, and homeownership. Hern's arguments seem to challenge the civil rights activism of Martin Luther King, Jr. by proposing separate and unequal neighborhoods. Martin Luther King, Jr.'s success was rooted in the morality of Bible verses with the founding fathers' view "that all men are created equal . . . with life, liberty and the pursuit of happiness." Dr. King held rallies and demonstrations at which he advocated that everyone should have access to safe housing, health care, and employment without the burden of discrimination. King argued that social change cannot happen without a vision for a better life. Hern's views are un-American; he wants to roll back the constitutional protections passed by civil rights leaders.

New investment is beneficial to homeowners. The key is not to stop gentrification, but to harness it to benefit the existing community. Black homeowners in Portland are delighted to see their home values increase. Neighborhoods are better off in general with gentrification and the investment of new dollars. Downtown minority and poor neighborhoods are worse off without new investment, experiencing high rates of foreclosures, abandonment, and racial segregation. Moreover, black leaders are also tired of half-way houses, homeless shelters, and drug rehab centers being the only kind of new developments in minority neighborhoods. Not all black neighborhoods are poor and dysfunctional. The belief that black neighborhoods are only for the poor is a stereotype. Black neighborhood leaders just want an ownership share in the success that comes with gentrification so that they can benefit equally from it.

Middle housing is a concept that represents the future of American cities, a shift away from economic and racial segregation and toward a diverse mix of incomes and people living in low-carbon neighborhoods that are fiscally solvent and environmentally sustainable.

THE STORY OF PORTLAND'S RESIDENTIAL
INFILL PROJECT

The long origin story of the most progressive new zoning and housing policy adopted by any major American city to date. (A personal story by Garlynn Woodsong.)

Origins in the Comprehensive Plan

At the end of 2014 and the beginning of 2015, Portland was working on updating its Comprehensive Plan, which hadn't been updated since the early 1980s despite a state mandate to do so every 20 years. An effort in 2008/9 was initiated under Mayor Tom Potter, but it transitioned into more of a visioning effort that did not lead to the adoption of a new Comprehensive Plan. By July of 2014, however, a discussion draft of a new Comprehensive Plan was out for public comment.

Meanwhile, as the city bounced back from the depths of the 2008/9 Great Recession, several builders that had formerly specialized in suburban subdivisions pivoted to focus on urban infill. Builders like Everett Custom Homes and Renaissance Homes were now in the business of buying small single-family homes, bulldozing them, and erecting large new single-family homes in their place (often, when lot size allowed, two or three at a time). They made money on the spread between the size of the small older home and the large new homes, multiplied by the spread between the cost of construction per foot and the sales price per foot from comparable sales of large new homes. Neighbors, however, were concerned. Very few people enjoy the process of watching the house next door get bulldozed and turned into a construction zone. Progressive Portlanders were also concerned that the resulting new homes were so expensive that neither they nor most people they knew could afford the high price tag. A sentiment was building in the city that the demolitions were bad, and something needed to be done.

My neighborhood, the Concordia Neighborhood Association, suggested a new path forward. In January of 2015, building from discussions that I led as chair of the Land Use and Transportation Committee during the latter half of 2014, it submitted a lengthy comment letter with testimony full of edits and suggestions for the draft Comp Plan. The letter included a recommendation to legalize "flats" in single-family zones served by high-quality transit, to ensure that it is economically feasible to build new homes affordable to median-income households. The letter called for multiple dwelling units, up to two per floor, to be allowed within structures that otherwise meet the form requirements for single-family homes, as regulated through minimum lot size, maximum lot coverage, building setbacks, height, protection of existing mature trees, and other issues relevant to neighborhood livability. It specifically called for the use of Community Design Standards to regulate all new development; these standards generally apply a form-based approach to regulation, requiring windows to be square or vertical rather than horizontal, for porches to orient to the street, and other aspects of form agreed upon based on conversations with the community in the early 1990s. Because Oregon is a by-right development state, projects could either choose to comply with the Community Design Standards to receive a building permit without discretionary review, or they could submit to Design Review if they wished to deviate from the standards.

The Residential Infill Project Stakeholder Advisory Committee

In August of 2015, under Mayor Charlie Hales, the city kicked off a new initiative, the Residential Infill Project, to re-examine its single-family zoning code. I was appointed to its Stakeholder Advisory Committee to represent the NE Coalition of Neighborhoods (NECN), of which Concordia was a member neighborhood. From the beginning, I pushed for fourplexes to be legalized as a part of the effort. The Residential Infill Project Stakeholder Advisory Committee (RIPSAC) met monthly through June of 2016.

In February of 2016, architect Dan Parolek, who coined the term "missing middle housing" to describe Accessory Dwelling Units (ADUs), duplexes, triplexes, fourplexes, courtyard housing, mansion apartments, live-work units, and townhomes, came to Portland. He spoke with the Planning & Sustainability Commission, as well as City Council and the Metro Council, presumably about the benefits of replacing single-family zoning with a form-based code that allows missing middle housing.

In June of 2016, Portland adopted its new Comprehensive Plan. Perhaps inspired by testimony from the Concordia Neighborhood Association, guidance from Dan Parolek, and guidance from the RIPSAC, this included a new goal, 5.A: Housing diversity, stating:

> Portlanders have access to high-quality affordable housing that accommodates their needs, preferences, and financial capabilities in terms of different types, tenures, density, sizes, costs, and locations.

This goal was backed up by a new Policy 5.4: Housing types.

> Encourage new and innovative housing types that meet the evolving needs of Portland households, and expand housing choices in all neighborhoods. These housing types include but are not limited to single-dwelling units;

multi-dwelling units; accessory dwelling units; small units; pre-fabricated homes such as manufactured, modular, and mobile homes; co-housing; and clustered housing/clustered services.

By itself, this policy would be sufficient to allow for middle housing types. But Portland didn't stop there. It also included Policy 5.6, Middle Housing:

> Enable and encourage development of middle housing. This includes multi-unit or clustered residential buildings that provide relatively smaller, less expensive units; more units; and a scale transition between the core of the mixed use center and surrounding single family areas. Where appropriate, apply zoning that would allow this within a quarter mile of designated centers, corridors with frequent service transit, high capacity transit stations, and within the Inner Ring around the Central City.

Interestingly, this policy provided fairly specific locational guidance for middle housing to be used as a transition type around centers and corridors.

Meanwhile, the Residential Infill Project was progressing on a parallel track. The staff wrote their proposal, focusing on legalizing two ADUs, duplexes, and triplexes, but not fourplexes, basically ignoring all input from the RIPSAC. The next step for the RIP was to go to City Council in December 2016, so that Mayor Charlie Hales could hold a council vote on it before leaving office at the end of his term. The council voted to instruct the staff to write up a code proposal.

Staff put their heads down in 2017, coming up in September with a discussion draft that focused on creating a new overlay zone focused around transit that would allow a house with two ADUs, duplexes, or duplexes plus one detached ADU on most lots, or a triplex on corner lots. This progressed to a Proposed Draft in April 2018 that included basically the same language, with some tweaks to the area for the new overlay zone.

Pivot to Advocacy and Coalition-Building

Meanwhile, I was working with advocates to try to change the situation. A small group of people associated with the Congress for the New Urbanism, including R. John Anderson, Mary Vogel, and myself, went around meeting with as many of the Planning & Sustainability Commission and City Council members as would give us an appointment, making the case that the RIP needed to at the very least legalize fourplexes in order to provide for a development type that might pencil out, and thus provide economic opportunity for property owners to become small developers, as well as housing opportunities in each of the new homes in a fourplex, which should be relatively cheaper than those in a duplex or triplex by being able to amortize the initial site acquisition cost over more units.

In the community, the long-standing group 1,000 Friends of Oregon hired a community organizer, Madeline Kovacs, to run a new community initiative called Portland For Everyone, which was largely focused on eliminating the overlay zone concept in favor of legalizing the entire package of middle housing types proposed under RIP on any formerly single-family zoned lot in the city. The inspiration for this was equity, the rationale being that the RIP represented an opportunity to build generational wealth through additional units that could be rented out to bring passive income; and that this opportunity could either be restricted to certain areas, which may have already been priced higher for a variety of reasons and thus would already be excluding from property ownership those who could not afford the price of admission; or it could be spread as widely as possible, including to areas with a history of under-investment where the city's most low-income property owners are concentrated.

I framed the position as the equity argument, in contrast to the city's proposal to focus the opportunity areas around transit, which I framed as the smart growth argument. For me, this was a new thing under the sun: equity versus smart growth, with strong arguments for both. In the end, however, it was very hard to argue against the equity argument, given that the city's other policies already concentrate higher-density growth around transit. There didn't seem to be a huge amount of downside to allowing greater opportunity everywhere within the City of Portland, which does not contain within its city limits any truly sprawling, ex-urban areas that could be so locationally inefficient as to justify completely restricting the growth of middle housing types.

The 2018 draft of the RIP went before the Planning and Sustainability Commission (PSC) in September of that year. The PSC heard testimony from a series of packed hearings that, thanks to the organizing of Portland 4 Everyone (P4E), strongly argued in favor of the equity position, with many people, including me, arguing on the record in favor of legalizing fourplexes as well. The PSC heard this message loud and clear and directed staff to revise the RIP to include fourplexes and to expand the area of the city within which all the RIP's middle housing types would be legalized.

Staff and the PSC went back and forth on code language through the spring of 2019, making tweaks to account for a large number of details, from what types to allow on historical lots of record, to which environmental or infrastructure deficiency zones to exclude from the area where the new types would be allowed.

By the summer of 2019, this resulted in an updated draft of the RIP. Meanwhile, the context for the RIP suddenly shifted.

HB 2001 and the Most Progressive Statewide Housing Policy in America

During the 2019 legislative session, which ended with the unprecedented act of Republican legislators walking out of the session rather than allowing a quorum to be met for the passage of a climate-focused cap & trade bill, there was the silver lining of HB 2001: the nation's most progressive housing and zoning policy reform bill. HB 2001 required all areas in the Portland metro region, and cities of 20,000 people or more statewide, to legalize duplexes, triplexes, fourplexes, cottage clusters, and townhouses in residential areas, including all single-family zones, by June 30, 2022. (It also required that smaller cities legalize duplexes in all single-family zones by June 30, 2021.)

This context-shifting legislative move obviously changed the goalposts for the RIP, by essentially requiring a certain outcome. Combined with the guidance from the PSC, it all but required staff to legalize fourplexes citywide, except for in specific areas with documented infrastructure deficiencies or environmental constraints.

During 2020, City Council engaged in a lengthy public comment period and amendment process, during which certain tweaks were made to the RIP. These included a focused allowance for six-plexes where at least half the units are affordable to households making no more than 60 percent of the Median Family Income (MFI), among other edits. This was a compromise: I had worked with advocates and non-profits to craft a proposal to allow eight-plexes if two of the units were affordable to households making no more than 80 percent of the MFI and one was affordable to a household making no more than 60 percent of the MFI, as this would allow for the market-rate units to cross-subsidize the affordable units so that the entire eight-plex could be built by a non-profit without additional subsidy. Council didn't see the benefit of this, however, and preferred the six-plex option, reasoning that plenty of subsidies would be available for non-profits or something. Or perhaps there was no reasoning; it was just one of those political compromises where nobody wins.

Regardless, this means that, when the RIP was finally adopted in August of 2020, it went beyond our original objective of allowing fourplexes and the strict requirements of HB 2001.

One area of concern around the RIP, as well as HB 2001, revolves around design. Looking back at the original testimony that the Concordia Neighborhood Association submitted, which called for the use of the existing Community Design Standards to regulate new multifamily types within existing neighborhoods, it is clear that the city has consistently ignored guidance to regulate design with the RIP. The state has similarly been blind to the role of design in its regulations, leaving this up to local jurisdictions if it is regulated at all. The danger here is that residents will see poorly designed multifamily structures replace aging homes in their neighborhood and will blame the fact of them being multifamily for becoming an ugly blight upon the neighborhood, without realizing that it is not this, but the lack of design regulation that leads to poor aesthetic outcomes. My hope is that somehow, good design regulations will be implemented before the blowback against poorly designed new structures becomes too severe.

After all, there's a lot to love with the RIP: not only did it legalize four-plexes in single-family zones citywide, making Portland the first major American city to do so, but it also eliminated minimum parking requirements for those same zones. More homes and less parking: these, at least, are necessary components for climate success.

The Future

The quirky, bold, and progressive city of Portland has firmly cemented its reputation as the green urbanist capital of North America, but what does its future look like? What does it need to do next to ensure that it continues to enjoy the social and economic benefits of green urbanism?

In recent years, equity has become the biggest topic of local "green growth" conversation. How do we ensure that the benefits of going green are spread equally among every Portlander? Portland has the potential to become one of the most equitable cities in the United States; but how does potential become realized, and how can that goal be instituted into policy? The city must ensure that minority/low-income communities are getting their fair share of the economic benefit from new public investments and private construction. This means improving management of the negative side effects of gentrification and continuing to support and expand programs that help minority neighborhoods.

The city must also continue to work on improving and "greening" its infrastructure, especially in its car-oriented suburbs. To do this, it must develop a unique toolkit for the various urban environments and topographies that exist in these areas. For example, there are several suburban communities that exist in the mountains just outside the city. A bike lane would not be remarkably effective in such an environment, so the city would have to develop a different strategy to encourage residents to reduce their car usage; an all-electric car-sharing program would be a good example. Transit plus e-bikes/scooters is a good solution even for areas with significant topography.

Portland, soon, will also have to deal with its "Falling Star" neighborhoods: first tier suburbs built in the 1940s, 1950s, and 1960s that are starting to decay and decline in value. Such neighborhoods are not old enough to have "historic" value, but their advanced age is starting to take its toll, and they are falling into disrepair. As the homes lose value, they are resold to lower-income owners who do not have the financial capacity to keep the homes maintained, which lowers the value further. This results in a negative feedback loop, which can cause the neighborhood to decline very quickly. Portland must devise new strategies to break this decline and revitalize these neighborhoods into healthy, stable communities.

Finally, Portland must continue to expand its internationally renowned bicycling and transit programs. It must learn to address public resistance resulting from the NIMBY attitude so that it can further expand its streetcar and light rail network. Additionally, it must create new and innovative strategies to increase bicycle ridership, which appears to have hit a ceiling, leveling out at about 7 percent bicycle mode share.

How does the city break this ceiling? Portland is now on the third generation of Biketown bike share, including on its second generation of bikes featuring electric assist. The city must focus on increasing minority ridership. Biketown now has a low-income option to do exactly that.

However, Biketown is currently just a municipal system, one whose service boundary is currently only a subset of the city limits. In order for access to this system to truly be equitable, it really should be managed by the regional government, Metro, rather than the city, so that the e-bikes can become available for suburban residents, including many people of color who had previously been priced out of the city and now reside in other nearby jurisdictions. Further, the city and the region need to concentrate many more resources on building out a regional bicycle and pedestrian trail system, to allow for same, comfortable, and enjoying bicycling and e-biking experiences to be available for more people to reach more destinations."

Portland: An Example of Green Urbanism in Action

The other 270 mid-sized cities in America should look to Portland as an example of how investing in alternative transit and dense, mixed-use neighborhoods can pay off handsomely; an example of how "being green" and economically thriving do not have to be competing ideas.

Portland has not only become a city with a steadily shrinking environmental footprint, but one that is experiencing an economic boom based on the "green" economy. Portland has the largest percentage of college-educated residents of any mid-sized city in America. Today, it continues to experience a massive influx of young and college-educated citizens and boasts steadily rising rates of walking, bicycling, and transit use. While visiting Portland, one can experience a vibrant craft economy, an active downtown, and fully revitalized urban neighborhoods.

The young, educated millennial loves the "green city" because it provides an opportunity to practice a healthier and more fulfilling lifestyle; nearly 40 percent of Portlanders choose to give up their cars and use alternative transportation. Indeed, Portland's charm is that it is a weird, unconventional, and exciting place to live, and today the city is living proof that investing in the "green" lifestyle pays off in a big way.

APPENDIX

Change in median home prices for 20 major metropolitan areas from June 2006 to June 2012 and average year-by-year change in median home values:

CITY	YEAR-OVER-YEAR CHANGE		SINCE PEAK		SINCE JUNE 2006	
Las Vegas	+28%		-49%		-48%	
San Francisco	+25%		-19%		-19%	
Los Angeles	+21%		-25%		-24%	
San Diego	+20%		-25%		-25%	
Phoenix	+19%		-39%		-39%	
Atlanta	+18%		-18%		-17%	
Detroit	+17%		-29%		-26%	
Miami	+14%		-40%		-39%	
Tampa	+13%		-37%		-36%	
Seattle	+12%		-17%		-10%	
U.S. 20-city index	+12%		-21%		-21%	
Portland	+12%		-16%		-12%	
Denver	+10%		+0%		+4%	
Minneapolis	+9%		-21%		-21%	
Dallas	+8%		+0%		+6%	
Chicago	+8%		-25%		-25%	
Charlotte	+8%		-9%		-2%	
Boston	+6%		-8%		-6%	
Washington, D.C.	+6%		-19%		-19%	
Cleveland	+4%		-14%		-14%	
New York	+4%		-21%		-21%	

Figure 16.1 Comparing Portland's Housing Costs with Other Cities—Housing Is Not More Costly in Green Cities. *Source:* Standard and Poor's Case-Shiller Home Price Index June 2006–12

REFERENCES

Abbott, C. (2011). *Portland in Three Centuries: The Place and the People.* Corvalis: University of Oregon Press.

Agyeman, J., McLaren, D., & Schaefer-Borrego, A. (2013). Sharing Cities. *Friends of the Earth Briefing*, 1–32.

Anderson, S. (July 24, 2013). Interview by C. P. Stevens and J. Gilderbloom.

Brian, H. (July 22, 2013). Interview by C. P. Stevens and J. Gilderbloom.

Carter, S., & Quealy, K. (July 30, 2013). *Housing's Rise and Fall in 20 Cities.* http://www.nytimes.com/interactive/2011/05/31/business/economy/case-shiller-index.html?_r=1&.

Casey, N. (January 21, 2018). Venezuela's Most-Wanted Rebel Shared His Story, Just Before Death. *The New York Times*, sec. Americas. https://www.nytimes.com/2018/01/21/world/americas/venezuela-oscar-perez-nicolas-maduro.html.

City of Portland. (2012). *The Portland Plan: Prosperous. Educated. Healthy and Equitable.* file:///C:/Users/shahb/Downloads/Adopted_PDXPLAN_Web.pdf.

citytoplists. (n.d.). *Top 101 Cities with the Most People below the Poverty Level, Excluding Cities with 15% or More of Residents in College and with the Median Age below 28 (Population 50,000).* http://www.city-data.com/top2/c3.html

Clifton, K. J., Morrissey, S., & Ritter, C. (2012). Catering to the Bicycling Market. *TR News*, (280), 26–32.

Cortright, J. (July 24, 2013). Interview by C. P. Stevens and J. Gilderbloom.

Cortright, J. (2007). *Portland's Green Dividend.* http://www.impresaconsulting.com/node/42.

Ehrenhalt, A. (2012). *The Great Inversion and the Future of the American City.* 1st ed. New York: Knopf.

Engels, F. (1973). *The Condition of the Working Class in England: From Personal Observation and Authentic Sources.* Moscow: Progress Publishers.

Fainstein, S. S. (2010). *The Just City.* Ithaca: Cornell University Press.

Florida, R. (2017). *The New Urban Crisis: How Our Cities Are Increasing Inequality, Deepening Segregation, and Failing the Middle Class—and What We Can Do About It.* 1st edition. New York: Basic Books.

Frederick, C., & Gilderbloom, J. (2018). Commute Mode Diversity and Income Inequality: An Inter-Urban Analysis of 148 Midsize US Cities. *Local Environment* 23, no. 1, 54–76.

Frederick, C., Riggs, W., & Gilderbloom, J. H. (2018). Commute Mode Diversity and Public Health: A Multivariate Analysis of 148 US Cities. *International Journal of Sustainable Transportation* 12, no. 1, 1–11.

Gilderbloom, J. (2014). Rebel Cities: From the Right to the City to the Urban Revolution by David Harvey, (book review) by John I. Gilderbloom. *Journal of Urban Affairs* 36, no. 5, 945–947.

Gilderbloom, J., & Appelbaum, R. P. (1987). *Rethinking Rental Housing.* Philadelphia: Temple University Press.

Gilderbloom, J. I. (2018). *Chromatic Homes: The Joy of Paint in Historic Places.* Lexington: University Press of Kentucky.

Gilderbloom, J. I., Hanka, M., & Lasley, C. B. (2009). Amsterdam: The Ideal City, Policy and Planning. *Local Environment: The International Journal of Justice and Sustainability* 14, no. 6, 373–392.

Gilderbloom, J., Hargrove, E., & Canfield, J. (2015). *From Blighted to Beautiful: Covington HOPE VI Final Report.* Louisville: University of Louisville: posted http://sun.louisville.org

Gilderbloom, J., Grooms, W., Mog, J., & Meares, W. (2016). The Green Dividend of Urban Biking? Evidence of Improved Community and Sustainable Development. *Local Environment* 21, no. 8, 991–1008.

Gilderbloom, J. I., & Squires, G. D. (October 20, 2015). Forget Red and Blue States: Go Green for Better Jobs, Health and Environment. *Courier Journal.*

Gilderbloom, J. I., & Squires, G. D. (2016). How Environmental Toxins Reduce Life Expectancy in Many American Cities. *Scholars Strategy Network.* Accessed February 4, 2018. http://www.scholarsstrategynetwork.org/brief/how-environmental -toxins-reduce-life-expectancy-many-american-neighborhoods.

Goodman, P. (1960). *Growing up Absurd.* London: Sphere Books.

Gotschi, T. (2011). Cost and Benefits of Bicycling Investments in Portland, Oregon. *Journal of Physical Activity and Health* 8, S49–S58.

Greenstone, S., & Van Wing, S. (May 19, 2016). *Are African-Americans Really Leaving Portland?* Accessed February 5, 2018. https://www.opb.org/radio/programs/ thinkoutloud/segment/oregon-portland-african-americans/.

Hale, N. (July 25, 2013). Interview by C. P. Stevens and J. Gilderbloom.

Heying, C. (2010). *Brew to Bikes: Portland's Artisan Economy.* Portland: Portland State University: Ooligan Press.

Kaufman, K. (July 25, 2013). Interview by C. P. Stevens and J. Gilderbloom.

Lehto, A. (July 23, 2013). Interview by C. P. Stevens and J. Gilderbloom.

Lord, B. (July 23, 2013). Interview by C. P. Stevens and J. Gilderbloom.

Marx, K. (2007). *The Economic and Philosophical Manuscripts of 1844.* New York: Dover Publications, Inc.

McCann, E., & Ward, K. (2011). *Mobile Urbanism: Cities and Policymaking in the Global Age.* Vol. 17. Minneapolis: University of Minnesota Press.

Montgomery, C. (2013). *Happy City: Transforming Our Lives through Urban Design.* New York: Macmillan.

Plato, Cooper, J. M., & Hutchinson, D. S. (1997). *Complete Works.* Indianapolis: Hackett Publishing.

Poole-Jones, C. (July 22, 2013). Interview by C. P. Stevens and J. Gilderbloom.

Reeves, M. (July 24, 2013). Interview by C. P. Stevens and J. Gilderbloom.

Sadowsky, R. (July 25, 2013). Interview by C. P. Stevens and J. Gilderbloom.

Swanson, K. (2012). Center for Disease Control, *Bicycling and Walking in the United States: 2012 Benchmarking Report.* Washington, DC: Library of Congress.

Vigdor, J. L., Massey, D. S., & Rivlin, A. M. (2002). Does Gentrification Harm the Poor? [With Comments]. *Brookings-Wharton Papers on Urban Affairs* 2002, 133–182.

Chapter 17

Amsterdam

Planning and Policy in the World's Most Livable City

John Hans Gilderbloom and Matthew J. Hanka

The urban revolution arrived in Amsterdam to ignite the change that other cities need to follow for the survival of modern civilization: dense and compact neighborhoods near work and school, biking and walking for over half the population, light and hard rail access, minimal personal usage of cars, historic preservation, personal freedom, and support for the arts to attract young people. Is Amsterdam the "ideal city"? Many of the social, economic, and environmental problems facing Amsterdam are considerably fewer than in cities in the United States and, in most cases, Western Europe. At this moment in history, Amsterdam might be the world's greatest city because of its ability to ensure necessities, freedom, and creativity. Tolerance of drugs and sexual freedom, along with the integration of different races, has reduced many of the "social problems" faced by most cities.

We have compared, on a per capita basis, differences between the Netherlands and the United States, and comparisons are made on the city level and per capita numbers are compared on the national level. Our data show that Amsterdam has lower rates of crime, murder, rape, drug usage (cocaine and marijuana), teenage pregnancy, diabetes, obesity, suicide, abortion, infant mortality, dependence on fossil fuels, and homelessness. There is also considerably less racial segregation. People live longer because of Amsterdam's walkability, bike usage, and access to parks. The Netherlands leads both Western Europe and the United States in walking and bicycling trips, significantly reducing car dependency. Ghettos are nearly non-existent compared to the United States with its discriminatory housing policy from the 1930s. Quality housing that gives residents pride of place is supplied to everyone compared to the stark, cold, and institutional "projects" provided by the U.S. federal government.

In the Netherlands, social housing organizations ensure that 4 million people live in 2.4 million houses. They are responsible for adequate and affordable housing, contribute to the quality of life in neighborhoods, and invest in the construction of new dwellings and sustainability. In the United Kingdom in 2021, around 4 million houses were occupied by households socially renting. The largest shares of social rented households in England during the period under observation were 2000 and 2001, when a total of almost 20 percent of all households were recorded as socially rented. People living in Amsterdam are more tolerant, secure, happier, and healthier compared to citizens in the United States. Great cities provide for everyone and enhance the lives of everyone and ensure an unrivaled level of freedom. This chapter demonstrates that social and environmental justice are tied together. Amsterdam is by no means perfect, but in comparison to many other democratic industrial cities, it is a far better place for citizens of all races, religions, and incomes.

A central concern of urban studies is to develop an understanding of social and spatial constraints on necessities that are distributed on a non-random basis, such as housing, environment, health, and transportation. British sociologist and neo-Weberian Pahl (1975) asserts that an individual's life opportunities are powerfully influenced by "managers" such as the government, banks, developers, landlords, and business owners, who determine the use of space (see also Molotch, 1976). In his book *Whose City,* Pahl (1975) argues that "access to resources is systematically structured in a local context" (p. 203). A person's life is not determined solely by his or her relationship to the means of production, but by his or her spatial location in the urban system. Inequalities are generated within and among cities. Why can certain cities within capitalism provide essential needs while others cannot?

Pahl (1975) attempts to understand how access to fundamental needs varies among urban areas and to identify why certain urban places have difficulty allocating necessities, while others do not. Pahl creates boundaries of what urbanists should and should not look at: "housing and transportation are elements in my view of the city, family allowances and pension schemes are not" (p. 10).

Pahl asserts that we must come to understand how urban managers and gatekeepers can increase or decrease these necessities for living in the city. Elected leaders, non-profit organizations, bureaucrats, social movements, and community organizers change, remake, rebuild, revitalize, and destroy the pattern and infrastructure of the city that can have an impact on the life opportunities of all citizens. The gatekeepers allocate resources, make policies, and enforce police actions (building code enforcement, planning approvals, in the criminal justice system, etc.). Pahl's thinking on cities was problematic because it focused only on basic human needs. It never explored or embraced the importance of cultural, creative, and personal freedoms.

METHODOLOGY

Like the founders of urban sociology (Louis Wirth, Ernest Burgess, and Max Weber), we conduct a comparative analysis that needs further refinement and testing. Our results are indicative of trends and not definitive. Ideally, we try to include large numbers of cities as our experimental variable, but alas, there is only one Amsterdam; after that everything gets diluted. We compare Amsterdam with the averages of other American cities. In terms of social sciences, ideally, it would be nice to have 50–75 Dutch cities to compare to 50–75 U.S. cities, but Amsterdam is unique. Amsterdam is the outlier, the extreme city, the wild city, the ideal city. What we're interested in is comparing social and environmental justice in the United States versus Amsterdam.

We believe using Amsterdam as a case study can yield some important findings (Feagin et al., 1991; Clavel, 1986; Durkheim, Weber, Capek, and Gilderbloom, 1992). French sociologist Durkheim (1895) argues that the essence of social science starts with comparison and should attempt to mimic the hard sciences as a method of comparing one sample that has the treatment to another sample without the treatment (see also Kantor and Savitch, 2005).

Few studies have been done across national boundaries, and if there are comparative studies across nations, few, if any, have been done across cities (Fainstein, et al., 1983; Kantor and Savitch, 2005; Abu-Lughod, 1999), particularly Van de Ven's (1997) book comparing Brussels to Amsterdam, and Musterd and Salet's (2003) book comparing Dutch cities with other European cities. Other comparisons across cities include studies between Houston and Santa Monica (Capek and Gilderbloom, 1992), Burlington, Berkeley, and Santa Monica in Pierre Clavel's (1986) *The Progressive City*, and rent regulation policies among U.S. cities (Gilderbloom and Appelbaum, 1988).

Comparing cities across international boundaries can be illuminating in measuring differences in social policy and planning. It is only through comparison and measuring relationships that we can achieve greater certainty (Kantor and Savitch, 2005). Durkheim's (1897) study of suicide was notable because he explained variations in suicide by what kind of religion was dominant in each area. His study was considered a landmark sociological study since it showed how taking one's life was a non-random phenomenon associated with religious norms tied to places. This chapter applies these themes to our framework of analyzing Amsterdam compared to American cities. We went to numerous government agencies in Amsterdam that collect data at the municipal, regional, and national levels, and gathered data for United States cities. We also used computer search engines to uncover data and create new tables that compare Amsterdam/Netherlands to the United States. While such data has been looked at for more than 100 years, we thought the comparison of Amsterdam to the United States was novel and the most comprehensive analysis to date. The data presented go well beyond previous studies in terms of scope and breadth. Our comparison would test whether public policy, political leadership, and public administration make a difference in the lives of people. Although this is not exhaustive, we think it's illustrative of how public policy can impact capitalist cities and their quality of life.

While there have been other studies that attempt to measure the quality of life in Amsterdam, and some provide comparisons with the United States and other European cities (Hajer, 1995; Terhorst and Van de Ven, 1997; Fischer and Hajer, 1999; Terhorst, Van de Ven, and Deben, 2003; Hoffman, Fainstein, and Judd, 2003), our more nuanced contribution uses a much wider variety of indicators that measure environmentalism, social justice, housing, transportation, health, crime, racial conflict, and sustainability. We did not cherry-pick these tables—everything we found on a comparative basis is presented—even if they go against our thesis.

While we've increased the number of indicators measuring social justice, many of these measures are done on a per capita national basis and are not broken down by city or region, either in the United States or in the Netherlands. It is important to note that Amsterdam's "social justice" is not isolated from the national government, which works hand in hand by adopting and passing the policies pushed by the Netherlands' largest city.

The differences observed would be sharper if they could be broken down by city. It is important to remember that the Netherlands is a tiny country, roughly about the size of Maryland. If it were a state, in terms of size it would rank 44th out of 50 U.S. states, followed by Massachusetts and Hawaii. Indeed, five out of eight tables/figures in this paper are comparisons done on a per capita basis.

What is commonly pushed in Amsterdam is passed on a national level, albeit on a more moderate level, so it is only fitting that national and local policies would be interconnected. While we believe this data suggests significant differences in measuring social justice in the United States and the Netherlands, we hope this will spark comparative analysis of the relationship between policies at the national and city levels and its impact on people, social justice, and the environment.

The Politics in Amsterdam—A Wide Range of Ideas

The Nazi occupation of the Netherlands left a permanent imprint on the Dutch. The Dutch experience under Nazi fascism, which led to horrific genocide, forced slavery camps, hunger and starvation, and a massive effort to exterminate nearly 100,000 Dutch Jews along with other "out" groups such as gay people, the disabled, and leftists, helped reinforce ideas of tolerance and guarantee essential rights and human necessities among the Dutch. This tolerance, along with numerous social programs fueled by the radicalism of the 1960s, is embedded in the Dutch identity and evident throughout post-World War II Dutch culture. This includes the Dutch political system, where Amsterdam and the Netherlands are represented by a plurality of parties aimed toward progressive goals.

Like some European nations, the Netherlands operates under a parliamentary form of government, with a monarch as the head of state. Parliamentary government usually consists of multiple political parties, which build a coalition government. There is a low threshold based on proportional representation, with a considerable number of parties winning a low percentage of seats (Musterd and Salet, 2003).

The national parties are represented in the Amsterdam municipal government. The municipal government has 45 seats, with each seat occupied by a council member elected to serve four-year terms. The mayor of Amsterdam is nominated by the municipal council but appointed by the Queen instead of being popularly elected, unlike the United States or even other parliamentary governments (World Mayor, 2006). In an interview, former Amsterdam mayor Job Cohen stated that "being elected by the inhabitants of the city would not give a mayor more authority—it would provide him with more powers because he would have obtained a mandate by the majority to execute his political programs" (World Mayor, 2006).

The recent municipal election in March 2006 gave the Labour Party (PvdA) 20 of the 45 seats. The Liberals (VVD) received eight seats, the Green Party (GroenLinks) and the Socialist Party (SP) each won six seats, the Democrats won two seats, and the Christian Democrats (the majority party in the Dutch parliament) have just 2 of the 45 seats on the Amsterdam City Council (www .iamsterdam.com, Council Election, 2006).

The 2006 municipal elections in Amsterdam resulted in a coalition among the three parties with the most seats in the council (the Labour (PvdA), the Liberals, and the Green Left) to form a "red-green" body for the first time. This new coalition has worked with the mayor and the rest of the Amsterdam municipal government to tackle issues such as poverty, housing, transportation, immigration, racial integration/segregation, economic development, promotion of a creative class business climate, and even some trivial issues such as "dogs pooping in public places" where parties take different stances.

Although the winner-take-all system on the municipal and local level in U.S. cities can still allow the election of progressive mayors, the structural limitations of the U.S. two-party system and the structure of federalism do not produce as many progressive outcomes as in the multi-party system of European nations such as the Netherlands. Unlike the U.S. system, the Dutch and Amsterdam governments have proportional representation, which forces them to build coalitions with other political parties to create a government. Many of these parties, particularly on the left, have helped reinforce the culture of tolerance that has come to define Dutch culture and Dutch identity.

Amsterdam: Decriminalization of Sex Trade

As an important port city, Amsterdam has been in the business of sex since the late thirteenth century. Approximately 25,000 prostitutes work in the Netherlands according to a survey by the A. de Graaf Foundation (Prostitution Information Center, 2007). Combined statistics show that on a given day, 6,000 prostitutes work in Amsterdam alone (Prostitution Information Center, 2007). Ideally, sex work should be free of middlemen, or pimps, and in the words of the Prostitution Information Center, "to end abuses in the sex industry" (Prostitution Information Center, 2007). In October 2007, Amsterdam's social housing organizations attempted to reduce the number of brothels by enforcing laws aimed at taking organized crime out of the sex trade, leading to a reduction in prostitution.

Despite the city's openness about the sex industry, Dutch teens have low pregnancy and abortion rates. Table 17.1 compares teen sexual behavior rates between the Netherlands and the United States. On average, Dutch teens wait longer to have sex, are less likely to become pregnant, are less likely to pursue abortion, and are more likely to use contraception. The abortion rate in the United States is nearly double that of the Netherlands. The rate in the Netherlands is growing, according to the National Ministry of Health, Welfare and Sport, largely due to a higher rate among women of foreign descent (Netherlands Ministry of Health, Welfare and Sport, 2007). Advocates for Youth, a U.S. organization that advocates for sexual education, highlighted the differences between European and American sexual education and media campaigns. Access to information, services, and contraception is confidential and free. American campaigns encourage abstinence, while Dutch campaigns deliver the message that safe sex is an issue of self- and partner-respect and is necessary for teen sex (Advocates for Youth, 1998).

Table 17.1 Comparisons of Sexual Behavior, Abortions, and HIV Rates between the United States and the Netherlands

	United States	Netherlands
Teenage birth rate[1]	20.3	4.5
Teenage abortion rate[1]	15 percent	0.3 percent
Age of first intercourse[2]	18.1	18
Contraceptives at first intercourse[1]	44 percent	93 percent
Abortions[3] (per 1000)	14.6	8.6
HIV[4,5]	0.3 percent	63

Sources: 1. Advocates for Youth (2006) European Approaches to Adolescent Sexual Behavior & Responsibility.
2. South Carolina Campaign to Prevent Teen Pregnancy (2003) Adolescent Sexual Health.
3. Sedgh G et al. (2007) "Induced abortion: rates and trends worldwide" in the *Lancet*, 370(9595):1338–1345.
4. U.S. Department of Health and Human Services, Centers for Disease Control and Prevention. (2007). Centers for Disease Control and Prevention. HIV/AIDS Surveillance Report, 2005. Vol. 17. Rev ed.
5. Van de Laar, M.J.W.; De Boer, I.M.; Koedijk, F. D. H. and E. L. M. Op de Coul. (2004) HIV and Sexually Transmitted Infections in the Netherlands in 2004. Ministry of Health, Welfare and Sports, Centre of Infectious Diseases Epidemiology.

In the Netherlands, sex and drugs is a health issue, not a criminal issue. In Amsterdam, there is a five-day waiting period between a request for an abortion and the procedure, giving a woman enough time for reflection (Netherlands Ministry of Health, Welfare and Sport, 2007). In the Netherlands, all late-term abortions are illegal (Netherlands Ministry of Health, Welfare and Sport, 2007). Until recently, U.S. law allowing abortions was determined by the state, but a 2003 law put a federal ban on third-trimester abortions. Prior to the legislation, most states forbade abortion in the final trimester of pregnancy.

Table 17.1 also shows HIV rates for both nations. Like the abortion rate, the HIV infection rate in the United States is higher than those in the Netherlands. Within these statistics, different groups have increased rates. Heterosexual infections are growing in the United States. In the Netherlands, growth has occurred among male homosexuals, a group that has leveled off in the United States (U.S. Center for Disease Control, Van de Laar, 2005). The Netherlands has a unique approach toward preventing the spread of AIDS among heroin addicts. The Dutch view that the establishment of a network, a consensus of preventive measures, and cooperation has been "essential in the successful implementation of an extensive and far-reaching national AIDS prevention program for drug users" (Leuw and Marshall, 1994, 67). This includes a wider needle-exchange program, which has not resulted in increased needle usage and indicates that 25 to 30 percent of all heroin addicts in the Netherlands use drugs intravenously (Leuw and Marshall, 1994, 67). Another study disputes this by finding that the percentage of intravenous drug usage is as low as 10.5 percent (Netherlands Ministry of Justice, 1999).

Amsterdam: Decriminalization of Drug Usage

Like sex, the Dutch consider drug use a health issue and focus on harm reduction. Soft drugs are decriminalized with penalties rarely enforced. Proprietors of coffee shops, which sell soft drugs in the Netherlands, are charged only under certain circumstances, such as selling of large amounts to a single user, selling to a child, or creating a nuisance. Hard drugs are restricted with several innovative programs to get people treatment. The Netherlands encourages health care prevention while cracking down on organized crime and major drug trafficking. This policy was developed in 1976, when Parliament established a commission to reconsider prohibition. Marijuana usage was considered a youth culture rebellion, but its use is typically short-lived, mostly recreational, and unlikely to lead to the use of more dangerous drugs (Dolin, 2001). In 1995, several government ministries combined to evaluate the policy and discovered that the Netherlands had some of the lowest rates of drug use in Europe for both hard and soft drugs (Dolin, 2001). Table 17.2 compares rates of drug use between the Netherlands and the United States, showing that U.S. citizens are twice as likely to smoke pot and four times more likely to try cocaine than in Holland. Cocaine usage is less in the Netherlands, while smoking and alcohol consumption are a little more. Table 17.2 finds that the annual prevalence of heroin or opioid use between ages 15 to 64 is much higher in the United States than in the Netherlands. In 1995, there were 2.4 drug-related deaths per million inhabitants in the Netherlands, compared to 9.5 for France, 20 in Germany, 23.5 in Sweden, and 27.1 in Spain (Netherlands Ministry of Justice, 1999).

Amsterdam: Lower Crime Rates

The Dutch believe that decriminalization of drugs and sex work has dealt a blow to organized crime networks that once controlled a large portion of

Table 17.2 Comparison of Drug Use (15–64 years old)

	United States	Netherlands
Annual prevalence of **cocaine** use[1]	2.4	0.6
Annual prevalence of **cannabis** use[1]	13.7	5.4
Annual prevalence of **amphetamines** use[1]	1.5	0.3
Annual prevalence of **ecstasy use**[1]	1.4	1.2
Annual prevalence of **opioids** use[1]	5.9	0.3
Annual prevalence of **drugs** use[1]	2.9	1.6

http://www.unodc.org/documents/data-and-analysis/WDR2011/StatAnnex-consumption.pdf [accessed 12/4/18].
Source: 1. Organization for Economic Co-operations & Development (OECD). World Drug Report 2011.

window prostitution (Brants, 1998). In fact, data shows a reduction in crime, which will be discussed in this chapter. The police can focus on more pressing concerns than prostitution and drug busts. The sale and use of soft drugs (cannabis and hash) are tolerated and not prosecuted for much the same policing principle. While all drugs are illegal according to the Opium Act, soft drugs are decriminalized. However, hard drug laws are still enforced with many innovative programs to get people the treatment they need, offered in lieu of penal sentences.

The criminalization of drugs and the sex trade is a factor in the increase of violent crime. Miron (2004) claims the U.S. homicide rate could be reduced by 25 to 75 percent by decriminalizing drugs. If that figure is accurate, decriminalizing drugs would reduce the American homicide rate to a number comparable to the Dutch. Decriminalization has resulted in lower crime rates, including lower rates of soft and hard drug use.

Table 17.3 shows criminal justice data collected by the United Nations Office on Drugs and Crime and the Organization for Economic Co-operations & Development (OECD). While police per capita are comparable, a closer look demonstrates critical differences. Overall, the United States experiences a

Table 17.3 Comparison of Crime Rates

Per 100,000 Population	United States	Netherlands
Police[2]	198	298
Total crime[3]	8,517	8,212
Per capita spending on criminal justice system (1998)[1]	379 euro	223 euro
Homicide[3]	5	0.6
Attempted homicide[3]	330	9
Assaults[3]	238	282
Rapes[3]	39	7
Robberies[3]	103	57
Thefts[3] (2016)	1773	3215
Automobile theft[3]	220	138
Burglaries[3]	491	1348
Frauds[3]	134	124
Total persons convicted[3]	23	520
Drug Offenses[2]	560	47
Juvenile convicted[3] (2013)	0.6	174
Incarceration rate[3]	638	116.8
Parole rate[2]	231	1
Guns[4]	220	2

Sources: 1. Van Dijk, Frans & Jaap de Waard, "Legal infrastructure of the Netherlands in international perspective: Crime control" (Netherlands: Ministry of Justice, June 2000).
2. United Nations Office on Drugs and Crime (UNODC). https://data.unodc.org/ [accessed 12/4/18]
3. CIVITAS Institute for the Study of Civil Society 2010–12. Comparisons of Crime in OECD Countries. *http://www.civitas.org.uk/crime/comments.php* [accessed 12/4/18].
4. Krug, E.G., Powell, K.E.; and L.L. Dahlberg. (1998) International Journal of Epidemiology, 27 (2):214–221.

higher crime rate. The incarceration rate per 100,000 is ten times higher in the United States than in the Netherlands, and per capita spending on the criminal justice system is nearly twice as high in the United States. The homicide rate is eight times higher in the United States than in the Netherlands. Attempted homicide includes any assault with a deadly weapon in the U.S. data but was not specified in Holland. There is a dramatic difference in these numbers. Violent person-on-person crime is 15 percent of all American crimes, compared to 5 percent of all crime in the Netherlands. Most Dutch crime is property and theft.

According to Table 17.3, citizens in the United States are much more likely to have a gun than a Dutch citizen. U.S. laws differ by state, but guns are easy to obtain and possess (Krug, Powell, and Dahlberg, 1998). In Holland, guns are nearly impossible to buy and keep. A Dutch person cannot even inherit a gun unless that person is a member of a licensed shooting club, hasn't committed a crime of any type, and is not known "to move in criminal circles." Guns must be locked in a special safe, separate from the ammunition (Netherlands Justice Ministry, 2005). Guns can get into the Netherlands via Belgium, where firearms are sold to the public.

From the data, we can also can see that more juveniles are convicted in the Netherlands by a greater margin. Alternatively, in the United States, the incarceration rate is increasing and 10 times that of the Netherlands (see Table 17.3). The trend toward violent crime explains some of this, but not all. A case can be made that the pursuit of violent criminals takes more police resources, so a low violent-crime rate, such as in the Netherlands, allows police more time to pursue thefts, burglaries, frauds, and embezzlements than their American counterparts. It would also inflate the Dutch crime rate, especially considering the lack of drug crimes. Rape cases in the United States are also increasing and almost six times higher than those in the Netherlands (see table 17.3).

Table 17.4 shows health and quality-of-life data (WHOSIS, 2007; U.S. CDC, 2005; OECD, 2018; Netherlands CBS, 2005). The Dutch are healthier despite a higher percentage of the GDP spent on health by the United States. The Dutch live longer. They maintain a healthier weight, get more exercise, and suffer less from diabetes. Infant mortality and cancer mortality are lower in the Netherlands compared to the United States; however, the infant mortality rate for both is decreasing. More people smoke tobacco in the Netherlands, though it is decreasing. Alcohol consumption is also decreasing in the Netherlands compared to the United States, where people drink more (see Table 17.4). All these factors are likely correlated. Unlike the United States, where many bars, restaurants, and entertainment establishments are going smoke-free, the Dutch still allowed smoking in bars until the summer of 2008. The suicide rate and GDP spent on health in both the United States and Netherlands are increasing as of 2016, with the United States having higher rates in both cases.

Table 17.4 Comparison of Health and Life Quality Statistics

Percentage Unless Noted	United States	Netherlands
Alcohol consumption[1]	8.8	8.3
Smoker[2]	11.8	18
Obesity (m)[1]	30.2	11.4
Obesity (f)[1]	30.1	15.7
Diabetes[1]	9.4	6
Gets enough exercise[2]	48	56
Suicide rate[1]	13.8	10.5
Life expectancy (m)[1]	76.1	80
Life expectancy (f)[1]	81.1	83.2
Healthy life expectancy (m)[2]	67	70
Healthy life expectancy (f)[2]	71	73
Infant mortality (per 1,000)[1]	5.9	3.3
Cancer mortality (per 100,000)[1]	318	304
Percent of GDP spent on health[1]	17.2	10.1

Sources: 1. Organization for Economic Co-operations & Development (OECD). 2018. Health Statistics 2018. http://www.oecd.org/els/health-systems/health-data.htm [accessed 12/3/18].
2. Netherlands Central Bureau of Statistics (2005).

Amsterdam: The Tolerant City

Amsterdam is the great liberal experiment in Europe. No part of the city announces that liberalism as boldly as the red-light district, where—in the Oudekerksplein—there coexists the Old Church (possibly the oldest building in Amsterdam, believed to be consecrated by the Bishop of Utrecht in 1306), all the prostitutes in their doorways and windows, and a kindergarten. What other city in the world would believe in the hopeful coexistence of religion, prostitution, and early childhood education? But it is both brave and original of the Dutch to celebrate human differences. (John Irving, November 2005; Majoor, 2005)

Before exploring the tolerance of Amsterdam, let's examine some definitions of tolerance. There are two kinds of tolerance, negative and positive. Negative tolerance is "the capacity to 'put up with' another's difference from self because the different other is simply not perceived and/or because the self and other do not intersect" (Lofland, 2000, 147). This can be done through either sharing a large, bounded space but not the smaller pieces of it or by sharing smaller spaces within the larger space, but separate from each smaller group (Lofland, 2000). Positive tolerance is "the capacity to put up with others' fully recognized differences from self . . . with a mild appreciation for, or enjoyment of, those differences" (Lofland, 2000, 146–149). Positive tolerance can be achieved by having diverse people forced into conflicts without recourse, developing a capacity for tolerance (Lofland, 2000).

Amsterdam touts its tolerance. The city's English-language website geared toward tourists emphasizes the ethnic, cultural, and sexual diversity of the city. The population is made up of 170 nationalities (de Boer, 2007). Amsterdam is a highly integrated city with a 45 percent ethnic minority population. Many of Amsterdam's minorities come from former colonies, such as Suriname, Indonesia, and the Netherlands Antilles. Others come from nearby largely Muslim nations, such as Turkey and Morocco.

Although the Surinamese population of Amsterdam has decreased slightly in past years (see Table 17.5), Amsterdam's Turkish population has increased by seven percent, the Moroccan population has increased by more than 10 percent, and the number of non-Western and Western foreigners who have moved to the city has increased (Cijfers, 2006). The location of these ethnic groups has increased the spatial inequality of these populations. Starting in 2000, the large Surinamese and Moroccan populations were deconcentrated out of the massive public housing in the Southeast, and in the Bijlmermeer section, away from the activity of Amsterdam Centrum (Deben, 2000). Today, the Bijlmermeer housing projects have been demolished and residents are being resettled and integrated throughout Amsterdam. The Surinamese settled in lower-income and less attractive neighborhoods but enjoy greater social mobility, while the Moroccans settled in higher-end areas but are less socially mobile, which can be explained by differences in cultural and economic origins (Musterd and Salet, 2003). Regardless, the ethnic groups experienced little segregation because of the tolerant culture.

Today, ghettos are unheard of in Amsterdam. This ethic of tolerance and diversity has resulted in a city policy that prevents the creation of ethnic ghettos and enhances exposure of immigrants to the ethnic Dutch population and culture. Unlike the United States, ghettos and/or highly segregated places, which are nearly all poor and made up of one race, do not exist in the Netherlands. This is because of the dominant role of the social welfare state in the

Table 17.5 Comparison of Population by Ethnic Origin

Ethnic Origin	2002	2003	2004	2005	2006	Index 2002 = 100
Surinam	71,464	71,471	70,717	70,446	69,645	97
Antillean	11,799	11,705	11,503	11,523	11,360	96
Turkish	35,806	36,594	37,333	37,943	38,337	107
Moroccan	58,809	60,767	62,691	64,370	65,426	111
Other non-western foreigners	63,063	65,791	68,159	70,049	70,401	112
Total non-western foreigners	240,941	246,328	250,403	254,331	255,169	106
Western foreigners	100,268	101,121	102,671	104,723	105,112	105
Native Dutch	394,119	388,596	385,689	383,897	382,746	97
Total	735,328	736,045	738,763	742,951	743,027	101

Source: Amsterdam in Cijfers. 2006.

Netherlands in fostering the integration of immigrants, as well as the use of the market and other "reciprocity networks" among various ethnic groups in the Netherlands (Aalbers and Deurloo, 2003, 197–208).

The world media has sensationalized the murder of two well-known Dutchmen (Theo Van Gogh and Pim Fortuyn) in the 2000s by two non-native Dutch, one of whom committed the crime in the name of the Muslim religion. These murders have given people the impression that the ethos of tolerance has been reduced since the Dutch have responded with stricter immigration and deportation laws requiring law-abiding foreigners to develop the ability to speak and write in Dutch and not be appalled by Dutch sexuality. Hanneke Gelderblom, a former Dutch Senator, has argued that Holland cannot allow everyone in the world to come to Amsterdam, especially those who don't want to embrace the Dutch culture of tolerance.

Racial unrest among Muslims in Paris was not present for the same group in Amsterdam. Leon Deben, a professor at the University of Amsterdam, argues that this kind of unrest would not happen in Amsterdam or any other major Dutch city because the minority population is integrated, as opposed to Paris, where Middle Eastern youth are isolated in large housing blocks (Expatica, 2005; Canadian Broadcasting Channel News, 2005; Deben and Bontje, 2006).

The ethic of tolerance is permanently embedded into the consciousness and identity of the Dutch people, and the experiences under Nazi occupation reinforced these values of tolerance and diversity. Heiner (2005) argues that it was their experience during World War II that helped the Dutch create a post-war society based on a commitment to a more humane criminal justice system in the Netherlands.

Also, trade with Indonesia and other former Dutch colonies meant greater diversity and tolerance. Many Dutch-speaking Indonesians arrived in the Netherlands during the post-war economic boom, as well as ethnic groups such as the Turks, Moroccans, Surinamese, and immigrants from the Antilles (Culture of the Netherlands).

While traditional Dutch society was built upon four "pillars," or social, political, and educational systems that were separated by religious and political beliefs, societal shifts after World War II and in the 1960s led to a rapid depillarization of society. Devoted religious practice was declining on the political left, and societal concepts of inclusion were on the rise; combined, these led to a dramatic societal shift. The injection of a fifth pillar in recent years, made up of a largely Muslim immigrant population, has further challenged the Dutch to accommodate the change. This has at times led to trouble, from flamboyant media to assassinations, but a prevailing societal commitment to inclusion demonstrates a Dutch commitment to working through social change to develop a new structure with a peaceful resolution (Halink and Carpman, 2003; Fainstein, 1999).

Tolerance and Creative Class

Richard Florida, who coined the term "creative class," believes that for cities to be successful they must possess the "three Ts" of economic growth, which are technology, talent, and tolerance (Florida, 2005; Deben and Bontje, 2006). Cities with a creative class have economic and social characteristics such as high levels of diversity and tolerance, a large gay population, and a highly educated and skilled workforce. The size and population of the city and the number of amenities the city offers also influence creativity. Tolerance toward drug usage and sex is a major magnet for the tourist industry as well as attracting creative young professionals to the city.

Critics claim that Florida's creative class would lead to more social exclusion, which one might think the city of Amsterdam doesn't want or doesn't have. By Florida's own admission, members of the creative class are "inward-looking and selfish" and have yet to develop a vision of society in which all citizens can participate and benefit. Individual mobility divides the United States into the "have regions" favored by the creative class, and the "have-not regions," which are largely abandoned (Dreher, 2003). The critical view of the creative class is much stronger in Europe than in the United States, where many cities have adopted Florida's ideas as gospel and argue that the creative class is the engine necessary for economic growth and development.

Statistical data on the sexual orientation of Dutch residents is difficult to find. These records are not kept due to privacy laws. However, Amsterdam's reputation as the "gay capital of Europe" is well known (Timmermans, 2006). The moniker is touted by the city and celebrated by the 1987 Homomonument and the annual Gay Amsterdam parade. The city made world history in 2001 when it became the first municipality to perform same-sex marriages (Municipality of Amsterdam, 2007). It also has a large immigrant population and is a world leader in technology. Amsterdam already has attracted a large creative class to the city, as a result of many of its residents and Western foreigners being attracted to the city's arts, culture, image, and history (Bontje and Crok, 2005). Amsterdam's lifestyle and tolerance are selling points to outsiders, something that Amsterdam reinforces as an economic development strategy. The producers of the creative class in Amsterdam, such as the creative industries, are there because of the location within the city, the role of the economy, and the city's image. The creative industries, as well as foreigners' and citizens' strong connection to the development of Amsterdam, have helped Amsterdam grow into a creative knowledge center, although the city hasn't developed a policy to become a creative knowledge city (Bontje and Crok, 2005).

The Dutch view of tolerance allows people to endure the beliefs or behaviors of others that differ from their own beliefs or behaviors, not just put up with them. This extends to the everyday norms of city life, including traffic. To the Dutch, the bicycle is not merely for transportation, but is a tool that promotes citizen solidarity (Reinarman, 2007). This is called pragmatic tolerance, where the different behaviors regarding drugs, sex, and lifestyle do not necessarily impact the sovereignty of others.

Amsterdam Housing Policies

The Netherlands leads all European nations (as well as the United States and Canada) in the production and maintenance of social housing. Social housing is essentially housing taken out of the free market and placed within a cooperative-run entity founded by a church or a political party. Most Dutch social housing is controlled by nonprofit organizations, not the government (Van der Veer and Schuling, 2005). As Harloe (1985) notes, "the problem is that the free market has, historically, been unable to provide housing to socially acceptable standards at a cost which is affordable by low-income households." Limited recognition of this fact provided the basis for the development of housing policy and subsidies in Europe. In the Netherlands, the percentage of social housing dwellings is the highest in Europe.

In Table 17.6, housing cost burden for homeowners and renters (both privatized and subsidized) is higher than in the U.S. However, about one-third of the Dutch housing stock was social in 2005. In the Amsterdam region, 34 percent of dwelling units are considered social, and 55 percent of the municipal units are social. The average rent share is 24 percent, much

Table 17.6. Housing Statistics Comparison

	United States	Netherlands
Housing cost burden[1] (Mortgage owners)	15.9 percent	19.5 percent
Housing cost burden[1] (Privatized & subsidized)	25.4 percent	29 percent
Dwellings[2] (per 1000)	419	449
Social rental dwellings[2]	4.3 percent	**34.1 percent**
Number of homeless[2]	564,708	31,000
Poor housing[1] (No flushing)	0.5 percent	0.0%
% GDP on housing Allowance[2]	0.1 percent	0.5 percent

Source: 1. Organization for Economic Co-operations & Development (OECD). 2018. Affordable Housing Database. http://www.oecd.org/social/affordable-housing-database.htm [accessed 12/3/18].
2. CBS/BE Bevolkingsstatistiek, CBS/WON Woningstastistiek, Ouwehand and Van Daalen.

lower than the rent for the American working class (Gilderbloom, 2008). Seventy-five percent of the housing stock, including all private rentals, is rent controlled. The waiting list for social housing in Amsterdam can be 9 to 12 years. Since social housing is a precious commodity, many people hang onto their apartments long after they can afford more expensive market-rate housing or rent them illegally to those who have little choice but to rent that unit.

Amsterdam residents consume about half as much housing as their American counterparts. They are creative in producing housing in attics, basements, boats, old warehouses, tree houses, silos, and even shipping containers. Housing policies are sound, with strong historic preservation practices that protect historic buildings. It is not uncommon to walk through Amsterdam and see a large collection of buildings that are 100, 200, 300, and even 400 years old. If any buildings are torn down, 90 percent of the materials (wood, brick, glass, and plumbing) are recycled and used again. Squatting laws also encourage landlords to fix up abandoned housing units or otherwise face losing these unused structures. The best green house is an old house.

While the United States has an estimated 744,000 homeless in shelters and on the streets, and nearly 322,000 unsheltered homeless in American cities, the Dutch have only a fraction of that number (National Alliance to End Homelessness, 2007). According to a Google search: Homelessness in the Netherlands is a growing social problem, affecting 39,300 people in 2018. The homeless population has risen between 2009 and 2019. We observe that homelessness is not nearly the problem in Amsterdam as it is in American cities (Gilderbloom, 2008; National Alliance to End Homelessness, 2007). Our own observations, along with four graduate classes doing fieldwork in Amsterdam, have yet to see a visible homeless population sleeping on sidewalks, park benches, and lawns, as you would find in a typical U.S. city.

In a new study, the Chicago Coalition for the Homeless (CCH) reports that 58,723 Chicagoans experienced homelessness in 2019, down 16 percent from the year prior. The decline is part of a trend seen during the latter half of the 2010s, with Chicago's homeless count decreasing by more than 30 percent between 2015 and 2019. In New York, about 45,300 people slept in NYC shelters, including 16,500 single adults and 14,600 children in 2022. However, the numbers reported by the DHS don't give the full scope of homelessness, as these exclude people in other facilities, such as faith-based centers, youth centers and overnight drop-in centers.

The 2020 Homeless Count & Survey shows that we can expect to see 3,974 people experiencing homelessness at any given moment in the Houston region. Throughout the entire Miami-Dade County, nearly 900 people are homeless, according to the Miami-Dade Homeless Trust. In 2019, volunteers counted more than 8,000 homeless people in San Francisco's shelters, jails, and living on the streets. It was a 17 percent rise from the prior count in 2017, a staggering jump that alarmed local officials who had been spending about $300 million a year on the issue. In January 2019, Los Angeles County had 58,936 people experiencing homelessness, but by January 2020 the number rose to 66,433. The city of Los Angeles counted 36,165 in 2019, and 41,290 in 2020.

Leon Deben (2007), the leading Dutch urban sociologist who has done counts of the homeless in Amsterdam, estimated that, according to the GGD, 500 to 1,000 people are homeless in Amsterdam. (Total inhabitants are 800,000.) He compared this number with Louisville, which has a homeless population of 6,986 unduplicated homeless people (both unsheltered and sheltered) between October 1, 2017, and September 30, 2018, which was a 4.3 percent increase from the 6,695 served in the previous year. The homeless living on the streets in Amsterdam rotate off the street and into either shelters, family homes, or detoxification programs with education and job training.

One concern is that the amount of social housing has fallen over the years. Moreover, the amount of space for the new Dutch middle-class is significantly less than that for Americans. For example, the Amsterdam middle class lives in two-bedroom, one-bath dwellings, while the typical American middle-class family lives in a three-bedroom, two-bathroom house, which occupies a much larger amount of land. Space matters.

Amsterdam: The Green City

With more bicycles than people, the lack of petroleum dependency is evidence that Amsterdam is an eco-friendly city and one of the 10 most eco-friendly cities, according to the Environmentor. Amsterdam is progressive and green in more ways than its famous "coffee shops." It's notorious for being the city where there are more bikes than people (bikes are piled over each other in the streets!) and holds the title of the most bicycle-friendly capital city in the world. Not only does biking ensure a healthier population, but it also reduces carbon emissions and pollution significantly. The municipality integrates actions to make it an example of green urbanism.

In Amsterdam, there are around 847,000 bikes in Amsterdam that belong to 442,693 households. There are four times more bicycles than cars in Amsterdam. 80 percent of Amsterdammers own a bike, and almost 60 percent use their bikes on a daily basis, more than most of its European neighbors and the United States. The Netherlands leads both post-industrialized North American and European countries in terms of the proportion of trips taken by walking or biking. In the Netherlands, half the trips are petroleum-free, while in the United States, it's 7 percent, while European countries like France and England lag far behind. In Amsterdam, 67 percent of the population uses a bike at least once a month. One-third of residents go to work by bike.

However, bike use among the Muslim immigrant community is far lower (de Boer, 2007). Intricate webs of bicycle lanes have improved comfort and safety for cyclists and helped increase use. Some parts of Amsterdam prohibit cars and allow only pedestrians and bikes. Bikeways also enhance mobility and access for the disabled. Tommy Clark, who directs the Louisville Metro Office of Disability Services, said that Amsterdam might be one of the most wheelchair-accessible major cities in the world.

Pucher (2007, 2003) has provided empirical data showing that urban residents who walk or ride bikes gain an extra hour of life for every hour of biking, or an average biking culture can increase the lifespan of a person from two and a half to four years in Holland and Germany. Moreover, the state saves on health cost payouts because people are healthier. He noted that people who ride bikes tend to be smarter and happier, and "we save money on commute times, give less money to countries that dislike us, and pollute the environment less." According to Pucher (2007:3):

> The image of the city is closely associated with cycling. In a new marketing campaign the city calls itself "Bike Capital of Europe." . . . 50% of them make daily use of the bicycle, 28% a few times per week and 8% weekly. . . . The city is flat and has a very compact semi-circular development pattern. Many bridges make bike transport very convenient.

He also makes this observation:

> Bike use reached a low point in the 1970s with only 25% of all trips. Then, bicycle advocates and environmentalists promoted the bike. Their views were shared by many citizens, as automobiles caused congestion and pollution in the city. The two possible scenarios were to either adapt the city to automobile traffic (building parking in city center, widening roads to ease congestion, etc.) or to improve public transport and cycling options. In 1978 a new city council took office, which promoted cycling. (Pucher 2007:2)

With a growing urban population and increased car and air traffic that accommodates it, Amsterdam knew it had a critical air pollution problem before the European Union established baseline levels for pollution. Improving air quality is one of the four pillars of the city government's current environmental policy, and the city is taking aggressive steps in monitoring and reducing emissions. Expected hot spots of poor air quality near the A-10 Highway and near Schiphol Airport exist. While Amsterdam's air quality suffers from typical urban pressures, it is still below CAFÉ standards and is aggressively attempting to stay there or improve. Cutting down on vehicle emissions, despite traffic that is growing at the rate of the economy, has kept ground-level ozone levels unchanged (Van Der Zee, Helmink, and De Jonge, 2007). Garbage incinerators in Amsterdam are among the cleanest in the world (Dare, 2007).

With 65 miles of ancient canals, water pollution can be a serious threat. When they were first constructed in the Middle Ages, the many working windmills kept water circulating. Today's urban environment offers additional threats, such as automobile pollution. To combat these threats, water is flushed two to four times a week, and special cleaning boats troll the waterways to clear them of surface litter. Converting the houseboats to the sewer system in 2005 made a huge difference and has turned the canals into a vibrant place of aquatic life with accompanying waterfowl. Herons and pelicans are common sights near the canals and on houseboat ledges.

In addition, dredging barges clean the canal bottoms, lifting about 10,000 discarded bicycles and 100 sunken boats each year (Amsterdam City Government, 2008). Amsterdam recycles soil and building supplies and encourages its boroughs to recycle material that it dredges from the canals. These and other efforts are outlined in its Environmental Policy Plan, which is updated every four years. The recently completed plan focuses on improving the environment in four areas that were believed could be improved in a short period of time. These are noise pollution, air quality, climate change, and the sustainable use of raw materials. These initiatives include aggressive mapping efforts, sustainable building practices, sustainable features in new large-scale developments, wind-generated energy, and centrally ordering tropical hardwood to minimize environmental and financial costs (Municipality of Amsterdam, 2007). Amsterdam even recycles its landfills. In 2002, efforts to clean up the Volgermeerpolder landfill began. Once rehabilitated, this area will serve as a nature preserve on the northern edge of the city (Municipality of Amsterdam, 2007). Amsterdam's city council recognizes a need for green places, if not for recreation, then just to keep people grounded. "The goal is that everyone is 15 minutes from looking a cow in the eyes," Democratic 66 Amsterdam councilman Sebastiaan Capel said about Amsterdam's efforts to incorporate meaningful green space into the community. "Our green spaces are not waiting for something else, but demands are high on space. Housing and public space are problems in the same areas" (Capel, 2007).

Despite pressure, policy mandates and enhances public green space. The method the city is using to meet this goal dates back to the 1935 General Expansion Plan. Although the original plan intended for green spaces with the city wedged between, the current look is quite the reverse: urban spaces with green fingers reaching in. No green space is in danger of being lost, not even to the car. Capel recalled plans for a new expressway, a ramp of which would cut through a green finger. City officials focused on the ramp, wanting to know what it was. They were told it was a proposed ramp, but it wouldn't make it to the final highway plans. The officials refused to consider the expansion until they were shown a plan without any lines through the green finger (Capel, 2007). The green spaces have survived through the 2002 plan with plans to expand the nine green fingers.

Part of the expansion will come as the city builds new islands on the north side of town. It's a necessary step because growing inland would put pressure on the agricultural Green Heart Dutch project. The Green Heart is a natural area surrounded by the Randstad consisting of landscapes, water, and cities. Holland's Green Heart is also not up for negotiation (Marchal, 2007).

With its dense population and tourism burden, public space in Amsterdam's Centrum is scarce. Even in this cramped city, there is room for green space in the city center. From the sloped mall of Museumplein that covers an underground parking garage, to the beloved Vondelpark, an art-imitating-nature urban park, to Westergasfabriek, a brownfield-turned-culture center, to Amsterdamse Bos, a post-World War II landscape park three times the size of Central Park in New York City, Amsterdam makes sure that while a cow may be more than 15 minutes away, an escape to nature is a short walk down the street.

Amsterdam: A City Engineered for Tomorrow

There's an Old Dutch saying: "God built the world, but the Dutch built Holland." They reclaimed their land from the waters of the North Sea, using ever-evolving civil engineering techniques to grab land from the seas, drain it, and raise it above sea level. This process of developing polders is what gives Holland its unique landscape. Flat peaty plateaus rise from one water table to the next highest. In the early days, the lifting of these polders was done by windmills pumping water from the land to enhance buoyancy. Pumps are now run by conventional electricity.

Today, population pressure has created a need for land near Amsterdam. Not to be deterred by something as simple as a lack of land, the Dutch dredge the bottom of the IJmeer, a lake that was once a shallow inlet of the North Sea. The dredged land is piled up and pressed out to develop islands used to expand the land area near Amsterdam (Marchal, 2007).

The Dutch are not the only nation to prepare for disasters, but they plan for much larger events. Protection is nationally mandated to defend the country from at least a 1,250-year flood. Massive storm barriers are engineered to defend the low-lying nation from storm surges. Dikes, storm walls, and dams must be built strong enough to prevent a significant disaster in the Netherlands (Deltawerken Online, 2004). Building to legal minimums is not common practice in the Netherlands. Many floodways are engineered for a 4,000-year flood. Levees and floodwalls in the United States are engineered at the 100-year level. As several recent American emergencies prove, such as Hurricane Katrina, the flooding of the Platte River, and the 1993 Great Midwest floods, this level of protection is often inadequate even for those beyond the 500-year floodplain. There were 34 federally declared flood disasters in the past decade in the United States. There were four in the past five decades in Holland, the most significant in 1953, before much of the current protections were developed (Kok, Vrijling, Van Gelder, and Vogelsang, 2002).

An estimated 70 percent of the Dutch population is at risk for flooding, something inherent in being a low-lying state with many water resources (Kok, Vrijling, Van Gelder, and Vogelsang, 2002). However, aggressive engineering and maintenance plans have made flood losses over the second half of the twentieth century a modest 1.58 billion euros with fewer than 2,000 deaths (Kok, Vrijling, Van Gelder, and Vogelsang, 2002). No conversion rate is needed to determine that the Dutch had less damage than the United States' $195 billion in flood damage over the same period (NOAA, 2007).

Since heavy rains and mountain melts in the heartland of Europe led to extreme flooding of the Rhine and Maas Rivers in 1993 and 1995, there has been a refocus on protecting Holland from its rivers. A re-evaluation of Holland's dike system has been performed. Changes are being made to allow a natural spillover floodway into natural and unpopulated floodplains. Another plan includes increasing the width of the IJssel River, which will disrupt some lives but is considered necessary to defend against global warming (Sterling, 2007).

Perhaps the greatest global warming challenge is the rising sea, threatening lands claimed hundreds of years ago. More than half of Holland lies below sea level. The Dutch solution: raise the land on the coast quickly. The plan is to inject the limestone foundation of the land with sulfuric acid, creating expanding gypsum with nowhere to go but up (Kroon, 1993).

From creating land to expanding floodways to defending against global warming, the Dutch understand that human/nature conflict is something that requires engineering entrepreneurship to avoid. While other nations, such as the United States and Japan, regulate against the hazards they can combat and help victims rebuild, the Dutch take emergency planning to the next level by employing their most creative ideas to protect Dutch lives and property.

CONCLUSION

There is much to learn from the Dutch about important solutions to policy and planning. Amsterdam provides us with valuable lessons. Cities should be remade for all to enjoy. Goodman (1956) observed that an individual "has only one life and if during it he has no great environment, no community, he has been irreparably robbed of a human right." This question is symbolic of a much larger debate about democracy, community, and the economic riches that will determine what kind of society the United States, the Netherlands, and other European and Western capitalist democracies will become in the twenty-first century. With the failure of socialist and communist societies, we need to work within the parameters of capitalism.

Marxism essentially argues that all cities in capitalism are wretched places filled with poverty, disease, fear, death, and crime. Amsterdam shows that a capitalist city can meet the essential needs of the people, such as health, housing, safety, individual freedom, sustainable living, and transportation. Amsterdam demonstrates how within the framework of capitalism, democracy and green thinking provide solutions for meeting the basic human needs of nearly all of Amsterdam's residents. Amsterdam is a place of freedom, not repression.

Amsterdam and the Netherlands rank the highest among the thirty-eight Organization for Economic Cooperation and Development (OECD) nations in providing the most thorough and effective welfare state by ensuring a higher standard of living for the dependent poor (even among the immigrant populations). They have achieved this by offering more employment options, housing options, and services to families and children than the United States (Mollenkopf, 2000). More importantly, policies ensure that every Amsterdammer has housing by providing housing for all income groups.

Most urban scholars can tell us what is wrong with a city, but few can provide a recipe or model for revitalizing inner-city neighborhoods with affordable housing, green urbanism, accessibility, and integration. Amsterdam is the ideal city to gain inspiration from because it is a city that solves problems instead of ignoring them. This chapter hopes to shed new light on the forces that shape our cities.

Can we have hope about the future of cities? Can a more comprehensive, more engaging, and perhaps more optimistic theory of the city be embraced? There is a lot we can learn from the Dutch about important solutions to policy and planning. Amsterdam is a city that does not ignore or shun the invisible populations (Gilderbloom, 2008). It has created progressive policies and initiatives that are designed to reach every section of society, thereby benefiting everyone. Learning about cities such as Amsterdam provides us with valuable insights and lessons, and more importantly, represents the greatest and most compelling example of sustainable living in our society. These are the kinds of questions we should be addressing as we seek to develop a new urban paradigm.

Revised and reprinted with permission from *Local Environment.*

NOTE: I still have family here, most notably Hanneke Gelderblom, known as the "Anne Frank that lived" and is subject of museum exhibits and documentary films. She survived the Nazis and thrived: becoming a city council member in The Hague, getting elected as a Senator, working on the World Court of war crimes, and co-founding a political party (D-67) that honors civil rights for Muslims, Jews, gays, and the elderly. She also helped pass strict gun control laws that have resulted in fewer weapons used in crimes. She serves on the European Commission. Her husband Hans Gelderblom is a noted architect in Amsterdam. They have asked me to carry on their work on human freedom. That is why when I met with my family in 1970s Holland, they asked me to use Hans as my name. More interestingly, she declined Steven Spielberg's offer to make a film about her life.

REFERENCES

Aalbers, M., & Deurloo, R. (2003). Concentrated and condemned? Residential patterns of immigrants from industrial and non-industrial countries in Amsterdam. *Housing, Theory, and Society, 20*(4), 197–208.

Abu-Lughod, J. (1999). *New York, Chicago, Los Angeles.* Minneapolis: University of Minnesota Press.

Advocates for Youth. (1998). *Differing European/U.S. approaches to teen sex show surprising results.* Advocates for Youth. http://www.advocatesforyouth.org/NEWS/PRESS/1998/100198.htm.

Amsterdam City Government. (2006). *Summary of environmental policy plan.* Government of Amsterdam (Gemeente Amsterdam).

Amsterdam in cijfers 2006. (2006). *Government of Amsterdam (Gemeente Amsterdam).* Dienst Onderzoek en Statistiek.

Amsterdam Tourism Bureau. (2008). *Amsterdam canals.* http://www.amsterdam.info/canals/.

Barclay, G., Tavares, C., Kenny, S., Siddique, A., & Wilby, E. (2003*). International comparisons of criminal justice statistics 2001.* London: Home Office Research, Development & Statistics Directorate.

Bontje, M., & Crok, S. (2006). Amsterdam, A creative knowledge city? The debate on the economic future of Amsterdam and its region. In L. Deben & M. Bontje

(Eds.), *Creativity and diversity: Key challenges to the 21st century city* (pp. 144–161). Amsterdam: Het Spinhuis.

Brants, C. (1998). The fine art of regulated tolerance: Prostitution in Amsterdam. *Journal of Law and Society, 25*(4), 621–635.

Canadian Broadcasting Channel News. (2005). *In depth: France riots. Understanding the violence.* CBC News Online, pp. 1–5.

Capek, S., & Gilderbloom, J. I. (1992). *Community versus commodity: Tenants and the American city.* Albany: State University of New York Press.

CIVITAS Institute for the Study of Civil Society 2010–2012. *Comparisons of crime in OECD countries.* Retrieved April 12, 2018, from http://www.civitas.org.uk/crime/comments.php.

Clavel, P. (1986). *The progressive city: Planning and participation: 1969–1984.* New Brunswick: Rutgers University Press.

Council Elections 2006. (2007). *Iamsterdam.* Retrieved August 12, 2007, from http://www.iamsterdam.com/introducing/government_politics/council_elections.

Culture of the Netherlands. (2007). http://www.everyculture.com/Ma-Ni/The-Netherlands.html.

Cunningham, M., & Henry, M. (2007). *Homelessness counts: Research reports on homelessness.* Washington, DC: National Alliance to End Homelessness.

Dare, P. (2006). Incineration debate stuck in past era, air quality expert says. *Ottawa Citizen,* March 7, 2006.

Deben, L., & Bontje. M. (2006). *Creativity & diversity: Key challenges to the 21st century city.* Amsterdam: Het Spinhuis.

Dolin, B. (2001). *National drug policy: The Netherlands.* Parliamentary research. Ottawa: Library of Parliament.

Dreher, C. (2002). Be creative or die: Interview with Richard Florida. *Salon.com.,* June 6.

Durkheim, E. (1895). *The rules of sociological method.* New York: The Free Press of Glenco.

Durkheim, E. 1897 [1951]. *Suicide: A study in sociology.* Translated by J. A. Spaulding & G. Simpson. New York: The Free Press of Glenco.

Fainstein, S., Fainstein, N., Hill, R., Judd, D., & Smith, M. (1983). *Restructuring the city. The political economy of urban redevelopment.* New York: Longman, Inc.

Federal Emergency Management Agency. (2007). *Significant flood events.* Retrieved August 16, 2007, from http://www.fema.gov/business/nfip/statistics/sign1000.shtm.

Fischer, F., & Hajer, M. A. (1999). *Living with nature: Environmental politics as cultural discourse.* Oxford: Oxford University Press.

Florida, R. (2002). *The rise of the creative class.* New York: Basic Books.

Florida, R. (2003). Cities and the creative class. *City and Community, 2*(1), 3–19.

Gilderbloom, J. I. (2008). *Invisible city: Poverty, housing, and new urbanism.* Austin: University of Texas Press.

Goodman, P. (1956). *Growing up absurd: Problems of youth in the organized system.* New York: Random House.

Hajer, M. A. (1995). *The politics of environmental discourse: Ecological moderniza-tion and the policy process.* Oxford: Oxford University Press.

Halinka, S., & Carpman, D. (2005). *Social capitalism: Challenging the monocultural fantasy of Dutch politics.* Amsterdam: Humanity In Action.

Harloe, M. (1985). Landlord/tenant relations in Europe and America—The limits and functions of the legal framework. *Urban Law and Policy, 7*(4–5), 359–383.

Heiner, R. (2005). The growth of incarceration in the Netherlands. *Federal Sentenc-ing Reporter, 17,* 227–230.

Hoffman, L. M., Fainstein, S. S., & Judd, D. R. (Eds.). (2003). *Cities and visitors: Regulating people, markets, and city space.* New York: Blackwell Publishing, Inc.

Iamsterdam. (2007). *Gay, lesbian and bisexual milestones.* Retrieved August 12, 2007, from http://www.iamsterdam.com/visiting_exploring/special_interest/gay _lesbian_bisexual/milestones.

Kantor, P., & Savitch, H. V. (2005). How to study comparative urban development politics: A research note. *International Journal of Urban and Regional Research, 29*(1), 135–151.

Kok, M., Vrijling, J. K., Van Gelder, P. H., & Vogelsang, M. P. (2002). *Risk of flood-ing and insurance in the Netherlands in flood defense.* New York: Science Press.

Kroon, R. L. (1993). *Dutch plan: Raise land above the seas.* International Herald Tribune [online]. Retrieved August 12, 2007, from http://www.iht.com/articles /1993/01/12/dike.php.

Krug, E. G., Powell, K. E., & Dahlberg, L. L. (1998). Firearm-related deaths in the United States and 35 other high- and upper-middle-income countries. *International Journal of Epidemiology, 27*(2), 214–221.

Leuw, E., & Marshall, I. H. (1994). *Between prohibition and legalization: The Dutch experiment in drug policy.* Amsterdam: Kugler Publications.

Lofland, L. H. (2000). Urbanity, tolerance, and public space: The creation of cosmo-politans. In L. Deben, W. Heinemeijer, & D. Van de Vaart (Eds.), *Understanding Amsterdam- Essays on economic vitality, city life, and urban form* (pp. 143–161) Amsterdam: Het Spinhuis.

Logan, J., & Molotch, H. (1987). *Urban fortunes: The political economy of place.* Berkeley: University of California Press.

Majoor, M. (2005). *De Wallen in beeld. Images of Amsterdam's red-light district.* Amsterdam: DeWallenwinkel/Prostitution Information Centre.

Marchal, M. (2007). Stad en Groen: Nieuw Groen voor Ijburg en Zuidas. Amsterdam: Municipality of Amsterdam, Municipal Project Summary.

Miron, J. (2004). *Drug war crimes: The consequences of prohibition.* Boston: Inde-pendent Institute.

Mollenkopf, J. (2000). Assimilating immigrants in Amsterdam: A perspective from New York. In L. Deben, W. Heinemeijer, & D. Van de Vaart (Eds.), *Under-standing Amsterdam- Essays on economic vitality, city life, and urban form* (pp. 197–218). Amsterdam: Het Spinhuis.

Molotch, H. (1976). The city as growth machine: Toward a political economy of place. *American Journal of Sociology, 82*(2), 309–332.

Moulaert, F., Martinelli, F., Swyngedouw, E., & González, S. (2005). Towards alternative model(s) of local innovation. *Urban Studies*, *42*(11), 1969–1990.

Musterd, S., & Deurloo, R. (2006). Amsterdam and the preconditions for a creative knowledge city. *Tijdschrift voor Economische en Sociale Geografie*, *97*(1), 80–94.

Musterd, S., & Salet, W. (2003). *Amsterdam human capital*. Amsterdam: Amsterdam University Press.

National Weather Service. (2007). *Hydrologic Information Center-Flood losses*. Retrieved August 15, 2007, from http://nws.noaa.gov/oh/hic/flood_stats/Flood _loss_time_series.htm.

Netherlands Central Bureau of Statistics. (2005). *Reported health and lifestyle data [online]*. http://www.cbs.nl/en-GB/menu/themas/gezondheid-welzijn/cijfers/ nieuw/default.htm.

Netherlands Ministry of Health, Welfare and Sports. (2007). *Abortion*. Netherlands Ministry of Health, Welfare and Sports web site. Retrieved August 10, 2007, from http://www.minvws.nl/en/themes/abortion/ default.asp.

Netherlands Ministry of Health, Welfare and Sport, Drug Policy in the Netherlands. (1999). *Progress report September 1997–1999*. The Hague: Ministry of Health, Welfare and Sport, p. 7.

Organization for Economic Cooperation & Development (OECD). (2011). *World drug report 2011*. Retrieved April 12, 2018, from http://www.unodc.org/documents/data-and-analysis/WDR2011/StatAnnex-consumption.pdf.

Organization for Economic Cooperation & Development (OECD). (2018a). *Affordable housing database*. Retrieved March 12, 2018, from http://www.oecd.org/ social/affordable-housing-database.htm.

Organization for Economic Cooperation & Development (OECD). (2018b). *Health statistics 2018*. Retrieved March 12, 2018, from http://www.oecd.org/els/health -systems/health-data.htm.

Pahl, R. (1975). *Whose city?* Middlesex: Penguin.

Pucher, J. (2007). *Case studies of cycling in Amsterdam: The Netherlands*. New Brunswick: Rutgers University, Center for Urban and Economic Research, Working paper.

Reinarman, C. (2007). *Signs and shapes of tolerance: Ethnographic reflections*. Draft Manuscript.

Scott, A. (2006). Creative cities: Conceptual issues and policy questions. *Journal of Urban Affairs*, *28*(1), 1–17.

Sterling, T. (2005). Dutch advisory panel: Divert the Rhine to prepare for global warming. Forecast Earth. Retrieved August 12, 2007, from http://climate.weather .com/articles/dutchrhineplans040507.html.

Summary Environmental Policy Plan Amsteram 2004–2006. (2006). *Policy summary*. Amsterdam: Municipality of Amsterdam.

Terhorst, P., Van de Ven, J., & Deben, J. (2003). Amsterdam: It's all in a mix. In L. M. Hoffman, S. S. Fainstein, & D. R. Judd (Eds.), *Cities and visitors: Regulating people, markets, and city space*. New York: Blackwell Publishing, Inc.

Terhorst, P. J. F., & van de Ven, J. C. L. (1997). *Fragmented Brussels and consolidated Amsterdam: A comparative study of the spatial organization of property rights* (pp. 75–90). Amsterdam: Netherlands Geographic Society.

Timmermans, A. (2006). *Trip out!* Amsterdam. Retrieved August 12, 2007, from http://www.logoonline.com/travel/destinations/europe/netherlands/amsterdam .jhtml.

Trimbos Institute. (November 2002). *Report to the EMCDDA by the Reitox national focal point, The Netherlands drug situation 2002.* Lisboa: European Monitoring Centre for Drugs and Drug Addiction, p. 28, Table 2.1.

United Nations Office on Drugs and Crime (UNODC). Retrieved April 12, 2018, from https://data.unodc.org/.

U.S. Center for Disease Control. (2007). *Cases of HIV infection and AIDS in the United States and dependent areas,* 2005, (17). Government annual report. Atlanta: U.S. Department of Health and Human Services.

U.S. Department of Health and Human Services (HHS). (2002). *Substance abuse and mental health services administration, national household survey on drug abuse: Volume I.* Washington, DC: Summary of National Findings.

Van de Laar, M. J. W., de Boer, I. M., Koedijk, F. D. H., & Op de Coul, E. L. M. (2005). HIV and sexually transmitted infections in the Netherlands in 2004: An update. Amsterdam: Ministry of Health, Welfare and Sport.

Van der Veer, J., & Schuiling, D. (2005). The Amsterdam housing market and the role of housing associations. *Journal of Housing and the Built Environment, 20*(2), 167–181.

Van Der Zee, S., Helmink, H., & De Jonge, D. (2007). *No decrease in PM10 concentrations in Amsterdam in the period 1999–2005.* World Health Organization Centre for Air Quality Management and Air Pollution Control Newsletter v. 39.

Van Dijk, F., & de Waard, J. (2000). *Legal infrastructure of the Netherlands in international perspective: Crime control.* Netherlands: Ministry of Justice.

"The Volgermeerpolder." (2007). *Dienst Milieu en Bouwtoezicht.* Retrieved August 12, 2007, from http://www.dmb.amsterdam.nl/ipp/main.asp?name=pagina&item _id=519&selected_balkitem_id=.

Walmsley, R. (2003). *World prison population list* (Fifth edition). London: Research, Development and Statistics Directorate of the Home Office.

Chapter 18

Historic Preservation as a Sustainability Strategy to Foster Pro-Environmental Cultures

John Hans Gilderbloom, Matthew J. Hanka, and Joshua D. Ambrosius

Preservation is environmentally friendly behavior because it involves the continued reuse of existing buildings and materials. One of the best ways to create affordable housing is by renovating historic properties. The cost of renovating an old, abandoned home is about one-half the cost of building new because you already have the sewer, water, plumbing, bricks, wood, and electrical infrastructure. Historic buildings should never be demolished. In Holland, it is against the law to demolish a building, and if it is allowed because of fires, floods, or explosions, you must recycle 98 percent of the materials. This reduces future energy consumption by taking advantage of the *embedded energy*—the energy used in the construction of these structures—already expended in the past. Preservation and reuse further assist in rejuvenating downtowns, making them more desirable and livable compared with suburbs, thereby reducing automobile usage and creating more walkable and bike-friendly communities within a metropolitan area. This chapter demonstrates the linkage between the preservation of historic structures and the creation of jobs, increases in property values, and the encouragement of sustainable attitudes and behaviors by residents.

HISTORIC DISTRICTS AND RESIDENTIAL PROPERTY VALUES

The bulk of literature about local and National Register historic districts shows that property values rise faster than in unprotected or undesignated neighborhoods. The value of each historic home is protected by controls on the exterior of the house or by mandating that the house be well-maintained using historic paint colors and materials. Property values are further protected by an assurance that other nearby properties will maintain their historic character and never be demolished, which limits negative externalities. Most studies have shown a positive correlation between property value increases and historic preservation (Gilderbloom & Hanka, 2007b; Gilderbloom, House, & Hanka, 2008; Rypkema, 1994; Coulson & Lahr, 2005; Coulson & Leichenko, 2001; Leichenko, Coulson, & Listokin, 2001; Ford, 1989; Shipley, 2000; Mason, 2005). Haughey and Basolo (2000) find a federal historic preservation district by itself has a positive impact on property. Rypkema's (1997b) study of a city in Indiana showed that five neighborhoods protected by local historic zoning ordinances in the state did better overall in property appreciation than similar, unprotected neighborhoods. Florida (2002) and Rypkema (2006) both focus on the powerful relationship between preservation and economic development. Incidentally, past research found that central cities do more historic preservation than suburban jurisdictions (Green & Fleischmann, 1991, 150). This results in greater sustainability if they see an investment is making a better return. The bonus with historic preservation is the sweat equity of the owners, which reduces rehab costs.

However, it should be noted that not all studies confirm a positive impact of preservation efforts as another strategy to reduce greenhouse gases. In some cases, local historic preservation ordinances have caused a loss in property appreciation (Asabere, Huffman, & Mehdian, 1994; Haughey & Basolo, 2000). Haughey and Basolo (2000) suggest that stringent local regulations, as opposed to federal designation, can cause property values to fall in historic districts. Another study of historic preservation in Charleston, South Carolina, found that housing of the lowest quality in a district experiences negative returns as a result of historic preservation (Lockard & Hinds, 1983). Lockard and Hinds (1983) do find that historic housing of the highest and medium quality tends to see a positive impact from historic preservation. They neglect to note it's good for the soul and can be costly if you replace the old with the new. Successful preservationists don't cover up the beautiful and thick hardwood floor or the less lovely but durable concrete floor with cheap toxic carpeting, or replace the metal roof with wood shingles (a thick white foam to push the sun rays back to the atmosphere instead of being absorbed inside the house), or cover up the brick with plaster when exposed brick provides warmth, or put in a false ceiling when the above 10 or 12 foot high ceiling helps cool the building, or worst of all, remove trees that cool the house in the summer.

Nevertheless, our research in Kentucky cities Louisville, Newport, and Covington, sponsored by the U.S. Department of Education and U.S.

Department of Housing and Urban Development, using rigorous data analysis consistently shows that nine historic districts in Louisville enhance the value and profitability of housing. Long-term real estate investments have four areas of profit: cash flow (which is not great at the start of the investment), appreciation, equity build-up, and taxable depreciation when paying taxes. Nearly all the properties appreciate in value (Gilderbloom, House, & Hanka, 2008).

Preservation of historic housing is strongly associated with the creation of affordable rental housing because it is profitable and increases property values. That is why 8 out of 10 preservationists in Kentucky claim it is a more profitable return on investment than other kinds of investments (Gilderbloom, House, & Hanka, 2008). Moreover, the cost of rehabilitating old buildings is not only more environmentally friendly (90 percent), but also costs less than constructing new buildings with the same amount of space, according to 78 percent of our respondents (Gilderbloom, Hanka, & House, 2008). While not always the case, these restoration and adaptive reuse strategies seem to be more affordable than building new units. Consequently, there is more money invested in new construction, and builders create the myth that renewal or preservation is too costly.

Preservation and Environmental Sustainability

Over 30 years ago, few recognized the connection between sustainable neighborhoods and historic preservation. The 1992 United Nations Earth Summit in Rio de Janeiro, Brazil, defined sustainable development as the means of providing for the basic necessities of life, such as food, education, jobs, worship, transportation, and safety, to meet our needs today while enabling future generations to meet their needs (United Nations, 1992; 2004).

A sustainable neighborhood is, by default, a historic neighborhood designed before the invention of the automobile or air conditioning. The layout of these neighborhoods placed stores, churches, schools, jobs, and recreation in close proximity to one another. These types of neighborhoods have lasted from past generations to the present and will allow future generations to live, work, and play there.

A sustainable neighborhood is one that preserves the past for the present and future generations. Restoring these beautiful buildings is an important environmental act. Historic preservation is a natural ally of environmentalism, which provides residents the opportunity to reduce their carbon footprint by refraining from excessive automobile and high-cost energy use. A historic neighborhood is a healthier neighborhood because many of its citizens are more active (Gilderbloom, House, & Hanka, 2008).

Older neighborhoods and newer housing have been compared in terms of the ease of commuting from home to school, work, recreation, shopping, or public transportation outlets (Rypkema, 2002, 7–9). Older neighborhoods are in closer proximity to work (i.e., CBD employment and other urban job centers) and places of recreation and leisure. According to the American Housing Survey (1999), historic house residents live within 5 miles of their work, compared to 23 percent of people living in new housing constructed within the past four years. Similarly, two-thirds of the people living in older neighborhoods were within one mile of an elementary school, with a 25 percent drop for those living in new houses (39 percent). The percentage of those that shop within a mile of their home was 62 percent for older neighborhoods versus 41 percent for new neighborhoods. In terms of the availability of public transportation, 59 percent had easier access in older neighborhoods versus 26 percent in newer developments. Finally, the amount of affordable housing was about 20 percent greater in older neighborhoods.

Houses built in the 1800s were designed without the need for air conditioning. In the nineteenth century, homes were designed with ten to 14-foot ceilings to allow hot air to rise and escape through door transoms, cooling the first floor on hot summer nights. Large attics, ranging anywhere from eight to 16 feet, were built to capture the hot air, and large basements were built to keep perishables cool in the summer. Also, working-class "shotgun" and "camelback" houses were built with raised floors and high walls that helped cool the buildings. The basements were often used for storage along with providing protection against inclement weather.

Kentucky is a national leader in preservation, ranking first in the White House´s Preserve America initiative with 73 recognized communities. Tax incentive programs have been an effective tool for creating positive changes in historic areas. Louisville has more affordable housing than any other city in the nation and that is largely due to having the oldest housing stock in the nation. There is more shotgun housing stock than post-Katrina New Orleans. In our previous books, looking at 144 medium-sized semi-isolated cities, we found over several decades that the greater amount of preserved older housing stock, the lower mean housing costs (Gilderbloom & Appelbaum, 1988; Gilderbloom, 2008).

With new advances in energy conservation, including insulation materials, fan and duct systems, and energy-efficient air conditioning, preservationists are able to create new spaces out of these unused spaces—whether it's an in-law apartment, a work-from-home space, a private refuge for either a "man's cave" or "woman's space," for pool playing, working out, a home office, or a rental unit to bring in revenue. Energy costs from these attic or basement spaces can be significantly lower if one uses passive solar design, fans, insulation, and proper ventilation. In our survey, we found that the majority of Kentucky preservationists (64 percent) believe owners should be allowed to convert basements, garages, and attics into additional housing in historic buildings (Gilderbloom, House, & Hanka, 2008).

Embodied Energy

Historic buildings are intrinsically valuable due to their architecture and construction, which capture the architectural technologies, styles, craftsmanship, and the integrity of the design of that time. But even beyond the value that is given to them due to their historic place, another value can be assigned to them when viewing their preservation in terms of environmentalism and sustainability. All existing buildings contain what is called "embodied energy," or the energy that is stored in them through their construction and throughout the building's life. Every year, approximately 1 billion square feet of existing buildings are demolished to be replaced by new construction, and it is estimated that approximately 82 billion square feet of existing buildings will be demolished between 2005 and 2030 (National Trust for Historic Preservation, 2011).

There are two types of embodied energy. The first is the initial embodied energy, which is the non-renewable energy that goes into the first stages of the construction process, including the consumption of raw materials and the processes needed to consume these materials. This type of energy has two components. First, the direct embodied energy is the energy that is consumed while materials are transported to the construction site and while the construction is occurring. Second, the indirect embodied energy is the consumption of energy that occurs during the attainment of raw materials and the production and manufacturing of construction materials. The other type of embodied energy is recurring, which is the energy that goes into the building's maintenance, restoration, or replacement of materials or systems that might occur over the life of the building (Measures of Sustainability).

The embodied energy that goes into a brick, for example, includes the non-renewable energy such as the burning of coal or gasoline that went into extracting the clay, the manufacturing of the brick from the clay, the transportation of the brick to the construction site, and the energy needed to make that brick part of the building. In addition, any energy used for maintenance, such as re-mortaring the brick. All the carbon that was released into our atmosphere during these processes cannot be taken back and therefore is a part of the building's embodied energy.

Embodied energy is the reason that "the greenest building is the one that already exists." This saying has become the slogan for many preservationists who now understand the environmental value of historic buildings. If an old building is torn down and replaced with an LEED-certified building with all the latest sustainable technology, it is still at a deficit of the embodied energy that had gone into the now-demolished original structure (www.asla.dirt.org, 2011).

The 2011 study by the Preservation Green Lab, an office of the National Trust for Historic Preservation, found many significant results on the sustainability of the reuse of buildings. The main finding of this study is that when comparing buildings of similar size and function, reuse almost always results in more environmental savings than demolition and new construction. In addition, it can take a new building that is 30 percent more energy efficient than average, 10 to 80 years to make up for the carbon emissions and climate change impacts of its construction. They also found that the scale of reuse could greatly impact and reduce carbon emissions. In the city of Portland, they found that if only 1 percent of the existing infrastructure is reused instead of demolished with new development in its place, 15 percent of the city's carbon emissions goal will be reached. The study also found that the amount and type of construction materials used in retrofitting could greatly impact the environmental benefits of reuse. These benefits can be reduced or even negated by choosing to use many new materials or the wrong types of materials.

We would add that the best greenhouse is an old house that lies within a functioning historic downtown neighborhood. Rypkema (2006) argues that every time a large historic house is demolished, the construction debris put in a landfill is equal to one million recycled aluminum cans. Rypkema argues the relationship between historic preservation and sustainability:

> Razing historic buildings results in a triple hit on scarce resources. First, we are throwing away thousands of dollars of embodied energy. Second, we are replacing it with materials vastly more consumptive of energy. What are most historic houses built from? Brick, plaster, concrete, and timber are among the least energy consumptive of materials. What are major components of new buildings? Plastic, steel, vinyl, and aluminum—among the most energy consumptive of materials. Third, recurring embodied energy savings increase dramatically as a building life stretches over fifty years. You're a fool or a fraud if you claim to be an environmentalist and yet you throw away historic buildings, and their components.

As Chiras (2004, 16) argues, the best kind of sustainable shelter is maintaining and enhancing historic housing. He states that renovating a historic home is the:

epitome of conservation and is arguably one of the most sustainable forms of construction . . . it uses existing resources such as lands, foundations, and walls. No new land must be bulldozed or cleared to make room for a new home: trees do not need to be cut down. Further benefits can be achieved if wastes generated from the project are recycled.

Many older houses can be saved at a cost substantially below market rate. An old house contains a great deal of "embedded" energy, which is wasted when it is demolished. Embedded energy describes the totality of energy used to build and create one house at one particular location, such as the sum result of energy needed to produce a house by cutting down trees in the forest for wood, hauling the wood back on trucks, manufacturing the steel and bricks, and creating the infrastructure of roads, sidewalks, gas, water, and sewer lines. According to Rypkema (2006):

When we throw away a historic building, we simultaneously throw away the embodied energy incorporated into that building. How significant is embodied energy? In Australia, they have calculated that the embodied energy in their existing building stock is equivalent to ten years of the total energy consumption of the entire country.

Richard Moe (2008), president of the National Trust for Historic Preservation, offers an example of preserving a historic building:

Boston City Hall has about 500,000 square feet of space. The amount of energy embodied in that building is about 800 billion BTUs. That's the equivalent of about 6.5 million gallons of oil and if the building were to be demolished, all of that embodied energy would be wasted. What's more, demolishing City Hall would create about 40,000 tons of debris. That's enough to fill more than 250 railroad boxcars a train nearly 2 ½ miles long, headed for a landfill that's probably almost full already. Finally, constructing a new 500,000-square-foot building on the City Hall site would release about as much carbon into the atmosphere as driving a car 30 million miles or 1,200 times around the world.

Preservation equals a commitment to sustainable practices. Governments can use an array of bold and innovative steps to enhance historic preservation efforts, such as raising the cap on state tax credits, establishing additional historic zoning overlays, providing soft second loans, providing grants for façade restoration, and expanding educational opportunities for historic property owners.

Environmental Findings

We find that urban residents are indeed more environmentally friendly in their behavior than their suburban counterparts, holding the control variables equal. This evidence leads us to conclude that those living within a central city are, on average, more pro-environment in their behavior and that those urbanites living in likely historic homes are even more pro-environment than those residing in higher-density, new-build apartment complexes and condos. This confirms our suspicions regarding the effect of historic preservation on environmentally sustainable behaviors and practices. Encouraging the renovation of historic structures not only preserves existing housing stock and conserves costs, energy, and materials, but it also encourages individuals to reside in neighborhoods that naturally foster more environmentally friendly behaviors.

Preservation Creates Jobs, Jobs, Jobs

We utilize a respected job multiplier simulation model to quantify the economic impact that determines the direct, indirect, and total effects of an external infusion of funds for historic preservation efforts. This simulation model, developed by Rutgers University for the National Park Service, is called the Preservation Economic Impact (PEI) model. Based on real case studies of job creation put into a computer simulation model, this software calculates the total economic impact of historic preservation, determining both the direct and multiplier effects of rehabilitation. The labor and materials used specifically to purchase or rehab a historic home would be considered a direct effect. On the other hand, the multiplier effect consists of any indirect impacts, meaning any money spent on goods and services by the construction industries that produce the rehabilitation materials (National Park Service, 2000).

Historic preservation results in more job creation than most other kinds of investments. According to Donovan Rypkema, investment in new construction creates 40 jobs per million dollars compared to an investment in historic rehabilitation, which results in anywhere from 43 per $1 million (Rypkema, 1997) to 49 new jobs per rehabilitation project (National Park Service Annual Report, 2006). We used a more conservative estimate derived from the PEI developed by Rutgers University Urban Planning Program and the National Park Service. According to the PEI model in Figure 18.1, 7,365 jobs were created as a result of Kentucky state tax credits program from 2005 to 2007, resulting from direct, indirect, and induced effects of the $171 million spent. From this investment, the PEI approximates that $229 million of income was generated and total gross domestic product was $356 million. The multiplier

	Economic Component		
	Employment (jobs)	Income (000$)	Gross Domestic Product (000$)
I. TOTAL EFFECTS (Direct and Indirect/Induced)*			
Private			
1. Agriculture	56	1,030.0	4,054.0
2. Agri. Serv., Forestry, & Fish	125	2,970.0	3,597.0
3. Mining	75	3,185.0	7,893.0
4. Construction	2,372	64,990.0	74,686.0
5. Manufacturing	1,166	41,679.0	69,088.0
6. Transport. & Public Utilities	414	15,114.0	32,524.0
7. Wholesale	313	13,079.0	22,020.0
8. Retail Trade	984	16,779.0	26,834.0
9. Finance, Ins., & Real Estate	568	29,223.0	52,992.0
10. Services	1,260	39,863.0	60,303.0
Private Subtotal	7,333	227,010.0	353,591.0
Public			
11. Government	32	1,945.0	2,818.0
Total Effects (Private and Public)	7,365	228,955.0	356,409.0
II. DISTRIBUTION OF EFFECTS/MULTIPLIER			
1. Direct Effects	3,027	89,023.0	117,032.0
2. Indirect and Induced Effects	4,339	139,932.0	239,377.0
3. Total Effects	7,365	228,955.0	356,409.0
4. Multipliers (3/1)	2.433	2.572	3.045
III. COMPOSITION OF GROSS STATE PRODUCT			
1. Wages—Net of Taxes			174,915.0
2. Taxes			
a. Local/State			31,426.0
b. Federal			
General			21,678.0
Insurance Trusts			17,649.0
Federal Subtotal			39,327.0
c. Total taxes (2a+2b)			70,753.0
3. Profits, dividends, rents, and other			110,741.0
4. Total Gross State Product (1+2+3)			356,409.0
EFFECTS PER MILLION DOLLARS OF INITIAL EXPENDITURE			
Employment (Jobs)			43.0
Income			1,338,038
Local/State Taxes			183,657
Gross State Product			2,082,890

Note: Detail may not sum to totals due to rounding.
*Terms:
Direct Effect (State)–the proportion of direct spending on goods and services produced.
Indirect Effects–the value of goods and services needed to support the provision of those direct economic effects.
Induced Effects–the value of goods and services needed by households that provide the direct and indirect labor.

Figure 18.1   PEI Model for Kentucky State Tax Credits, 2005–2007. *Source:* Center for Sustainable Urban Neighborhoods, formerly University of Louisville, transferred to Neighborhood Associates, Washington, D.C.

effect of state tax credits is 43—so for every $1 million dollars spent on state tax credits, 43 jobs were created (Figure 18.2).

The Main Street program in Kentucky has produced 4,720 jobs, resulting in approximately $149 million in income and a total gross domestic product of over $237 million in 2006. According to the model, every $1 million spent on Main Street reinvestment results in approximately 29 new jobs. Out of the $292 million spent on Main Street reinvestment in 2006, including approximately $128 million in private investment, $70 million in public improvements (e.g., streetscape), and nearly $95 million in new construction, 8,468 jobs were created.

The PEI job creation estimates are just the tip of the iceberg, since there are hundreds who are employed maintaining, restoring, and upgrading thousands of historic homes around Kentucky. For this reason, most Kentucky preservationists see a large economic return (87 percent), and another 81 percent see

	Economic Component		
	Employment (jobs)	Income (000$)	Gross Domestic Product (000$)
I. TOTAL EFFECTS (Direct and Indirect/Induced)*			
Private			
1. Agriculture	33	607.0	2,434.0
2. Agri. Serv., Forestry, & Fish	32	797.0	936.0
3. Mining	31	1,300.0	3,397.0
4. Construction	1,410	38,046.0	44,407.0
5. Manufacturing	781	29,197.0	47,023.0
6. Transport. & Public Utilities	261	10,101.0	22,845.0
7. Wholesale	231	9,708.0	16,206.0
8. Retail Trade	641	10,830.0	17,412.0
9. Finance, Ins., & Real Estate	408	20,490.0	40,281.0
10. Services	866	26,423.0	40,383.0
Private Subtotal	4,695	147,501.0	235,326.0
Public			
11. Government	25	1,494.0	2,173.0
Total Effects (Private and Public)	4,720	148,995.0	237,498.0
II. DISTRIBUTION OF EFFECTS/MULTIPLIER			
1. Direct Effects	1,877	56,321.0	79,286.0
2. Indirect and Induced Effects	2,843	92,674.0	158,212.0
3. Total Effects	4,720	148,995.0	237,498.0
4. Multipliers (3/1)	2.515	2.645	2.995
III. COMPOSITION OF GROSS STATE PRODUCT			
1. Wages--Net of Taxes			114,544.0
2. Taxes			
a. Local/State			21,758.0
b. Federal			
General			14,570.0
Insurance Trusts			11,683.0
Federal Subtotal			26,252.0
c. Total taxes (2a+2b)			48,010.0
3. Profits, dividends, rents, and other			74,944.0
4. Total Gross State Product (1+2+3)			237,498.0
EFFECTS PER MILLION DOLLARS OF INITIAL EXPENDITURE			
Employment (Jobs)			28.7
Income			907,365
Local/State Taxes			132,502
Gross State Product			1,446,345

Note: Detail may not sum to totals due to rounding.
*Terms:
Direct Effect (State)--the proportion of direct spending on goods and services produced.
Indirect Effects--the value of goods and services needed to support the provision of those direct economic effects.
Induced Effects--the value of goods and sevices needed by households that provide the direct and indirect labor.

Figure 18.2   PEI Model for Kentucky's Main Street Program, 2006. *Source:* Center for Sustainable Urban Neighborhoods, formerly University of Louisville, transferred to Neighborhood Associates, Washington, D.C.

potential job growth from preserving the physical built environment heritage of our ancestors (Gilderbloom, House, & Hanka, 2008).

In sum, this translates to:

- State Tax Credits: $171 million investment results in 7,365 jobs in 3 years (2005–2007)
- Federal Tax Credits: $52 million investment results in 2,236 jobs in 1 year (2006)

- Main Street Program: $292 million investment results in 4,720 jobs in 1 year (2006)
- Heritage Tourism: Total budget of $96 million and employs 2,700 Kentuckians (2002)

So, who gets the jobs? Impian Software, which is a more conservative, measures estimates that 83 percent of the jobs go to workers with only a high school diploma or less. Moreover, it's not just job creation at the site but increased demand for plumbing, wood, cement, appliances, doors, windows, roofing, gables, kitchen and bathroom supplies, and administration costs, as well as local eateries. Also, more jobs are created in repurposing buildings because it is labor-intensive. New housing is less labor-intensive. In my experience as a private developer, renovating an old home for occupancy means many of the materials there can be reused and are half the cost.

Newport, from Vice to Nice

Once known as Cincinnati's dumping ground for organized crime activity, prostitution, and gambling, Newport, Kentucky, has become one of the greatest comeback cities in America. Newport was a notorious den of corruption that then-Attorney General Robert Kennedy brought national attention to by cracking down on the illegal activity that made the city infamous. Some 40 years ago, Newport was known as the "vice capital of the south" according to *Time Magazine*. Just a decade ago, the downtown was a depressed area littered with sex-oriented businesses dominating historic Monmouth Street. Since then, the city has gone from vice to nice and from hips to hip. How did this miracle happen?

In an urban world beset with defeat, Newport is a feel-good story about how one small Ohio River city turned itself around with strong leadership, entrepreneurship, forward-thinking federally funded programs, and effective partnerships between non-profit organizations, the higher education and business community, city government, and neighborhood leaders. This resulted in destroying the outdated public housing and replacing it with programs designed to lift people out of welfare dependency with a second chance for education and job opportunities.

HOPE VI is a bipartisan urban revitalization program created in 1992 by HUD that eliminates inadequate and poorly designed public housing that was a product of the urban renewal style of the 1940s and 1950s. Public housing projects of this era perpetuated a culture of poverty that contributed to an increase in crime, drugs, vandalism, and hopelessness.

This old public housing was replaced with single-family homes in mixed integrated communities. HOPE VI played a role in improving the overall quality of life by reducing crime and increasing business and economic development in the downtown. Property values are up, and former public housing residents are getting off welfare and finding jobs with the promise of better housing. Some of these residents have purchased attractive homes and have become homeowners. Deconcentrating the poor and creating greater economic opportunities have positively impacted the crime rate in Newport's neighborhoods. As a result, overall crime in the four police districts in and around the old public housing decreased by 19 percent from 2000 to 2007. No mega-homeless shelters in this river town.

The city of Newport has shown not only the initiative to pursue federal housing grants such as HOPE VI, but also to pursue a wide range of opportunities to increase economic development in downtown Newport. Businesses in Newport's downtown since HOPE VI are of higher quality than in the 1990s. The site of the old public housing has been cleared to make way for a $1 billion investment over the next 10–15 years. The Ovation development will offer first-class, mixed-use development, including retail, office, condominiums, a hotel, and additional lodging, entertainment, and recreational amenities in close proximity to downtown. The economic revitalization of Newport has occurred through property value changes in HOPE VI and non-HOPE VI neighborhoods. The HOPE VI areas have experienced a higher appreciation in property value than those in non-historic and non-HOPE VI areas. Historic Preservation districts have had the largest increase in property values.

Much of the success in Newport is the result of the steady and consistent leadership of Mayor Thomas Guidugli, who will complete his third and final term and 16 years in office as mayor at the end of 2008. This consistency has enabled the mayor and his staff to set goals and implement a common-sense strategy for revitalizing Newport by aggressively pursuing property acquisition on the riverfront for the Newport on the Levee development and the Newport Aquarium. These developments have attracted new businesses, tourists, jobs, and economic development in the downtown area. The city has pursued accreditation for the City of Newport police department to create a professional and progressive department. Additionally, Newport received a big assist from Cincinnati by reinvigorating downtown Cincinnati through the development of new museums and professional sports stadia for the Cincinnati Reds and Bengals.

Newport still has untapped potential to encourage civic engagement and economic development. One strategy is converting one-way streets to two-way streets in the city. Converting one-way streets to two-way streets with parking, trees, and bike lanes to calm traffic will make neighborhoods more livable for families, young urban professionals, and the elderly, who want to live closer to medical care downtown. Newport's resurgence from "vice" to "nice" has played a major part in this successful comeback story. HOPE VI has been a positive force in the renewal, restoration, and revitalization of Newport, and the Newport model provides inspiration and hope.

What's Up with Anti-Gentrification Hysteria?

Preservation is a great tool for environmental stewardship, yet there have always been naysayers who make the bogus claim that preservation takes away housing from the poor. Anti-gentrification is the new racism. It is unconscious racism reflecting a belief that Blacks are not deserving of first-class neighborhoods and black neighborhoods are second-class places. Those who oppose gentrification argue that any kind of neighborhood improvement would raise the price of housing: transit stops, parks, calm streets, bike lanes, two-way streets, and historic preservation are all considered verboten in the Marxist playbook of planning. Anti-gentrification is a code word for racism: keep people who are not like us out. The origin of the anti-gentrification movement was rooted in fear—fear of gays (the true gentry) or whites.

Please gentrify me! This was the slogan of Andres Duany, co-founder of Congress of New Urbanism. Three cheers for those investing in West Louisville. But rather than blame white people as the problem for daring to move into black neighborhoods, why not focus on the real villains? Pollution, redlining, and racist ideology. Ruth Glass, who coined the term "gentrification," was an East German member of the Communist Party. She found an academic job in London, blamed capitalism as the problem and communism as the solution, and proudly pointed to the achievements of the Soviet Union. She was upset to see racial integration in an East London neighborhood where white gentry (mostly gay) were moving into a neighborhood of immigrants from India. She wrote about the cruelty of whites moving into an inner-city poor neighborhood and living side-by-side with Indians. She might have claimed to be a member of the Communist Party, but her childhood was rooted in Nazi ideology, and she believed in the separation of races. What was their crime? Investing in dilapidated housing without hot water, heat, or sanitation, and with horrible rat infestations. The anti-gentrifiers are social engineers trying to promote disillusionment, anger, and rebellion against capitalism.

In terms of Civil Rights, Ruth Glass was the opposite of Martin Luther King, Jr. Anti-gentrification efforts equal racism: only people of a certain race or economic class are welcomed in neighborhoods. She argued that racial integration caused displacement. The Communist Party was hoping this would cause resentment and racial division. Just like Marx and Engels, she never wanted the poor to have a comfortable home or neighborhood; otherwise, they would lose their revolutionary spirit.

Is the enemy really the "gentry" of young middle-class people who are teachers, policemen, firemen, nurses, supervisors, and mailmen who don't want to live far away in all-white conformist suburbs? These folks are not monsters, but good people who want to find peace and harmony by living with others. Is this a rebuttal of Jesse Jackson's Presidential campaigns in the 1980s where he told adoring crowds of the need for good role models, black and white, in poor black neighborhoods? As he said, the drug pusher needs to be replaced by a person who works a nine to five job. This was also the dream of Martin Luther King, Jr.

This is what is known as gentrification hysteria, and it's simply not real. An examination of U.S. Census reports shows that the displacement of Blacks is rare in the 14 neighborhoods we have partnered with in three different cities (Gilderbloom & Mullins, 2005; Gilderbloom, 2005). How bad of a problem is gentrification? Governing Magazine conducted a comprehensive survey of every city in America and found that it was only happening in a small number of cities and neighborhoods. In fact, more minorities tend to move into these neighborhoods where people are investing, renewing, and building.

Poor people concentrated by one race—whether black or white—is an economic disaster. Our study of 174 Louisville neighborhoods that was published in the highly respected scholarly journal, Housing and Society, found that poor neighborhoods die without investment, and investors flee when they see no return on investment. When people see others investing with their sweat, they get inspired to fix up their own homes. This is documented in the book, *Chromatic Homes: The Joy of Color in Historic Places*. Whether it is the spectacular renewal in Louisville's Original Highlands, Newport, Covington, downtown Cincinnati, New Orleans, Miami, or internationally in Lisbon, Amsterdam, Barcelona, or Cardiff.

What is worse than gentrification? No gentrification. Why? These are neighborhoods that have been neglected by our city leaders. Polluting industries have caused multiple problems such as reduced life expectancy, lower housing values, cancer, lung disease, poor school performance, and lots of crime. What bank is going to make a loan when they see so many houses sinking in value and disappearing? No investment in a neighborhood is death. In Louisville, you have over 5,000 abandoned housing units in neighborhoods with little or no investment. Many of these homes are going for $10,000 to $20,000.

In a recent HUD survey, two-thirds of low-income Blacks said they would welcome upper-income folks to their neighborhood, whether Black or White, gay or straight (Gilderbloom, House, & Hanka, 2008). What do they have in common with the middle-class? They are more likely to call the police, fire, or EMS to keep the neighborhood safe, healthy, and prosperous. People do better in integrated neighborhoods. Do Blacks want to live separately from Whites? In a study done of African American public housing residents, two-thirds found racially mixed housing more desirable than wholly black housing (see chapter 15 on Hope VI). White people are not pushing Blacks out of West Louisville, especially with the problems of pollution, unlivable neighborhoods, and redlining.

The consequences of gentrification or regeneration are mostly good. Badly needed investment in abandoned housing and buildings would spark current homeowners to invest money in their own homes and ride the wave of prosperity. In other words, more investment means more housing, not less, and results in a neighborhood going up in value and getting a return on investment. In the Highlands, prices have skyrocketed and the percentage of minorities has increased, according to the U.S. Census. Moreover, it is a job stimulator for the largest group of unemployed persons: those without or only with high school diplomas. Gentrification means good-paying jobs. Construction is much more labor intensive, creating 17 jobs per million dollars invested, and 82 percent of these jobs go to people with the highest unemployment levels and roughly one-third go to minorities. This is a much greater job return than building freeways, which creates about five jobs per million.

What is so wrong with prosperity? Sam Watkins, the former head of Louisville Central Community Center, once asked, "What's wrong with Black people making money?" In a partnership with the University of Louisville Center for Sustainable Urban Neighborhoods, developers, bankers, and residents, Sam pushed to turn abandoned homes into beautiful first-class homes. In East Russell, he was hailed as a hero, visionary, and brave. A published study found that his nearly 1,000 homes built in East Louisville did not cause Blacks to be displaced. In fact, his efforts saved housing by investment. This was followed by successful upper-middle-class Blacks building $500,000 homes, which signaled even more investment. The 200 or so homeowners saw their bargain investment triple in value from a $50,000 home to a $150,000 home. Regeneration promotes the preservation of older and more affordable housing. Our studies of preservation activity show that this is one of the best ways to preserve and create affordable housing. It also promotes a greener environment that keeps the embedded energy.

Discussion

A historic property owner feels that their investment is more secure in historic districts because their neighborhoods are not only preserved but are also well maintained. They recognize that a historic preservation district protects their investment and makes it more profitable. That is why neighborhoods that have enacted historic preservation districts throughout Kentucky overwhelmingly do not repeal them, and in several cases have increased the boundaries of the district. Stephen Roosa, a major owner of historic buildings in Louisville, agrees:

Real estate can be a super long-term investment, especially in historic neighborhoods. When I first decided to focus my efforts in the Old Louisville and Highlands neighborhoods, most real estate professionals I talked with discouraged me from investing in these neighborhoods. I persisted and now have several properties in my portfolio in these areas. As these neighborhoods have improved over the years, these investments have provided much greater returns than if I had invested elsewhere in the county (Gilderbloom & Hanka, 2007).

Investors do not want to see the home next door or across the street demolished and replaced with a cinderblock house that looks like it came out of the hills of Costa Rica, or see a 1920s bungalow covered with vinyl siding, with the original windows replaced and the original wooden doors replaced by a cheap, manufactured door bought from a big-box store. Removing these important architectural details and modernizing them with inexpensive materials is devastating to the value of the defaced home, and it also hurts nearby property values.

CONCLUSION

This chapter argues that historic preservation has a positive impact on job creation, property values, and environmental stewardship. We demonstrated empirically via an economic stimulation model that historic preservation creates more jobs than most other investments. Our research also supports policy recommendations at the local, state, and federal levels such as higher tax incentives for historic preservation, including façade restoration and forgivable loans for historic rehabilitation. These types of measures will not only encourage economic development and increase revenues but will do so in a way that protects the natural environment. Restoring an older home or building is more labor-intensive than other kinds of public investments. Because of their proximity to downtown and consumer preferences for housing with reduced energy costs, historic buildings experience higher appreciations in property values than newly constructed homes and buildings. Many of these historic buildings were designed to be lived in year-round before electricity and air conditioning. This chapter also shows that environmentalism and historic preservation are linked together and complement one another. Residents of historic neighborhoods exhibit more environmentally friendly behavior, particularly those living in single-family homes. Saving one home, including its pipes, wires, brick, wood, and metal, means one less house built in suburbia. More research needs to be conducted in other countries, states, and cities on the valuable contributions of historic preservation policies.

REFERENCES

Ambrosius, J.D., & Gilderbloom, J.I. (2015). Who's Greener? Comparing Urban and Suburban Residents' Environmental Behaviour and Concern. *Local Environment, 20*(7), 836–849.

Asabere, P.K., Huffman, F.E., & Mehdian, S. (1994). The Adverse Impacts of Local Historic Designation: The Case of Small Apartment Buildings in Philadelphia. *Journal of Real Estate Finance and Economics, 8*, 225–234.

Chiras, D.D. (2004). *The New Ecological Home: A Complete Guide to Green Building Options*. White River Junction: Chelsea Green Publishing, Inc.

Coulson, N.E., & Lahr, M.L. (2005). Gracing the Land of Elvis and Beale Street: Historic Designation and Property Values in Memphis. *Real Estate Economics, 33*(3), 487–507.

Ernie Fletcher's Communications Office. (May 17, 2007). http://kentucky.gov/Newsroom/travel/parksntw.htm.

Florida, R. (2002). *The Rise of the Creative Class*. New York: Basic Books.

Ford, D.A. (1989). The Effect of Historic District Designation on Single-Family Home Prices. *American Real Estate and Urban Economics Association (AREUEA) Journal, 17*(3), 353–362.

Gilderbloom, J. (1988). *Rethinking Rental Housing*. Philadelphia: Temple University Press.

Gilderbloom, J. (2008). *Invisible City: Poverty, Housing and New Urbanism*. Austin: University of Texas Press.

Gilderbloom, J., & Hanka, M. (2007a). Interview with Stephen A. Roosa, President, Energy Systems Group, Louisville, KY, September 24.

Gilderbloom, J., & Hanka, M. (2007b). Louisville's New Urbanism. *Courier Journal*, Louisville, KY, June 13.

Gilderbloom, J., & Hanka, M. (2009). Interview with Scot Walters, Tax Credit specialist, Kentucky Heritage Council, March 11.

Gilderbloom, J.I., House, E.E., & Hanka, M.J. (2008). *Historic Preservation in Kentucky*. Hodgenville: Preservation Kentucky, Inc.

Gilderbloom, J., & Mullins, R. (2005). *Promise and Betrayal: Universities and the Battle for Sustainable Urban Neighborhoods*. Albany, NY: SUNY Press.

Green, G.P., & Fleischmann, A. (1991). Promoting Economic Development: A Comparison of Cities, Suburbs, and Non-metropolitan Communities. *Urban Affairs Quarterly, 27*(1), 145–153.

Green, M. (2007). Urban Areas See Biggest Gains in Property Value. *Courier Journal*, Louisville, KY, June 5.

Haughery, P., & Basolo, V. (2000). The Effect of Dual Local and National Register Historic District Designations on Single-Family Housing Prices in New Orleans. *The Appraisal Journal, July*, 283–289.

Historic Confederation of Kentucky, the Kentucky Association of Museums, & the Kentucky Historical Society. (2002). *The State of Museums and History Organizations in Kentucky*, Frankfort, KY.

Historic Old Louisville. (2008). *Historic Old Louisville Visitors Center.* http://www
.oldlouisville.org/.

Jacobs, J. (1961). *The Death of and Life of Great American Cities.* New York: Random House.

Jones, R.E., Fly, J.M., & Cordell, H.K. (1999). How Green Is My Valley? Tracking Rural and Urban Environmentalism in the Southern Appalachian Ecoregion. *Rural Sociology, 64,* 482–499.

Kentucky Department of Agriculture, Office of Agricultural Marketing and Product Promotion. (2007). *PACE: Farmland Preservation Program.* http://www.kyagr
.com/marketing/farmland/index.htm.

Kentucky Department of Tourism. (2006). *Economic Impact of Travel on Kentucky: State Totals From 2003–2006.* http://www.kentuckytourism.com/NR/rdonlyres
/3CAC4E9A-4E7B-4AB3-987F-AD354C908B97/0/EconomicImpactStatewide.pdf

Kentucky Department of Tourism. (2007). *Tourism Spending in Kentucky Hits $10 Billion.*

Kentucky Heritage Council. (1988). *Economic Benefits of Historic Preservation in Bowling Green, Kentucky.* Frankfort, KY.

Kentucky Heritage Council. (2003). *Planning to Preserve: The 2004–2009 State Historic Preservation Plan for the Commonwealth of Kentucky.* Frankfort, KY.

Kentucky Heritage Council. (2007). *Ida Lee Willis Memorial Foundation Preservation Awards Program Project Award Nomination: Henry Clay Hotel,* delivered by Stephen Collins, Ida Lee Willis Foundation Chair, May 22. Frankfort, KY.

Kentucky Heritage Council. (2007). *Ida Lee Willis Memorial Foundation Preservation Awards Program Service to Preservation Award Nomination: Bardstown,* delivered by Stephen Collins, Ida Lee Willis Foundation Chair, May 22. Frankfort, KY.

Kentucky Heritage Council. (2007). *Survey to Kentucky State Tax Credit Users.* Frankfort, KY.

Kentucky Heritage Council. (1979–2006). *Kentucky Main Street Reinvestment Statistic Reports, 1979–2006.* Frankfort, KY.

Kentucky Heritage Council. (2005–2007). *Kentucky State Tax Credit Project Monthly and Annual Reports, 2005–2007.* Frankfort, KY.

Leichenko, R.M., Coulson, N.E., & Listokin, D. (2001). Historic Preservation and Residential Property Values: An Analysis of Texas Cities. *Urban Studies, 38*(11), 1973–1987.

Lockard, W.E., & Hinds, D.S. (1983). Historic Zoning Considerations in Neighborhood and District Analysis. *The Appraisal Journal, October,* 485–497.

Louisville Metro/Jefferson County Government. (2007). *Historic Landmarks and Preservation Districts.* http://www.louisvilleky.gov/PlanningDesign/Historic
+Landmarks+and+Preservation+Districts+Commission.htm

Mason, R. (2005). *Economics and Historic Preservation.* Washington, DC: The Brookings Institution.

Moe, R. (2008). *Featured Speech—Historic Preservation & Green Building: Finding Common Ground.* National Trust for Historic Preservation Conference.

Morrissey, J., & Manning, R. (2000). Race, Residence and Environmental Concern: New Englanders and the White Mountain National Forest. *Research in Human Ecology, 7*, 12–24.

National Center for Preservation Technology and Training. (2001). *Preservation Economic Impact Model User Guide*. Natchitoches: National Park Service.

National Highway Traffic Safety Administration. (2005). *Traffic Safety Facts 2003 Data: Pedestrians*. Washington, DC. http://wwwnrd.nhtsa.dot.gov/pdf/nrd30/NCSA/TSF2003/809769.pdf. 7/10/2007.

National Park Service, Heritage Preservation Services Division. (2003–2006). *Federal Tax Credits Annual Reports, 2003–2006*. Washington, DC. National Park Service.

National Park Service, Heritage Preservation Services Division. (2003–2006). *Federal Tax Credits Statistical and Report and Analysis, 2003–2006*. Washington, DC. National Park Service.

National Park Service. (2007). *List of Preserve America Communities and Neighborhoods in Kentucky*. http://www.preserveamerica.gov/PAcommunities.html#k.

National Trust for Historic Preservation. (2007). *Average Reinvestment from Main Street Programs*. Washington, DC. http://www.mainstreet.org/content.aspx?page=2080.

National Trust for Historic Preservation. (2007). *What is the Main Street Approach to Commercial District Revitalization?* Washington, DC. http://www.mainstreet.org/content.aspx?page=47&section=2.

Nooney, J.G., Woodrum, E., Hoban, T.J., & Clifford, W.B. (2003). Environmental Worldview and Behavior: Consequences of Dimensionality in a Survey of North Carolinians. *Environment and Behavior, 35*, 763–783.

Old Louisville Visitors Center. (2007). *Old Louisville Facts and Statistics*. www.oldlouisville.com.

Public Sector Consultants. (2006). *RHDI Stakeholder and Local Leader Interview Guide*. Lansing: Report prepared for Preservation Kentucky, Inc.

Rypkema, D. (1997a). *Historic Preservation and the Economy of the Commonwealth*. Kentucky Heritage Council and Commonwealth Preservation Advocates.

Rypkema, D. (1997b). *Preservation and Property Values in Indiana*. Historic Landmarks Foundation of Indiana.

Rypkema, D.D. (2002). *Historic Preservation and Affordable Housing: The Missed Connection*. Washington, DC: National Trust for Historic Preservation.

Rypkema, D.D. (2006). *Smart Growth, Sustainable Development and Historic Preservation*. Presentation at the Bridging Boundaries—Building Great Communities Regional Smart Growth Conference, Louisville, KY.

Samdahl, D.M., & Robertson, R. (1989). Social Determinants of Environmental Concern: Specification and Test of the Model. *Environment and Behavior, 21*, 57–81.

Shipley, R. (2000). Heritage Designation and Property Values: Is There an Effect? *International Journal of Heritage Studies, 6*(1), 83–100.

Steitzer, S. (2007). Coal-to-Gas Plant Likely in Kentucky—Peabody Was Offered $250 Million in Incentives. *Courier-Journal*, Louisville, KY, October 26.

St. James Art Fair Foundation. (2007). *Quick Facts about the St. James Court Art Fair: America's #1 Fine Art & Craft Show.* http://www.stjamescourtshow.com/QuickFacts.asp.

United Nations 2004. (1992). The Rio Declaration on Environment and Development. United Nations Conference on Environment and Development, Rio de Janiero, Brazil. Reprinted in: *Sustainable Development Reader*, edited by S.M. Wheeler and T. Beatley. New York: Routledge.

U.S. Census Bureau. (2000). *Census of Population and Housing, Washington, D.C.* Compiled by E. Schneider, Kentucky State Data Center, Louisville: University of Louisville.

Walton, T. (2006). *Environmental Concern and Behavior in an Urban Social Structural Context.* Paper presented at Mid-South Sociological Association annual meeting, Lafayette, LA.

Part IV

PUT A MASK ON POLLUTION

Chapter 19

Climate of Hope

Cities Leading the Way

John Hans Gilderbloom and Chris Nolan

It's going to get worse. Look around, and you'll see that no place is safe from the ravages of climate chaos. Every state has experienced the pain of climate chaos—Hawaii, Florida, New York, Arizona, California, and, of course, Kentucky. East to West, North to South, we are witnessing the slow destruction of the Earth as we once knew it. Or look around the world: North Pole, Africa, Middle East, China, Russia, and South America. We have a climate crisis that is destroying our way of life with record high temperatures. We are facing a catastrophic crisis that will destroy the Earth as we know it in eight years, according to the United Nations.

One response to climate chaos is doing nothing, followed by a rapid increase in millions of people immigrating to places of greater opportunity for food, water, health, and work. It's either migrate or die. Climate chaos, if it goes unchecked, will result in more wars and catastrophic environmental damage. This trajectory of worsening conditions can be reversed. We have the means to immediately reduce greenhouse gases by 80 percent; that's why we are changing the mainstream view of a climate of doom to a climate of hope.

Climate change is a life-and-death struggle. Studies show that 70 percent of people worldwide fear for the future, believing they are helpless to change or solve the climate crisis, and most are young people under the age of 40. The United States is one of the biggest emitters of greenhouse gases in the world—27 percent of greenhouse gases come from United States. Yet, the population of the United States makes up only 6 percent of the world's population (EPA, 2015:12-29). The United States causes more damage than any other country in the world, nearly five times as much damage compared to its population size. Greenhouse gases create holes in the Earth's fragile atmosphere causing the sun to heat the Earth. This year, record high temperatures were recorded on Earth and next year (2025) will be even hotter. With these future trends, the Earth will be completely unlivable. There is both good and bad news about solving the climate crisis. This is the bad news.

The good news is that we have the tools and technology to dramatically reduce these greenhouse gases here and everywhere around the world. Throughout this book we have relied on the Toxics Release Inventory which provides estimates from industry officials who downplay the amount of dangerous toxins released into our air, water, and soil. The European Space Agency (ESA) can provide a more reliable estimate using satellite technology that can on a block-by-block and building-by-building basis accurately identify the extent of greenhouse gas emissions. Our next research effort with the Center for Sustainable Urban Neighborhoods is to give citizens accurate information on toxins invading their precious neighborhoods so they can fight back with facts, not guesstimates.

We can reverse the climate change catastrophe. This is part of the theme of this book and our documentary: our cities can save us. At the local level of cities and neighborhoods, we need people to plead to world leaders to make a difference and demand accountability.

The United States can be a leader in showing how the climate crisis can be solved. This book shares a message of optimism and highlights tested policies from around the world that will reduce greenhouse gases by 80 percent. We have the brainpower to reverse the catastrophic climate crisis, not just slow it down by three to four years.

Why do so many books and documentaries have a hard time convincing people of the dangers present and ahead of us? Because the human mind struggles with the abstract, the chaotic, and large-scale concepts that don't directly touch upon our local daily experiences. Climate change, for many, feels distant, intangible, staggeringly random, and complex. It's a collection of intricate data points, statistics, models that blur the communication for the general populace. A documentary that features the research in this book is under development. It is called *Climate of Hope: Cities Saving the World*. Unlike any other film, this documentary will take a different approach to climate change by describing how cities, communities, and citizens around the world are leading the way to create more livable, healthier, and more prosperous life. This book delves deeper into the overall vision for the documentary and the storylines of families in Louisville, Portland, Amsterdam, Los Angeles, and other places where greenhouse gases are being reduced.

We want to create a realistic international movement to roll back the destruction caused by climate chaos. We have put together an amazing team of successful and proven change makers to save the Earth from decimation.

It started innocently enough. The roots of sociology were discovering the impact of "social forces" shaping our lives and using large data sets. We are connected to neighborhood organizations worldwide. Our mission is to inspire and inform more front-line neighborhood crusaders to advocate for city-driven climate change mitigation worldwide. We aim to create a powerful grassroots movement for change that will amplify the distribution and momentum of the film's message.

The film aims to inspire hope and catalyze change through stories of families, children, and neighborhoods taking bold, transformative actions. It champions the idea that with technology, will, and heart, our cities can not only survive but thrive. At its heart, the film is about urban transformation, challenges faced by cities/neighborhoods, and examples of cities leading the way with climate change action—ultimately highlighting the potential for other cities to transform into success stories. Filmmaker Chris Nolan is already in talks with PBS to ensure wide audience awareness and garner engagement in the movement. There are also plans in place to feature videos from the film on social media platforms such as YouTube and TikTok with the intent of reaching millions of viewers to create momentum and advocacy for this mission-critical pursuit (http://www.sunlouisville.org and http://www .climateofhopefilm.com).

Our best chance to reverse climate change is our cities. In 2050, 70 percent of the population will live in cities. Imagine our world in 2050, where seven out of ten people reside in cities. The world's population is projected to then be around 9.7 billion, which means an astonishing 6.6 billion people will be living in cities. The relentless urban migration is up 55 percent since 2018, which means the brunt of climate change will be experienced in metropolises. There have been best-selling books on how cities have taken control of the climate crisis, such as *Climate of Hope, How Cities, Businesses, and Citizens Can Save the Planet* and *Solved: How the World's Great Cities Are Fixing the Climate Crisis*. They have chronicled the stories of progressive cities around the world—large cities like Los Angeles, Amsterdam, New York, Toronto, Oslo, Shenzhen, and Sydney, and small towns such as Georgetown, Texas. However, since *Climate of Hope* was written in 2017, artificial intelligence (AI), technology, and even more city initiatives are making an incredible difference in reducing global emissions and implementing sustainable solutions.

Seventy percent of people worldwide fear the future, believing we are helpless to change or solve the climate crisis. A significant portion of this group consists of young people under the age of 40 who feel the planet is doomed. This book will help change that narrative. It aims to inspire hope and catalyze change. It highlights the urgency and feasibility of urban transformation, championing the idea that with technology, will, and heart, our cities can not only survive but thrive.

We cannot wait for national governments to agree on how to reduce greenhouse gas emissions. We aim to inspire a new conversation that will spur more action by cities to increase the pace and scale of climate change solutions that will make the world healthier and more prosperous. We possess the means to curb climate change. Every block, every neighborhood, and every family matters. Climate is about more than data; it's about human lives. Cities, at their core, are dynamic organisms. They evolve, adapt, and with the right nudge, can heal. Look at how these great cities have been rebuilt even stronger from earthquakes, fires, and war: San Francisco 1906, Chicago 1871, Tokyo 1923, Warsaw 1944, Dresden 1945, Beirut, Berlin 1945, Lisbon 1755, Hiroshima 1945, London 1940, and Rotterdam 1940. This book and the film will provide a roadmap for tackling the most complicated challenge the world has ever faced and inspire those who want to take positive action and make a significant impact on the world.

Louisville (which likely is an extreme case) is only one of 1,000 polluted places in the United States. Environmental justice has been defined as the pursuit of equal justice and equal protection under the law for all environmental statutes and regulations without discrimination based on race, ethnicity, and/or socioeconomic status. This concept applies to governmental actions at all levels—local, state, and federal—as well as private industry activities. Lower-income communities and minority populations have historically been the target of many sources of pollution. Air pollution from industrial sites, toxic contamination from incinerators and brownfields, contamination of ground and source water, and lead exposure from aged housing structures are just a few of the environmental hazards that disproportionately affect low-income communities.

The EPA has had an effective impact in places where citizens, science, and government demand healthy air, water, and soil. California environmental protections work to expand lifespan, improve quality of life, and preserve lovable places. But this is sporadic because we know that there are 1,000 places where the EPA turns a blind eye to environmental disasters, creating great harm to all forms of life (humans, animals, fish, and vegetation) necessary for sustainable life that allows us to thrive, not just survive. (Shaw and Younes, 2021). Local EPA offices are political bodies that answer first to powerful and deadly corporations that destroy the Earth.

Louisville is a primary example where corporations ask for exemptions to release more deadly pollution. How do they justify one pound of toxins landing on the bodies of 60,000 residents who are mostly poor and minority and have a lifespan twelve years shorter than folks outside the sacrifice zones (Gilderbloom, 2023)? In a study my research team conducted, this difference in lifespan was not explained away by the lifestyle of poor and minority residents since the comparable demographic group lived longer in non-sacrifice zones. We exposed the lifestyle explanation as disinformation and pseudo propaganda blaming the victim instead of the industries that pollute. Fuzzy math from the EPA justifies increasing deadly pollution despite hearing testimony from residents who live and work there (Gilderbloom, 2023).

The call for greater racial equity means cleaning up the air, water, and soil. Poor people needlessly suffer more in Louisville than low-income people in West Coast cities. If Louisville adopted the same tough environmental regulations as its West Coast counterparts, West Louisville would surely bloom instead of slowly die. The unfairness between black and white neighborhoods is stark and vivid. Science and public health officials can show Louisville how to solve some of its most pressing problems, and other cities can learn from its example. If Louisville would find the will to address the pollution of West Louisville, it could prove to be a case study in best practices on how cities can confront environmental and health injustice. Deadly pollution is Louisville's most urgent problem, making many Westside neighborhoods unlivable, unsustainable, unhealthy, and unprosperous. It is the number one cause of environmental racial injustice.

The powerful want others to believe that pollution is not a problem. In other words, if they cannot see it, it must not exist. Science, data, facts, and truth tell a much different story. But President Trump wanted the EPA to hide this data from scientists and the public (Gilderbloom et al., 2020b). Indeed, pollution is largely invisible because the deadliest pollution, PM2.5, is microscopic, about 1/16th of a human hair, and gets clogged up in a person's lungs, heart, brain, and liver. Louisville might be among the worst examples of the extreme negative impact of pollution in the city. Compared with 144 other mid-size cities, Louisville has some of the worst and deadliest pollution in the United States (Gilderbloom et al., 2020).

This book adds a human dimension to how pollution impacts our brains, hearts, and lungs in a case study of Louisville. We challenge the mainstream corporate view that pollution is harmless to humans and that it just "dissipates into the air." It's hard to dismiss the impact of one pound of toxic pollution per resident or student entering the nose, ears, eyes, mouth, and skin. It shortens lifespans. But they do dismiss it. And few challenge that view, including taxpayer-funded universities, for fear of alarming donors. We hope this research will stimulate even more rigorous research as we try to obtain more research funds via the Center for Sustainable Urban Neighborhoods. High levels of pollution reduce lifespans, reduce housing equity, damage the fragile atmosphere, and increase the chance of getting COVID-19.

The map by Beelen et al. (2014) showing deaths per 100,000 attributable to pollution was published by *The Lancet Commission*, the leading medical journal in the world (Beelen et al., 2014). Notice that North America, Australia, and Northern Europe get a "pass" for hardly any deaths being caused by pollution (Beelen et al., 2014). As we have shown throughout this book, this claim is preposterous, lazy, and maddening.

Chetty et al. (2014) found that in some parts of the USA, some places have reduced lifespans, but he discounted pollution, which is consistent with *Lancet*. We found otherwise in this book.

ProPublica was able to identify 1,000 places around the United States in which dangerous toxins are released into the air, and this is what their map showed. The majority are in the Rust Belt area and the unregulated South (Shaw and Younes, 2021). The map by Shaw and Younes (2021) shows concentrations of chemical releases mostly in the Rust Belt and sun belt but few along the East and West Coast.

As we showed in chapter 3, my research team discovered an error in air pollution measuring: two of the four EPA measures of pollution were missing. We looked at a sample of 144 semi-isolated cities with air quality ranging from clean to toxic. We wanted to correct the mistakes made by *Lancet* and ProPublica and develop a clear and unbiased representative map to understand how high levels of toxins have a devastating impact on communities.

KIESD (Kentucky Institute for the Environment and Sustainable Development) partnered with American Synthetic Rubber, which alone was responsible for the production of roughly 90 percent of a deadly cancer-causing

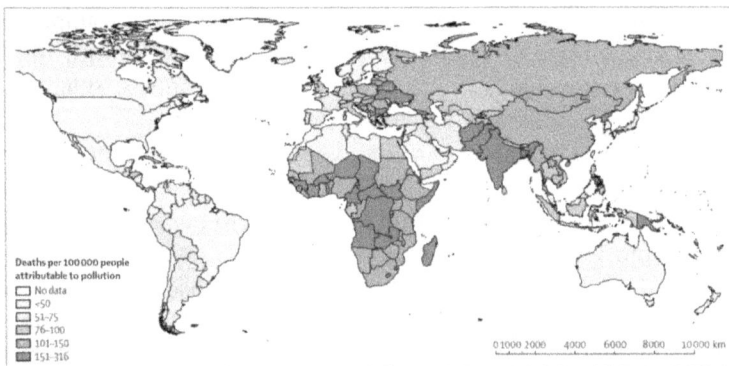

Figure 19.1 Lancet Commission World Map Showing Concentrated Pollution Deaths. *Source*: Beelen, R. et al. (2014). Effects of long-term exposure to air pollution on natural-cause mortality: an analysis of 22 European cohorts within the multicentre ESCAPE project. The Lancet Vol. 383 No. 9919: 785–795. DOI: 10.1016/S0140-6736(13)62158-3.

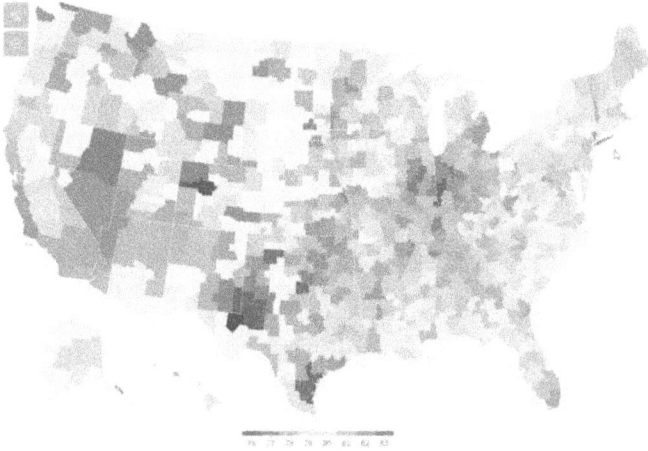

Figure 19.2 Chetty's 2014 Map of Shortened Lifespans Throughout the United States. *Source*: Chetty, R., Cutler, D., Stepner, M., & Abraham, S. (2016). The Association Between Income and Life Expectancy in the United States; 2001–2014. *Journal of the American Medical Association*, 315(16), pages 1, 750–751, 7666.

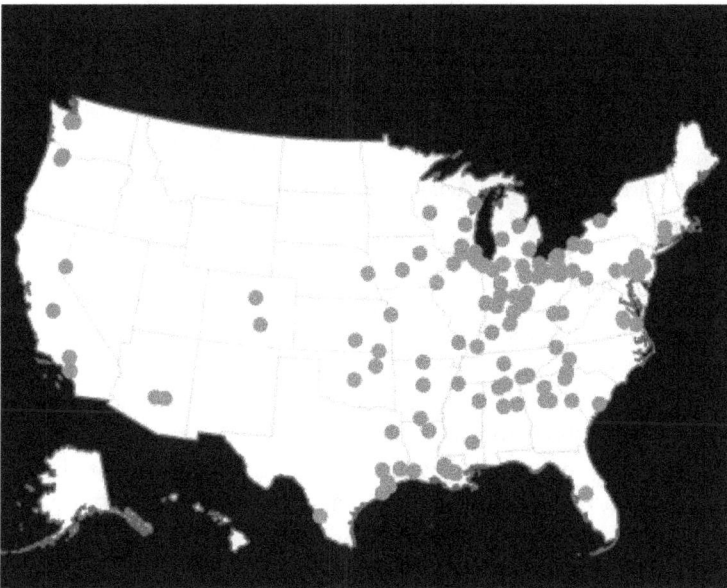

Figure 19.3 ProPublica Identified 1,000 Places with Dangerous Toxins. *Source*: Shaw, A., & Younes, L. (2021). The Most Detailed Map of Cancer-Causing Industrial Air Pollution in the United States. ProPublica, November, 2.

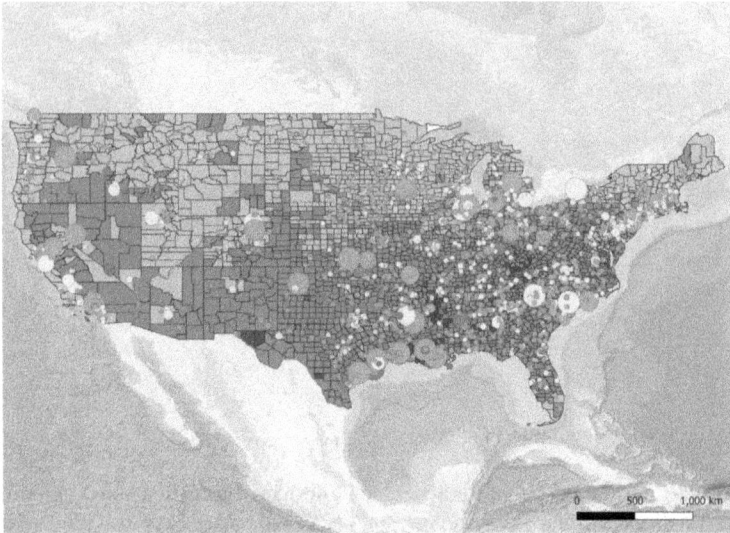

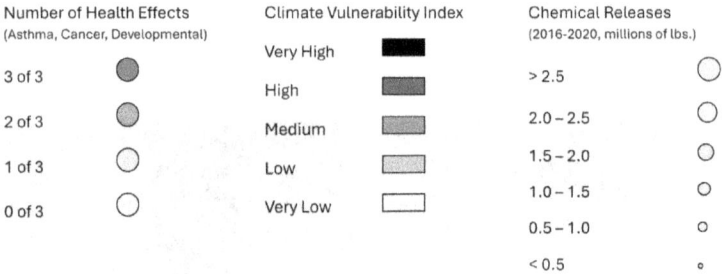

Figure 19.4 Chemical Exposure Action Map for USA. *Source: Environmental Defense Fund, Collins & Proville, 2023*

chemical: 1,3-butadiene. Russ Barnett found that the chemical was mostly non-detectable in the air when American Synthetic Rubber shut down for two weeks during the summer. KIESD proposed a $3 million dollar anti-pollution device that would stop 1,3-butadiene from being released into the air, water, and soil. This deadly chemical is highly correlated with cancer. Russ made a "good neighbor" proposal to have American Synthetic Rubber install a $3 million dollar device to forestall this dangerous pollutant from going into the air. Imagine the positive impact if the chemical companies that produce 75,000 pounds of chemical pollutants were equipped with such devices. Moreover, our research pushed the EPA to reduce the amount of pollution by at least 6,000 pounds (Giffin, 2024; see also Gilderbloom, 2023). EPA officials tipped their hats to my series of articles on pollution in the local newspapers and academic journals, along with my research cited by the U.S. Department of Justice (2023; see also Gilderbloom, 2023, 2020c, 2020b, 2020a). Good measures of data help move officials to act.

How deadly are the toxins in Louisville? One accident, one angry employee, or one terrorist attack could wipe out thousands of lives in West Louisville with a chemical leak or explosion. This is a ticking time bomb. Given the relative density of West Louisville, roughly 60,000 residents could die instantly, according to former officials from the Commonwealth of Kentucky Energy and Environmental Cabinet, the recently closed Kentucky Institute for the Environment and Sustainable Development, and the Center for Sustainable Urban Neighborhoods.

Heavy pollution not only affects health and housing in industrial neighborhoods, but it also creates an even bigger hole in the fragile atmosphere, making Louisville one of the highest emitters of greenhouse gases per capita. High levels of industrial pollution explain why Kentucky has a disproportionate number of man-made disasters: flooding in Eastern Appalachia and dangerous tornados in the South. For example, in their 2021 article in the Courier Journal, environmental writer James Bruggers and his colleague Phil McKenna estimate that just one chemical plant out of the 45 in Louisville emits enough dangerous greenhouse gases to equal the pollution of 671,000 cars (McKenna and Bruggers, 2021a). Several weeks later, they amended the article to read it was now 750,000 cars with two vicious greenhouse gases (McKenna and Bruggers, 2021b).

> Chemours most harmful climate super pollutant is the byproduct, dydroflurocarbon-23 (HFC-23), a potent greenhouse gas that produces 12,400 times more warming than carbon dioxide, the main chemical compound responsible for climate change. . . . In addition to being a climate super pollutant that is 1,720 times more effective at warming the atmosphere than carbon dioxide, HCFC-22 also destroys atmospheric ozone that helps protect the Earth from harmful ultraviolet rays. (McKenna and Bruggers, 2021b)

In the previous chapters, we talked about the correlation between pollution and its impact on humans: proficiency scores, lifespan, and rates of COVID-19, as well as neighborhood impacts such as abandoned housing, housing equity, and crime. Deadly toxic pollution occurs in over 1,000 places around the United States, according to ProPublica. It's a problem that has not been well-documented due to significant pushback from chemical companies, corrupt local governments, chambers of commerce, and even universities joining in on the deceit.

Reducing Greenhouse Gas Emissions and the Need for Renewable Energies

In this chapter, Roosa states that many of our cities have evolved to become "post-industrial" and are unlike anything that our forbearers might have imagined. Cities in Brazil, Mexico, China, and India are reeling from population growth, poor development choices, and environmental damage. Access to energy is key to sustainability. Roosa notes that hope for solutions can be found in policies, programs, and technologies. Governments are creating policies, corporations are reconstructing strategic plans, and institutions are redefining their missions. There is an evolving consensus on the horizon, one that will change how we prioritize our efforts to become more sustainable.

Historic Preservation as a Sustainability Strategy to Foster Pro-Environmental Cultures

This chapter examines the impacts of historic preservation on jobs, property values, and reducing fossil fuel consumption. Louisville, Kentucky, is a national leader in preservation, ranking first in the White House's Preserve America initiative with 73 recognized communities. Tax incentive programs have been an effective tool for creating positive changes in historic areas. Historic preservation results in more job creation than most other public investments. In the presence of escalating gas prices and assorted environmental practices, we show that neighborhoods containing historic districts exhibit higher increases in median neighborhood housing values than undesignated neighborhoods. This chapter also shows that environmentalism and historic preservation are linked together and complement one another. Residents of historic urban neighborhoods exhibit more environmentally friendly behavior, particularly those living in single-family homes. More importantly, they recycle materials that reduce energy consumption with embedded energy.

How to Make Our Schools Greener and Our Students Smarter

The transition from energy-intensive schools to green schools can result in enormous energy savings; higher grades and test scores, less violence, and happier students (because of daylighting in classrooms); lower building costs; and reduced water costs (from water conservation techniques and rainwater collection systems). Plus, students' proficiency scores will rise in a clean air, solar environment coupled with nearby chemical and liquor distillery companies shutting down the deadly toxins.

Will Planting Eight Billion More Trees Solve Climate Chaos? No!

Trees have the ability not only to filter our air but also to add beauty to our cities. As a result of the many benefits associated with trees and their ability to improve urban areas, it is imperative that city officials and planners start incorporating trees into cities and urban areas on a much larger scale. They help cool the streets and houses, make walking more pleasant and enjoyable, and create beauty which leads to a more lovable and sustainable neighborhood. They help reduce the threat of flooding and reduce energy costs as well. However, they don't do much to clean up toxic pollution.

Just one tree can make a difference, but it is not a miracle cure to address the issue of heavy chemical pollution. However, planting eight billion trees will not significantly lower the warming of the Earth or clean up the pollution. It's a corporate decoy designed to take the focus away from pollution and dodge having to focus on pollution regulation. As we show in 145 semi-isolated cities, there is no difference between cities with high levels of tree canopy versus low levels of tree canopy in terms of lifespan and cancer.

Does Walkability Matter? Yes, It Reduces Foreclosure, Improves Health, and Reduces Greenhouse Gases

The ideal neighborhood is one that is walkable. The more people walk or bike or use buses/trains, the less pollution there is in the air. In general, inner-city neighborhoods built before the mass production of cars are more walkable than sprawling "residential only" suburban neighborhoods that are isolated from the basic necessities of everyday life. We find that walkability is statistically significant in predicting an increase in neighborhood housing values and has a significant negative correlation with foreclosures in the neighborhoods of Louisville. Finally, walkability is also associated with reduced crime in several measures.

Biking Is the Best Choice for Health, Safety, and Zero Emissions

Biking is one of the most viable transportation choices for healing a broken Earth. Compared to cars, it releases none of the dangerous greenhouse gases that wreak havoc on the environment. Choosing a bike over a car means better health for bikers, longer life spans, decreased medical expenditures, and an average of $8,000 to $12,000 per year in savings. A bike-friendly neighborhood also increases community and housing values. The local economic benefits are significant in shifting car-centered transportation toward a more bike-friendly culture in urban neighborhoods. Fewer cars and more bikes and bike-oriented infrastructure could translate into higher property values, more jobs, decreased traffic congestion, and, ultimately, more money in the consumer's pocket. This spending on housing repairs, installation of community gardens, and locally operated grocery stores and restaurants (expected results of the cost savings created by the reduction or discontinuance of car-based commuting) represents the potential green dividend of investments in urban bicycling. Another important finding is that people who ride bikes will be healthier due to increased physical activity and improved air quality. Most importantly, the net zero impact of carbon emissions from bicycles means a reduction of greenhouse gases and, hopefully, a way to heal the Earth back to normalcy. However, the primary problem is the lack of safe bicycling infrastructure in most cities today, requiring major public investment before a significant number of people will choose this option.

Portland, The Best Green City in America: Equity, Health, and Safety

Portland is a model for one of the most livable and sustainable mid-sized cities in the United States and is different from most other U.S. cities in one crucial way: it has reduced its per capita carbon emissions. Portland has not only become a city with a steadily shrinking environmental footprint, but one that is experiencing an economic boom based on the "green" economy, a massive influx of young and college-educated residents, and steadily rising rates of walking, bicycling, and transit use. How has the city accomplished all of this? The answer is: by embracing new urbanism. Over the last 40 years, the city of Portland has instituted policies that incentivize the construction of dense, mixed-use development. Combined with generous investments in streetcars, light rail, and bicycling infrastructure, Portland has created a dense, walkable urban environment that has been attracting new residents and new investment ever since. Portland's charm is that it is a weird, unconventional, and exciting place to live, and the city is living proof that investing in the "green" lifestyle pays off in a big way. In addition, they have banned new single-family housing construction and mandated new housing must at a minimum have an accessory unit. Single-family housing is just too expensive in terms of taxpayer infrastructure cost and moreover, they tend to be more concentrated in the outlying areas, which means longer commute times.

Amsterdam: Planning and Policy in the World's Most Livable City

Amsterdam includes many components that make it an ideal city: dense and compact neighborhoods near work and school, biking and walking for over half the population, light and hard rail access, minimal personal use of cars, historic preservation, personal freedom, and support for the arts to attract young people. Many of the social, economic, and environmental problems facing Amsterdam are considerably fewer than in cities in the United States, and in most cases, Western Europe. Amsterdam, at this moment in history, might be the world's greatest city because of its ability to ensure necessities, freedom, and creativity. The data show that Amsterdam has lower rates of crime, murder, rape, drug usage (cocaine and marijuana), teenage pregnancy, diabetes, obesity, suicide, abortion, infant mortality, dependence on fossil fuels, and homelessness. There is also considerably less racial segregation. People live longer because of Amsterdam's walkability, bike usage, and access to parks. People living in Amsterdam seem more tolerant, secure, happier, and healthier compared to citizens in the United States. Amsterdam is by no means perfect, but in comparison to many other democratic industrial cities, it is a far better place for citizens of all races, religions, and incomes. There is much to learn from the Dutch about important solutions to policy and planning and Amsterdam provides us with valuable lessons.

Academic Freedom and Science Under Attack

Throughout history, there has been an attack on science. Who stands up for science? Not the cowardly university administration that bows down to the wishes of those who donate millions of dollars to the university and uses the same tactics that were used to attack Rachel Carson, author of the award-winning book, *Silent Spring*. Other scientists have been silenced; Galileo, the greatest scientist of all time, was put in home incarceration for 12 years. When he argued that the Earth circled the Sun and conflicted with the teachings of the Catholic Church, his microscope and telescope were taken from him, and he was told that he could not write for the last 12 years of his life. Despite being a gunshot survivor and having my life threatened, I remain a fearless truth teller who informs people that we are not being good stewards of the Earth.

How are 3,000-plus places of higher education seemingly silent about environmental crimes against humanity and reducing greenhouse gases? There are only a few academic universities willing to speak out without interference from the polluters: Arizona State University, University of California (UC) Santa Barbara, UC San Diego, and Harvard are doing cutting-edge environmental scholarship and activism. Way too many are afraid to speak out and be good stewards of the Earth, especially those in the South and the Rust Belt where pollution levels are very high. Even the federal EPA under the Democrats is too timid.

CONCLUSION

Climate change is not just a scientific phenomenon; it's an intimate dance of survival, adaptation, and hope. Every block, every neighborhood, and every city add a note to our global symphony. The stakes are high and the challenges profound, but woven within are tales of resilience, transformation, foresight, and progress. We feel the weight of why this matters, and we've seen how our fight is down to individual city blocks. We've glimpsed the future of super cities and the next fifty years. Our children are the inheritors of dire world challenges, but they are also the heartbeat of promise for a more abundant, brighter, better future.

End toxic pollution and watch a magical transformation happen in the 1,000 toxic neighborhoods in the USA: a jump in elementary school student proficiency scores, higher home values, longer lifespans, reduced greenhouse gases, fewer COVID-19 cases, and more business startups. We need to stop the unnecessary loss of life. We need to fight for clean air, water, and soil without compromise. Universities have millions of dollars to identify the problem and the solutions to solve it. The University should be a fearless defender of science and free speech, and an unassailable beacon of truth. Universities need to get on the side of science instead of climate denialism. We need to have honest and uncomfortable debates. We can solve this problem before it's too late.

Our planet is facing catastrophic destruction in just seven or eight years, according to the United Nations. The doomsday clock is ticking; it's minutes before midnight. We are talking about saving civilization. We are talking about not just trying to slow down climate change by three years but actually reversing it. We need to educate citizens and leaders about existing tools to save civilization. We have access to the best data once hidden and not accessible to scientists: EPA, CDC, Toxics Release Inventory, and advanced satellite technology like the Copernicus Sentinel, which can zoom in on a city, neighborhood, or a building to measure what kinds of toxins are being released. The omnipresent eye of the Eurpean Space Agency's Copernicus Sentinel program captures data and highlights the profound connection between climate, cities, and neighborhood communities.

At the UN Climate Change Conference COP 27, António Guterres, Secretary-General of the United Nations, said in a speech, "We are on the highway to climate hell, and we've got to take our foot off the gas pedal." This book is about hope, a climate of hope. It gives people workable tools to reverse climate change. We have provided examples of what might work to empower citizens to not accept defeat.

REFERENCES

Beelen, R., Raaschou-Nielsen, O., Stafoggia, M., Andersen, Z. J., Weinmayr, G., Prof Hoffmann, B., Wolf, K., Samoli, E., Fischer, P., Nieuwenhuijsen, M., Vineis, P., Xun, W. W., Prof Katsouyanni, K., Dimakopoulou, K., Oudin, A., Forsberg, B., Modig, L., Havulinna, A. S., Lanki, T., Turunen, A., Oftedal, B., Nystad, W., Nafstad, P., De Faire, U., Pedersen, N. L., Östenson, C.-G., Fratiglioni, L., Penell, J., Korek, M., Pershagen, G., Eriksen, K. T., Overvad, K., Ellermann, T., Eeftens, M., Petra, P., Peeters, H., Meliefste, K., Wang, M., Bueno-de-Mesquita, B., Sugiri, D., Krämer, U., Heinrich, J., de Hoogh, K., Prof Key, T., Peters, A., Hampel, R., Concin, H., Nagel, G., Ineichen, A., Schaffner, E., Prof Probst-Hensch, N., Künzli, N., Schindler, C., Schikowski, T., Adam, M., Phuleria, H., Vilier, A., Clavel-Chapelon, F., Declercq, C., Grioni, S., Krogh, V., Tsai, M.-Y., Ricceri, F., Sacerdote, C., Galassi, C., Migliore, E., Ranzi, A., Cesaroni, G., Badaloni, C., Forastiere, F., Tamayo, I., Amiano, P., Dorronsoro, M., Katsoulis, M., Trichopoulou, A., Brunekreef, B., & Hoek, G. (2014). Effects of Long-Term Exposure to Air Pollution on Natural-Cause Mortality: An Analysis of 22 European Cohorts within the Multicentre ESCAPE Project. *The Lancet, 383*(9919), 785–795. DOI: 10.1016/S0140-6736(13)62158-3.

Chetty, R., Cutler, D., Stepner, M., & Abraham, S. (2016). The Association Between Income and Life Expectancy in the United States; 2001–2014. *Journal of the American Medical Association, 315*(16), 1, 750–751, 7666.

Giffin, C. (2023). New EPA Rule EPA Rule Would Effect these Four Chemical Plants, Provide Air Monitoring in West Louisville. *Courier Journal.*

Gilderbloom, J., Gregory Squires, G., & Meares, W. (2020a). Mama I Can't Breathe: Louisville's Dirty Air has Steep Medical Costs. *Local Environment: The International Journal of Justice and Sustainability, 25*(8), 619–626. https://doi.org/10.1080/13549839.2020.1789570.

Gilderbloom, J. H. (2023). Where Is the Public Outrage? The Reckless Decision to Keep Polluting West Louisville. *Courier Journal* op-ed April 3, 2023.

Gilderbloom, J. H., Meares, W. L., & Squires, G. D. (2020b). Pollution, Place, and Premature Death: Evidence from a Mid-Sized City. *Local Environment: The International Journal of Justice and Sustainability, 25*(6), 419–432. https://doi.org/10.1080/13549839.2020.1754776.

Gilderbloom, J. H., Riggs, W., Frederick, C., Squires, G., & Quenichet K. (2022). The Missing Link of Air Pollution: A Closer Look at the Association Between Place and Life Expectancy in 146 Mid-Sized Cities. *International Journal of Strategic Energy and Environmental Planning, 4*(4), 31–51. https://papers.ssrn.com/sol3/papers.cfm?abstract_id=3506217.

Gilderbloom, J. H., & Squires, G. (2022). Put a Mask on Pollution: Connect the Dots Between COVID-19 and Unhealthy Air. *Social Policy, 51*(4), 15–18.

Gilderbloom, J. H., Washington, C. B., Quenichet, K., Manella, C., Dwenger, C., Slaten, E., Sarr, S., Altaf, S., & Frederick, C. (2020c). What Cities Are the Most Dangerous to Your Health? Ranking the Most Polluted Mid-Size Cities in the United States. Pre-prints with *The Lancet*, January 9, 2020.

Gilderbloom, J. I. H., Kingsberry, I., & Squires, G. D. (2021). How Many More Children Must Be Hurt by Pollution? *Harvard Medical School Primary Care Review*. https://info.primarycare.hms.harvard.edu/review/children-hurt-pollution.

McKenna, P., & Bruggers, J. (2021a). Louisville Chemical Plant Emits Super-Pollutant Equal to Emissions from 671,000 Vehicles. *Courier Journal*, March 10, 2021. https://www.courier-journal.com/story/news/2021/03/10/global-warming-louisville-plant-emits-super-pollutant-contributors/6932934002/.

McKenna, P., & Bruggers, J. (2021b). Louisville's Super-Polluting Chemical Plant Emits Not 1, but 2 Potent Greenhouse Gases. *Courier Journal,* April 5, 2021. https://www.courier-journal.com/story/news/2021/04/05/super-polluting-louisville-chemical-plant-emits-more-greenhouse-gases-than-thought/7075408002/.

Meares, W., Gilderbloom, J., H., Squires, G., & Jones, A. (2021). Pollution and the Pandemic: Explaining the Differences in COVID-19 Rates Across 146 U.S. Communities. *International Journal of Strategic Energy and Environmental Planning,* 4(2), 9–25.

Shaw, A., & Younes, L. (2021). The Most Detailed Map of Cancer-Causing Industrial Air Pollution in the US. *ProPublica,* November 2.

U.S. Department of Justice. (2023). *Investigation of the Louisville Metro Police Department and Louisville Metro Government.* United States Department of Justice Civil Rights Division and the United States Attorney Office Western District of Kentucky Civil Division March 3, 2023.

Al Gore's Speech at the Opening of the World Leaders Summit at the United Nations Climate Change Conference (COP27) in Sharm el-Sheikh, Egypt, on November 7, 2022

We are all here today because we continue to use the thin blue shell of atmosphere surrounding our planet as an open sewer. Today as every day, we are spewing 162 million tons of man-made, heat trapping, global warming pollution into the sky. It adds up and accumulates and lingers there. On average, each molecule lingers 100 years, and the accumulated amount now traps as much extra heat as would be released by 600,000 Hiroshima class atomic bombs exploding every day on our planet. That's why we're seeing these disasters, and the pattern is very clear. It's getting steadily worse.

We have a credibility problem—all of us. We're talking and we're starting to act but we're not doing enough. It is a choice to continue this pattern of destructive behavior. We have other choices. In my faith tradition, I learned a teaching that is common to all three of the Abrahamic religions: that God has set before humanity a choice between blessings and curses, between life and death. We face that choice today. We can continue the culture of death that surrounds our addiction to fossil fuels by digging up dead life-forms from eons ago and burning them recklessly in ways that create more death, including 8.7 million people every year that die from the air pollution principally caused by the burning of fossil fuels.

The curses that we are continuing to choose are evermore apparent. We saw them on that amazingly powerful film. One third of Pakistan underwater, 1,700 people killed. A heatwave in China that lasted 70 days with the heat above 40 degrees, 104 Fahrenheit over a vast area of China. Stronger storms, bigger downpours, rain bombs, worse floods. Another million displaced in Nigeria two weeks ago. Another million in Chad a few days after that. The Nile Delta being salinated just as Hannibal's fields were salted. We are

459

causing the salting of the greatest agricultural area in Egypt as well as in the Mekong Delta and elsewhere. Droughts are drying up the mighty Mississippi River in my country, The Tigris and the Euphrates in the cradle of civilization, the Poe River in Italy, Loire in France, the Rhine in Germany. All over the world, the greatest reservoirs in North America, Lake Mead and Lake Powell, are reaching what is called dead pool status. Starvation and famine in the Horn of Africa. Devastation in Madagascar. And the areas of our world that are presently considered by doctors uninhabitable because of the combination of heat and humidity are relatively small today but they are due to expand to the point where experts are predicting as many as one billion climate migrants crossing international borders in the balance of this century. Think of the millions that are crossing borders now and the xenophobia and the authoritarian populism that is caused by a large surge of refugees and then imagine if you will what a billion climate refugees would do. It would end the possibilities of self-governance.

We have to act. We know that tragically across the nations of Africa these impacts are even worse and in the Pacific islands and elsewhere, and the poor suffer the most. The late Archbishop Desmond Tutu said, "Climate change is the apartheid of our times." We don't have to choose curses. We can choose blessings, including the blessings of renewable energy. We are now in the early stages of a sustainability revolution that has the magnitude of the industrial revolution and the speed of the digital revolution. If we invest in it and stop subsidizing the culture of death, we can save ourselves. Dollar for dollar, each dollar spent in renewable energy creates three times as many jobs as dollars spent in fossil energy. And this sustainability revolution is massively deflationary at just the time when the world is trying to find new ways to battle inflation.

The International Energy Agency has called solar energy the cheapest energy in the entire history of the world and Africa can be the renewable energy superpower. Africa has 40 percent of the entire world's global potential for renewables. The potential for solar and wind in Africa is 400 times larger than their total fossil fuel reserves. And these are real possibilities. This is not pie in the sky. If you look at all of the new electricity generation installed worldwide last year, 90 percent of it was renewable because it is cheaper almost everywhere and in two years it will be cheaper in 100 percent of the world. But it is not enough to simply increase our support for clean energy if we are simultaneously backtracking on our commitments against developing and financing fossil fuels.

The world's leading scientists and energy experts have told us that any new fossil fuel development is incompatible with 1.5 degrees as the limit to the temperature increase. Mr. Secretary General, you said we are on a highway to climate hell; we need to take our foot off the gas. We need to obey the first law of holes: when you're in one, stop digging. We have to stop making this crisis worse. We must see the so-called "dash for gas" for what it really is: a dash down a bridge to nowhere leaving the countries of the world facing climate chaos and billions in stranded assets, especially here in Africa. At a time of turbulence in the global energy markets, the wealthy nations of the world should not confuse the short term with the long term and should not be fooled by the absolute need to backfill the shortage of fossil energy caused by the cruel and evil war launched by Russia in Ukraine as an excuse for locking in long term commitments to even more dependence and addiction on fossil fuels.

We have to move beyond the era of fossil fuel colonialism and that's what it is. The dash for gas in Africa is a dash for gas to be sent to wealthy countries. We have to remove the barriers to the trillions needed for climate finance and the scale of what is needed can only be provided by the private sector. I support governments paying money for loss and damage and adaptation but let's be very clear that that's a matter of billions or tens of billions. We need four and a half trillion per year to make this transition and that can only come by unlocking access to private capital. In the United States and in Canda, if you look at all of the financing of renewable energy, 96 percent of it comes from the private market. If you look at Africa, much lower amounts overall but what percentage of it comes from the private markets? Not 96 percent, 14 percent; 86 percent comes from governments. Why is that? Because, for example, if you want to build a solar field in Nigeria, even though it's profitable, you have to pay interest rates that are 7 times higher than the interest rates paid by OECD countries. That is unjust. It is insane. That is what the World Bank and the multilateral development banks are supposed to address. We need to reconvene Bretton Woods and completely revamp and reform the World Bank system and make access to private capital available for developing countries.

This is a moment for a global epiphany. It is not time for moral cowardice and reckless indifference to the future of humanity. If you want an example of reckless indifference just recall the banker who said earlier this year in discussing the impacts of the climate crisis, "Who cares?" When he was fired it reminded me of the famous writer who said that in the highest reaches of business and politics a gaffe is when someone accidentally tells the truth. The truth is that all of us have limitations and weaknesses and vulnerabilities but as human beings we also have the God given ability to rise above those limitations. The greatest leader of my country, Abraham Lincoln, once said, "The occasion is piled high with difficulty, and we must rise with the occasion. As our case is new, we must think anew and act anew. We must disenthrall ourselves and then we shall save our country." And that is the way we can save our world.

More is needed but we have the basis for hope. Just days ago, the people of Brazil chose to stop the destruction of the Amazon. Earlier this year, the people of Australia chose to start leading the renewable energy revolution. Earlier this year, the people of my country chose leaders who enacted at long last the biggest and most ambitious climate legislation in the history of the world. The world has passed and enacted the Kigali Amendment. We have the basis for hope. If we actually do reach true net zero, the scientists tell us temperatures will stop going up with a lag time of as little as three to five years. And if we stay at true net zero, half of all of the manmade CO_2 will fall out of the atmosphere in as little as 25 to 30 years. But we have to choose blessings instead of curses. We have to choose life over death.

If we are truly to make this COP the one that drives implementation of the pledges made in Glasgow, we must equip ourselves with every tool and resource to facilitate our path to true net zero and work together to drive impact and increase accountability. I want to applaud the Secretary General for making transparency around these net zero goals a priority, and I'm proud that two days from now Climate Trace, a nonprofit coalition of artificial intelligence experts and data scientists and researchers and NGOs, will announce the first inventory of site specific facility sources of greenhouse gas pollution, the most detailed inventory ever assembled that can be used to identify exactly where all this pollution is coming from. There is a path that we can make from here to a future with hope. As a final point, I started by saying that we have a credibility problem, but we also have the ability to address it and if anyone doubts that we can summon the political will to do what is necessary and to save our future, always remember that political will is itself a renewable resource. Thank you.

United Nations Public Domain—https://www.youtube.com/watch?v=qLTcC-7srnLw&t=231s

A Prayer for Our Earth

All-powerful God,
you are present in the whole universe
and in the smallest of your creatures.
You embrace with your tenderness all that exists.
Pour out upon us the power of your love,
that we may protect life and beauty.
Fill us with peace, that we may live
as siblings, harming no one.
O God of the poor,
help us to rescue the abandoned
and forgotten of this earth,
so precious in your eyes.
Bring healing to our lives,
that we may protect the world and not prey on it,
that we may sow beauty,
not pollution and destruction.
Touch the hearts
of those who look only for gain
at the expense of the poor and the earth.
Teach us to discover the worth of each thing,
to be filled with awe and contemplation,
to recognize that we are profoundly united
with every creature
as we journey towards your infinite light.
We thank you for being with us each day
Encourage us, we pray, in our struggle
for justice, love and peace.

Pope Francis, Laudato si' (246): A joint initiative of the Holy See and the Stockholm Environment Institute. (2023). *Our Common Home: A guide to caring for our living planet* [Booklet].

Index

A. de Graaf Foundation, 398
AAA. *See* Automobile Association of
    America (AAA)
Abbott, C., 373
Abraham, S., 13, 24
Abramson, Jerry, xxviii, 164, 310
academic freedom, 454–55
acetylene, 157
acrylic latex emulsions, 158
acrylonitrile, 159
active solar, 341
Adams, Sam, 375
AEE. *See* Association of Energy
    Engineers (AEE)
Aerial Tram, 370
aerosol propellant, 159
affordability, 289–90, 335–56, 369
affordable housing, 335–36
African Americans, 3, 6, 94, 100,
    107–8, 135–36, 171, 214
Agency for Toxic Substances and
    Disease Registry (ATSDR), 190
agriculture, xviii–xix, xxi–xxii,
    115
Agyeman, Julian, 190
air, xx, 9–10; clean, 9; PAH levels in,
    49; poor quality of, 10, 49–51, 62,
    94, 173, 175, 198, 250, 252, 254,
    267, 408; quality, 267–68

air conditioning systems, 126, 239, 241,
    270–72, 332, 342, 346, 423–24, 435
air filtration, trees planting for, 267–69
air pollution, 3, 8, 49, 195, 267–68; air
    quality analysis and, 32–33; case
    selection criteria for study on, 33, 53;
    cleanest and dirtiest cities, analysis
    to, 35, 56–60; control variables
    for, 34; dependent variable for, 33;
    discussion on, 39–40; educational
    achievement, impacts on, 14–15; and
    educational performance, 247–48;
    emissions from coal-fired generation,
    249–51; EPA's measures for, 9–10;
    excluded cases, 53–54; facts over
    politics on, 62–63; fine particle,
    3–4; human health, impacts on,
    30, 50–52; human mortality rates
    and, 9; impact on public school
    achievement, 247–56; implications
    of, 60–62; from industrial sites, 12;
    infants health and, 30, 51; levels
    of, 9–10; on life expectancy (low
    vs. high emission cities), impacts
    on, 9–10, 29, 36; limitations of
    methodology applied, 63; literature
    review on, 50–52; local conditions
    matter in, 31; negative effects on
    quality of life, 50–52; pollutants,

About the Contributors

Joshua D. Ambrosius is an associate professor of political science and is the director of the Master of Public Administration program at the University of Dayton. He holds a PhD in Urban & Public Affairs from the University of Louisville and a Master's in Public Policy from Johns Hopkins University. He regularly teaches courses related to applied quantitative research methods; urban, housing, and faith-based social policy; and space exploration policy. His diverse research agenda addresses unique questions related to the social science of space exploration, religious institutions, sustainability, community development, and social capital. Dr. Gilderbloom was his PhD chair.

Bobby William Austin is an academician, non-profit executive director, author, and editor. Most recently, he was the executive director of Neighborhood Associates. He is the executive director of the Public Kinship Institute, focused on sharing his ideas and philosophy about Americans' interconnectedness. He is the author of five books. As the director of the W.K. Kellogg Foundation's National Task Force on African American Men and Boys, chaired by Andrew J. Young, he co-authored and edited the 1996 report, *Repairing the Breach: Key Ways to Support Family Life, Reclaim Our Streets and Rebuild Civil Society in America's Communities*. Based on that report, he developed the first major philanthropic and social infrastructure network for African American men and boys, through which 30 national organizations reached 17,000 participants. Bobby Austin and the legacy of the Task Force were highlighted at the Harvard Graduate School of Education's 2014 conference, "Revisiting *Repairing the Breach*."

He serves as the Mahatma M.K. Gandhi fellow of the American Academy of Political and Social Science and an advisory member of the Southern Reconstruction Foundation. He was recently chosen as a member of

489

Evolutionary Thinkers. He is a resident fellow of the Institute of Politics, Policy and History, is included in *HistoryMakers,* and is a member of the Sovereign Order of St. John of Jerusalem, Knights Hospitaller.

Russell Barnett was formerly the director of the Kentucky Institute for the Environment and Sustainable Development and the co-editor of the *Journal of Sustainability.* Previously, he worked in the Kentucky Energy and Environmental Cabinet. Russ was a mentor to John Hans Gilderbloom and to many PhD students who graduated from the University of Louisville Urban and Public Affairs.

Christopher Bird is a structural engineer working in building design in Washington, D.C. At this firm, he is a leader in analyzing the embodied carbon impacts of structural materials and construction strategies. He studied at the University of Louisville (UofL) Speed School of Engineering, earning a Bachelor's and Master's degree in structural engineering. He served as the UofL Services Vice President for the Student Government Association, advocating for sustainable construction and planning on campus.

Leah Callahan works in multimedia and computer graphics. She designed two of Dr. Gilderbloom's books: *Chromatic Homes: The Joy of Color in Historic Places* and *Chromatic Homes: The Design and Coloring Book.* She worked on the cover design of this book and does computer graphics for business, education, and non-profits.

Chad Frederick is a professor of urban geography and sustainable planning at Grand Valley State University in Allendale, Michigan. He holds a PhD in Urban Affairs from the University of Louisville, a Master's degree in Urban and Environmental Planning from Arizona State University, and a Bachelor's in Interdisciplinary Studies from Metropolitan State University in St. Paul, Minnesota. His interests include education for sustainable development, systems properties, the built environment, and social science pedagogy. He lives in Grand Rapids, Michigan, with his wife and children. Dr. Gilderbloom was his PhD chair.

Robert Friedland is a professor at the University of Louisville Medical School, where he conducts research on aging and dementia. He has published multiple books and articles on these subjects.

John Hans Gilderbloom is an urban planning legend. Dr. Gilderbloom's insights and research underscore the importance of cities—they are not just structures, they are living, breathing souls with dreams, despair, hopes, and aspirations. They are truly the future of the planet. Dr. Gilderbloom is ranked as one of the world's top urban thinkers and was an adviser to Vice President Al Gore's White House Sustainability Commission, the Livable Communities Initiative. He was also invited to sit in the stands with President Clinton at his second inauguration parade, was invited to his inauguration party, and was later asked to provide input to a draft State of the Union address which, had no mention at the time of tax cuts resulting in greater homeownership opportunities or clear language on environmental stewardship. During President Clinton's first term of office, he was an active consultant and advised HUD Secretary Henry Cisneros, who later wrote the introduction to Dr. Gilderbloom's book, *Promise and Betrayal.* During Clinton's second term, he was hired by HUD Secretary Andrew Cuomo to calculate job creation through housing and preservation investment—these were later shown to be superior job creators.

In 2022, he received the American Sociological Association Lifetime Achievement Award in Urban and Community Studies. He is the Director of the Center for Sustainable Urban Neighborhoods at Neighborhood Associates in Washington, D.C. He is hated by the radical right and left. He is a gunshot survivor who learned to thrive by embracing excellence in his research, teaching, and service in Louisville and Phoenix. His work in urban affairs includes significant contributions to cities worldwide in urban development, sustainability, housing, health, and transportation. Among the UPA faculty, he is the most cited author.

Beyond being a renowned urban planner, Dr. Gilderbloom is a visionary seeking sustainable solutions to concrete chaos, a luminary whose expertise illuminates the crux of our urban dilemma, and a modern-day steward of the future of cities. Dr. Gilderbloom has taught since 1980 at the University of California (SB), University of Wisconsin (GB), University of Houston, and University of Louisville (UofL), where he is Professor of the Graduate Planning, Public Administration, Sustainability, and Urban Affairs program. He is a former professor of Economics in the College of Business at the UofL, where he directed the Center for Sustainable Urban Neighborhoods for 34 years before the University closed it and moved it to Neighborhood Associates, Washington D.C. His research in sustainability, health, housing, and green renewal has appeared in seven books, 68 scholarly journals, 30 chapters in edited books, 11 monographs, and opinion pieces in *The Wall Street Journal, The Washington Post, Los Angeles Times, The New York Times, Chicago Sun-Times, San Francisco Chronicle, Courier-Journal, Houston Chronicle,* and *USA Today* magazine. His work has also appeared in prominent academic journals such as *Harvard Medical Review Primary Care, Journal of* the *American Planning Association, Urban Affairs Quarterly, Journal of Urban Affairs, Social Policy, Local Environment,* and *Lancet* (prepublication, which had the most downloads of any article according to SSRN and published again here and in a peer-reviewed journal).

He won the UofL Presidential Medal for Distinguished Research and is a two-time winner of the University of Louisville Arts and Sciences Research and Creativity Award. Dr. Gilderbloom was named one of the top urban thinkers in an international survey by Planetizen and has been featured in the documentary, Rubbertown, and on *CNN*. He was featured in a lengthy *New York Times* Sunday piece on university-community partnerships and on short-term rentals before the advent of Airbnb. He is regularly cited in major newspapers in the United States, Europe, and Japan. He has won multiple awards in planning, preservation, sustainability, and housing.

His latest book, *Climate Chaos: Killing People, Places, and the Planet,* will be released in 2025. The book includes writings from Pope Francis, Nobel Prize Winner Albert Gore, and a host of major thinkers. We believe that at the grassroots level, cities can save the world. This book inspired three-time Emmy Award winner Chris Nolan to create his latest film, Climate of Hope: Cities Saving the World http://www.climateofhopefilm.org. Dr. Gilderbloom plans to finish several more books, including *Burn This Book: How Universities Stifle Scientific Research for the People* and *Fight Back: How to Win Progressive Legislation with Statistics.* He is available for speaking engagements to promote the film and book.

Email: john.gilderbloom@louisville.edu / jgilde02@sprynet.com
   Website: http://www.sunlouisville.org

KEY BOOKS:

Rethinking Rental Housing by John Hans Gilderbloom and Richard Appelbaum

Chromatic Homes: The Joy of Color in Historic Places by John Hans Gilderbloom

Invisible City: Poverty, Housing, and New Urbanism by John Hans Gilderbloom

Promise and Betrayal: Universities and the Battle for Sustainable Urban Neighborhoods by John Hans Gilderbloom and Robert L. Mullins

Community Versus Commodity: Tenants and the American City by Stella Capek and John Hans Gilderbloom

Rent Control: A Source Book edited by John Hans Gilderbloom

KEY ARTICLES:

"Does Walkability Matter? An Examination of Walkability's Impact on Housing Values, Foreclosure, and Crime" by John Gilderbloom, William Riggs, and Wesley L. Meares. *Cities* 42, 13–24 2015.

"Scapegoating Rent Control: Masking the Causes of Homelessness" by Richard Appelbaum, Michael Dolny, and John Hans Gilderbloom. *Journal of the American Planning Association* 57 (2) 153–164.

Elliott Grantz has leveraged her expertise in sustainability to drive positive change in developing communities. After graduating with a Master of Science in Sustainability from the University of Louisville, she joined the United Nations World Food Programme (WFP) as an international development consultant focused on government partnerships and program development. Elliott's work with WFP has focused on school meals, resilience building, and emergency response programs across diverse regions, including Latin America, West and Southern Africa, and Asia. Currently based in Lao PDR, she's actively involved in developing the School Lunch Program, ensuring its long-term sustainability.

Dr. Matthew J. Hanka is a professor of political science at the University of Southern Indiana (USI) in Evansville, Indiana. Hanka earned a BA in History and Politics from The Catholic University of America in Washington, DC, and an MA in Political Science and a PhD in Urban and Public Affairs, both from the University of Louisville. Dr. Hanka was hired at USI in 2010 to serve as the Director of the Master of Public Administration (MPA) program, a position he held for eight years. His research interests include housing policy, community development, urban policy and governance, state and local government, strategic planning, and social capital. He has co-authored 12 peer-reviewed journal articles and nine technical reports. He is the author of the book *What Is Happening in Your Community? Why Community Development Matters* published by Lexington Books in 2021 and is writing a book on the Louisville-Jefferson County, Kentucky merger titled *Merger Dreams* slated to be published in 2025 by University Press of Kentucky. Dr. Hanka served as Director of the Commission on Homelessness for Evansville and

Vanderburgh County, was the former co-chair of Housing Organizations United Serving Evansville (HOUSE) and a member of the Evansville Affordable Housing Trust Fund Advisory Committee. He lives in Evansville, Indiana with his wife Ann and their two sons, MJ and David. Dr. Gilderbloom was his PhD chair.

Bunny Hayes is a social scientist working to help solve social, ecological, and environmental justice issues in her surrounding communities. She is a graduate of the University of Louisville's Geography and Environmental Sciences Department. She holds a Master of Science degree in Sustainability and a Master of Arts degree in Communication.

Antwan Jones, PhD, is an associate professor of epidemiology and sociology at George Washington University.

Avery Kahl is a graduate student in sustainability at the University of Louisville. She is Dr. Gilderbloom's research assistant.

Zachary E. Kenitzer is a Market Intelligence Manager for the Columbus Division of PulteGroup (Pulte Homes) and is based in Columbus, Ohio. He graduated with a PhD in City and Regional Planning from The Ohio State University in 2016 and completed his post-doctoral work at the Kirwan Institute for the Study of Race and Ethnicity at The Ohio State University in 2018.

Isaiah Kingsberry is a graduate research assistant in the Department of Geography and Geosciences at the University of Louisville. Isaiah's research aims to understand the gaps between our physical and experienced environments. Email: Isaiah.kingsberry@louisville.edu.

Wesley L. Meares is an associate professor of public administration and political science in the Department of Social Sciences at Augusta University, where he serves as the graduate program director for the Master of Public Administration program. His research interests include housing policy, community development, sustainability, and local government cybersecurity. Dr. Meares' research has appeared in journals such as *Cities, Public Administration Review, Public Administration Quarterly, Local Environment, Journal of Urban Affairs, Housing Studies, Journal of Public Affairs Education* and *Journal of Urbanism*. He received his BA in Political Science from Lagrange College, a Master of Public Administration from Western Kentucky University, and a PhD in Urban and Public Affairs from the University of Louisville. Dr. Gilderbloom was his PhD chair.

Justin Mog is a director of the University of Louisville sustainability program. He holds a PhD from the University of Wisconsin. He has been a lifelong advocate of conservation and reducing greenhouse gas emissions.

Lisa Murray is an undiscovered but talented "street artist" who came up with the cover art of the smokestacks. Dr. Gilderbloom saw the art and thought it illustrates the pollution problem in 1,000 mostly poor and minority neighborhoods in the United States. It also captures the signature slogans used in protests against the chemical factories in Louisville. Lisa added the sunflowers as a dark joke about the Russians responding to the Chernobyl nuclear accident with sunflowers. In West Louisville, the chemical companies have emitted 75,000 pounds of dangerous toxins, causing 12 years of shortened life for the 60,000 residents. But instead of doing their due diligence by putting in pollution controls, they focus on hiding the problem by making the smokestacks invisible to the eyes and removing the smells, thereby reducing climate activism.

Chris Nolan is a multiple Emmy Award-winning director and screenwriter. He has written and directed documentaries such as "It's VUCA, The Secret to Living in the 21st Century" and the recent film on AI, "Look Up Now," as well as feature-length movies, television, and hundreds of hours of content and commercials. Chris has received countless industry awards and has been nominated for the Humanitas Prize, an award for film and television writing that promotes human dignity.

Karrie Ann Quenichet is a PhD student in the College of Education and Human Development at the University of Louisville. She headed an outreach program designed to improve the health and education of poor West Louisville residents.

William (Billy) Riggs, PhD, AICP, LEED AP, is a global expert and thought leader in the areas of automated mobility and future transportation, urban technology, transport economics and business models, and sustainable urban development. He is a professor and program director at the University of San Francisco School of Management, and the director of the Autonomous Vehicles and the City Initiative. He has worked in venture capital and management consulting and has been an advisor to multiple companies and start-ups on technology, engineering, smart mobility, and urban development. This follows two decades of experience working as an urban planner, economist, and engineer. He has over 100 publications and has been featured in multiple global media outlets such as *The Economist, The Wall Street Journal, The New York Times, Washington Post, PBS, TF1,* the *Atlantic,* and more. Billy is the author of the books *End of the Road: Reimagining the Street as the Heart of the City* and *Disruptive Transport: Driverless Cars, Transport Innovation and the Sustainable City of Tomorrow.* Dr. Riggs has been a fellow with the National Science Foundation, sits on the Transportation Research Board (TRB) Committee on Policy and Law, the Committee on Emerging Technology and Law, and the Committee on the Landscape and the Built Environment.

Stephen A. Roosa, PhD, MBA, CEM, BEP, CSDP, REP, CBCE, CMVP, LEED AP, is an energy engineering professional focused on developing solutions that bridge the gaps between energy and sustainability. He is a former senior process engineer and energy manager for a Fortune 100 company. He is presently the President of RPM Asset Management and a senior energy consultant for Ameresco, LLC. He is the Association of Energy Engineer's (AEE) Director of Sustainable and Local Programs. He has taught over 125 seminars and workshops on the topics of energy management, sustainable development, renewable energy, and microgrids. Dr. Roosa has written six published books, over 70 published journal articles, and edited over 250 published papers on topics such as energy policies, planning and program development, energy conservation, engineering, management, renewable energy, nuclear energy, and sustainable development. He is the Editor-in-Chief of the *International Journal of Strategic Energy and Environmental Planning* and the *Alternate Energy and Distributed Generation Journal.* He serves on the Editorial Review Board of the *International Journal of Energy Management.* Dr. Roosa has received numerous awards during his career, including the AEE International Energy Manager of the Year Award, the AEE Energy Project of the Year, the U.S. Federal Energy and Water Management Project Award, the U.S. Army Corps of Engineers National Energy Engineers Systems Technology Award, the U.S. Department of Defense Joint Chiefs of Staff Citation for Energy Management, the AEE Distinguished Service Award, and the Honorable Order of Kentucky Colonels Medal of Distinction. He is a past AEE president, an AEE Fellow, a Legend in Energy, and a member of the Energy Manager Hall of Fame. Email: roosa.stephen@yahoo.com. Dr. Gilderbloom was his Ph.D. Chair.

Ellen Slaten holds a PhD in sociology from the University of Texas at Austin. She worked at the University of Texas Health Science Center in San Antonio for seven years, coordinating several federally funded research projects. She holds a master's degree in sociology from the University of Texas at Austin and a bachelor's degree in sociology from the University of Houston. She works as an editor for Dr. Gilderbloom.

Carla J. Snyder has a master's degree in Leadership from Grand Canyon University. A renowned public speaker on the national/international stage, Ms. Snyder has a long history in academia, corporate training, and not-for-profit leadership. She taught in the health sciences arena in the Maricopa Community College District. She has written several publications, including one book. In 2022, she was awarded International Woman of the Year for Promoting Peace, and in 2024, she won the Businesswoman of the Year award in Phoenix and the International HERS Award for Promoting Peace Worldwide.

She is currently working with three-time Emmy Award winner Chris Nolan on the film *Climate of Hope: Cities Saving the World* (http://www .climateofhopefilm.org). The film is based mostly on this book, on which she provided valuable feedback. She also has a successful consulting business (Website: https://www.carlajsnyder.com/ Email: Carla@carlajsnyder.com). She has been the spark that pushed Dr. Gilderbloom to write this book. They grew up together attending the same nursery school, place of worship, summer camps, and high school in San Mateo, California, south of San Franciso. They didn't cross paths for many years until the summer of 2018 when she attended a presentation he did on his popular book, *Chromatic Homes: The Joy of Color in Historic Places*, and they have been together ever since.

Gregory D. Squires is a research professor and professor emeritus in the Department of sociology at George Washington University. He has served as a board member, consultant, or advisor to several government and non-profit organizations, including the Consumer Advisory Council of the Federal Reserve Board, Housing and Urban Development's Office for Fair Housing and Equal Opportunity, National Association of Insurance Commissioners, U.S. Commission on Civil Rights, Social Science Advisory Board of the Poverty and Race Research Action Council, Fair Housing Task Force of the Leadership Conference on Civil and Human Rights, and several others. He is currently a community development columnist for *Social Policy*. Email: squires@gwu.edu.

Jennifer Stekardis is a Louisville native and recent graduate from Vanderbilt University, earning BA degrees in Economics and Public Health. She currently resides in Raleigh, North Carolina, where she works at KPMG US as an Advisory Associate in the firm's State and Local Solutions practice. In this position, she helps state governments strategize and execute policy plans related to healthcare, electric vehicle infrastructure, and other pertinent issues. Additionally, she stays connected with the Kentucky community by working part-time with Canopy Kentucky. Canopy is a nonprofit start-up whose mission is to grow Kentucky businesses to positively impact people, the planet, and the future.

Porter Stevens is originally from Louisville, Kentucky, where he was fortunate to grow up in an older neighborhood where he could walk, ride bicycles, and take public transit. His interest in sustainable cities was ignited by a college semester abroad in Europe, where he experienced how urban places can be beautiful, healthy, and thriving communities. He earned his master's degree in Urban Planning at the University of Louisville in 2014 and has been working as a professional urban planner for almost 10 years. He currently lives in Lancaster, Pennsylvania, with his wife and two children.

Ra'Desha Williams graduated from the University of Louisville Planning Program and is now a Senior Planner in Phoenix, Arizona, and the organizer of the group, Black Women in Business.

Garlynn Woodsong With over 28 years of expertise in regional planning, urban analytics, and transportation policy, Mr. Woodsong is dedicated to guiding communities toward a low-carbon, resilient, and equitable future. His passion lies in steering communities toward a future that is low-carbon, resilient, and equitable. Renowned for his proficiency in scenario planning, Mr. Woodsong played a pivotal role in the development and deployment of cutting-edge tools RapidFire and UrbanFootprint. His impactful contributions include the management of projects such as Vision California and Honolulu TOD, as well as geospatial analysis for the influential Bay Area TOD Study.

Taking a brief hiatus from planning after 2012, Mr. Woodsong delved into real estate development. During this period, he delivered the first adaptive re-use conversion of a single-family residence into a four-plex under the modern commercial building code in Portland, OR.

Since returning his attention to planning in 2018, his consultancy has been dedicated to the convergence with climate action (including mitigation and adaptation), public health, economic development, development feasibility, and natural and working land conservation and restoration. His strategic approach seamlessly integrates planning, regulatory codes, and multidisciplinary outcomes, emphasizing effective communication tailored to resonate with the values and aspirations of local communities. Mr. Woodsong's expertise lies not only in envisioning sustainable futures but also in translating these visions into tangible, community-driven realities.

Charlie Zhang is a professor at the Department of Geographic and Environmental Sciences, University of Louisville. He is an urban geographer with research interests and expertise in crime mapping, school segregation, environmental pollution, and health disparities. Dr. Zhang's publications have appeared in premier academic journals including *Environmental Pollution, Environmental Research, Environmental Science and Technology, Exposure and Health, Journal of Exposure Science and Environmental Epidemiology, Journal of Rural Health, Journal of Urban Affairs,* and *Urban Geography.* Email: c.zhang@louisville.edu.

www.ingramcontent.com/pod-product-compliance
Lightning Source LLC
Chambersburg PA
CBHW031807270326
41932CB00008B/335